Safety and Security Issues in Technical Infrastructures

David Rehak
VSB - Technical University of Ostrava, Czech Republic

Ales Bernatik
VSB - Technical University of Ostrava, Czech Republic

Zdenek Dvorak
University of Zilina, Slovakia

Martin Hromada
Tomas Bata University in Zlin, Czech Republic

A volume in the Advances in Information Security, Privacy, and Ethics (AISPE) Book Series

Published in the United States of America by
IGI Global
Information Science Reference (an imprint of IGI Global)
701 E. Chocolate Avenue
Hershey PA, USA 17033
Tel: 717-533-8845
Fax: 717-533-8661
E-mail: cust@igi-global.com
Web site: http://www.igi-global.com

Library of Congress Cataloging-in-Publication Data

Names: Rehak, David, 1978- editor. | Bernatik, Ales, 1972- editor. | Dvorak, Zdenek, 1961- editor. | Hromada, Martin, 1983- editor.
Title: Safety and Security Issues in Technical Infrastructures / David Rehak, Ales Bernatik, Zdenek Dvorak, and Martin Hromada, editors.
Other titles: Safety and security issues in technical infrastructures
Description: Hershey, PA : Information Science Reference, an imprint of IGI Global, [2020] | Includes bibliographical references and index. | Summary: "This book presents a current overview and new trends of the safety and security issues in technical infrastructures"-- Provided by publisher.
Identifiers: LCCN 2019047052 (print) | LCCN 2019047053 (ebook) | ISBN 9781799830597 (hardcover) | ISBN 9781799830603 (ebook)
Subjects: LCSH: Industrial safety.
Classification: LCC T55 .H2764 2020 (print) | LCC T55 (ebook) | DDC 363.1--dc23
LC record available at https://lccn.loc.gov/2019047052
LC ebook record available at https://lccn.loc.gov/2019047053

This book is published in the IGI Global book series Advances in Information Security, Privacy, and Ethics (AISPE) (ISSN: 1948-9730; eISSN: 1948-9749)

British Cataloguing in Publication Data
A Cataloguing in Publication record for this book is available from the British Library.

For electronic access to this publication, please contact: eresources@igi-global.com.

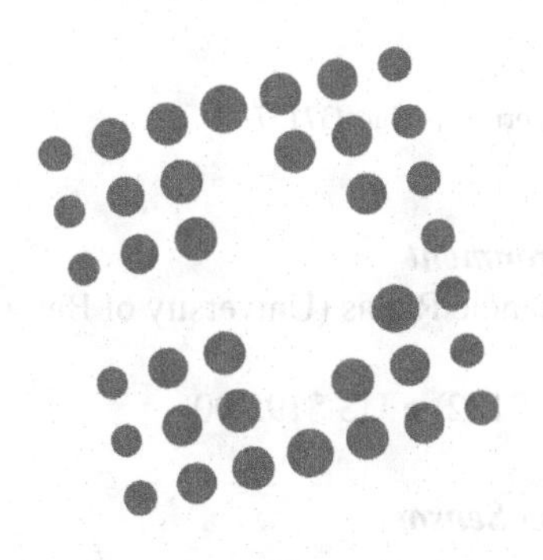

Advances in Information Security, Privacy, and Ethics (AISPE) Book Series

Manish Gupta
State University of New York, USA

ISSN:1948-9730
EISSN:1948-9749

Mission

As digital technologies become more pervasive in everyday life and the Internet is utilized in ever increasing ways by both private and public entities, concern over digital threats becomes more prevalent.

The **Advances in Information Security, Privacy, & Ethics (AISPE) Book Series** provides cutting-edge research on the protection and misuse of information and technology across various industries and settings. Comprised of scholarly research on topics such as identity management, cryptography, system security, authentication, and data protection, this book series is ideal for reference by IT professionals, academicians, and upper-level students.

Coverage

- Cyberethics
- Telecommunications Regulations
- Security Information Management
- Electronic Mail Security
- Device Fingerprinting
- Security Classifications
- Privacy Issues of Social Networking
- Risk Management
- Access Control
- IT Risk

IGI Global is currently accepting manuscripts for publication within this series. To submit a proposal for a volume in this series, please contact our Acquisition Editors at Acquisitions@igi-global.com or visit: http://www.igi-global.com/publish/.

The Advances in Information Security, Privacy, and Ethics (AISPE) Book Series (ISSN 1948-9730) is published by IGI Global, 701 E. Chocolate Avenue, Hershey, PA 17033-1240, USA, www.igi-global.com. This series is composed of titles available for purchase individually; each title is edited to be contextually exclusive from any other title within the series. For pricing and ordering information please visit http://www.igi-global.com/book-series/advances-information-security-privacy-ethics/37157. Postmaster: Send all address changes to above address.

Titles in this Series

For a list of additional titles in this series, please visit:
http://www.igi-global.com/book-series/advances-information-security-privacy-ethics/37157

Cyber Security of Industrial Control Systems in the Future Internet Environment
Mirjana D. Stojanović (University of Belgrade, Serbia) and Slavica V. Boštjančič Rakas (University of Belgrade, Serbia)
Information Science Reference • © 2020 • 374pp • H/C (ISBN: 9781799829102) • US $195.00

Digital Investigation and Intrusion Detection in Biometrics and Embedded Sensors
Asaad Abdulrahman Nayyef (Sultan Qaboos University, Iraq)
Information Science Reference • © 2020 • 320pp • H/C (ISBN: 9781799819448) • US $235.00

Handbook of Research on Intrusion Detection Systems
Brij B. Gupta (National Institute of Technology, Kurukshetra, India) and Srivathsan Srinivasagopalan (AT&T, USA)
Information Science Reference • © 2020 • 400pp • H/C (ISBN: 9781799822424) • US $265.00

Applied Approach to Privacy and Security for the Internet of Things
Parag Chatterjee (National Technological University, Argentina & University of the Republic, Uruguay) Emmanuel Benoist (Bern University of Applied Sciences, Switzerland) and Asoke Nath (St. Xavier's College, University of Calcutta, India)
Information Science Reference • © 2020 • 315pp • H/C (ISBN: 9781799824442) • US $235.00

Internet Censorship and Regulation Systems in Democracies Emerging Research and Opportunities
Nikolaos Koumartzis (Aristotle University of Thessaloniki, Greece) and Andreas Veglis (Aristotle University of Thessaloniki, Greece)
Information Science Reference • © 2020 • 200pp • H/C (ISBN: 9781522599739) • US $185.00

Quantum Cryptography and the Future of Cyber Security
Nirbhay Kumar Chaubey (Gujarat Technological University, India) and Bhavesh B. Prajapati (Education Department, Government of Gujarat, India)
Information Science Reference • © 2020 • 343pp • H/C (ISBN: 9781799822530) • US $235.00

Impact of Digital Transformation on Security Policies and Standards
Sam Goundar (The University of the South Pacific, Fiji) Bharath Bhushan (Sree Vidyanikethan Engineering College, India) and Vaishali Ravindra Thakare (Atria Institute of Technology, India)
Information Science Reference • © 2020 • 300pp • H/C (ISBN: 9781799823674) • US $215.00

701 East Chocolate Avenue, Hershey, PA 17033, USA
Tel: 717-533-8845 x100 • Fax: 717-533-8661
E-Mail: cust@igi-global.com • www.igi-global.com

Editorial Advisory Board

Paul Swuste, *Delft University of Technology, Netherlands*
Kari Telaranta, *Finnish National Rescue Association (SPEK), Finland*
Ivana Turekova, *Constantine the Philosopher University in Nitra, Slovak Republic*
Jan Valouch, *Tomas Bata University in Zlin, Czech Republic*

Table of Contents

Detailed Table of Contents

projects at European level. As a case study, the state of security and safety in the transport infrastructure of the Slovak Republic will be presented. The following will be a generalized set of recommendations to improve security and safety in the transport infrastructure. The chapter will be supplemented by relevant sources of information on the issues addressed.

The healthcare sector has been considered a part of critical infrastructure (CI) of society and has faced numerous physical-cyber threats that affect citizens' lives and habits, increase their fears, and influence hospital services provisions. The two most recent ransomware campaigns, WannaCry and Petya, have both managed to infect victims' systems by exploiting existing unpatched vulnerabilities. It is critical to develop an integrated approach in order to fight against combination of physical and cyber threats. In this chapter, key results of the SAFECARE project (H2020-GA787005), which aims is to provide solutions that will improve physical and cyber security, to prevent and detect complex attacks, to promote incident responses and mitigate impacts, will be presented. More specifically, healthcare critical asset vulnerabilities; cyber-physical threats that can affect them; architecture solutions, as well as, some indicative scenarios that will be validated during the project will be presented.

There is an identified and elevated threat level to water services by modern terrorists, in consideration of increasing levels of observed violence in recent terrorist attacks across Europe. This chapter raises significant aspects related to the security of the water critical infrastructure (water CI). Initially, dependencies and interdependencies of water CI, with other CIs, are highlighted as a potential incubating risk, which may well be hidden within the complexities of the modern water value chain of logistics and services. Threats to water CI including single points of failure are further described, followed by terrorist water attack planning methodologies and strategies. Finally, the water CI protection that may be considered to reduce any future threat levels from acts of terrorism is discussed.

This chapter deals with the issue of process safety in industrial companies and major accident prevention. In the present-day technologically advanced world, industrial accidents appear ever more frequently, and the field of major accident prevention has become a dynamically developing discipline. With accelerating technical progress, risks of industrial accidents are to be reduced. In the first part, possible approaches to quantitative risk assessment are presented; and continuing it focuses on the system of risk management in industrial establishments. This chapter aims at providing experiences, knowledge, as well as new approaches to the prevention of major accidents caused by the implementation of the Seveso III Directive.

The new industrial revolution will encompass massive change. Manufacturing Companies are pursuing digitalization and trying to figure out how to implement collaborative robots, all the while trying to manage data safety and security. It is a big challenge to deal with all the needed infrastructures to handle the big data digitalization provides whilst having to account for the shielding of it. Even more so when one has to succeed at it while taking care of the workers, the sustainability of their jobs, the implementation of safe practices at work, based on the contributions of the whole, through efficient vertical communication, imbued with Safety Culture and aiming the sustainability of the Company itself. This chapter proposes to address the role of standardization in managing industry 4.0, where culture, Risk Management and Human Factors are key, and how the tools provided by these norms may contribute to nimbly balance each Company's needs.

Accidents and near-miss accidents in chemical industries are widespread. Most of the incidents occurred due to combinations of organizational and human factors. To identify the causes for an incident of an accident analysis is needed, because it reveals the possible causes behind the accidents. Accident analysis shows the human and organizational factors that support learning from the events. Literature review shows that human error plays an important role of accidents in process industries. The chapter discusses some case studies which are received very little media publicity and also no proper assessment. At first reports on the incidents were collected from newspapers and then the place was visited to conduct an interview with local people and present and past workers with the help of the PESO (M/S Petroleum and Explosive Safety Organization, Eastern Region, Govt. of India).

Explosions can be considered to be the most devastating events in industry. Reasons that lead to such an event are often very complex. Nonetheless, the basic phenomenon is generally simple. To understand what leads to an explosion and how this can be prevented, the underlying physical and chemical processes as well as the basic steps that lead to an explosion are clarified. Practically a system of standards and regulations ensures a framework to avoid unwanted events. This together with typical sequence of events will be given and lead to a general overview of explosion process safety.

Actual and future concerns in fire safety in buildings and infrastructure are challenging. Modern technologies provide rapid development in area of fire safety, especially in education, training, and fire-engineering. Modelling as a tool in fire-engineering provides possibility to design specific fire scenarios and investigate fire spread, smoke movement or evacuation of occupants from buildings. Development of emerging technologies and software provides higher possibility to apply these models with interactions of augmented and virtual reality. Augmented reality and virtual reality expand effectivity of training and preparedness of first (fire wardens) and second (firefighters) responders. Limitations such as financial demands, scale and scenarios of practical training of first and second responders are much lower than in virtual reality. These technologies provide great opportunities in preparedness to crisis in a safety way with significantly limited budget. Some of these systems are already developed and applied in safety and security area e.g. XVR (firefighting, medical service).

The construction of road tunnels is an important part of road infrastructure. The operation of road tunnels has historically been accompanied by a number of extraordinary events. Fires are among the most dangerous ones. Individual countries create their own safety standards that mutually differ to a large extent. Some of the differences of the requirements for safety devices, including the requirements for their functionality, are compared and commented on in this chapter. Moreover, attention is paid to the use of asphalt surfaces on roads and sidewalks in tunnels. This chapter also describes the approach to fire ventilation in tunnels, one of the most significant safety devices. Special attention is paid to the choice of the strategy of longitudinal ventilation, which has been the subject of many discussions. This chapter outlines the possible directions for a solution in the future.

This chapter deals with materials used in safety and security engineering. The most commonly used materials in this field include shielding materials, materials for protective suits, electrically insulating materials and materials for fire protection. The first part of the chapter describes the properties of materials used in the above applications. The second part of the chapter focuses on characteristics of materials that accurately describe their fire risk. The fire risk of a material is quantified by its resistance to ignition (determined generally by critical heat flux and ignition temperature) and by the impact of the fire on the environment. The impact of fire is usually determined by the heat release rate, toxicity of combustion products (primarily determined by carbon monoxide yield and for materials that contain nitrogen, also through the hydrogen cyanide yield) and the decrease of visibility in the area (depending on the geometry of the area and the smoke production rate).

Because of their unique characteristics, research and development of new nanomaterials is one of the major disciplines of the twenty-first century, and examining their special properties, especially toxicity, is therefore necessary. As well as their benefits (technological improvements, specific material properties, improved resistance to natural effects), new materials also bring new risks requiring assessment in terms of occupational health and safety and their abuse as potential biological carriers or other materials. The study presents general information about nanoparticles and their distribution and properties in relation to entering aquatic, soil, and atmospheric environments. The study describes and cites examples of measurements conducted on the exposure of different nanomaterials to the work environment. Risk assessments of nanomaterials according to the available methodologies, measures to protect against nanoparticles, and importantly, the abuse of nanoparticles as a potential threat to the CZ population are also described.

The security of infrastructure systems is increasingly associated and ties to ensuring a company's basic functional continuity. Increasing security and ultimately the resilience of infrastructure systems is significantly linked to the process of infrastructure security assessment. It is obvious that the basic pillar of ensuring the required level of security and resilience of infrastructure systems is the level of physical security. Therefore, the chapter will discuss the methods for physical security assessment with a link to the different nature of selected infrastructure systems. The basic logic will be the exploitation of qualitative-quantitative methods, assessing an existing or proposed security system, based on certain measurable values such as probability of detection, response force time, delay time and probability of correct and timely guard communication, where based on this data, the probability of interruption is estimated.

In today's world, critical infrastructure encompasses facilities vital to the economy, politics, and population. Their maintenance and safe operation can ensure the supply for the population. These facilities are at risk due to climate change, natural disasters, terror attacks, or wars which are increasingly affecting countries around the world. In addition, the human factor can also cause uncertainty and damages. The function

Foreword

Critical infrastructure describes the physical and cyber systems, networks and assets that are essential for the functioning of a society and its economy. The industrial world has always relied on critical infrastructure to provide essential services such as communications, transport, power for light and heat and water for drinking and cleaning, but with dawn of the electronic age the dependence on technical infrastructures has increased exponentially.

Developed in the 60's, SCADA systems are used to monitor and control plant and equipment in industries such as transport, telecommunications, water and waste control, energy, oil and gas refining. IoT, big data, and blockchains are already revolutionising the efficiency and delivery of services to our homes, towns, cities, and industries. That increasing dependency on computers and electronic systems to deliver and manage essential services creates new vulnerabilities and interdependencies, some obvious and some more difficult to identify.

This book offers an overview of the safety and security issues; present current approaches, methods, techniques and trends in the application of actual safety and security measures in technical infrastructures. Safety and security are examined from different aspects, e.g. fire, explosive, transport, process, occupational, construction, material, technical, but also in terms of risks, uncertainties or the decision-making process.

The International Association of Critical Infrastructure Protection Professionals (IACIPP) is delighted to be supporting and contributing to this publication, which will make a significant contribution to the comprehensive awareness raising of experts in individual infrastructure safety and security.

John Donlon
Chairman, International Association of Critical Infrastructure Protection Professionals, UK

Preface

In relation to urbanization trends, the population is increasingly dependent on technical infrastructures. These infrastructures provide for population irreplaceable services in the form of energy supply, provision of information and communication technologies, construction and maintenance of transport networks, industrial production, etc. It is therefore essential to ensure a permanent high level of safety/security for these infrastructures as well as the safety/security of dependent infrastructures and populations in case of disruption or the failure of these technical infrastructures. High level of safety and security for these infrastructures can be achieved in a variety of ways and means that are constantly evolving and modernizing in the context of current safety/security threats. Based on the above, this book focuses on actual safety and security issues in technical infrastructures.

The subject of the book is the elaboration of cross-cutting publication presenting a current overview and new trends of the safety and security issues in technical infrastructures. Safety and security are examined from different aspects, e.g. fire, explosive, transport, process, occupational, construction, material, technical, but also in terms of risks, uncertainties or decision-making process. The aim of the publication is to present current approaches, methods, techniques, and trends in the application of actual safety and security measures in technical infrastructures. The publication significantly contributes to the comprehensive awareness raising of experts in individual infrastructures safety and security. The impact of the publication is related to and consists in concentrating the knowledge and experience of leading experts from selected European countries into a single monograph.

The book is suitable for both academics and practitioners, especially safety and security managers. It is also suitable for students of safety and security disciplines. The book is addressed to all experts involved in safety and security engineering, i.e. fire protection, civil protection, occupational and process safety, persons and property security engineering.

The book consists of fifteen capitols that focus on safety and security in the area of electricity infrastructures, transport infrastructures, cyber-physical infrastructures, water infrastructures, industry infrastructures and other relevant infrastructures. Safety and security in these infrastructures are discussed especially in the context of process safety, occupational safety, explosion safety, fire safety, safety and security materials, nanotechnology safety and security, risk assessment, security systems, safety and security uncertainties, and business continuity.

The first chapter is "Electricity Infrastructures Technical Security." In general, energy infrastructure is a basic but very complex system of elements, interconnections, functional inputs and outputs, which creates the need to decompose subsystems, systems and infrastructure areas. The aim of this chapter is therefore to discuss the possible implementation of approaches to risk assessment and risk management in relation to the application of technical security measures. The chapter therefore discusses risk analysis methods where the transition from general approaches to risk analysis, through risk identification methods and procedures and the assessment of major industrial and technological risks, to specific risk analysis methodologies for electricity infrastructures, are presented. An important part of the chapter is also the introduction of practical approaches and methodologies that are accepted as 'best practices' in connection with ensuring the technical security of electricity infrastructures.

The next chapter, "Transport Infrastructures Safety & Security," is focused on explaining the causal links between security and safety within transport infrastructure. The aim of the authors is to present the current state of protection and resilience of transport infrastructures in Europe. The introductory part is focused on comparative analysis of the latest information on transport infrastructure. In addition, an overview of current European transport infrastructure directives and legal acts are included. This is followed by an analysis of the results of scientific research projects at European level. As a case study, the state of security and safety in the transport infrastructure of the Slovak Republic are presented. The following is a generalized set of recommendations to improve security and safety in the transport infrastructure.

The third chapter is "Cyber-Physical Security in Healthcare." The healthcare sector that has been considered as a critical infrastructure (CI) has faced numerous physical-cyber threats that affect citizens' lives and habits, increase their fears and influence hospital services provision. The two most recent ransomware campaigns, WannaCry and Petya, have both managed to infect victims' systems by exploiting existing unpatched vulnerabilities. It is prevalent that it is critical to develop an integrated approach in order to fight against combination of physical and cyber threats. In this chapter, key results of SAFECARE project (H2020-GA787005), which aims to provide solutions that improve physical and cyber security, to prevent and detect complex attacks, to promote incident responses and mitigate impacts, are presented. More specifically, healthcare critical assets' vulnerabilities; cyber-physical threats that can affect them; architecture solution as well as, some indicative scenarios that are validated during the project, are presented.

In addition to the energy, transport and information technology sectors, another important issue is the and the title "Security of Water Critical Infrastructure." In this chapter is an identified and elevated threat level to water services by the modern terrorists, in consideration of increasing levels of observed violence in recent terrorist attacks across Europe. This chapter raises significant aspects related to the security of water critical infrastructure. Initially, dependencies and interdependencies of water CI, with other CIs, are highlighted as a potential incubating risk, which may well be hidden within the complexities of the modern water value chain of logistics and services. Threats to water CI including single points of failure are further described, followed by terrorist water attack planning methodologies and strategies. Furthermore, water CI security and protection that may be considered to reduce any future threat levels from acts of terrorism is discussed. Finally, this chapter is considered significant and important for European Union Water CI owners, as there is also an observed generational change within the terrorist environment and those who would harm the community. This factor leads to differing perspectives related to terrorist actions directed at potable water and water services.

The next chapter is "Industry Process Safety." This chapter deals with the issue of process safety in industrial companies and major accident prevention. In the present-day technologically advanced world, industrial accidents appear ever more frequently, and the field of major accident prevention has become a dynamically developing discipline. With accelerating technical progress, risks of industrial accidents are to be reduced. In the first part, possible approaches to quantitative risk assessment are presented; later, it focuses on the system of risk management in industrial establishments. This chapter aims at providing experiences and knowledge as well as new approaches to the prevention of major accidents caused by implementation of the Seveso III Directive. The aim of major risk prevention is to operate process installations at an acceptable societal level. Process safety focuses on preventing fires, explosions, and accidental chemical releases at chemical process facilities or other facilities dealing with hazardous materials. The chapter describes benefits of carrying out the risk assessment.

The next chapter is "Industrial Occupational Safety." The new industrial revolution (Industry 4.0) encompasses massive change. Manufacturing companies are pursuing digitalization and trying to figure out how to implement collaborative robots, all the while trying to manage data safety and security. It is a big challenge to deal with all the needed infrastructures to handle the big data digitalization provides while having to account for the shielding of it. Even more so when one has to succeed at it while taking care of the workers, the sustainability of their jobs, the implementation of safe practices at work, based on the contributions of the whole, through efficient vertical communication, imbued with a safety culture and aiming the sustainability of the company itself. This chapter proposes to address the role of standardization in managing industry 4.0, where culture, risk management, and human factors are the key, and how the tools provided by these norms may contribute to nimbly balance each Company's needs.

The next chapter describes another important factor for industry process safety in the next chapter "Effects of Human Factors in Process Safety." Industrial accidents directly effects on both the public and the environment. Accidents may damage the industrial economy and also results in injuries, pain or even death of workers. Chemical process plants are more accident prone. Due to rapid development of chemical technology, there is a continuous growth of ever more complex installations with extreme and critical process conditions. In general, as the chemical process plant handled dangerous substances and their plant complexity, they can be characterized with a high accident potential. Accident analysis revealed that most of accidents occurred in chemical plants due to combinations of organizational and human factors and errors. Accident analysis also recommends some future preventive plan which protects plants in future similar accidents. The chapter discussed about human factors is one of the major causes for accidents. Human factors occurred due to lack of knowledge, lack of motivation, lack of attention, lack of perception, carelessness/negligence, lack of memory, etc. The introductory section of the chapter contains an overview of chemical industry, process safety and human factors then discussed three accidental case studies. The case studies also find the probable human factors behind the incidents. This chapter may be very useful for students of engineering disciplines, safety personnel, workers and all other persons related with plant safety.

In the context of Industry Process Safety, it is also important to pay attention to the chapter "Explosion Process Safety." An explosion is one of the most destructive events that can occur in industry. To assure explosions process safety a set of regulations are in force. To understand these, the phenomena behind explosions are explained in detail. Starting from the very basic requirements to get a combustible material ignited the influencing factors are explained. To predict and avoid unwanted explosion events always the entire systems has to be taken into consideration. Consequently, the systematic behind explosions protection is derived, dividing the measures in primary, secondary and tertiary explosions protection. To each section, typical procedures and safety characteristics are explained. These are briefly described as well as how they are reflected in the European ATEX regulations. Limits of application are discussed as well as the deviations in application that need to be accounted for when other national regulation have to be applied. Typical examples and research activities in the field of Explosion Process Safety are given.

The ninth chapter of this book is "Simulation and Modelling in Fire Safety." Actual and future concerns in fire safety in buildings and infrastructures are challenging. Modern technologies provide rapid development in area of fire safety, especially in education, training and fire-engineering. Modelling as a tool in fire-engineering provides the possibility to design specific fire scenarios and investigate fire spread, smoke movement, or evacuation of occupants from buildings. Development of emerging technologies and software provides higher possibility to apply these models with interaction of augment and virtual reality. Augment reality and virtual reality expand effectivity of training and preparedness of first (fire wardens) and second (firefighters) responders. Limitations such as financial demands, scale and scenarios of practical training of first and second responders are much lower than in virtual reality. These technologies provide great opportunities in preparedness to crisis in a safety way with significantly limited budget. Some of these systems are already developed and applied in safety and security area, e.g. XVR (firefighting, medical service etc.).

The area of fire safety is also dealt within the chapter "Current Safety Issues of the Road Tunnel Constructions." The construction of road tunnels is an important part of road infrastructure. The operation of road tunnels has historically been accompanied by a number of emergencies. Fires are among the most dangerous ones. Individual countries create their own safety standards that mutually differ to a large extent. Some of the differences of the requirements for safety devices, including the requirements for their functionality, are compared and commented on in this chapter. Differences are caused by a number of influences and does not have to be necessarily perceived as negative. Moreover, attention is paid to the use of asphalt surfaces on roads and sidewalks in tunnels. The possible contribution of asphalt to the fire intensity and asphalt softening are considered problematic. The results of experiments cannot be evaluated as clearly positive and the authors do not recommend the use of asphalt surfaces in tunnels longer than 1000 meters. This chapter also describes the approach to fire ventilation in tunnels, one of the most significant safety devices. Special attention is paid to the choice of the strategy of longitudinal ventilation, which has been the subject of many discussions too. This chapter outlines the possible directions for a solution in the future.

An important issue in technical infrastructures is outlined in the chapter "Materials for Safety & Security." This chapter deals with materials used in safety and security engineering. The most commonly used materials in field of safety and security include shielding materials, materials for protective suits, electrically insulating materials and materials for fire protection. The chapter is divided into two parts. The first part describes the properties of materials used in the above-mentioned applications. The second part of the chapter focuses on characteristics of materials that accurately describe their fire hazard. The fire hazard of a material is described by its resistance to ignition (mainly determined by critical heat flux) and by the impact of the fire on the environment. The impact of fire is most often determined by the heat release rate, toxicity of combustion products (primarily determined by carbon monoxide yield and hydrogen cyanide yield) and the decrease of visibility in the fire compartment (depending on the geometry of the fire compartment and the extinction area of smoke released from burning fuel). This chapter is mainly important and intended for both researchers and professionals in fields of occupational safety and health, safety of technologies and materials and also in fire engineering.

In connection with modern technical advances in the field of material engineering is the chapter titled "Nanotechnology Safety & Security." Because of their unique characteristics, research and development of new nanomaterials is one of the major disciplines of the twenty-first century, and examining their special properties, especially toxicity, is therefore necessary. As well as their benefits (technological improvements, specific material properties, improved resistance to natural effects), new materials also bring new risks requiring assessment in terms of occupational health and safety and their abuse as potential biological carriers or other materials. The study presents general information about nanoparticles and their distribution and properties in relation to entering aquatic, soil and atmospheric environments.

It also notes the ambivalent attitude towards developing nanotechnology (futurists versus fanatic opponents). The study describes and cites examples of measurements conducted on the exposure of different nanomaterials to the work environment. Risk assessments of nanomaterials according to the available methodologies, measures to protect against nanoparticles, and importantly, the abuse of nanoparticles as a potential threat to the Czech population are also described.

The chapter "Security of Infrastructure Systems" is increasingly associated today and ties to ensuring the company's basic functional continuity. Increasing security and ultimately the resilience of infrastructure systems is significantly linked to the process of infrastructures security assessment. It is obvious that the basic pillar of ensuring the required level of security and resilience of infrastructure systems is the level of physical security. Therefore, the chapter discusses the methods for physical security assessment with a link to the different nature of selected infrastructure systems. The basic logic is exploitation of qualitative/quantitative methods, assessing an existing or proposed security system, based on certain measurable values such as probability of detection, response force time, delay time and probability of correct and timely guard communication, where based on this data, the probability of interruption is estimated.

The chapter "Uncertainties in Safety & Security" describes another important factor in the functioning of technical infrastructures. In this chapter, the authors want to introduce the importance of the uncertainties effects in critical infrastructure protection with the context of the human factors. This century is the information technology age, but all infrastructures are supported by human operators. The critical infrastructure includes all infrastructure what is essential for everyday life. This is the reason why protection is so important. This chapter goal is to present the process of critical infrastructure protection in the case of uncertainty event. The human operations can increase the risk in the uncertainty situation; therefore, it needs to understand the human factors. Between the human factors and the critical infrastructure protection can be find a relationship. Due the fact, that the protection is supported by the human source, the chapter includes selected human factors and risk.

The last important area of this book is the chapter "Business Continuity of Critical Infrastructures for Safety and Security Incidents." This chapter provides an overview of important and timely concepts with a focus on the safety and security for critical infrastructures. The content is a result of triangulation of sources from the fields of academia, best practices and lessons learned, legislation and scientific research. The protection of CIs has been a popular topic of discussion through recent years but also a topic for initiative towards the undisrupted function, prosperity and well-being of nations in a world of interconnections and dependencies. In respect to that, the following chapter provides input which will assist in the understanding of the concepts surrounding the safety and security of critical infrastructure while combining theoretical approaches with practical guidelines for the composition of a business continuity plan. The chapter also discusses the factors contributing to the criticality of technical infrastructures as part of a nation or a cross-border network and the threats to which a CI can be exposed whether these are natural or man-made.

Finally, let me say a few words about how the book affects both the practical and scientific fields and how it contributes to the development of the issues. This is a unique set of topics contributing to the issues of safety and security in technical infrastructures. The book provides a comprehensive view of this area in the context of significant critical infrastructures, i.e. energy, transport, information technology, water, and industry. The authors, from various aspects such as process safety, occupational safety, explosion safety, fire safety, safety and security materials, nanotechnology safety and security, risk assessment, security systems, safety and security uncertainties, and business continuity, and examine these infrastructures.

The book brings views and opinions of experts from around the world and presents current trends in the field of improving Safety & Security in Technical Infrastructures. I firmly believe that the book will benefit a wide range of readers and provide to them answers to a number of questions that the title raises. I wish you pleasant reading.

David Rehak
VSB – Technical University of Ostrava, Czech Republic

Acknowledgment

The editors would like to thank the parent institutions for their support:

- VSB – Technical University of Ostrava, Faculty of Safety Engineering, Czech Republic
- University of Žilina, Faculty of Security Engineering, Slovak Republic
- Tomas Bata University in Zlín, Faculty of Applied Informatics, Czech Republic

The institutions that supported writing process of this book:

- International Association of Critical Infrastructure Protection Professionals (IACIPP)
- European Commission's Joint Research Centre (JRC)
- Ministry of Interior – General Directorate of Fire Rescue Service of the Czech Republic
- Technology Platform Energy Security of the Czech Republic (TPEB)
- Czech Transmission System Operator (ČEPS)

The institutions where the authors work:

- Center for Security Studies (KEMEA), Greece
- Fire Rescue Brigade of the Moravian-Silesian Region, Czech Republic
- International Association of CIP Professionals, Great Britain
- Mettle Crisis Leaders, Australia
- Norwegian Institute of Bioeconomy Research, Norway
- Óbuda University, Hungary
- Occupational Safety Research Institute, Czech Republic
- Otto von Guericke University Magdeburg, Germany
- Prasanta Chandra Mahalanobis Mahavidyalaya, India
- Security Risk Watch, Netherlands
- Slovak University of Technology in Bratislava, Slovak Republic
- Spanish Fire Protection Association (CEPREVEN), Spain
- Tomas Bata University in Zlín, Czech Republic
- University of Calcutta, India
- University of Guanajuato, Mexico
- University of Ljubljana, Slovenia
- University of Minho, Portugal
- University of Žilina, Slovak Republic
- VSB – Technical University of Ostrava, Czech Republic

And last but not least to all authors for their conscientious approach and willingness to participate in the elaboration of this significant book.

Acknowledgment

Chapter 1
Electricity Infrastructure Technical Security:
Practical Application and Best Practices of Risk Assessment

Martin Hromada
https://orcid.org/0000-0003-0347-7528
Tomas Bata University in Zlin, Czech Republic

David Rehak
https://orcid.org/0000-0002-4617-0553
VSB – Technical University of Ostrava, Czech Republic

Neil Walker
International Association of CIP Professionals, UK

ABSTRACT

In general, energy infrastructure is a basic but very complex system of elements, interconnections, functional inputs and outputs, which creates the need to break down subsystems, systems, and infrastructure areas. The aim of this chapter is therefore to discuss the possible implementation of approaches to risk assessment and risk management in relation to the application of technical security measures. This chapter of the book will therefore discuss risk analysis methods where the transition from general approaches to risk analysis, through risk identification methods and procedures and the assessment of major industrial and technological risks, to specific risk analysis methodologies for electricity infrastructures, will be presented. An important part of the chapter is also the introduction of practical approaches and methodologies that are accepted as "best practices" in connection with ensuring the technical security of electricity infrastructures.

DOI: 10.4018/978-1-7998-3059-7.ch001

INTRODUCTION

Critical infrastructure (CI) must be seen as a complex system of highly vulnerable elements, whose protection is based on reducing vulnerability and increasing their resilience to the effects of emergencies and crisis situations (Rehak et al., 2013).

The Council of the European Union (CEU, 2008) defines critical infrastructure as "an asset, system or part thereof located in Member States which is essential for the maintenance of vital societal functions, health, safety, security, economic or social well-being of people, and the disruption or destruction of which would have a significant impact in a Member State as a result of the failure to maintain those functions". A similar perception of critical infrastructure is also given in the National Infrastructure Protection Plan of 2013 (DHS, 2013) where the U.S. Department of Homeland Security defined the critical infrastructure as "systems and assets, whether physical or virtual, so vital to the United States that the incapacity or destruction of such systems and assets would have a debilitating impact on security, national economic security, national public health or safety, or any combination of those matters".

Energy is the most important sector of critical infrastructure. This sector was labelled as uniquely critical by the presidential directive PPD-21 (The White House, 2013). This unique criticality lies mainly in the high dependence of other critical infrastructure sectors on the supply of individual types of energy. In the event of disruption or failure of these supplies, cascading effects[1] are initiated in the critical infrastructure system (Rinaldi et al., 2001; Rehak et al., 2018), resulting in widespread impact on society. These mainly impact on state protected interests, i.e. state security, economy and basic living needs of the population (CEU, 2008).

Every year, energy infrastructure elements are exposed to the negative impact of a wide range of interacting anthropogenic and natural hazards. Current global risks affecting the energy industry include Global Warming and Climate Change, Catastrophic Events, Cyber Threats, Rapidly Changing Industry, Regulation and Public Policy, Tariffs and Trade Tension, and Talent Retention and New Hires (Heisler, 2018). In contrast, the most significant non-global risks are mainly technological accidents (BBC, 2005), cascading effects (Rehak et al., 2018) or terrorist attacks (Tichy, 2019).

The importance of critical infrastructure protection was first highlighted by the United States in 1995. Over the years, critical infrastructure protection activities were initiated by other countries – Canada in 1998 and the United Kingdom, Sweden and Switzerland in 1999. Since the infamous attacks of September 11/2001, many European countries have defined their critical infrastructure assets and launched critical infrastructure protection efforts (Rehak et al., 2016). Currently, the most important factors in the management of critical infrastructure protection are the risk assessment and management process and the process of assessing and strengthening resilience. For this reason, attention will be paid in the following to risk analysis and to ensuring the efficient and effective technical security of electrical infrastructures.

ENERGY INFRASTRUCTURE DESCRIPTION

Energy infrastructure reference objectives can be perceived from several aspects. The first aspect is the formulation of objectives according to type plans and municipal approaches to ensure optimal energy supply. The second is the definition of infrastructures, which importance is identified as critical and which have a major impact on ensuring the functional continuity of society. This would be critical infrastructures as amended by Act No. 240/2000 Coll. on Crisis Management and on Amendments to

Certain Acts (Crisis Act) (Act 240, 2000) and Government Regulation No. 432/2010 Coll. (Government Regulation 432, 2010) on the criteria for determining the critical infrastructure element. In relation to the above facts and legal standards, the following groups of energy infrastructure reference objects may be considered: Electricity, Oil and Gas, Heat production and supply, and Petroleum and petroleum products.

Electricity

According to the Type Plan for the Crisis Situation of the Disruption of Large-scale Electricity Supply (MIT, 2014a), the power system is considered a nationwide area system, with a high level of links to the electricity systems of neighbouring states. It follows from the document that the electricity system consists of:

1. Production parts producing electricity at different sources,
2. 400 kV, 220 kV transmission lines and equipment (substations - transformer stations) and selected 110 kV lines and equipment,
3. Medium voltage distribution systems of 3 kV, 6 kV, 10 kV, 22 kV, 35 kV and 110 kV,
4. Low voltage distribution systems 0.4 / 0.23 kV,
5. Technical control rooms hierarchically arranged to manage the whole system.

Given the structure of the power system, it can be stated that this system is very sensitive to the correct functioning and required interaction of its individual elements, which are closely related and interacting, which increases its vulnerability and importance in relation to the resilience of critical infrastructure as a whole. Given that it is not possible to store electricity on a larger scale, the balance between production and consumption must be consistently maintained.

Therefore, the electricity system as a whole, must continually meet the requirements for securing the amount of electricity consumption that change over time.

In the context of critical infrastructure, these are (Government Regulation 432, 2010):

- Electricity plants (with a total installed electrical capacity of at least 500 MW, generators providing ancillary services with a total installed electrical capacity of at least 100 MW, power generation and supply lines, control room of the electricity producer).
- Transmission system (transmission system lines with a voltage of at least 110 kV, transmission system electrical stations with a voltage of at least 110 kV, technical control centre of the transmission system operator).
- Distribution system (distribution system electrical stations, 110 kV lines, and their associated lines, assessed according to their strategic importance in the distribution system, the technical control room of the distribution system operator).

Gas

It can be stated that the gas system is nationwide, and the network system is almost entirely dependent on gas supplies from abroad. In general, the structure of this sector is made up of (MIT, 2014b):

1. Production plants (gas production or mining facilities)
2. Transmission systems (interconnected set of high-pressure gas pipelines and compression stations)
3. Distribution systems (interconnected sets of high-pressure, medium-pressure and low-pressure gas pipelines not directly connected to compression stations)
4. Direct gas pipelines (not part of the transmission or distribution system - additionally set up to supply gas to eligible customers)
5. Underground gas storage facilities
6. Gas connections
7. Gas dispatching centres - workplaces ensuring the balance between sources and gas demand and safe and reliable operation of the gas system

In the context of critical infrastructure, these are (Government Regulation 432, 2010):

- Transmission system (high-pressure transit gas pipeline with a nominal diameter of at least 700 mm, high-pressure domestic gas pipeline with a nominal diameter equal to or less than 700 mm, compressor station, transfer station technical control centre).
- Distribution system (high-pressure and medium-pressure gas pipeline, transfer and control stations, technical control centre).
- Gas storage (underground gas storage with a storage capacity of at least 50 million m3 of gas, technical control room).

An important advantage of natural gas is the possibility of its storage in underground storage facilities.

Heat

The heating system according to the Type Plan for the Crisis Situation of the Disruption of Large-Scale Thermal Energy Supply (MIT, 2014c) is characterized as an interconnected set of equipment for production, distribution and consumption of thermal energy, including heating networks and connections. The basic components of the heating system are (MIT, 2014c):

1. Thermal energy sources
2. Distribution heat equipment (heating networks and transfer stations)
3. Thermal networks (transport of thermal energy or interconnection of sources)
4. Thermal connections (devices that carry the heat transfer medium from a source or distribution equipment only to one customer)
5. Heat consumption equipment (equipment connected to a source or distribution of thermal energy intended for internal distribution and consumption of thermal energy in the building or its part or in the set of the customer's premises.

In the context of critical infrastructure, these are (Government Regulation 432, 2010):

- Heat plant (plant with a total installed capacity of at least 800 MW, heat output from the heat source, heat producer's control centre).

- Heat distribution (thermal energy supply system with a capacity of at least 500 MW, technical control room of the distribution system operator).

It can be stated that district heating systems are predominantly local/municipal and are not interconnected.

As in previous cases, the heating industry is an important part of the energy complex, which is characterized by a huge amount of supplied energy, a wide range of used fuels and types of sources.

A specific feature of the heating infrastructure is the fact that heat supplies are provided by a number of heating companies that operate independent, not interconnected heating systems. It can therefore be stated that this infrastructure has a local network character (MIT, 2014c).

Petroleum and Petroleum Products

Thus, the production of fuels and other petroleum products in refineries is often and directly dependent and related to the supply of oil by oil pipelines from abroad, both in the quantity and quality of the oil processed in the refineries. The basic structure of the elements of the oil infrastructure may include (ASMR, 2014):

1. Transit and domestic oil pipelines
2. Pumping stations
3. Storage tanks for crude oil and petroleum products
4. Technical control room

In the context of critical infrastructure, these are (Government Regulation 432, 2010):

- A transit pipeline with a nominal diameter of at least 500 mm, including entry points
- A national pipeline with a nominal diameter of at least 200 mm, including entry points
- Technical control room
- Pumping stations
- Terminal equipment for the transfer of oil
- The beginning and the end of the doubling of the oil pipeline and the branching line - the pig chamber
- A pipeline with a nominal diameter of at least 200 mm, including entry points
- Technical control room
- Lifting stations
- Container and container complex with a capacity of at least 40.000 m3
- Technical control room
- Refineries with an atmospheric distillation capacity of 500.000 t / year or more.

The Most Vulnerable Elements of the Electricity Infrastructure or their Specific Parts

In connection with the process of risk analysis and evaluation, assets for a selected critical infrastructure area were defined and categorized, within the scope of Act No. 240/2002 Coll. on Crisis Management and on Amendments to Certain Acts (Crisis Act) (Act 240, 2000) and Government Regulation No. 432/2010 Coll. on the criteria for determining the critical infrastructure element (Government Regulation 432, 2010). In relation to the above-mentioned legislative standards, the following groups of assets in the production, transmission and distribution of electricity can be considered:

1. Thermal power plants and heating plants
2. Hydropower plants
3. Electrical stations
4. Dispatching building
5. Power line

For comprehensive understanding there was a need to classify information systems according to the level of information security. It can be stated that this classification is based on a unified concept of defining security levels for information assets in Czech Electricity sector, which was extended and supplemented by security classes for information assets within the Production Division. This definition was created in accordance with the requirement to use the currently used practical approaches within the owner or the CI subjects. We enclose a detailed description of each class (Deloitte Advisory, 2012):

- IB-I systems perform important safety/security functions. Their malfunctioning or failure can have a significant impact on nuclear safety, can lead to significant technological damage, long-term shutdowns of production facilities, harm to health or life threats to many people and environmental impacts.
- IB-II systems fulfil minor or only supplementary safety/security functions, provide management of important or extensive technological units, or some selected important information systems, the outputs of which are essential for decision-making and manual control. Their malfunction may have a significant effect on the technical safety and operability of important equipment, short-term shutdowns or injury to operating personnel.
- IB-III systems perform mainly monitoring and diagnostic functions without direct influence on technical safety and operability, eventually they provide control of only minor technological units or only minor technological parts. Their malfunction does not have a major impact on the technical safety and operability of the equipment, and failures will not result in significant damage or personal injury.

Due to the sensitivity of the presented data, it is not possible to go into more detail and thus get to the elementary level of the individual components of systems forming a selected group of electrical critical infrastructure.

GENERAL APPROACHES TO CRITICAL INFRASTRUCTURE RISK ANALYSIS

Risk analysis is based on improving risk understanding. It provides inputs for risk assessment and decision-making on whether identified risks need to be managed and what are the most appropriate risk management strategies and methods. Risk analysis takes into account the causes and sources of threats, their positive and negative impacts (consequences) and the likelihood that these impacts may occur. Factors that affect the severity of the impact and the likelihood of an event occurring should also be identified. The result of the risk analysis is to determine or estimate the levels of individual risks. (Rehak, 2012)

Risk analysis can be carried out in varying degrees of detail, depending on the specific risk, the purpose of the analysis and the available sources of information. The analysis may be carried out in a qualitative, semi-quantitative or quantitative manner, or a combination thereof, depending on the requirements and availability of the information. In practice, qualitative analysis is often used first to obtain general risk level data and to identify the main risks. If possible, a more objective semi-quantitative or quantitative risk analysis should be performed in the next step.

The qualitative analysis uses verbal evaluation to describe the severity of the potential impacts and the likelihood of these impacts occurring (see Table 1). The qualitative scales may be adjusted or adjusted to suit the circumstances, e.g. different descriptions can be used for different risks. Qualitative analysis can be used as an initial activity to identify the risks that require a more detailed analysis, where appropriate, where this type of analysis is appropriate for decision making, or where numerical data or resources are not adequate for quantitative analysis.

Table 1. Example of expressing values in individual types of risk analysis

	Qualitative analysis	Semi-quantitative analysis	Quantitative analysis
Probability of occurrence	high	4,5	79%
Severity of impact	medium	3	3 120 000, - €

Legend:
For simplicity, the severity of the impact is only expressed by the value of the asset analysed.
Qualitative analysis - expression through verbal evaluation.
Semi-quantitative analysis - expression through a point scale e.g. from 1 to 5.
Quantitative analysis - expression through actual values (i.e. the probability percentage and the magnitude of the impact in money).

In the semi-quantitative analysis, the corresponding values are assigned to the qualitative scale, i.e. Scale point scale (see Table 1). The aim is to create a wider scale of assessment than is usually used in qualitative analysis, but not to propose realistic values for risk calculation, as is the case with quantitative analysis. Particular attention should be paid to the semi-quantitative analysis, as the figures selected may not adequately reflect reality, which may lead to inconsistent, deviating or disproportionate results. Semi-quantitative analysis may not correctly distinguish risks, especially when the impact or probability reaches extreme levels.

Quantitative analysis uses numerical values that are much more accurate than descriptive scales used in qualitative and semi-quantitative analysis (see Table 1). Data from various sources are used to reflect the severity of the impact and the likelihood of occurrence. The value of the critical infrastructure element (if it is expected to be totally destroyed) or the costs necessary to repair the damage (i.e. the reconstruction or renewal of the CI element) are most often used to quantify the severity of the impact. In the case of probability, we can speak of quantitative expression if the actual frequency or probability of the occurrence of an event is known. The quality of the analysis then depends on the accuracy and completeness of the numerical values and the validity of the models used.

However, impacts can also be determined by modelling the outcome of an event or a set of events or by extrapolating experimental research or available data, and can be expressed in terms of tangible and intangible impacts. In some cases, it is required that the specification of impacts for different times, places, groups or situations be realized by more than one numerical value or descriptive element.

CIs vulnerability and risk must be analysed and assessed in order to prepare to address them by design, operation and management. Risk analysis as a formalized subject has existed for about three decades, and has reached a wide range of applications with the purpose of revealing and identifying potential failure modes and hazards in our systems so that they may be corrected before they manifest. (Zio, 2016)

A number of simple but also complex methods are currently used to analyse the risks in the critical infrastructure system. The most commonly used simple methods include Brainstorming, Checklist, What if, Delphi method, and Structured or semi-structured interviews (IEC 31010, 2019). These methods are mainly based on qualitative analysis. In contrast, complex methods are designed to perform semi-quantitative or quantitative analysis. The most commonly used these methods for risk analysis in the CI system are mainly Failure Mode, Effects, and Criticality Analysis (FMECA), Fault Tree Analysis (FTA), Event Tree Analysis (ETA), Hazard and Operability Study (HAZOP), and Risk and Vulnerability Analysis (RVA).

The FMEA / FMECA method was developed in the 1960s as a tool for systematic and highly organized analysis of failure modes of system elements and for assessing their consequences for individual subsystems and the system as a whole. The impetus for the emergence of these methods was the difficulty of ensuring the reliability of new technical systems, characterized by great complexity, the failure of which could have led to catastrophic consequences to a large extent. The FMEA method is a structured, qualitative analysis used to identify ways of system failures, their causes and consequences. If the analysis includes an estimate of the criticality of failure consequences and the probability of their occurrence, it is an analysis of the modes, consequences, and criticality of the failures, known as the FMECA method. (IEC 60812, 2018)

Fault Tree Analysis (FTA) is used to evaluate the probability of failure, respectively reliability of complex systems. Due to its versatility, it finds its application in many areas, especially in the area of risk management and quality management, or safety management. It is applicable as a preventive method, as well as a method of analysing an existing problem (such as an accident). The FTA method usually follows FMEA and is designed for complex systems. The FTA method is based on an analysis of a peak event or problem (generally a negative phenomenon such as accident, breakdown, poor quality, high cost) and helps to systematically identify factors that cause or negatively affect system functionality. Its aim is a detailed analysis - finding the causes of the negative phenomenon and further reduces the probability of its occurrence. For a simple system, it is preferable to use FMECA or HAZOP. (IEC 61025, 2006)

Event Tree Analysis (ETA) is a causal analytical technique that is used to evaluate the progress of a process and its events leading to a possible accident. The ETA method was developed at the request of the nuclear industry following the Three Mile Island accident. The principle of the ETA method is similar to that of the FTA method, except that failure events are monitored, not just failures as in the case of FTA. It is used especially in the area of risk management and quality management or safety management. The ETA method is based on an analysis of a sequence of activities and events in a process leading to an accident, which it displays using a graphical logical model. The ETA shall also consider the possible responses of the safety system and the human operator (operators). The ETA analysis results in different accident scenarios. (IEC 62502, 2010)

Hazard and Operability Study (HAZOP) is one of the simplest and most widespread approaches to risk identification. The HAZOP method is based on an assessment of the likelihood of a hazard and the resulting risks. Its main goal is to identify potential risk scenarios - thus it allows to identify dangerous conditions that may occur on the examined facility. The method looks for critical points and then evaluates the potential risks and dangerous states. It is a team expert multidisciplinary method where team members look for scenarios at a joint meeting using, for example, the brainstorming method. The results are formulated in a final recommendation to improve the process or system. (IEC 61882, 2016)

Risk and vulnerability analysis (RVA) is an important first step in the comprehensive civil preparedness planning process. Danish Emergency Management Agency (DEMA) has therefore developed a generic scenario-based model for risk and vulnerability analysis (the RVA Model). The model is developed for government agencies with responsibilities for society's critical functions, and consists of tables which make the user identify and evaluate threats, risks and vulnerabilities. The underlying focus is to formulate suggestions for countermeasures. The RVA Model consists of four parts, which each come in a MS Word document. In Part 1, the purpose and scope of the analysis is defined. In Part 2, the scenarios are developed. The users generate their own scenarios based on what is most relevant to them. In Part 3, users are asked to assess the risks and vulnerabilities associated with each scenario. Risks are assessed according to likelihood and consequences. Vulnerabilities are assessed according to existing capacities to prepare for, respond to, and recover from the type of incident in the scenario. In Part 4, the analysed scenarios are presented graphically in a risk and vulnerability profile. (RVA, 2016)

Other suitable methods can be found in the publications "Risk assessment methods for critical infrastructure protection. Part I: A State of the Art" (Giannopoulos et al., 2012), "Risk assessment methods for critical infrastructure protection. Part II: A new approach" (Theocharidou and Giannopoulos, 2015) or "Risk analysis and risk management in critical infrastructures" (Motaki, 2016).

RISK ASSESSMENT METHODS IN THE CONTEXT OF MAJOR INDUSTRIAL AND TECHNOLOGICAL HAZARDS IN ENERGY CRITICAL INFRASTRUCTURE

The methods presented in the previous section of this chapter should be seen as general approaches for risk analysis that can also be applied to critical infrastructure. However, in addition to these approaches, there are also methods that are already specifically applicable for risk assessment in a critical infrastructure system. These methods are intended not only for risk analysis but also for their identification and assessment and in some cases for subsequent strengthening of resilience. In view of the focus of this chapter, this part of the article is devoted to risk assessment methods in the context of major industrial and technological hazards in energy critical infrastructure. Incident reporting, which is an important supporting element of risk assessment, is also part of this chapter.

Risk Assessment Methodologies

Effective risk assessment methodologies are the cornerstone of a successful Critical Infrastructure Protection programme. The extensive number of risk assessment methodologies for critical infrastructures clearly supports this argument. Risk assessment is indispensable in order to identify threats, assess vulnerabilities and evaluate the impact on assets, infrastructures or systems, taking into account the probability of the occurrence of these threats. This critical element differentiates a risk assessment from a typical impact assessment methodology.

There is a significant number of risk assessment methodologies for critical infrastructures. In general, the approach that is used is rather common and linear, consisting of some main elements: Identification and classification of threats, identification of vulnerabilities and evaluation of impact. This is a well-known and established approach for evaluating risk and it is the backbone of almost all risk assessment methodologies (Giannopoulos et al., 2012). The following methodologies (see Table 2) were analysed in relation to the project focus.

Table 2. Risk assessment methodologies (Giannopoulos et al., 2012)

Risk Assessment methodologies	Sector/Hazards	Resilience	Objectives
Better Infrastructure Risk and Resilience	All sectors/All Hazards	Yes	Vulnerabilities assessment, risk reporting
BMI	All sectors/All Hazards	No	Vulnerabilities and risk assessment, Foster collaboration between policy makers and private sector
CARVER 2	All sectors/All Hazards	Yes (partially)	Risk evaluation, evaluation of alternatives, allocation of protective measures
CIMS (Critical Infrastructure Modeling Simulation)	All sectors/All Hazards	Yes (implicitly)	Rapid decision making, prioritization of emergency operations
CIPDSS (Critical Infrastructure Protection Decision Support System)	All sectors/All hazards	No	Risk informed design
CIPMA (Critical Infrastructure Protection Modeling and Analysis)	Energy, Communications, banking and finance/All hazards	Yes (implicitly)	Prevention, preparedness
Counteract (Generic Guidelines for Conducting Risk Assessment in Public Transport Networks)	Transport, Energy/ Terrorist threats	No	Risk reporting, protection measures effectiveness evaluation
FAIT (Fast Analysis Infrastructure Tool)	All sectors/All hazards	No	Interdependencies assessment and disruption impact
NSRAM (Network Security Risk Assessment Modeling)	All sectors/All hazards	Yes	What if analysis under malicious attacks
RAMCAP Plus	All Sectors/Terrorism, man-made	Yes (implicitly)	Risk Assessment

The following text describes a risk assessment methodology for Critical Infrastructures (CI) based on two staff working documents, one from DG ECHO on "Risk Assessment and Mapping Guidelines for Disaster Management" (Theocharidou et al., 2015; EC, 2010) and one from DG HOME on "A new approach to the European Programme for Critical Infrastructure Protection. Making European Critical Infrastructures more secure". As a result of the DG ECHO staff working document, several Member States (MS) have provided an overview of risks where the risk of "loss of critical infrastructure" has been identified as a man-made risk (EC, 2013).

A new approach: CRitical Infrastructures & Systems Risk and Resilience Assessment Methodology (CRISRRAM) takes into account the methodological findings and gaps identified in the previous chapters (see Figure 1). The main characteristics of the proposed methodology is that it adopts a system of systems approach and aims to address issues at asset level, system level and society level. In addition, it follows an all-hazards approach, which is an important element considering that DG ECHO policy is more focused on natural disasters while DG HOME policies embark on the security and man-made hazards. These various direct or indirect impacts are reflected in the layered approach which is proposed (Theocharidou et al., 2015).

Society layer. The proposed methodology starts with the definition of a hazard scenario that may directly have an impact on the society (e.g. flooding, earthquake) but at the same time it may impact a critical infrastructure. This layer complies with the national risk assessment guidelines, as risk is calculated according to a risk matrix, based on threat likelihood and (societal) impact assessment. However, this approach also considers impacts due to the failure of a CI or other dependent CIs (cascade impact). These are assessed based on the direct impact of the threat on a CI (asset layer) or due to the indirect impact of the hazard to other CIs (system layer) (Theocharidou et al., 2015).

Asset layer. It is possible to provide an estimation on the direct impact on one or more directly affected CI, on the basis of historical data, the results of vulnerability assessment of the CI or the presence of resilience mechanisms. This is usually assessed in terms of inoperability level or economic loss per each asset. This direct effect to each CI - a service degradation, a disruption or a failure - is related to an impact at societal level. If this is not the case then this infrastructure should not be considered as a CI at first hand. This assessment links asset level disruptions with societal impact (Theocharidou et al., 2015).

System layer. When the selected hazard scenario indirectly affects other CI, then we have to consider the dependencies among these CIs. Interdependencies are a key issue in modern critical infrastructures. As a consequence, dependency assessment should be introduced in our risk assessment framework. Modelling, simulation and analysis tools provide a good idea of the rippling effects in interconnected systems that finally lead to societal impact due to these indirect effects. However, this may be a continuous loop since interdependencies among infrastructures may lead to cyclic increase of indirect effects. It is clear that this can lead to augmented impact at societal level (Theocharidou et al., 2015).

Risk Assessment and Incident Reporting

To increase the relevance of the theoretical analysis of the integral security system, it is necessary to analyse a document that summarizes the security incidents of the selected critical infrastructure sector in a concrete way. Such a document is 10th Report of the European Gas Pipeline Incident Data Group (period 1970 – 2016) (EGIG, 2018). This report describes the structure of the EGIG database and presents different analyses and their results. The results of the analyses are commented on and give the most interesting information that can be extracted from the database. Linking of results of different analyses is provided where possible. Anyone who would like to combine different results should be very careful before drawing conclusions.

It can be stated that The EGIG database contains general information about the European gas transmission pipelines system as well as specific information about the incidents. Table 3 describes main information and criteria that were for this report used.

The analysed document summarizes the occurrence and percentage of the causes of the incident and presents these facts in a wider spectrum of understanding. To illustrate the distribution of incidents for the period 2007-2016 is figured (see Figure 2).

Figure 1. Proposed CI-rich NRA methodology (Theocharidou et al., 2015)

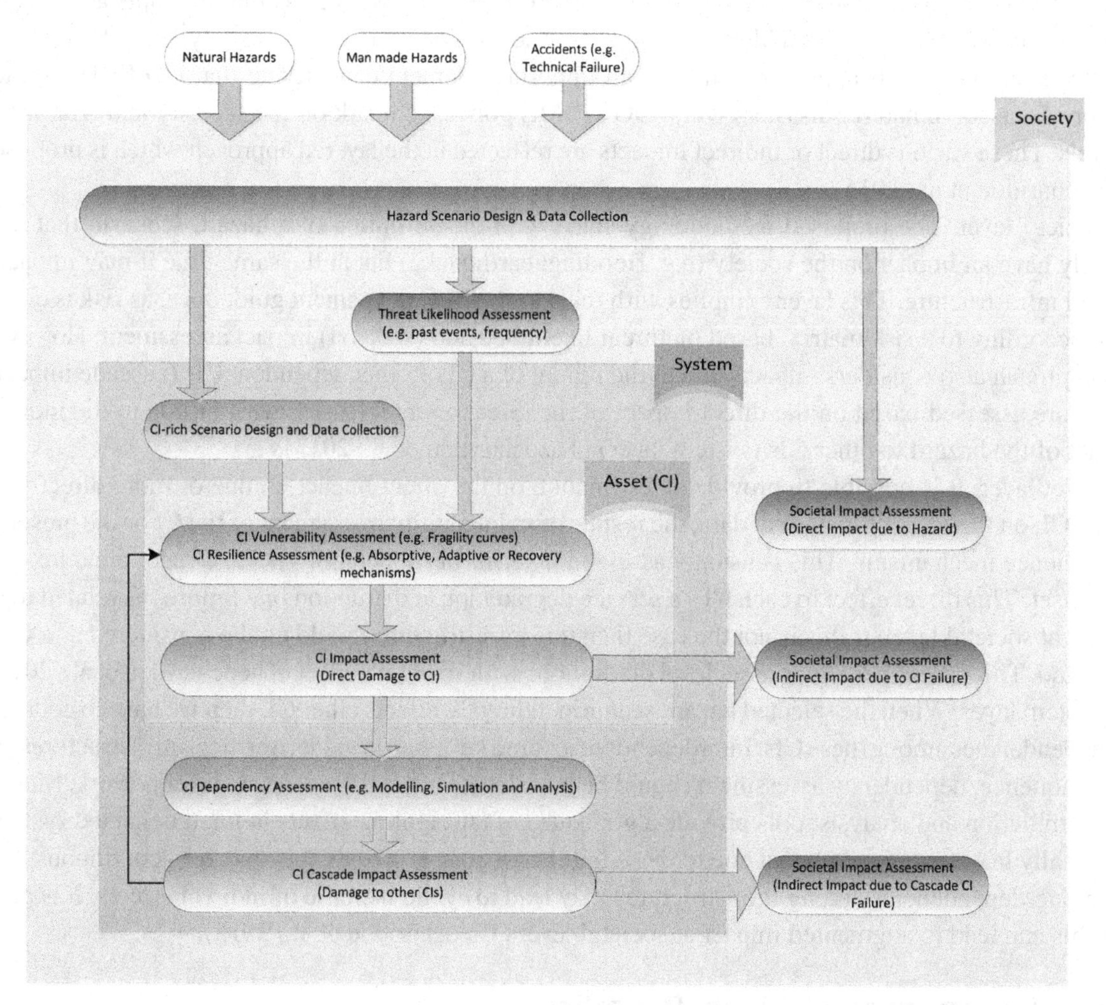

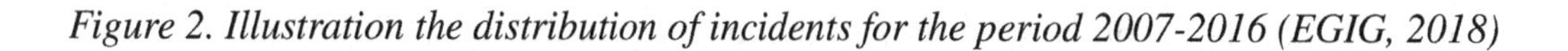

Figure 2. Illustration the distribution of incidents for the period 2007-2016 (EGIG, 2018)

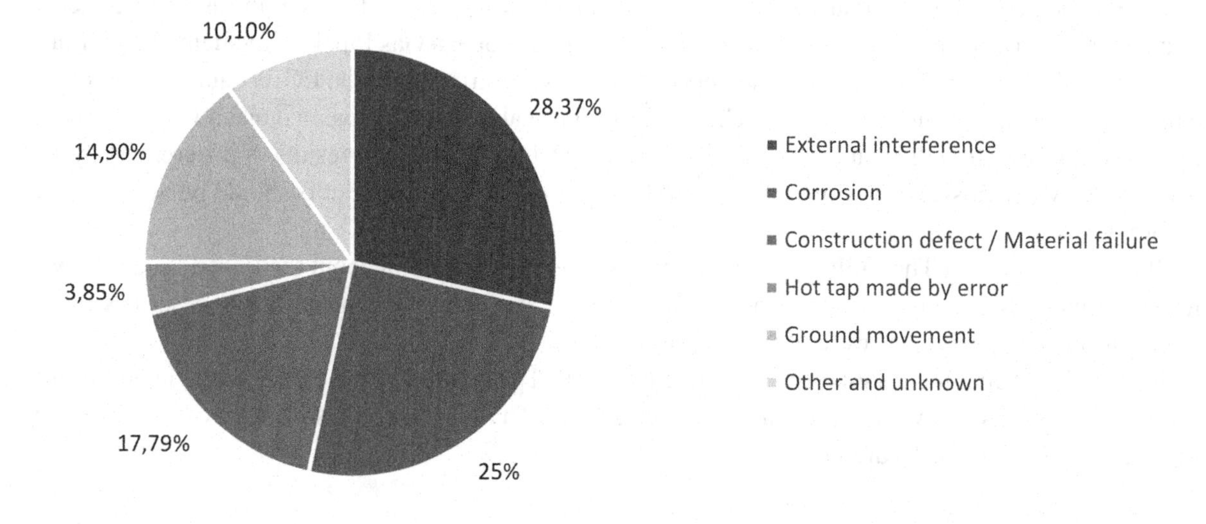

Table 3. Selected information and criteria for incident report (EGIG, 2018)

Pipelines parameters	Specific information about incidents	Additional information
• Diameter • Pressure • Year of construction • Type of coating • Depth of cover • Grade of material • Wall thickness.	The characteristics of the pipeline on which the incident happened, namely the general information listed above. The leak size: a) Pinhole/crack: the effective diameter of the hole is smaller than or equal to 2 cm b) Hole: the effective diameter of the hole is larger than 2 cm and smaller than or equal to the diameter of the pipe c) Rupture: the effective diameter of the hole is larger than the pipeline diameter. The initial cause of the incident a) External interference b) Corrosion c) Construction defect/material failure d) Hot tap made by error e) Ground movement f) Other and unknown. The occurrence (or non-occurrence) of ignition. The consequences. Information on the way the incident has been detected (e.g. contractor, landowner, patrol). A free text for extra information.	External interference: a) The activity having caused the incident (e.g. digging, piling, ground works). b) The equipment involved in the incident (e.g. anchor, bulldozer, excavator, plough). c) The installed protective measures (e.g. casing, sleeves). Corrosion: a) The location (Internal, External, Unknown). b) The appearance (General, Pitting, Cracking). c) In line inspected (yes, no, unknown). Construction defect/material failure: a) The type of defect (construction or material). b) The defect details (hard spot, lamination, material, field weld or unknown). c) The pipeline component type (straight, field bend, factory bend). Ground movement: a) The type of ground movement (dike break, erosion, flood, landslide, mining, river or unknown). Other and unknown: a) The sub-causes out of category such as design error, lightning, maintenance error.

The authors consider the findings and outputs of the document to be a significant motivation and starting point for achieving the main objectives and outputs of the project. The following text will distort and theoretically describe the role of the integrated security system in the issue of protection and resilience of critical infrastructure.

SPECIFIC RISK ANALYSIS METHODOLOGIES FOR ELECTRICAL INFRASTRUCTURE

In connection with the risk analysis and assessment process implementation, the selected critical infrastructure area assets were defined and categorized in the extension of the crisis law (Act 240, 2000) and the Government Decree on criteria for identification and designation of critical infrastructure element (Government Regulation 432, 2010).

Design and Optimization of selected critical infrastructure area protection management system is conditioned to a process of threat analysis and risk assessment, and selection of relevant methodology. The following risk assessment methodology is semi-quantitative approach that works with the three risk components (asset, threat, and vulnerability), where:

- **Asset:** Part of the system or its data with relevant value for company,
- **Threat:** Any activity using intentionally or unintentionally vulnerability with a negative impact on the confidentiality, availability and integrity of assets and is expressed as a probability of threat occurrence,

- **Vulnerability:** Expression of asset or group of assets weak points, under certain assumption will be used, resulting in a threat of injury or assets loss and any processes / functions that support these assets.

Detailed description of mentioned risk components is presented in Table 4.

Table 4. Risk components assessment (Rehak et al., 2016; Deloitte Advisory, 2012)

Numerical value	Asset value	Threat level	Level of vulnerability
0	none or no unrated	unlikely or unrated	none or no unrated
1	low	very unlikely	low
2	insignificant	unlikely	insignificant
3	middle	moderately likely	middle
4	high	very likely	high
5	very high	highly probable or certain	very high

To finalize the process of risk assessment in relation to selected areas of critical infrastructure has been defined equation (1):

$$R = A \cdot T \cdot V \qquad (1)$$

where R = risk level; A = asset value; T = threat level; V = level of vulnerability.

After the risk components determination, it is possible to quantify the risk value and the resulting value is divided into groups reflecting the increasing level of risk. To express the risk level the following categories were identified: 1-40 low risk; 41-70 intermediate risk, 71-125 high risk.

For the next risk analysis and determination of protection management system process, it is important to express correlation between the identified risks. For this purpose, the KARS methodology was used (Pacinda, 2010) which enables the risk quantitative analysis, through their correlation. The importance of this method is seen especially in the context of risk diversification based on the level of risks activity and passivity in relation to other risks (whether the selected risk has the potential to cause other risks or whether it may be caused by other risks - domino effect). The process of implementation KARS analysis is a multi-step, in a first step, an inventory of risks was defined, that is for the area (the area of generation, transmission and distribution of electricity) specific, and performs the correlations assessment (see Table 5).

The next step is to express the above-mentioned activity - which reflects the overall risk potential to cause additional risks; or whose passivity expressions indicate a number of risks which may cause the risk. Equations (2) and (3) are used for the calculation of the coefficients.

$$K_A R_i = \frac{\sum_{i=1}^{x} R_i}{x-1} \qquad (2)$$

Table 5. Table for correlations assessment (Rehak et al., 2016; Deloitte Advisory, 2012)

	Index	1	2	3
Index	Risk	High temperature	Lightning	Tree fall
1	High temperature		1	0
2	Lightning	0		1
3	Tree fall	0	0	

Legend:
To express the relationships between the risk is the risk correlation table filled with values where:
1 – Is the real possibility that the risk *Ri* may cause risk *Rj*
0 – Expresses a condition where there is no real possibility that the risk *Ri* may cause risk *Rj*

$$K_P R_i = \frac{\sum_{i=1}^{x} R_i}{x-1} \quad (3)$$

where $K_A R_i$ = activity coefficient of i-th risk; $K_P R_i$ = passivity coefficient of i-th risk; $\sum R_i$ = the sum of the risk; x = total amount of risks.

The last step of the analytical part of protection management systems development process is the assessed risks prioritization. In this step, the assessed risks are divided according to their significance into the following three segments: (1) primarily significant risks, (2) secondary significant risks, and (3) tertiary significant risks. In accordance with the Pareto rule (Koch, 2008) it is assumed that in the first segment will be 80% of the most significant risks. In this part of the process can also be used Criteria Risk Analysis of Facilities for Electricity Generation and Transmission (Rehak et al., 2014a) or Preferences Risk Assessment of Electric Power Critical Infrastructure (Rehak et al., 2014b).

The output of the application of the specified risk analysis is presented in the Tables 6 and 7, prioritizing the risks into individual groups.

The risks thus prioritized subsequently enabled the objective development of an integrated security system, which will to some extent be an inspirational source of knowledge for formulating relevant requirements for the operational level of security.

IMPORTANCE OF TECHNICAL SECURITY OF ELECTRICAL INFRASTRUCTURE

Technical security can be understood in the context of electrical infrastructure as a set of essential requirements and measures to ensure the required level of protection of infrastructure elements. The essence of technical security is the continuous application of security measures, which can be classified into three basic groups: regime measures, physical security and technical means. If the electrical infrastructure element is protected by all of the above-mentioned safety measures, we can speak of the so-called integrated security system.

Table 6. Risk prioritization table – physical security (Rehak et al., 2016)

Primary important risks	Secondary important risks	Tertiary important risks
Fire Backup power supply failure Server failure ACS element failure CCTV element failure Lock system failure Software failure Working procedures failure Lack of material resources Lack of human resources Security service failure Unauthorized access of a third party to the area Unauthorized access to Technical security system elements (TSS) Unauthorized manipulation of TSS elements Theft by third parties Destruction of cooling equipment Destruction of space or its part Destruction or retirement of the permanent service	Flood Heavy rain Lightning Electromagnetic radiation Power cut Interruption of water supply Water leakage from water. row in space Failure of lighting in the assessed area Air contamination with dangerous gas Nuclear accident Operating fault Workstation failure Wiring malfunction Inappropriate working procedures Operational error of employees Operational error of third party workers Incorrect handling of TSS elements Ignorance, unpreparedness of employees Negligence, ignorance of workers Obtain space protection information Impersonation of user identity Intentional damage to the area by the employee Intentional damage to space by a foreign person Staff sabotage Forcible intrusion of a stranger into space Burglary into space Destruction of overhead lines – columns Use weapons Demonstration near space Blackmail of employees Holding hostages Abduction of employees Destruction or disposal of technological premises for data processing and transmission Destruction or disposal of the dispatching centre	Snow Heating failure Air conditioning failure Impersonate physical identity by third-party workers Destruction of overhead lines – ropes Threats of bomb placement - by phone, e-mail Bomb Placement Threats – Writing Other threats Mail and parcel items with dangerous content Use of annoying agents

Table 7. Risk prioritization table – information security (Rehak et al., 2016)

Primary important risks	Secondary important risks	Tertiary important risks
Falsifying user identity by identifiable persons Falsifying user identity by contractors Falsifying user identity by third parties Unauthorized use of the application Introduction of destructive and harmful programs Infiltration of communication Manipulation of communication Incorporate malicious programs Wrong routing Network gateway technical fault Computer fault in network management / management Network interface technical fault Network service technical fault System or network software failure Application software failure	Misuse of system resources Communication interception Communication failure Computer malfunction Memory device technical failure Printing device technical failure Operating error Hardware maintenance error Software modifying Error User error	Lack of personnel

Integrated Security System

The identified and prioritized risks of unacceptable level have to be corrected to acceptable level using respective measures. The method of their correction and used tools and measures are contained in the security plan of electricity infrastructure elements. Quality of measures for lowering the evaluated security risks is positively correlated by the quality of the security plan. It is influenced by the used methodology, scope or complexity of proposed measures and their synergy, as well as by their effectiveness and purposefulness. Taking into account the importance of CI element (regional, national, European), it is necessary to lower, as far as possible, the riskiness of the element, given by potential security risks, threats or hazards in effect. It is possible to achieve effective lowering of security risks through complex of measures that act both preventively and repressively. The essence lies in simultaneous use of technical security devices, deployment of persons tasked with execution of physical protection (guarding) and organizational and regime measures. These elements constitute integrated security system (hereinafter "ISS") (Vidrikova et al., 2017).

The protection of the CI elements should be ensured by the means of permanent security measures, mainly using (Act 45, 2011):

- Technical security devices (i.e. mechanical barriers, alarm systems, other technical devices)
- Security services providing physical protection
- Organization and regime measures
- Vocational training of persons that provide the security of element
- Control measures for adherence to permanent security measures

The purpose of the technical security devices is to provide a combination of mechanical barriers and alarm systems (see Figure 3) with the early detection of intruders and slowing down of their actions, thus enabling the intervention unit to stop them from unlawful activity. The technical means of security are, in particular, an access control system, an electronic security system, a CCTV system, an electric fi re alarm system and other devices (e.g. devices for detecting substances and items or devices for the physical destruction of information carriers. (Lovecek et al., 2010)

Subsystem of physical protection (hereinafter "PP") consists of persons tasked with execution of physical protection (guarding) and of normative acts that regulate execution of physical protection. Its main tasks are the execution of security surveillance and control of adherence to the regime and organizational measures, acquirement, collection and analysis of security information on intrusion into the protected object using resources of an alarm system, evaluation of deviations of the occurred state from the security standard, ensuring or carrying out active intervention against intruder and acceptation of necessary measures for setting the of whole ISS into the defined state according to the security standard.

Subsystem of organizational and regime measures (hereinafter "ORM") is a sum of administrative, normative and regime measures for securing the protected interests and values. It is characterized by the system of order and regime, its provision and periodic control.

At the same time, this very brief description of the basic components of the integrated security system creates a framework for understanding the next subchapter dealing with the specific application of selected processes of the operational level of security of the critical infrastructure sector.

Figure 3. Classification of technical security devices (Vidrikova et al., 2017)

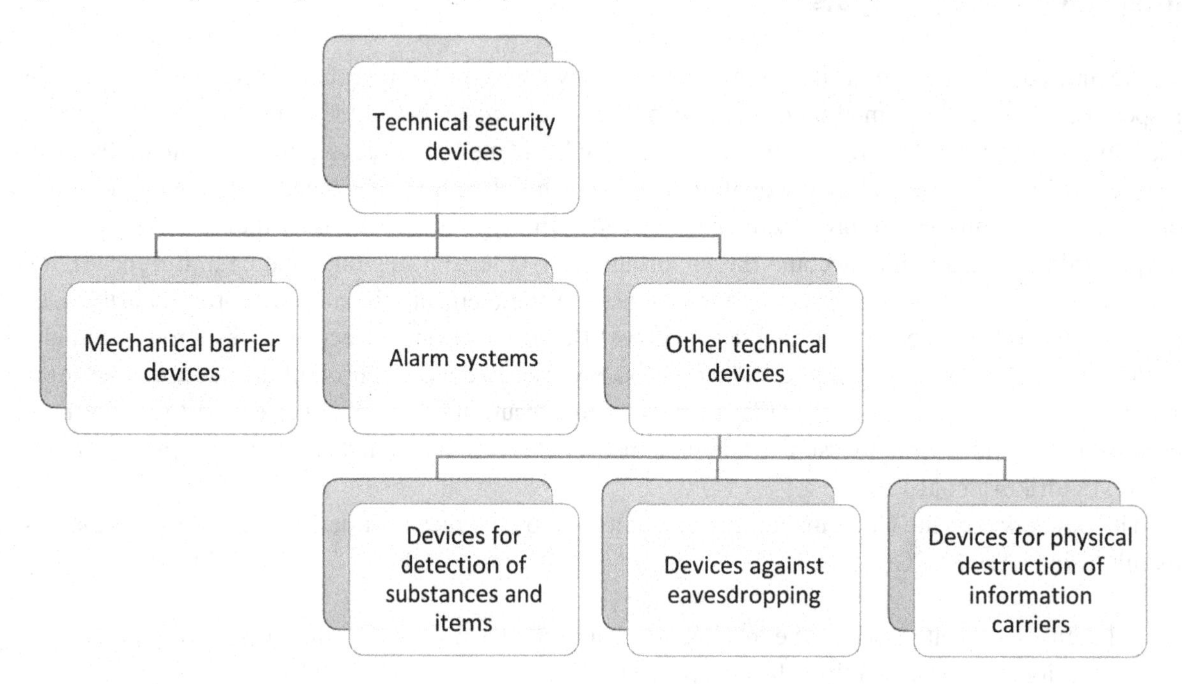

Best Practices to Ensure Technical Security of Electrical Infrastructure as an Implementation of Selected Methods and Approaches

The establishment of requirements for the integrated security system (ISS) in selected sector of critical infrastructure results from the need to maintain functional continuity of electricity supply. Within the process of formulating the requirements and parameters for the ISS and its individual parts, the specifics of the selected critical infrastructure area, the outputs of the analytical and pilot operations phases and the verification of the methodology were taken into account (Deloitte Advisory, 2012).

The methodical procedure itself is divided into complex, interdependent units. The outcome of the methodology is to specify selected security areas in relation to the optimization of the critical infrastructure ISS in the selected area in order to ensure continuity of electricity supply. The security areas, which will be described in a structural manner in the following text, can be divided into the following units:

- Physical security
- Information security
- Administrative security
- Personnel security
- Crisis management and planning

Physical Security

This part of the methodological procedure sets out structural and qualitative requirements for individual parts of the physical protection system as a means of ensuring the physical security of critical infrastructure, which to a certain extent reflect the identified risks and their potential to degrade functional continuity of the subject (see Tables 8-10).

Another relevant area of the critical infrastructure subject's physical protection system in the context of the implementation of a comprehensive system of critical infrastructure protection management is the regime measures and physical surveillance, for which detailed formulation of selected principles and requirements for specific setting of these areas is required protection, resp. response to the intruder activity and response.

Table 8. Mechanical barriers (Rehak et al., 2016; Deloitte Advisory, 2012)

Areas of application	Location - area, object - building, space parameters
Perimeter area	External fencing
	Entrances (gateway)
	Entrances (entrance and siding gate)
	Buildings in the perimeter
Exterior	Outdoor power station energy equipment
	Parking areas inside the building with stored property
	Inputs to through cable channels
Interior spaces and buildings	Entrance (outside) doors and gates in the building envelope incl. emergency exits and entrances from through or through cable ducts
	Locking system or padlock in the entrance (outside) door and gate to the building
	Self-closing mechanism on the main entrance door to the building
	Glazed parts (doors, windows) in the building shell
	Glazed parts (cellar windows) in the building shell, which are below the level of the surrounding terrain
	Other technical openings in the building shell
	Solid ladders on the building shell leading to the roof

The general definition of basic structural requirements for individual parts of the physical protection system is an important input activity within the process of creating a comprehensive system of critical infrastructure protection management in a selected area. It can be stated that the set requirements reflect the outputs of the risk and fact analysis, which were identified in the process of physical verification of selected objects and elements of critical infrastructure.

Information Security

Setting and formulation of Information Security Management is an essential aspect of setting uniform standards to ensure the functionality of the information systems of critical infrastructure entities. The complexity of information systems and current trends in the development of ICT increase the emphasis on building information security standards. Information (ICT) security is one of the fundamental pillars of critical asset protection, in particular by providing protection to information managed within databases and information systems and to processes that process data through application equipment. In addition, the ICT area provides a communication platform to other security systems, such as physical protection

Table 9. Alarm systems, CCTV, ACS (Rehak et al., 2016; Deloitte Advisory, 2012)

Areas of application	Location - area, object - building, space parameters
Perimeter area	External fencing
	Entrances (gateway)
	Entrances (entrance and siding gate)
	Buildings in the perimeter
	Other penetrations
Exterior	Outdoor power station
	Parking areas inside the building with stored property
	Inputs to through cable channels
	Evidence of entry in the outdoor premises of the building
	Transmission of alarm and other functional states of AS to the power control centre
	Transmission of alarm and other functional states of AS to the regional supervisory workplace
Interior spaces and buildings	Entrance (outside) doors and gates in the building shell incl. emergency exits and entrances from through or through cable ducts
	Glazed parts (doors, windows) in the building envelope
	Glazed parts (doors, windows) in the building envelope accessible from accessible places (walkable cornices and roofs, ladders, balconies)
	Glazed parts (cellar windows) in the building envelope, which are below the level of the surrounding terrain
	Other technical openings in the building shell
	Solid ladders on the building shell leading to the roof
	Outgoing or through cable duct outlet (to / from the interior of the building)
	Entrance (internal) doors to the premises, respectively. rooms related to the operation of the building
	Spaces, respectively. rooms related to the operation of the building
	Spaces, respectively. rooms related to the operation (management) of the building and continuous presence of persons (permanent service)
	Interior spaces at the entrance door to the building (vestibule) and other common areas (corridors, staircases)
	Space or room with installed AS
	Evidence of entry into the building and evidence of entry into selected areas or rooms related to the operation of the building (in AS or ACS)
	Transmission of alarm and other functional states of AS to the power control centre
	Transmission of alarm and other functional states of AS to the regional supervisory workplace

systems. The following sections identify standard IS / IT security management areas in which identified risks can be minimized in a consistent manner and are relevant to the critical infrastructure protection management needs (see Table 11).

Formulation of basic information security requirements is another cornerstone of critical infrastructure protection management in the field of electricity generation, transmission and distribution. As in the previous sections, we assume that the basic set of requirements in the Annexes are considered to be a set of standard parameters that can be drawn and adapted according to the individual needs of individual critical infrastructure entities.

Table 10. Regime measures and physical protection (Rehak et al., 2016; Deloitte Advisory, 2012)

Distribution of parameters	Principles of regime measures and physical security
Parameters related to regime measures	Determination of designated entrances for persons and vehicle entrances to the building
	Determination of the scope of persons' entry and means of transport to enter the building
	Regime of movement of persons, vehicles in the building
	Regime of movement of material in the building (introduction, removal of property)
	Key handling and TSS regime
	Emergency handling (security incidents)
Parameters related to physical protection	Performance of permanent service on the premises
	Guarding on the building without installed TSS (regular or irregular)
	TSS operation at the supervisory workplace (non-stop)
	Patrol activity on the object (regular or irregular)
	Evaluation of TSS states and reactions to them (continuous)
	Mobile intervention on the building (without delay)

Administrative Security

Another important part of the protection management system for the selected area of critical infrastructure is also the aspects of administrative security, which ensures the sufficient protection of documents in paper and electronic form during their creation, receipt, registration, processing, sending, transport, transfer, storage, shredding, archiving etc. They are considered as essential areas of administrative security (see Table 12).

Personnel Security

Based on the analysis and consultations with the responsible authorities in the area of critical infrastructure, another part of the comprehensive protection management system was defined as personnel security, which is perceived as a system of selection of individuals in relation to access and access to information assets of the organization access to, education and protection of information. The requirements focus on minimizing the impact of human error, potential theft, fraud, or misuse of the organization's information resources. In relation to the formulation of the principles of personnel security the methodology defines eight areas (see Table 13).

Crisis Management and Planning

Crisis management and planning is considered to be a very important part from the point of view of critical infrastructure protection management, which should identify the requirements and needs to ensure the continuity and renewal of critical infrastructure functionality in electricity generation, transmission and distribution. The proposed structural and functional requirements for individual areas of crisis management and planning optimize the process of dealing with an emergency / situation in the system, while maintaining the basic functions of the selected area and their recovery.

Table 11. Principles of Information Security Management (Rehak et al., 2016; Deloitte Advisory, 2012)

Required parameter of the security management system	Related processes
Identification and authentication	Password distribution
	Password length
	Password complexity
	Using a password
	Frequency of password change
	User identifiers
	Sharing accounts
Logical access control	Principles of access control
	Restricting access to information
	Workstation timeout / password protected screensavers
	Blocking default accounts
	Access to audit records
Event recording and audit	Event recording
	Period of retention of the logbook
	Monitoring
	Regular checks
	access permissions
	Analysis of event records
	Investigation of the incident
Software Integrity	Software integrity checks
	Updating of operating systems, databases and network components
	Turn-key SW update
Data backup and shredding	Back up traffic
	Data Backup
	Safe disposal of media
	Logic shredding
Network resilience	Redundancy of network devices
	Protocols and services
	Encrypted connections
	Firewall; IDS / IPS; Proxy server
	Powers of network administrators
	Back up routing tables
System testing	Security testing
	Test backup recovery
	Testing for updates
Protection against harmful programs	Measures to protect against harmful programs
	Removing malicious software
System management control	Reduce changes in the commercial application package
	Control access to system administrator accounts
Operational controls	Operational procedures
	Backup logs
Public Key Infrastructure	Using certificates for selected IS
Software change check	Emergency software repairs
Customer Authorization	Authentication services
	Customer management services
	Secure remote access
Vulnerability analysis	Identification and evaluation of assets
	Identifying and assessing threats and vulnerabilities
	Identification and evaluation of protective measures
Document / media checks	Save documents, media.
Virtualization	Communication and traffic management
	Classification and management of assets
	Access control
Security organization	Information security infrastructure
	Third party access security
Capacity planning	Software capacity planning
	Check of sufficient capacity
	System acceptance

Table 12. Administrative security (Rehak et al., 2016; Deloitte Advisory, 2012)

Areas of administrative security	
Administrative security	Responsibilities, duties and powers
	Marking and classification of documents
	Document handling
	Loss of documents and their media - media
	Administrative security in personnel changes

Table 13. Personnel security (Rehak et al., 2016; Deloitte Advisory, 2012)

Areas of personnel security	
Personnel security	Responsibilities, duties and powers
	Examination of employees
	Information protection agreements
	Conditions of work performance
	Staff training
	Responding to security incidents and failures
	Disciplinary process
	Termination of working relationship

The Table 14 sets out standard areas of responsibility and action at each level of management and planning in a selected critical infrastructure area.

In the event of an emergency, crisis situation (hereinafter referred to as EM / CS), increased emphasis is placed on formulating the optimal form of crisis management and dividing the management process into several levels. The Table 15 sets out the essential requirements of the crisis management process at each level, and it is clear that in order to maintain functional continuity of electricity generation, transmission and distribution, individual levels and activities will be tailored to the real requirements of each entity.

It is important to prioritize activities and levels of crisis management and to determine the degree of importance of EM / CS in relation to the availability of key system functions. Therefore, the Table 16 formulates the areas, resp. degree EM / CS (see Tables 16 and 17).

Table 14. Personnel structure of CMS (Rehak et al., 2016; Deloitte Advisory, 2012)

Personnel structure of CMS	
Personnel structure of CMS	Organization Representative for Emergency Management / Liaison Security Officer
	Organization crisis management manager
	Guarantor of crisis management organization for the section
	Solver
	Employees

Table 15. Crisis team management level (Rehak et al., 2016; Deloitte Advisory, 2012)

Crisis team management level	
Crisis team management level	Operational level
	Tactical level
	Strategic level

Table 16. Degree of emergency / crisis situations (Rehak et al., 2016; Deloitte Advisory, 2012)

Degree of emergency / crisis situations	
Degree of emergency / crisis situations	Degree 1 (ACTIVATION OF CP)
	Degree 2
	Degree 3
	Degree 4

Table 17. Activities of the crisis team of the organization (Rehak et al., 2016; Deloitte Advisory, 2012)

Activities of the crisis team of the organization	
Activities of the crisis team of the organization	Emergence and announcement of MS / KS
	Activation of crisis organization team
	Emergency Management
	Assessment and termination of emergency

In the following part of the process, attention is paid to the Critical Infrastructure Protection Management System (CIPMS) (it can be understood as Integrated Security System), which connects the various security areas into a managed deployment process that improves measures aimed at protecting the assets of a pre-defined Critical Infrastructure area. The CIPMS is designed according to the PDCA (Plan-Do-Check-Act) continuous process. Inputs are individual requirements defined by the CIPMS´ areas of interest. The output is a continuous CIPMS in selected security areas or other areas that will be later included in the CIPMS. The individual steps of the CIPMS can be depicted as a continuous cycle (see Figure 4).

Due to the different needs and conditions of individual subjects, it is necessary to clarify and consequently perform a differential analysis of the current and proposed states of identified safety/security areas. The results of the Difference Analysis are inputs for the subsequent steps of the CIPMS process. The activities that should be implemented step-by-step in the CIPMS process are presented in Table 18.

The PDCA Cycle should not be regarded only as a summary stage for all security areas´ progress during the selected period but, if it is possible, to also simultaneously monitor multiple cycles in different security areas, or even by means of individual measures, because deploying measures does not usually take place simultaneously for all measures in all areas, but rather during the year, according to the needs and possibilities of individual entities.

Figure 4. Graphical representation of PMS continuous cycle (Rehak et al., 2016)

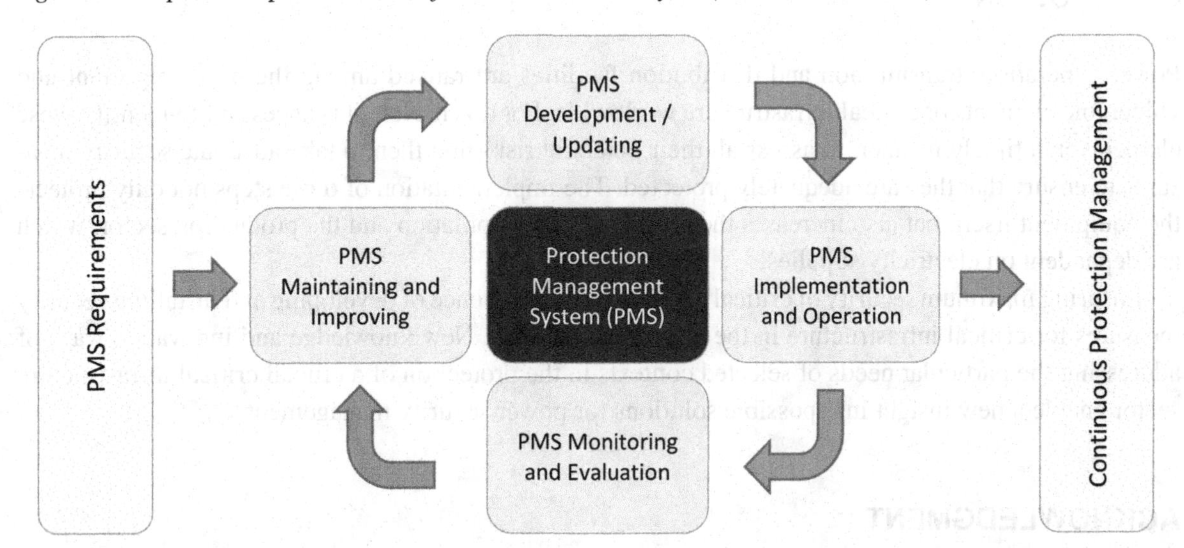

Table 18. Setting requirements for the PMS (Rehak et al., 2016; Deloitte Advisory, 2012)

PMS Phases	Description of continuous cycle
Plan - Create	Security manager should, in cooperation with the responsible employees to make regular planning of protective measures, their implementation, regular internal and independent controls and other activities in the field of security. The schedule should be established and capacity should be allocated, which will enable to implement the planned activities.
Do - Implementation and operation	In cooperation with system administrators, physical protection staff, the administrator or deputy of administrative security and other selected persons responsible for security thereof should implemented and operated security measures, for which these managers are responsible.
Check - Monitoring and evaluation	Control measures should take place both at the level of security events assessment in these areas, but also directly by individual measures settings control in relation to their projections to the success in the prevention and detection of security incidents.
Act - Maintenance	Based on the findings gained through regular monitoring and assessment of PMS, including consideration of the current situation in the legislative area and other requirements for the security. It is necessary to be ready to modify the current requirements / update the PMS in all selected areas.

The methodology presented here is an integrated approach, in relation to the comprehensive definition and the structural and qualitative requirements definition of selected safety and security aspects, in relation to the needs of the selected critical infrastructure area.

As was mentioned in the introduction to the subchapter, operational requirements from security chapter will do the literature review in context of integrated security systems, even given that the operational level is an elementary component of the system. The introduction was conceived as a theoretical and general analysis of the issue, followed by a presentation "best practices" linked to the previous text and its practical dimension.

The conclusions drawn from this text are considered as a starting point for the production of the main results and outputs of the project.

CONCLUSION

Power generation, transmission and distribution facilities are ranked among the most important and vulnerable elements of critical infrastructure worldwide. For this reason, it is necessary to identify these elements in a timely manner, to assess all their potential risks and then to take adequate security measures to ensure that they are adequately protected. The implementation of these steps not only protects the equipment itself, but also increases the security of the population and the production sector, which are dependent on electricity supplies.

Ensuring maximum security of critical equipment is the essence of developing and installing security measures for critical infrastructure in the electricity industry. New knowledge and innovative ways of addressing the particular needs of selected contexts in the protection of a crucial critical infrastructure sector enable a new insight into possible solutions for power security management.

ACKNOWLEDGMENT

The chapter was supported by the European Commission [Project SecureGas, Grant agreement 833017] and the Ministry of the Interior of the Czech Republic [Project Resilience2015, Grant number VI20152019049].

REFERENCES

Act No. 240/2000 Coll. on Crisis Management as amended. (2000).

Act No. 45/2011 Coll. on the Critical Infrastructure as amended. (2011).

ASMR. (2014). Type plan for Large-Scale Oil and Petroleum Products Disruption. *Administration of State Material Reserves*. Retrieved from http://www.sshr.cz/pro-verejnou-spravu/ropna_bezpecnost/ropna_bezpecnost/Typovy%20plan.pdf

BBC. (2005). Massive Power Outage in Indonesia. Retrieved from http://news.bbc.co.uk/2/hi/asia-pacific/4162902.stm

CEU. (2008). *Council Directive 2008/114/EC of 8 December 2008 on the identification and designation of European critical infrastructures and the assessment of the need to improve their protection*. Brussels: Council of the European Union.

Deloitte Advisory. (2012). *Methodology to ensure of critical infrastructure protection in the area of electricity generation, transmission and distribution*. Prague: Deloitte Advisory. (in Czech)

DHS. (2013). *National Infrastructure Protection Plan: Partnering for Critical Infrastructure Security and Resilience*. Washington, DC: U.S. Department of Homeland Security.

EC. (2010). *Commission staff working paper: Risk assessment and mapping guidelines for disaster management*. Brussels: European Commission.

EC. (2013). *Commission staff working paper on a new approach to the European programme for critical infrastructure protection - making European critical infrastructures more secure.* Brussels: European Commission.

EGIG. (2018). *Gas Pipeline Incidents: 10th Report of the European Gas Pipeline Incident Data Group (period 1970 – 2016).* Groningen: European Gas Pipeline Incident Data Group.

Giannopoulos, G., Filippini, R., & Schimmer, M. (2012). *Risk assessment methodologies for critical infrastructure protection. Part I: A state of the art.* Ispra: European Commission, Joint Research Centre; doi:10.2788/22260

Government Regulation. (2010). Coll. on Criteria for Determination of the Critical Infrastructure Element.

Heisler, A. (2018). Seven Critical Risks Impacting the Energy Industry. *Risk & Insurance*. Retrieved from https://riskandinsurance.com/7-risks-impacting-energy-industry

IEC 31010. (2019). *Risk management -- Risk assessment techniques*. Geneva: International Organization for Standardization.

IEC 60812. (2018). *Failure modes and effects analysis (FMEA and FMECA).* Geneva: International Electrotechnical Commission.

IEC 61025. (2006). *Fault Tree Analysis*. Geneva: International Electrotechnical Commission.

IEC 61882. (2016). *Hazard and operability studies (HAZOP studies) - Application guide*. Geneva: International Electrotechnical Commission.

IEC 62502. (2010). *Analysis techniques for dependability – Event tree analysis*. Geneva: International Electrotechnical Commission.

Koch, R. (2008). *The 80/20 Principle: The Secret to Achieving More with Less*. London: Crown Publishing Group.

Lovecek, T., Ristvej, J., & Simak, L. (2010). Critical Infrastructure Protection Systems Effectiveness Evaluation. *Journal of Homeland Security and Emergency Management*, *7*(1), 34. doi:10.2202/1547-7355.1613

MIT. (2014a). Type plan for large-scale electricity supply disruption. *Ministry of Industry and Trade*. Retrieved June 10, 2019, from http://download.mpo.cz/get/26093/58202/615552/priloha007.doc

MIT. (2014b). Type plan for large-scale gas supply disruptions. *Ministry of Industry and Trade*. Retrieved June 10, 2019, from http://download.mpo.cz/get/26093/58202/615554/priloha005.doc

MIT. (2014c). Type plan for large-scale disruption of thermal energy supply. *Ministry of Industry and Trade*. Retrieved June 10, 2019, from http://download.mpo.cz/get/26093/58202/615556/priloha003.doc

Motaki, K. (2016). *Risk Analysis and Risk Management in Critical Infrastructures* (Master Thesis). Piraeus: University of Piraeus.

Pacinda, S. (2010). Network Analysis and KARS. *The Science for Population Protection*, *2*(1), 75–96.

Rehak, D. (2012). Introduction to risk management. In L. Lukas (Ed), Security technologies, systems and management II (pp. 74-95). Zlin: VeRBuM. (in Czech)

Rehak, D., Cigler, J., Nemec, P., & Hadacek, L. (2013). *Critical Infrastructure in the Energy Sector: Identification, Assessment and Protection*. Ostrava: Association of Fire and Safety Engineering. (in Czech)

Rehak, D., Danihelka, P., & Bernatik, A. (2014a). Criteria Risk Analysis of Facilities for Electricity Generation and Transmission. In R. D. J. M. Steenbergen, P. H. A. J. M. van Gelder, S. Miraglia, & A. C. W. M. Vrouwenvelder (Eds.), *Safety, Reliability and Risk Analysis: Beyond the Horizon: ESREL 2013* (pp. 2073–2080). Boca Raton, FL: CRC Press.

Rehak, D., Hromada, M., & Novotny, P. (2016). European Critical Infrastructure Risk and Safety Management: Directive Implementation in Practice. *Chemical Engineering Transactions*, *48*, 943–948. doi:10.3303/CET1648158

Rehak, D., Markuci, J., Hromada, M., & Barcova, K. (2016). Quantitative evaluation of the synergistic effects of failures in a critical infrastructure system. *International Journal of Critical Infrastructure Protection*, *14*, 3–17. doi:10.1016/j.ijcip.2016.06.002

Rehak, D., & Senovsky, P. (2014b). Preference Risk Assessment of Electric Power Critical Infrastructure. *Chemical Engineering Transactions*, *36*, 469–474. doi:10.3303/CET1436079

Rehak, D., Senovsky, P., Hromada, M., Lovecek, T., & Novotny, P. (2018). Cascading Impact Assessment in a Critical Infrastructure System. *International Journal of Critical Infrastructure Protection*, *22*, 125–138. doi:10.1016/j.ijcip.2018.06.004

Rinaldi, S. M., Peerenboom, J. P., & Kelly, T. K. (2001). Identifying, Understanding and Analyzing Critical Infrastructure Interdependencies. *IEEE Control Systems Magazine*, *21*(6), 11–25. doi:10.1109/37.969131

RVA. (2016). *DEMA's model for risk and vulnerability analysis (the RVA Model). Birkerød.* Danish Emergency Management Agency.

The White House. (2013). *Presidential Policy Directive - Critical Infrastructure Security and Resilience*. Washington, DC: The White House.

Theocharidou, M., & Giannopoulos, G. (2015). *Risk assessment methodologies for critical infrastructure protection. Part II: A new approach*. Ispra: European Commission, Joint Research Centre; doi:10.2788/621843

Tichy, L. (2019). Energy infrastructure as a target of terrorist attacks from the Islamic state in Iraq and Syria. *International Journal of Critical Infrastructure Protection*, *25*, 1–13. doi:10.1016/j.ijcip.2019.01.003

Vidrikova, D., Boc, K., Dvorak, Z., & Rehak, D. (2017). *Critical Infrastructure and Integrated Protection*. Ostrava: Association of Fire and Safety Engineering.

Zio, E. (2016). Critical Infrastructures Vulnerability and Risk Analysis. *European Journal for Security Research*, *1*(2), 97–114. doi:10.100741125-016-0004-2

ADDITIONAL READING

Hromada, M., & Lukas, L. (2012). Multicriterial evaluation of critical infrastructure element protection in Czech Republic. In *Computer Applications for Software Engineering, Disaster Recovery, and Business Continuity* (pp. 361–368). Jeju Island: Springer-Verlag Berlin. doi:10.1007/978-3-642-35267-6_48

Hromada, M., & Lukas, L. (2013). The status and importance of robustness in the process of critical infrastructure resilience evaluation. In *Proceedings of the IEEE International Conference on Technologies for Homeland Security (HST 2013)* (pp. 589-594). Waltham, MA: The Institute of Electrical and Electronics Engineers.

National Infrastructure Advisory Council. (2009). *Critical Infrastructure Resilience Final Report and Recommendations*. Washington, DC: U.S. Department of Homeland Security.

Rehak, D., Danihelka, P., & Bernatik, A. (2014a). Criteria Risk Analysis of Facilities for Electricity Generation and Transmission. In R. D. J. M. Steenbergen, P. H. A. J. M. van Gelder, S. Miraglia, & A. C. W. M. Vrouwenvelder (Eds.), *Safety, Reliability and Risk Analysis: Beyond the Horizon: ESREL 2013* (pp. 2073–2080). Boca Raton, FL: CRC Press.

Rehak, D., & Senovsky, P. (2014). Preference Risk Assessment of Electric Power Critical Infrastructure. *Chemical Engineering Transactions*, *36*, 469–474. doi:10.3303/CET1436079

Rehak, D., Senovsky, P., & Hromada, M. (2018). Analysis of Critical Infrastructure Network. In Z. Chen, M. Dehmer, F. Emmert-Streib, & Y. Shi (Eds.), *Modern and Interdisciplinary Problems in Network Science: A Translational Research Perspective* (pp. 143–171). Boca Raton, FL: CRC Press. doi:10.1201/9781351237307-6

Rehak, D., Senovsky, P., Hromada, M., & Lovecek, T. (2019). Complex Approach to Assessing Resilience of Critical Infrastructure Elements. *International Journal of Critical Infrastructure Protection*, *25*, 125–138. doi:10.1016/j.ijcip.2019.03.003

Rehak, D., Senovsky, P., Hromada, M., Lovecek, T., & Novotny, P. (2018). Cascading impact assessment in a critical infrastructure system. *International Journal of Critical Infrastructure Protection*, *22*, 125–138. doi:10.1016/j.ijcip.2018.06.004

Vichova, K., Hromada, M., & Rehak, D. (2017). The Use of Crisis Management Information Systems in Rescue Operations of Fire and rescue system in the Czech Republic. *Procedia Engineering*, *192*, 947–952. doi:10.1016/j.proeng.2017.06.163

Vidrikova, D., Boc, K., Dvorak, Z., & Rehak, D. (2017). *Critical Infrastructure and Integrated Protection*. Ostrava: Association of Fire and Safety Engineering.

KEY TERMS AND DEFINITIONS

Critical Infrastructure: An asset or system which is essential for the maintenance of vital societal functions.

Electricity Infrastructure: Infrastructures and facilities for generation and transmission of electricity in respect of supply electricity.

Infrastructure Protection: The ability to prevent or reduce the effect of an adverse event.

Infrastructure Resilience: The ability to reduce the magnitude, impact, or duration of a disruption. Resilience is the ability to absorb, adapt to, and/or rapidly recover from a potentially disruptive event.

Integrated Security System: The essence lies in simultaneous use of technical security devices, deployment of persons tasked with execution of physical protection (guarding) and organizational and regime measures.

Risk Analysis: Consideration of relevant threat scenarios, in order to assess the vulnerability and the potential impact of disruption or destruction of critical infrastructure.

Risk Assessment: Overall process of risk identification, risk analysis and risk evaluation.

Technical Security: A set of essential requirements and measures to ensure the required level of protection of infrastructure elements.

ENDNOTE

1 Cascading effects are impacts caused by disruptions or failures in an element/sub-sector/sector of the CI that continue to spread across the CI and cause a failure of dependent elements/sub-sectors/sectors - which results in escalating further impacts.

Chapter 2
Transport Infrastructures Safety and Security:
State of the Art of Transport Infrastructure Protection and Best Practices From Research Projects

Zdenek Dvorak
https://orcid.org/0000-0002-8320-1419
University of Zilina, Slovakia

Bohus Leitner
https://orcid.org/0000-0001-5314-5666
University of Zilina, Slovakia

Lenka Mocova
University of Zilina, Slovakia

ABSTRACT

The chapter focuses on explaining the causal links between security and safety within the transport infrastructure. The chapter presents the current state of protection and resilience of the transport infrastructure in Europe. The introductory part will focus on comparative analysis of the latest information on transport infrastructure. In addition, an overview of current European transport infrastructure directives and legal acts will be included. This will be followed by an analysis of the results of scientific research projects at European level. As a case study, the state of security and safety in the transport infrastructure of the Slovak Republic will be presented. The following will be a generalized set of recommendations to improve security and safety in the transport infrastructure. The chapter will be supplemented by relevant sources of information on the issues addressed.

DOI: 10.4018/978-1-7998-3059-7.ch002

INTRODUCTION

The historical development of the human population and individual state units has been closely connected with the development of transport infrastructure since ancient times. Successful empires expanded their territories thanks to the ability to transport troops and logistically secure them. In the history of wars, battles took place on the ground and water. The beginning of the 20th century saw the expansion of fighting into the airspace. Later, submarines were deployed which allowed effective combat even under the sea. With the advent of the cosmic era around 1960, the military started to discuss another domain - space combat. With the rapid development of ICT and global connectivity on the Internet, another domain has opened - cyberspace.

Social security research is a challenge of the beginning of the 21st century. The society is being globalized, modern information and communication technologies (ICT) penetrate all areas, and the social perception of security is changing continuously. During the 20th century, the greatest threats to the human security were the wars, but, currently (2019) the greatest threats are the climate change, hybrid threats, cyberspace threats, uncontrolled mass migration, and terrorism.

The society relies on the functioning of key infrastructure systems - drinking water, sewerage, electricity, gas, oil and heat supply, road, air and water transport, cloud storage, servers, key optical networks, etc. In 2008, the European Union issued a directive – Council directive 2008/114 / EC of 8 December 2008 on the identification and design of European Critical Infrastructures and the assessment of the need to improve their protection. In L 345/81, it defines two sectors (energy and transport) and eight sub-sectors (1. Electricity, 2. Oil, 3. Gas, 4. Road transport, 5. Rail transport, 6. Air transport, 7. Inland waterways transport and 8. Ocean and short-sea shipping and ports), (Council Directive, 2008).

At present, modern society is extremely dependent on the functioning of networks of various kinds. Besides interpersonal relationships and social networks, which have always been important, people are increasingly dependent on the reliability of network systems such as pipelines, transport networks or the Internet, and related applications that use information flow and sharing.

During many years of research, researchers have come to generally valid conclusions that need to be communicated to the scientific and professional public in this form. The publication aims to extend the current level of knowledge in the field of protection of point, line and area objects of transport infrastructure. Furthermore, using appropriate methods to unify the approach to security management, safety management and resilience of selected transport objects. Another objective is to contribute to improving cooperation between the various entities involved in the preparation and management of extraordinary and crisis events and provide a basic orientation to all students of security management, safety management, and critical infrastructure protection.

The authors fulfill the objectives in question by describing the state of protection and security of transport infrastructure. Furthermore, an overview of current European directives and legal acts is presented, followed by a presentation of results and recommendations from scientific and research projects at Central European and national level. The next part presents the researchers' opinion on the direction of science development in the field of protection and safety of transport infrastructure objects.

BACKGROUND

The definition of basic terms is a prerequisite for understanding the issue of "transport infrastructure safety & security". Some of the terms used are generally established and are used almost identically in several countries, while others differ from one country to another. For this reason, in this subchapter, the researchers present the definitions, which are used later in the publication.

Security - according to the Cambridge dictionary "is a protection of a person, building, organization, or country against threats such as crime or attacks by foreign countries". According to researchers, security is the property of the system (object, activity, etc.) not to threaten either persons or the environment (especially the environment) (Cambridge, 2019).

Technical safety - Safety according to the Cambridge dictionary: "is a state in which you are safe and not in danger or at risk". According to safety researchers, it is a system setting to prevent losses caused by the unintentional activity of benevolent actors (Cambridge, 2019).

The concepts of safety and security are complementary to each other, today's research is aimed at creating comprehensive security solutions that require strong safety and security measures.

The Cambridge dictionary defines risk as "the possibility of something bad happening", according to ISO 31000: 2009, it is the "effect of uncertainty of intent", which means that the word "risk" refers to both positive and negative options (Cambridge, 2019).

Acceptable risk is defined as the risk that stakeholders, taking into account all operational and humane conditions, are willing to bear.

According to the Cambridge dictionary "transport is the movement of people or goods from one place to another," according to researchers, transport is the movement of means of transport by road or the activity of transport facilities carrying out transport (Cambridge, 2019).

A transport system is an ordered set comprising transport infrastructure, means of transport and technology, management and information systems, the legal framework and people.

An infrastructure according to Cambridge dictionary can be defined as "the basic systems and services, such as transport and power supplies, that a country or organization uses in order to work effectively" (Cambridge, 2019).

Transport Infrastructure is a collection of all facilities necessary for comprehensive transport provision.

Transport policy is defined at the European level by the European Commission, at the national level by governments. In the field of rail transport, the current objectives of the European transport policy are aimed at opening up the transport market and free competition, increasing interoperability and security of national networks, and developing railway infrastructure.

CURRENT EUROPEAN TRANSPORT INFRASTRUCTURE LEGAL FRAMEWORK

The legal framework of safety and security aspects of transport infrastructure is determined by many international agreements, technical standards, European regulations and recommendations, national laws and regulations at the sectoral level and internal regulations of specific companies that manage transport infrastructure. Given the complexity of the issue, the researchers focused only on land transport - road and rail transport infrastructure. This basic limitation excluded air transport, which has historically been set at a different quality level in safety and security aspects. Water transport (on navigable rivers, lakes, and canals) is relatively underrepresented in Central European countries.

Transport systems are a fundamental part of our modern social and economic infrastructure. In the case of the European Union, transport systems enable the free movement of goods and services, generation of economic wealth across all member states, and the import and export of goods and services to and from the region.

For transportation to take place, there are four key components - Modes, Infrastructures, Networks, and Flows. From the point of view of the chapter, the key component of transport is the area of transport infrastructure, see Figure 1 - Components of Transport Modes.

Figure 1. Components of transport modes (Transportgeography, 2019)

Modes
Mobile elements of transportation. Conveyances (vehicles) used to move passengers or freight.

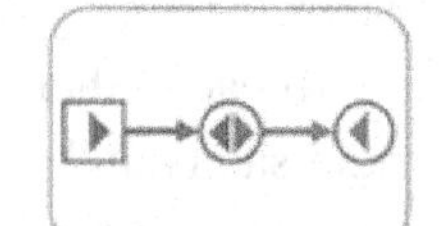

Infrastructures
Fixed elements of transportation. Physical support of transport modes, such as routes and terminal.

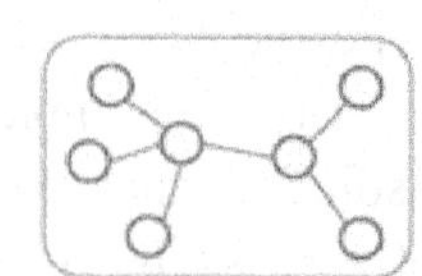

Networks
Functional and spacial organization of transportation. System of linked locations (nodes).

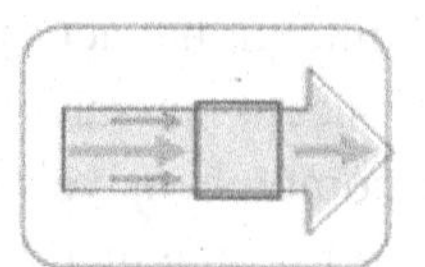

Flows
Flows have origins, intermediary locations and destinations. Movements of people, freight and information over network.

Modes. They represent the conveyances, mostly taking the form of vehicles that are, used to support the mobility of passengers or freight. Some modes are designed to carry only passengers or freight, while others can carry both.

Infrastructures. The physical support of transport modes, where routes (e.g. rail tracks, canals or highways) and terminals (e.g. ports or airports) are the most significant components. Infrastructures also include superstructures which are movable assets that usually have a shorter lifespan. So, for an airport, the infrastructure would be assets such as the runways while the superstructure would be the terminals and control equipment. For a port, the infrastructure would be piers and navigation channels while the superstructure would be cranes and yard equipment. In the area of assessing the resilience/vulnerability of transport infrastructure, infrastructure objects are usually divided into point, line and surface objects.

Networks. A system of linked locations that are, used to represent the functional and spatial organization of transportation. This system indicates which locations are connected and how they are serviced. Within a network, some locations are more accessible (more connections) than others (fewer connections).

Flows. Movements of people, freight and information over their respective networks. Flows have origins, intermediary locations, and destinations. An intermediary location is often required to go from an origin to a destination. For instance, flying from one airport to another may require a transit at a hub airport. Source: (Transportgeography, 2019)

The aim of the researchers is to present real knowledge that arises in the framework of research tasks at universities and research institutes, especially in the Czech Republic and Slovakia. Each task in the first part contains an analysis of the legal framework. The aim of the research teams is to discuss existing legal acts to contribute to the updating, amendment, and precision of both technical standards and norms as well as legal documents - laws, and regulations.

On transport infrastructure protection, researchers cite the EU Directive 2008/114 (Council Directive, 2008) as a key legal act directive has launched a Pan-European debate on critical infrastructure protection. The individual Member States gradually issued their own laws and decrees in 2009 -2012, fully implementing the Directive into life. A Pan-European debate on the content and need to amend EU Directive 114/2008 in 2018 was a breakthrough, see (Evaluation study, 2019)

In the context of the focus of the EU Directive 114, the possible impacts can be classified into:

- The Fatalities Criterion, i.e., life casualties or those with the need of subsequent hospitalization.
- The Economic Effects Criterion, assessed in terms of the significance of the economic loss threshold.
- The Public Effects Criterion, assessed in terms of their impacts on public confidence, physical suffering.
- The Disruption of daily life including the loss of essential services (i.e. cross-cutting criteria).

The values of these criteria for European Critical Infrastructure aspects have been defined in the Directive.

Further relevant European juridical documents were:

- European Commission. (2011). White Paper. Road Map to a Single European Transport Area – towards a competitive and resource-efficient transport system.
- European Commission. (2011). White Paper. Road Map to a Single European Transport Area – towards a competitive and resource-efficient transport system.
- European Commission. (2012). Research and innovation for Europe's future mobility. Developing a European transport-technology strategy.
- World Economic Forum. (2014). Strategic Infrastructure. Steps to Operate and Maintain Infrastructure Efficiently and Efficiently.
- European Commission. (2016). The road from Paris: assessing the implications of the Paris Agreement and accompanying the proposal for a Council Decision on the signing, on behalf of the European Union, of the Paris Agreement adopted under the United Nations Framework Convention on Climate Change. Source: (Eur-lex, 2019)

The following European documents were gradually issued for the safety area:

- Directive (EU) 2001/14/EC on the allocation of railway infrastructure capacity and the levying of charges for the use of railway infrastructure and safety certification (Railway Safety Directive, 2001).
- Directive (EU) 2004/49/EC of the European Parliament and of the council of 29 April 2004 on safety on the Community's railways and amending Council.

- 2009/460/EC: Commission Decision of 5 June 2009 on the adoption of a common safety method for assessment of achievement of safety targets, as referred to in Article 6 of Directive 2004/49/EC of the European Parliament and of the Council.
- Commission Regulation (EU) No 1078/2012 of 16 November 2012 on a common safety method for monitoring to be applied by railway undertakings, infrastructure managers after receiving a safety certificate or safety authorization and by entities in charge of maintenance.
- Commission Implementing Regulation (EU) No 402/2013 of 30 April 2013 on the common safety method for risk evaluation and assessment and repealing Regulation (EC) No 352/2009
- Directive (EU) 2016/798 of the European Parliament and of the Council of 11 May 2016 on railway safety.
- Commission Regulation (EU) No 1303/2014 of 18 November 2014 concerning the technical specification for interoperability relating to 'safety in railway tunnels' of the rail system of the European Union.

TRANSPORT INFRASTRUCTURE NOWADAYS AND FUTURE

Social changes, globalization, and informatization of society, economic development and information explosion are bringing significant changes to the modern European society. As the economy grows, there is a gradual settlement change in individual states. In 2008, there was a historical change, when, on a global scale, the number of inhabitants living in cities outgrew the number of inhabitants in rural areas. Rural areas are gradually being depopulated and several urban centers are emerging in each country, concentrating industry, trade and employment opportunities. This trend is also visible in Eastern European cities. Paradoxically, new homes, flats, and industrial buildings are being built on the outskirts of urban agglomerations, but without prior construction of transport, information and energy infrastructure. Individual city satellites and industrial parks are created without a real connection to new capacity infrastructure. As a result of not building new transport infrastructure, large traffic congestions are generated every day, causing major financial losses. Currently, in the Central European countries, (the Czech Republic and Slovakia), congestion losses are twice as high as losses caused by accidents.

The development of state-of-the-art transport infrastructure is also evolving thanks to the massive deployment of computer technology. There are dozens of software tools that allow efficient planning and design of transport infrastructure. If historically the development of motoring in the countries of Central and Eastern Europe was artificially hampered until 1989, then in recent years it is growing uncontrollably. In addition to a significant shift in the area of transport infrastructure, there was a shift in car technology to the possibility of an autonomous vehicle. Transport management and information systems are improving significantly, in some parts of the world it is referred to as telematics systems, and elsewhere they are referred to as intelligent transport systems. With the advent of 4G and 5G networks, the new transport infrastructure will be further interconnected with modern telematics devices and moving vehicles. The concepts of car-car, car-infrastructure, infrastructure-car and infrastructure-infrastructure are gradually becoming a reality.

Transport as a system is now understood as a deliberate movement of a means of transport along a transport path. There are more aspects to the transport classification. One basic division divides transport into passenger and freight, others into individual and bulk transport, or land, sea, and air transport. In this chapter, the authors will pay attention to road and rail transport as traditionally these are the two most important modes of transport in Central European countries.

For years, researchers at the University of Zilina have been among the leaders in solving a wide range of transport problems. Within the research and application projects, they create new methods, procedures, and technologies for the entire transport system. In the field of transport infrastructure, the aim is to build roads that will serve decades without a significant need for further investment.

On a global scale, the transport policy of developed countries focuses on the distribution of transport needs between individual and mass public transport. Individual transport in Central Europe is dominated by passenger cars and walking in case of short distances. The less represented are taxis, motorcycles, bicycles, scooters and other modern means of transport. A significant problem of individual transport is static transport - parking lots, parking is a frequent problem in big conurbations. One of the possibilities for modern cities is the use of telematics devices and Internet applications to guide the driver to a free parking space.

According to Drdla, public transport in the regions consists mainly of rail transport, bus transport, air transport, water transport, in cities and conurbations metro, trams, trolleybuses and unconventional modes of transport. Cog railways and cableways, maglev railways and moving walkways are used exceptionally. The dramatic increase in performance is apparent from data on the dynamic population mobility in the Czech Republic; in 1955 the average distance traveled in thousands of km per year was 2.75 and in 1985 it was 6.75. In 2015, dynamic mobility exceeded 10,000 km/year. The statistics show that it has grown more than threefold over the past 60 years. (Drdla, 2018)

Along with the history of transport, road operation rules and regulations have also evolved. In different countries around the world, the culture of driving behavior varies. Historically, safety has focused on defining rules and technical requirements for vehicles and the quality of transport structures. The development of informatization has brought new possibilities to the transport system to further increase the capacities of individual infrastructures. Gradually, security divided from safety. The term safety is understood as technical safety, whereas security mainly as organizational security. These terms have become very prominent in aviation, and this process is ongoing in other modes of transport. Researchers at the Technical University of Kosice study the issue of aviation security - e.g. (Kelemen et al., 2018).

In relation to transport infrastructure, the term safety is understood as the safety of technical facilities. An example is a road tunnel, where safety refers to those technologies, technical equipment, whose task is to increase the safety of road users. The examples include variable traffic signs, warning and emergency systems, passenger evacuation systems, etc. When applying the term security, the focus is on organizational measures in relation to the tunnel, such as reducing the speed before entering the tunnel, increasing the distance between vehicles before entering the tunnel, preventing the entry of oversized vehicles or vehicles carrying dangerous goods, etc.

Recently, the issue of cybersecurity, which affects all fields of human activity, seems to be crucial. In the Czech Republic, this problem is studied by the University of Defense in Brno and the Brno University of Technology. Petr Hruza is one of the most prominent authors (Hruza et al., 2014). The timeliness and breadth of cyber protection and cybersecurity are becoming dominant topics also in the processes of information and control systems of transport and transport infrastructure. At the University of Zilina, the problem of cybersecurity in transport was already addressed by a dissertation thesis in 2005-2008 (Dolnak, 2008). The issue of cyber protection is facing a problem with the confidentiality of information, or the classification of some information as sensitive. Therefore, global authors are only marginally concerned with this topic in the field of transport. Any published piece of information may be misused in an attack on information security. According to recent research, the attention of the attackers is also focused on transport. The results of a questionnaire survey carried out in 2018 in the Czech Republic indicate that 2/3 of companies do not detect a cyberattack on their own information system within 200 days after the attack, fewer than 1/3 find it within 10 days after the attack and only a few percent of companies discover the attack on the day of its performance.

When analyzing current knowledge in the field of transport infrastructure security, it is important to focus attention on important national and international institutions that manage processes in the field of transport.

At the global level these are:

The United Nations Economic Commission for Europe (UNECE) was established in 1947 and one of the important agendas is transport, especially:

The UNECE Convention related to Border Crossing Facilitation was created to enhance transport facilitation and security. The first significant step was the creation of the Framework of Standards to Secure and Facilitate Global Trade (SAFE) by the World Customs Organization. The Inland Transport Committee set up an expert group on security that gradually develops the following agenda:

- Transport safety and transport security.
- Inventory of regulatory initiatives at the national and international level.
- Inventory of standards, initiatives, guidelines, best practices by the private sector.

In 2013-2018, seminars, workshops, and discussion forums on these topics were organized – a workshop on rail security, Securing Global Transport Chains - a workshop on security of transport infrastructure.

Road Traffic Safety focuses on safety agenda in road transport and pays close attention to the following issues: traffic rules, toad signs and signals, construction and technical inspection of vehicles, road infrastructure, driving times and rest periods for professional drivers, safe transport of dangerous goods and hazardous materials. In respect to the addressed issue, the key activities are in the field of transport infrastructure. The Convention on Road Signs and Signals has been in force since 1968. In 2006, the Convention was amended and published as Convention on Road Signs and Signals of 1968 European Agreement Supplementing the Convention and Protocol on Road Markings, Additional to the European Agreement.

At European Union level, this includes:

Within the European Union, the most important institution is The EU Agency for Railways, whose main mission is to create the Single European Railway Area and ensure the functioning of the interoperable European Rail Traffic Management System. As part of the sub-activities, EU ERA focuses on safety culture, Rail Accident Investigation, Technical Specifications for Interoperability, Safety Management System, Conformity Assessment, Common Safety Methods, Transportation of Dangerous Goods (EU ERA, 2019)

Concerning significant supranational organizations, dozens of organizations could be listed in this section of the publication, authors cite those that have an impact on the functioning of rail transport. A key organization is The World Railway Organization (UIC), whose agenda addresses Freight, Information and Control Systems / Signaling, Environment, Safety / Security and Standardization (UIC, 2019)

An important international organization is the Organization for Security and Co-operation in Europe (OSCE) and its cross-sectional panels, e.g. in 2012 it was the Inland Transport Security Discussion Forum (OSCE, 2012)

Important relevant national institutions in the Czech and Slovak Republics include:

Ministry of Transport of the Czech Republic/Ministry of Transport (MD CR, 2019), Ministry of Transport and Construction of the Slovak Republic (MDV SR, 2019), both ministries are conceptual, strategic institutions with the mission of creating legal environment in the field of transport. Another central authority in the Czech Republic is the Railway Authority whose mission is to manage rail transport and railways. In the Slovak Republic, there is a Transport Authority with competence in the field of railway transport, civil aviation and inland navigation.

At the national level in the Slovak Republic, there are state organizations created for the administration of transport infrastructure for individual transport modes. Their task is to manage the railway infrastructure of the Slovak Republic/Slovak Railways, road transport/ National Motorway Company and District Road Administration, water transport - river basin enterprises / Slovak Water Management Enterprise of the Danube, Váh, etc. In air transport, there is the Air Traffic Services of the Slovak Republic, managing air traffic in Slovakia.

In addition to the above, there are some associations and non-profit organizations in the field of transport infrastructure and its protection, the examples include:

- International Association of Public Transport (next UITP) is a non-profit advocacy organization for public transport authorities and operators, policy decision-makers, scientific institutes, and the public transport supply and service industry. UITP has 1,600 member companies coming from 99 countries. The main mission of the organization is enhancing the quality of life and economic well-being by supporting and promoting sustainable transport in urban areas worldwide. Source (UITP,2019)
- The European Conference of Transport Research Institutes (ECTRI) is the leading European Research Association for Sustainable and Multimodal Mobility. Its main mission is promoting transport research and enhancing its scientific quality and effectiveness, providing independent, evidence-based advice to decision-makers in Europe, incorporating and representing the foremost European transport research institutes and universities. Six strategic objectives of ECTRI can be defined as:
    - Shaping the future European Research Area (ERA)
    - Promoting participation in R&D projects
    - Servicing its Members through information, exchange, and representation
    - Fostering education and training
    - Disseminating research results
    - Joining research capabilities and infrastructures Source: (ECTRI, 2019)

TRANSPORT INFRASTRUCTURE SAFETY MANAGEMENT

Safety management of transport infrastructure objects is a specific activity, which is based on the responsibility of the top management of the organization. Creating a safety policy is key to the functioning of safety management. In it, safety is characterized as the highest priority and value of the organization and must become one of the organization's strategic goals. The practical goal of safety management is to create a functional system and use proper security management methods. The system must ensure the safety of all processes and activities while at the same time avoiding the risks of causing damage to life and human health, property damage and environmental damage. The principles of safety management and their real application have become part of the corporate environment and a priority of the safety community since 2010. They are currently part of the win-win-win-win philosophy, where the first win is the customer win, the second win is the win of a supplier of goods or services, the third win is a win for a company/state, and the fourth win is a win for the environment.

The basic principles of safety management were gradually defined by researchers and tested in practice:

- Commitment to the Safety Principle - includes a statement of commitment by the senior management. Safety is declared as the highest priority of the organization.
- Responsibility for safety - the principle requires that all employees of the organization have personal responsibility for their activities with regard to safety and that management is responsible for the safety processes and activities within the organization.
- Safety planning - is an important prerequisite for the active implementation of the Safety Management System (SMS).
- Setting safety standards - the principle ensures compliance with applicable safety requirements and the organization's efforts to adopt internationally recognized safety standards and best practices in safety management.
- Safety management - confirmed by the commitment to a clear and effective approach to safety through the formal SMS structure.
- Achieving security - the principle requires that the means, processes, procedures, and resources used (e.g. risk assessment, incident reporting, and investigation, etc.) ensure a high level of organization's safety objectives and their achievement.
- Safety assurance - includes the means, processes, procedures, and resources to demonstrate compliance with safety standards and provides the required evidence of the level of safety achieved (e.g. safety review, safety records, etc.).
- Promotion of safety - the principle ensures that safety instructions and key safety information are disseminated throughout the organization, communication on safety issues is promoted and changes systematically aim to improve safety.

In the specific conditions of companies responsible for the development and maintenance of transport infrastructure, safety management systems (SMS) are gradually being introduced. Within the corporate environment, this represents a systematic approach to safety management, starting with a change in organizational structure, determining responsibility, defining policies and procedures. The established SMS of an organization forms a part of the overall organization management system. The basis for creating an SMS in each organization is:

- Internal safety legislation/standards.
- Described external and internal safety risks.
- The size of the organization's assets and the willingness to invest in their safety.

The real basis of a functioning SMS in a company is the determination of available means and forces, as well as the determination of regime measures to ensure, prevent and effectively protect it against events with negative consequences. A set of basic institutional and systemic safety tools are identified (Belan, 2015):

- Management elements of the organization - top management, line managers.
- Elements of the organization's safety management - safety manager, professional safety managers.
- Safety staff, both own and outsourced.
- Safety information on the external and internal safety environment and possible safety risks.
- Identification of owners of decision-making and communication processes.
- A list of related safety legislation.
- A set of regulations and restrictions in the form of regime measures.
- Forces and means to protect persons, property and the environment.
- Mutual links and relationships.

For a graphical representation of the security management of transport infrastructure managers, it is describing by authors graphically to summarize the basic pillars of security management, see Figure 2.

Figure 2. System of security management of the organization

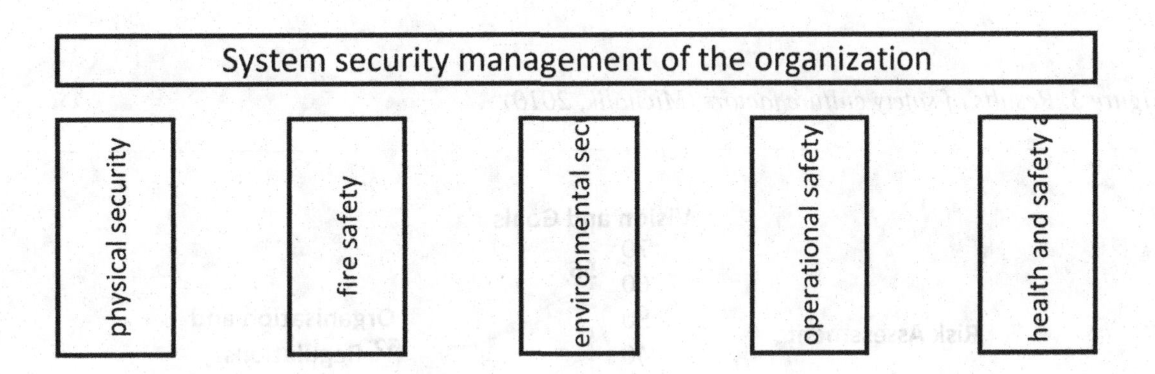

Safety and Security Culture

The concept of safety and security culture in transport is relatively new, it was first used in the IAEA report in 1986. Currently, it is understood as a complex of cultures in relation to both an individual and an organization. Within the safety and security culture, personal attitudes and rules applied by the organization are presented. The aim is to create and change environments to prevent accidents. In the Central European area, the safety and security culture is understood as a set of standards and values, knowledge, words and symbols that are continuously communicated between co-workers in the company.

If safety and security cultures are introduced into a business, then their following features will be manifested:

- Safety and security have become common values for all members of the organization.
- Each member of the organization acts independently, actively, and deliberately and safely supports safety and security culture processes.
- Safety and security standards are accepted as everyday normal behavior.

Several factors influence the state of the safety and security culture, such as:

- Values
- Standards and regulations
- Knowledge
- Assumptions
- Official statements
- Symbols
- Leadership

Within the available resources, several measurements of safety and security culture factors have been conducted. Figure 3 presents the status of the measured values in the company responsible for the maintenance and recovery of the transport infrastructure. It is clear from the above that the factors of risk assessment, safety talks and instruction and leadership and participation are not at the expected level. Therefore, the measures that a company must take in the safety and security culture processes will address these three lowest-rated factors.

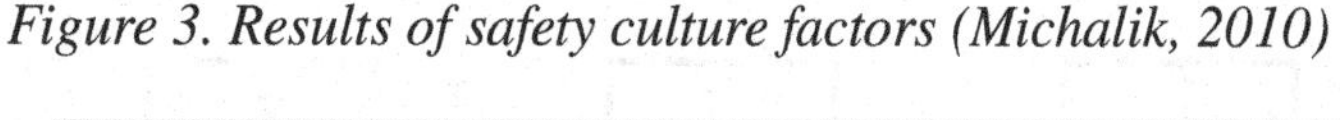

Figure 3. Results of safety culture factors (Michalik, 2010)

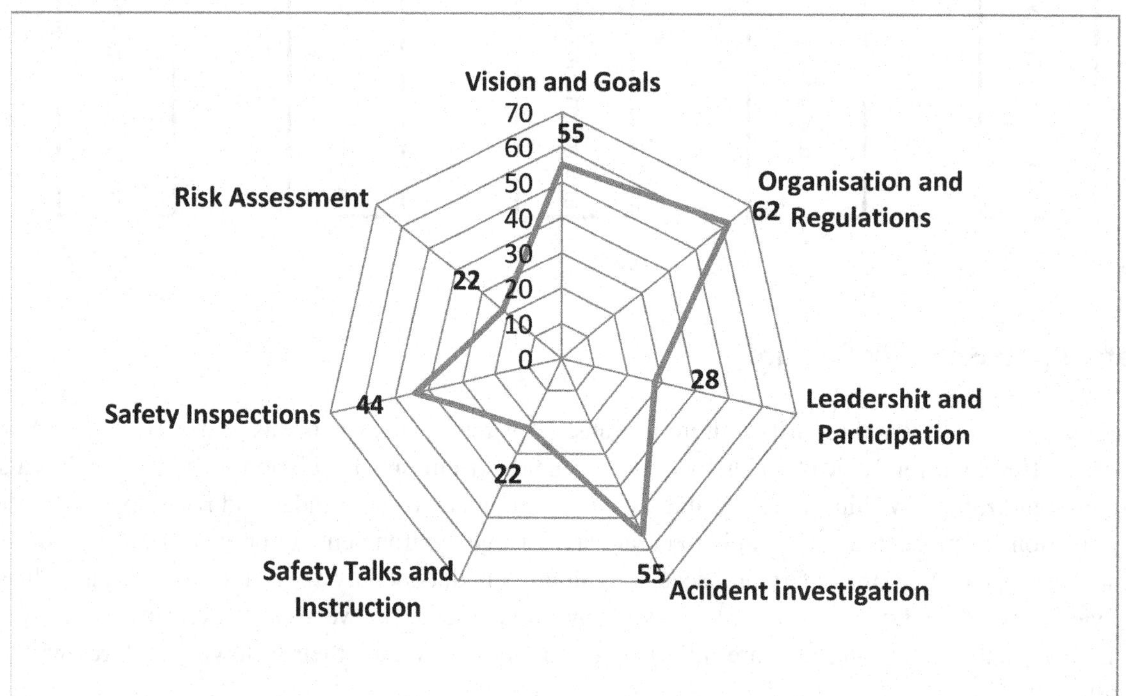

Three levels of safety culture have been described as part of the implementation of the safety culture in the corporate environment within the transport infrastructure. The first are manifestations of behavior - corporate rituals, and visible manifestations - clothes, design manual. The second level is represented by values that can be derived from the company philosophy and strategic goals. The third level is human prerequisites - nature, sense of responsibility, motivational aspects, subconscious and conscious behavior. It is advisable to implement a simplified model of safety culture during the implementation of safety and security culture. This model is graphically presented in Figure 4. Its goal is to harmonize what we believe in with what we say and what we do.

Figure 4. A simplified model of safety culture (Michalik, 2010)

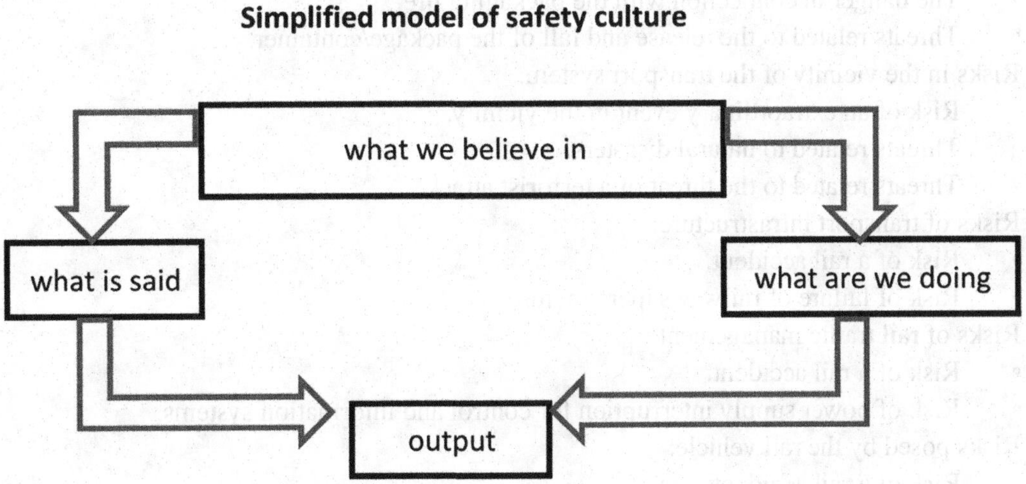

Simplified model of safety culture

what we believe in

what is said

what are we doing

output

SAFETY MANAGEMENT

Researchers have long been involved in risk management and transport safety management. These topics overlap. The ISO standard 01 0380 was issued in the Slovak Republic in 2003 (ISO 310000:2018, 2018). In parallel with its publication, researchers solved a number of research projects focused on risks in railway transport, e.g. Management of Transport in Case of Crisis Situations (2002-2003), Organization of rail and road transport in the event of disruption of the transport routes (2003-2005), Analysis of forces and means to ensure construction and remediation activities in rail transport in crisis situations (2006-2007). The technical standard 01 0380 was replaced by the technical standard ISO 31000, issued in 2009. In the active application of technical standards in solving scientific and research projects, researchers used the following risk classification:

- **Group 1:** Risks dependent on human activity:
  - Technogenic risks.
  - Sociogenic risks.

    - Combined anthropogenic risks.
- **Group 2:** Risks independent of human activity:
    - Natural hazards - meteorological, tectonic, tellurian and topological.
    - Cosmogenic risks.
- **Group 3:** Combined risks.

More detailed risk classification was used for rail transport (Dvorak et al., 2010):

- Risks associated with the transportation process:
    - Risk of an extraordinary event.
    - The threat in connection with the packaging failure.
    - The danger in connection with the packaging fire.
    - Threats related to the release and fall of the package/container.
- Risks in the vicinity of the transport system:
    - Risk of an extraordinary event in the vicinity.
    - Threats related to natural disasters.
    - Threats related to the threat of a terrorist attack.
- Risks of transport infrastructure:
    - Risk of a rail accident.
    - Risk of failure of railway superstructure.
- Risks of rail traffic management:
    - Risk of a rail accident.
    - Risk of power supply interruption for control and information systems,
- Risks posed by the rail vehicle:
    - Risk of a rail accident.
    - Risk of a technical failure of the wagon or a drive unit.
    - risk of power supply interruption for the drive unit.
- Risks caused by humans:
    - Risk of a rail accident.
    - Risk of a technical failure of the rail vehicle.
    - Risk of a strike of operating employees.
    - Threats related to the threat of a terrorist attack.
- Risks of an economic nature:
    - Risk of power supply interruption for the drive unit.
    - Risk of a strike of operating employees, etc.

The issue of transport risks has also been studied by (Iromuanya et al., 2015) a (Crnčevič, 2015), further development of knowledge was related to the application of ISO standard 31000, within the project - Automated comprehensive information system for remote management of crisis situations in rail transport with an emphasis on critical infrastructure, theoretically possible risks were summarized in an extensive database. Researchers have found that in theory there are 2400 different risks of rail transport. In the real world, it makes sense to consider about 600 risks. In the process of preparation and testing of the proposed information system, the aim was to create an information system that would enable automated activation of procedures for resolving the railway crisis event based on the remote control code sent. The main method was to send a notification to the designated crisis managers alongside a list of all the tasks for which they are responsible and which they have to conduct after activation.

In the subsequent phase of crisis event solution, the information system was supposed to determine the real state of performed tasks by smartphones of crisis managers. On the basis of the actual situation, its task was to modify the relevant crisis operational plan in real-time and send updated tasks to all crisis managers.

The set of methods and principles for the remote management of crisis events that was developed during the project was abbreviated KISDIS. All the principles described in the KISDIS methodology were implemented in the KISKAN information system. Source: (ANAKAN, 2019)

A basic task for remote crisis management is a set of well-prepared measures. In the process of defining the appropriate measure, it is necessary to know the context of its operation. Risk research is a multidisciplinary issue, always requiring a multidisciplinary approach, typically involving security, technical, scientific, environmental, social and economic aspects. Risk can be quantified and some quantification is necessary to determine its acceptability. The level of risk is specified by quantitative assessment. It is a numerical value (e.g. the estimated number of deaths caused by events per year) or a numerical function that describes the relationship between the probability and consequence of the selected source of risk. The source of risk may be equipment, activities or technologies, but also other objects or processes that endanger human or the environment. The project used the following basic relationship to calculate risk: $R = P * I$ (1)Where R is risk, P is probability and I is impact, more detail in ISO 31000.

This relationship represents the basic principle. The risk assessment process, based on the technical standards for risk management and the Common Security Method, consists of the following activities: determining the context, identifying threats, analyzing risks, assessing risks, proposing measures to mitigate threats and monitoring the current threat status; see Table 1.

Table 1. Risk assessment by KISDIS methodology (KISDIS, 2019)

Risk Assessment	1. Context - relations with the environment	Examining a specific object in relation to the surrounding environment - multi-sectoral view of the transport system
	2. Identify threats and define assets	Select a specific transport infrastructure location - e.g. tunnel, identify in detail all theoretically possible threats
	3. Risk analysis	Create a risk map - qualitative, semi-quantitative and quantitative methods
	4. Risk assessment and draft measures and monitoring	Measures towards active risk policy, risk prevention and monitoring

Minimizing risks and increasing safety levels is possible in three ways, specifically by:

- Organizational measures that represent various oral or written instructions, regulations and orders. They are essential in a comprehensive approach to security.
- Technical means (road transport equipment + road transport infrastructure + intelligent transport systems), which must be clearly preferred. With them, the risk can be significantly reduced, sometimes eliminated.
- Training measures that have limited effectiveness but are of great importance for building a safety culture.

To simplify the entire risk assessment system, the methodology of threat types has been applied. The type threat has similar characteristics for multiple risk locations and events and can, therefore, be seen as a template to facilitate risk assessment. The basis for identifying type threats is to determine their name.

The threat type names contain four basic pieces of information:

- **Event:** How the operation of the system was interrupted.
- Activation of the source of risk - an action that causes an event at the source of risk.
- **Location:** The initial specification of the location of the source of risk.
- **Cause:** The primary fact activating the source of risk in the most general form.

In order to create the names of the type threats, it is necessary to clarify whether there are enough characteristics in the field of activity for which the type threats are processed to create the type. This means that the elements under consideration are repeated in the system with such a degree of similarity they can be considered type-specific. An example of the type is prefabricated houses, which are built according to type projects and practically differ only by a specific location. The significance of the type is that the emergency procedure applicable to one type (prefabricated house) can also be related to other representatives of the given type (prefabricated houses).

For a proper understanding of the type selection, limitations for the name creation have been defined.

The system of naming the type threats is based on the fact that a risk is understood as a single description of the risk, which includes everything that can happen and could endanger the functionality of the system under investigation (e.g. railways). The following procedure is used to create the name of the type threats:

- The type of threat locations in the structure corresponding to the location of the source of the risk (where this may happen).
- The sources of risk and how they are triggered (who or what is the potential bearer of the risk and what can happen to it).
- Causes of crisis events, such as the primary activation (why this can happen).
- Event (action) disturbing the essence of the system balance (change of the monitored process).
- The consequences of crisis events (which events they may subsequently trigger - cascade, synergy).

The possible threat locations in rail transport were divided into the open line, the railway station with a track lead, the train and the surrounding area. In exploring the risk sources and the methods of their activation, a total of 50 activation mechanisms of risk sources for railway transport were defined, e.g. railway line bombing, turnout failures, fire in the line vicinity, rail line explosion. Four types of events that may occur were defined – line closure, change in train schedule, change in technology and change in the scope of services. The key result for further computer processing was the creation of a template for recording the name of the type threats, see Table 2.

The real name of the type threat is formed by the combination of aspects: Immobilization of the towing vehicle on the train caused by human error. Theoretically, the number of types threat names is given as the product of a number of events * a number of risk source activations * a number of locations * a number of causes. Under the conditions of rail transport in the Czech Republic, the total number of types threats were 627.

Table 2. Threat type template (KISDIS, 2019)

Event	Activation of risk source	Place	Cause
change in operating conditions due to	the immobilization of the towing vehicle caused	on the train	caused by a human error
change in operating conditions due to	theft of operating equipment	on the line	caused by crime

When assessing risks, it is also important to take into account the degree of consequences. In the real project, the researchers considered the primary consequences - loss of property of citizens, businesses and public administration, loss of life and health and environmental damage. The secondary losses included the restrictions on the scope of public transport, changes in train schedule, road traffic exclusion, losses on the infrastructure manager's assets. Last but not least, the tertiary consequences were also defined - damage to the traveling public, changes in the routing of public transport lines, endangering the economic interests of stakeholders, etc.

RESILIENCE OF TRANSPORT INFRASTRUCTURE

From the information and context presented so far, it is clear that transport is vital to a well-functioning economy. Physical infrastructure is needed for its implementation. The most common types of transport infrastructure are roads (roads, railways, waterways) and ports (dry ports, seaports, airports, terminals, etc.). These are parts of the physical so-called hard infrastructure transport systems that include soft infrastructure such as policies and regulations and institutions responsible for the planning, financing, operation, and maintenance of such systems. All these components are interconnected in transport systems.

However, if sudden adverse events occur – e.g. major technical failures, serious accidents, terrorist attacks, natural disasters (floods, earthquakes, and other naturally occurring adverse events) - the transport systems must have the required level of resilience in order to avoid serious disruption and loss of functionality. The increasing degree of interdependence of hard and soft infrastructure components can significantly increase the vulnerability of such systems. This can make negative events more serious, longer-lasting and with greater impact on society.

In the past, individual transport infrastructure systems and services were less complex. The causes of the disturbance were relatively easy to identify and repair, (e.g. railroad switches were operated manually and repairs were easier to implement, as opposed to the current remote-controlled operation via communication lines and power supply). The physical elements of the infrastructure were not very interdependent. Infrastructure is now a key element for increasingly complex global supply chains and transport operations. Any disruption to the continuity of its operation can spread more easily in the form of a cascading effect along the entire length of the supply chain. Transport systems have become more vulnerable to disasters, mainly because they are usually spread over a wider geographical area and depend on the functionality of several sectors. Possible examples of cross-sector interconnectivity are e.g. dependence on energy supplies, telecommunications services, but also the availability of specialist repairers after the system disruption. The resilience of the transport system can be increased by systematically assessing the threats and risks posed by inter-subsystem dependencies and ensuring the ability of transport systems to absorb their failures and loss of functionality quickly and effectively.

Both research and practical solutions should aim at procedures and measures to improve the resilience of transport systems in terms of the number of dependencies arising along with such systems, in particular of a linear nature. The literature review shows that a common framework for building the resilience of transport systems is not commonly available. While there are ongoing discussions on how to capture and assess resilience, there is a perceptible increase in the interest of the academic community in objectifying the assessment of the resilience of infrastructure systems. The works of Rehak (Rehak et al., 2016) and (Rehak et al., 2019) can serve as an example.

Regarding infrastructure as a technical system, the approach of Rourke (2007), which is a follow-up to the work of Bruneau (2003) is frequently cited. According to this approach, more detailed parameters for assessing resilience have been identified through document analysis to adapt them to the specific conditions of transport systems. The presented framework was divided into several parameters, which are particularly relevant for large infrastructure systems where transport infrastructure occupies an important place, see Table 3.

Table 3. Resilience parameters for transport systems (KISDIS, 2019)

Robustness	**Redundancy**
Status of assets Capacity to withstand climatologic Physical interdependency with other systems Geographical interdependency Logical interdependency	Local workarounds and substitutions Availability and capability of alternative routes Availability and capability of alternative modalities
Resourcefulness	**Rapidity**
Availability of emergency funding Availability of expertise and manpower Availability of equipment and materials for restoration and repair	System downtime Restoration / repair time

Robustness

It is the ability of the system to withstand threats or deviations without the loss of functionality. In this context, the technical condition and the operational parameters and properties of the elements play a significant role. Due to wear or neglected maintenance (e.g. corrosion, cracks in concrete), the object gradually loses its initial ability to withstand dynamic loads. The ability to withstand climate change is usually captured in the original design of the building, but the exposure may have changed over time. Such events occur increasingly frequently.

Redundancy

Regarding local failures and their solutions, an important question is whether alternative routes are available in the infrastructure networks. For example, whether the goods can be transported to their destination on another road or by rail. This is usually problematic for waterways since river systems do not usually have the possibility of diverting.

Resourcefulness

If natural disasters disrupt the infrastructure or the transport process there, the flexible transport system has the ability to restore its functionality, e.g. by temporary restoration, temporary bridging, etc. Unforeseen costs needed to cover costs and manage emergencies need to be continually planned so that they are available at all times and system downtime is minimized. Expertise and a skilled workforce are also needed. Sometimes it can be difficult to find the right staff at a particular place and time. As with the workforce, the availability of appropriate technology, equipment and material can be a similar challenge.

Rapidity

If the infrastructure is resilient, it means that it is designed and operated in a manner that can be restored relatively quickly after a loss of functionality and that the effects of the disruption will not be unduly restrictive. By using systems, assets, and features that are easy to be replaced, involving recovery teams that are trained and have real experience in responding to emergencies and defining likely scenarios and plans to quickly resolve the expected problem, the recovery time can be significantly reduced.

The characteristics of the main dimensions of transport system resilience make it possible to define measures to reduce the impact of negative events in the transport system. Infrastructure managers and cooperating organizations are constantly facing new infrastructure challenges. Disasters regularly affect transport systems and, when they occur, their nature and context are often significantly different in causes, conditions, space, and impacts. It is, therefore, more difficult to prepare to cope with all possible combinations of the causes and consequences of a system failure related to a disaster. Organizations responsible for the smooth operation of transport systems must find appropriate recommendations in the areas of:

- Reducing the impact of a disaster
- Shortening the duration of the interruption
- Resource efficiency to achieve the desired level of resilience

These groups of measures may be particularly useful for existing systems and their elements but are essential for the development of new infrastructure or the expansion and modernization of existing systems. Building an economically and technically adequate level of resilience of transport infrastructure systems and services is currently the primary trend in the design and implementation of modern and secure transport systems.

RESULTS OF RESEARCH PROJECTS - SOLUTIONS AND RECOMMENDATIONS

Research in the field of transport and transport safety is supported at EU levels (programs, 5FP, 6FP, 7FP and H2020) where several successful projects have been solved at the University of Zilina. Within 5FP it was the CETRA project. CETRA - Centre for Transport Research at the University of Zilina is one of two Centers in the Slovak Republic chosen by the European Commission to support research infrastructures in the countries joining the EU. The main objective of its establishment was to improve the research capability of the University and to put it at the service of the social and economic needs of Slovakia, in conformity with the interest of the European Union as a whole.

In 2015-2019 it was the ERAdiate project - Enhancing Research and Innovation Dimensions of the University of Zilina in Intelligent Transport Systems, which aimed to deepen cooperation in the field of Intelligent Transport Systems, intermodal ITS solutions, decarbonisation and mobility, sustainable smart city solutions and other multidisciplinary topics (ERACHAIR, 2019)

Historically, research at the University of Zilina has been directed towards transport, with the aim of being a key institution in the field of applied research in transport. The individual faculties and units have always been largely devoted to technical, technological, software solutions for the needs of all modes of transport. In the area of security and transport infrastructure, the faculties – of Security Engineering, Electrical Engineering, Civil Engineering and the Faculty of Management Science and Informatics - work very closely together. In the field of protection and recovery of transport infrastructure, it builds on the rich experience of the former military faculty. A methodological, technological and technical framework for solving the problems of technical protection and recovery of the railway and road network was being created there for 55 years. The basic tasks in the operational preparation of the transport system were defined in solved projects of science and research. Researchers participated in the following projects:

- Analysis of forces and means to ensure construction and remediation activities in rail transport in crisis situations
- Management of Transport in Case of Crises Events
- Methodology of Railway Transport Organization in a Restricting Track Section, which has arisen by Disruption of the Transport Routes
- Organization of Railway and Road Transport in Case of Transport Route Disruption
- Traffic security in emergencies and the creation of a crisis management system at the Ministry of Transport.

With the development of the Critical Infrastructure Protection Agenda, researchers' attention has focused on developing a methodological framework for addressing critical infrastructure issues, starting with terminology, establishing a legal framework, researching methods and tools to objectify criteria for categorizing objects among critical infrastructure elements, tools for improving the object protection and researching the issue of cybersecurity. Within these activities, the following projects were solved:

- Critical Infrastructure Protection in Sector Transportation, project 2011-2014
- A process model of critical infrastructure safety and protection in the transport sector, project 2015-2018
- Risk Analysis of Infrastructure Networks in response to extreme weather, project 2014-2017
- A comprehensive model of risk management of critical infrastructure, project proposal

As part of their own research, researchers have studied European international research projects in the field of security and transport infrastructure. Substantial parts of relevant projects are presented here:

FP7: MODSAFE - Modular Urban Transport Safety and Security Analysis Source: (MODSAFE, 2019)

The European Urban Guided Transport sector (Light rails, Metros, but also Tramways and Regional Commuter trains) is still characterized by a highly diversified landscape of Safety Requirements, Safety

Models, Responsibilities, Roles, Safety Approval, Acceptance and Certification Schemes. While a certain convergence in architectures and systems can be observed, the safety life cycle still differs from country to country and in some cases even within one country. Furthermore, security items occur more and more as vital for the urban transport sector. In some cases, these items are linked to the safety of the urban transport systems. In this context "safety" is seen as everything dealing with the methods and techniques to avoid accidents. "Security" is concerned with the protection of persons and the system from criminal acts. (MODSAFE, 2019)

FP7: RAIN - Risk Analysis of Infrastructure Networks in Response to Extreme Weather Source: (Rain, 2019)

The RAIN vision is to provide an operational analysis framework that identifies critical infrastructure components impacted by extreme weather events and minimize the impact of these events on the EU infrastructure network. The project has a core focus on land-based infrastructure with a much wider consideration of the ancillary infrastructure network in order to identify cascading and inter-related infrastructure issues. The outputs from the project will result in enhanced safety and reliability of critical infrastructure networks in the case of major weather-induced disruptions and will address European policy in the areas of safety and security, inter-modality and emergency response planning. (RAIN, 2019)

FP7: INFRARISK - Novel Indicators for identifying critical Infrastructure at RISK from Natural Hazards Source: (INFRARISK, 2019)

The project will develop reliable stress tests on European critical infrastructure using integrated modeling tools for decision-support. It will lead to higher infrastructure networks resilience to rare and low probability extreme events, known as "black swans". INFRARISK will advance decision-making approaches and lead to better protection of existing infrastructure while achieving more robust strategies for the development of new ones. INFRARISK proposes to expand existing stress test procedures and adapt them to critical land-based infrastructure which may be exposed to or threatened by natural hazards.

FP7: INTACT On the Impact of Extreme Weather on Critical Infrastructures Source: (INTACT, 2019)

The resilience of critical infrastructures (CI) to Extreme Weather Events (EWE) is one of the most salient and demanding challenges facing society. Growing scientific evidence suggests that more frequent and severe weather extremes such as heatwaves, hurricanes, flooding, and droughts are having an ever-increasing impact, with the range and effects on society exacerbated when CI is disrupted/destroyed. EWE causing (cascaded) CI outages frequently cause major social and economic loss in disasters. Obviously, there is a need to build anticipatory adaptation and organizational resilience to these relatively unforeseen and unexpected EWE impacts on CI.

FP7: DEMASST - Security of Critical Infrastructures Related to Mass Transportation Source: DEMASST, 2019)

To develop adequate and well-accepted security for mass transportation in Europe and the citizens affected by it is a formidable task. The malicious threats, particularly those posed by terrorists, require a comprehensive approach: if security improvements are patchy, perpetrators are likely to find the loopholes left. With their open access points and interconnections, surface mass transportation systems are highly vulnerable, while it is technically and economically, impossible for the multiple operators to employ security measures similar to those used at airports. The project develops a highly structured approach to the demonstration program built on identifying the main security gaps and the most promising integrated solutions, utilizing sufficiently mature technologies, for filling them. DEMASST proposes to build the methodological infrastructure for this. But an optimal demo project design does not stop with finding scientific answers: the issue of turning the demonstration into innovation is top on DEMASST's agenda. And this approach will have utility also beyond transportation.

FP7: CIPRNET - Critical Infrastructure Preparedness and Resilience Research Network Source: (CIPRNET, 2019)

The Critical Infrastructure Preparedness and Resilience Research Network or CIPRNet establishes a Network of Excellence in Critical Infrastructure Protection (CIP) R&D for a wide range of stakeholders including (multi)national emergency management, critical infrastructure (CI) operators, policymakers, and the society. CIPRNet builds a long-lasting, durable virtual center of shared and integrated knowledge and expertise in CIP and CI MS&A (Modelling, Simulation, and Analysis) by integrating part of the resources of the CIPRNet partners and their R&D activities acquired in more than 50 EU co-funded projects.

FP7: SMART RAIL - Smart Maintenance and Analysis of Transport Infrastructure Source: (SMART RAIL, 2019)

The SMART Rail project brings together experts in the areas of highway and railway infrastructure research, SME's and railway authorities who are responsible for the safety of national infrastructure, The goal of the project is to reduce replacement costs, delay and provide environmentally friendly maintenance solutions for aging infrastructure networks. This will be achieved through the development of state of the art methods to analyze and monitor the existing infrastructure and make realistic scientific assessments of safety. These engineering assessments of the current state will be used to design remediation strategies to prolong the life of existing infrastructure in a cost-effective manner with minimal environmental impact.

FP7: SERON - Security of Road Transport Networks Source: (SERON, 2019)

The European road network, particularly TERN highways and TENT projects, is of major importance for the European economy and the mobility of the European citizens. A major task of highway owners and operators is to ensure high availability of all-important links. Even smaller disruptions due to traffic restrictions or failure of road network elements lead to severe traffic interferences resulting in high economic follow-up costs and negative environmental impacts. The project focuses on the development of a methodology which is to help owners and operators to analyze critical road transport networks or parts hereof with regard to possible terrorist attacks. It will evaluate planned protection measures for critical road transport infrastructures concerning their impact on security and cost-effectiveness.

FP7: STRESST - Harmonized approach to stress tests for critical infrastructures against natural hazards Source: (STERSST, 2019)

Critical Infrastructures (CIs) provide essential goods and services for modern society; they are highly integrated and have growing mutual dependencies. Recent natural events have shown that cascading failures of CIs have the potential for multi-infrastructure collapse and widespread societal and economic consequences. Moving toward a safer and more resilient society requires improved and standardized tools for hazard and risk assessment of low probability-high consequence (LP-HC) events, and their systematic application to whole classes of CIs, targeting integrated risk mitigation strategies. Among the most important assessment tools are the stress tests, designed to test the vulnerability and resilience of individual CIs and infrastructure systems.

FP7: RASEN - Compositional Risk Assessment and Security Testing of Networked Systems Source: (RASEN, 2109).

The expected result of RASEN is an approach to security assessment that consists of methods and techniques to support the following. Compositional security assessment: How the security assessment can be broken down into smaller parts and systematically composed to obtain the global assessment. Risk-based security testing: How to derivative security test cases from security risk assessment results. Test-based security risk assessment: How to verify and update of the security risk assessment based on security test results. Legal security risk assessment: How to assess and understand compliance with legal norms related to information security. Continuous security assessment: How reuse results from previous security assessments and to rapidly update the security risk assessment based on passive testing (also called monitoring).

FP7: FORTESS - Foresight Tools for Responding to Cascading Effects in a Crisis Source: (FORTESS, 2019)

FORTRESS will identify and understand cascading effects by using evidence-based information from a range of previous crisis situations, as well as an in-depth analysis of systems and their mutual interconnectivity and (inter-)dependency. FORTRESS will seek to intervene in current crisis response practices by bridging the gap between the over-reliance on unstructured information collection on one side and a lack of attention to structural, communication and management elements of cross-border and cascading crisis situations on the other.

H2020: PLASA - Smart Planning and Safety for a Safer and More Robust European Railway Sector Source: (PLASA, 2019)

he PLASA project aims at significantly increasing customer experience and system robustness in the European rail sector. On the one hand, it improves planning activities of various stakeholders in the railway system by means of precise railway simulation and, on the other hand, it provides a methodology to manage the safety of the railway system based on risk assessment.

This will be achieved by a holistic approach involving partners of the rail industry, the operators and universities.

H2020: SAFURE - Safety and Security by Design for Interconnected Mixed-Critical CYBER-Physical Systems Source: (SAFURE, 2019)

"SAFURE targets the design of cyber-physical systems by implementing a methodology that ensures safety and security ""by construction"". This methodology is enabled by a framework developed to extend system capabilities so as to control the concurrent effects of security threats on the system behavior. The current approach for security on safety-critical embedded systems is generally to keep subsystems separated, but this approach is now being challenged by technological evolution towards openness, increased communications and use of multi-core architectures.

H2020: RESIST - Resilient Transport infrastructure to Extreme Events Source: (RESIST, 2019)

The overall goal of RESIST is to increase the resilience of seamless transport operation to natural and man-made extreme events, protect the users of the European transport infrastructure and provide optimal information to the operators and users of the transport infrastructure. The project is addressed to extreme events on critical structures, implemented in the case of bridges and tunnels attacked by all types of extreme physical, natural and man-made incidents, and cyber-attacks.

H2020: SAFE-10-T - SafetSource: (SAFE-10-T, 2019)

The SAFE-10-T project developed a Safety Framework to ensure high safety performance while allowing longer life-cycles for critical infrastructure across the road, rail and inland waterway modes. Moving from considering critical infrastructures such as bridges, tunnels, and earthworks as inert objects to being intelligent (self-learning objects) the project provides a means of virtually eradicating sudden failures.

Researchers obtain information about the protection of transport objects from the world mainly thanks to books. From these sources, the authors mention the following publications - (Quayle, 2006), (Crnčevič, 2015), (Vidrikova et al., 2017), (Postranecky et al., 2018) (Moridpour et al., 2019), and journals (ECTRI, 2019), (Komunikacie, 2019), (Transport, 2019).

In the framework of the projects under investigation, researchers referred to the following publications – (Boile, 2000), (Rinaldi et al., 2001), (Bogdevičius et al., 2004), (O'Connor, 2007), (Prentkovskis et al., 2010), (Trucco et al., 2012), (Kadri et al., 2014), (Ouyang, 2014), (Young et al., 2014), (Andersson et al. 2015), (Alcaraz and Zeadally, 2015), (Zagorecki et al., 2015), (Margaritis, 2016) and (Rehak et al., 2018).

RECOMMENDATIONS

Researchers, in cooperation with partner research organizations, have long been involved in the issue of the resilience of the state transport system. Attention is paid mainly to important infrastructure objects of railway transport, road transport, and airports. In recent research projects, it was possible to generalize as one of the conclusions that old transport infrastructure objects, especially outside the Trans-European Transport Networks (TEN-T), with neglected maintenance, are a major threat to the transport services and the regional development. Building and reconstructing transport infrastructure is very important for the sustainable future of modern society. According to Hofreiter, a new understanding of human security will change in the following aspects:

- Security increasingly focuses on human beings and human society.
- Protecting territorial sovereignty is no longer paramoun.
- Human security is understood as a dynamic system.
- The company emphasizes cooperation processes and cooperation (Hofreiter, 2004).

Around 2000, the researchers solved research projects focused mainly on intelligent transport systems (implementation of telematics applications) in road and rail transport (CONNECT, EASYWAY, Intelligent Transport Technologies, and Services) projects. The functional architecture of intelligent transport systems has been defined as one of the significant benefits; see Fig. 5.

Figure 5. Functional architecture of intelligent transport systems (Dado et al., 2007)

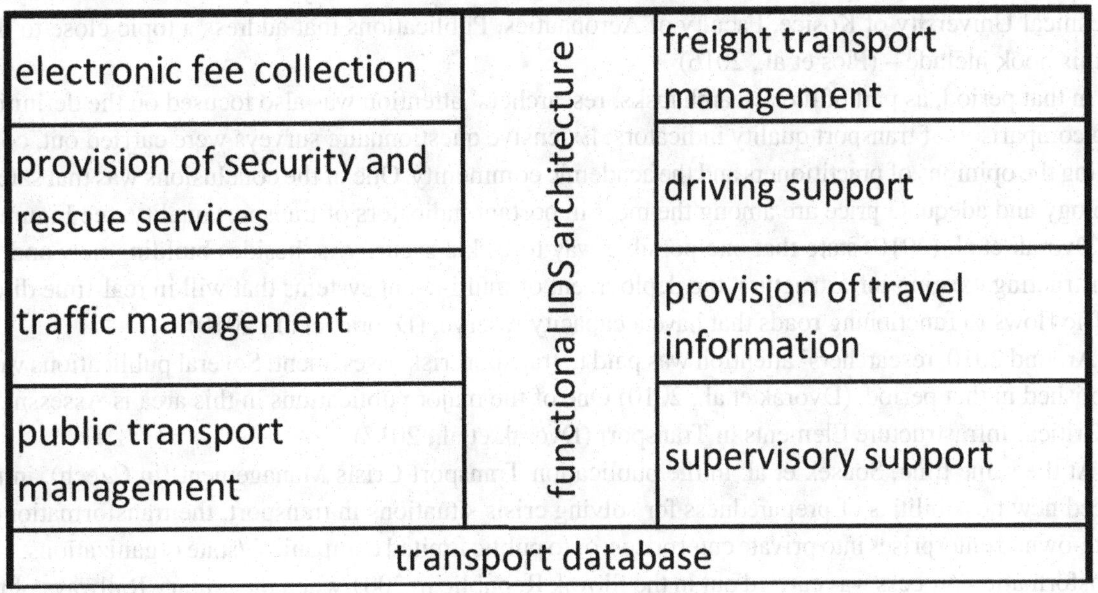

electronic fee collection	functional IDS architecture	freight transport management
provision of security and rescue services		driving support
traffic management		provision of travel information
public transport management		supervisory support
transport database		

The research focused on all modes of transport, but mainly on road and rail transport. In the area of safety and security, air transport is the most sophisticated of all transport modes. The research results

Figure 6. Transport quality indicators (Dvorak et.al. 2010)

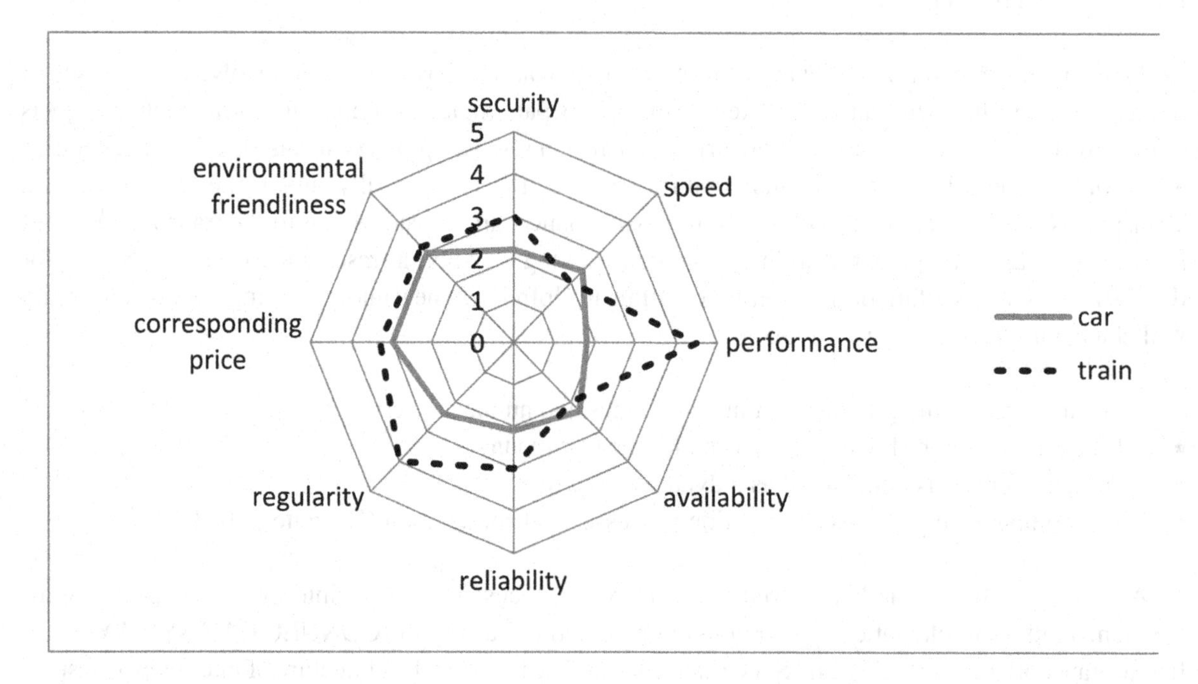

are of high quality as in the Czech Republic (the Czech Technical University in Prague, the Faculty of Transportation Sciences and the University of Pardubice, the Faculty of Transport Engineering. In the Slovak Republic, civil aviation is addressed by the University of Zilina in Zilina, Faculty of Operation and Economics of Transport and Communications. The issue of military aviation is dealt with by the Technical University of Kosice, Faculty of Aeronautics. Publications that address a topic close to that of this book include – (Plos et al., 2016)

In that period, as part of the research tasks, researchers' attention was also focused on the definition and comparison of transport quality indicators. Extensive questionnaire surveys were carried out, combining the opinions of practitioners and the academic community. One of the conclusions was that safety, ecology and adequate price are among the most important indicators of transport quality, see Figure 6.

Dvorak et.al. (2010) state that one possible way to tackle security is, besides building new and reconstructing existing infrastructure, the deployment of multi-agent systems that will in real-time direct traffic flows to functioning roads that have a capacity reserve. (Dvorak et al., 2010)

Around 2010, researchers' attention was paid to transport risk assessment. Several publications were published in that period. (Dvorak et al., 2010) One of the major publications in this area is Assessment of Critical Infrastructure Elements in Transport (Dvorak et al., 2017).

At the same time, Sousek et al. in the publication Transport Crisis Management (in Czech), introduced new possibilities of preparedness for solving crisis situations in transport, the transformation of state-owned enterprises into private enterprises, or to public limited companies' /state organizations. The transformation process was carried out in the Slovak Republic in 2001 when the unitary Railways of the Slovak Republic were divided into Railways of the Slovak Republic (railway infrastructure manager) and Railway Company Slovakia (passenger transport) and Railway Company Cargo Slovakia (freight transport). In the Czech Republic, the process started in 2003, when the original Czech Railways were divided into the Railway Infrastructure Administration (state organization) and Czech Railways, a.s. (CD) (passenger transport). In 2007, CD cargo, a.s. (freight) separated from Czech Railways, a.s.

At the European level, these fundamental organizational changes were launched in 1996 – by the European Commission document focusing on the transformation of rail transport - White Paper 1996 - A strategy for the recovery of public rail transport. Another groundbreaking document was the strategy - White Paper 2011 - Roadmap to a Single European Transport Area - Creating a competitive and resource-efficient transport system and the Transport 2050 itinerary. The objectives of the E.C. transport Themes were:

- no conventional fuel vehicles in cities - 40%
- the use of sustainable low carbon fuels in aviation
- at least a 40% reduction in emissions from shipping
- 50% shifting of passenger and freight transport from road to rail and waterborne transport. Source: (E.C. transport Themes, 2011)

CONCLUSION

The chapter Transport infrastructure safety & security was intended to extend the current level of knowledge in the field of protection of point, line and area objects of transport infrastructure. Research in this field at the University of Zilina has a tradition of more than 30 years. The original focus was on the technical protection and recovery of roads and railways. The original system of technical protection and recovery was condemned to extinction by the change of social system in 1989. New social facilities and new information and communication technologies have brought new possibilities to change the system of safety and protection of transport infrastructure.

Another aim was to unify the approach to security management, safety management and resilience of selected transport infrastructure objects using appropriate methods. A holistic approach to safety issues and the pursuit of building-integrated safety solutions in transport and transport infrastructure is a very important social issue. Everyday accidents and transport congestion require a multi-sectoral and multi-level approach. A significant increase in ICT applications is visible in both transport and transport infrastructure. Today's science and research deals with topics that will determine the state of life in 10 to 50 years. Investments in the recovery and construction of new transport infrastructure are made with a perspective of one hundred years or more.

Another aim of the researchers was to contribute to improving cooperation between the various actors involved in the preparation and management of emergencies and crisis situations. Managers of individual infrastructure objects have relatively well-prepared recovery plans within one mode of transport. They lack a certain perspective which would take into account the cooperation of several modes of transport.

The last goal was to focus on students of security management, safety management, and environmental infrastructure protection and the dissemination of project results.

Other challenges in the field of critical infrastructure protection and its safety and security aspects in transport infrastructure arise from the tasks defined in (Evaluation study, 2019):

- Refine the definition of 'protection' so as to include/account for the concept of resilience and its current use in relevant sectoral legislation.

- Set minimum elements to be covered when performing the risk analysis, and review the JRC risk assessment guidelines.
- Develop criteria and methodologies to be used during the identification process so as to assess 'cross-sector dependencies.
- Provide guidelines so as to assess the 'availability of alternatives'.
- Develop a competency framework for the SLO function.
- Organize training sessions for security officers in relation to ECI and NCI.
- Revise the template for the biannual report by the MS to the EC on risks, threats, and vulnerabilities.

The issue of security and protection of transport infrastructure as a multidisciplinary and multilevel topic requires a comprehensive approach and real application of the research results into practice.

ACKNOWLEDGEMENT

Publication of this chapter was supported by the Ministry of the Interior of the Czech Republic, grant No. VI20152019049, entitled 'Dynamic Resilience Evaluation of Interrelated Critical Infrastructure Subsystems'.

REFERENCES

Alcaraz, C., & Zeadally, S. (2015). Critical Infrastructure Protection: Requirements and Challenges for the 21st Century. *International Journal of Critical Infrastructure Protection*, *8*(1), 53–66. doi:10.1016/j.ijcip.2014.12.002

ANAKAN. (2019). Retrieved from http://anakan.cz/

Andersson, A., O'Connor, A., & Karoumi, R. (2015). Passive and Adaptive Damping Systems for Vibration Mitigation and Increased Fatigue Service Life of a Tied Arch Railway Bridge. *Computer-Aided Civil and Infrastructure Engineering*, *30*(9), 748–757. doi:10.1111/mice.12116

Belan, L. (2015). *Safety management – security management.* Zilina, Slovak Republic: EDIS.

Bogdevičius, M., Prentkovskis, O., & Vladimirov, O. (2004). Engineering solutions of traffic safety problems of road transport. *Transport*, *19*(1), 43–50. doi:10.3846/16484142.2004.9637952

Boile, M. (2000). Intermodal transportation network analysis - a GIS application. In *Proceedings of the 10th Mediterranean Electrotechnical Conference - MELECON* (vol. 2, pp.660-663). Academic Press.

Cambridge. (2019). Dictionary. Retrieved from https://dictionary.cambridge.org/

CIPRNET. (2019). Retrieved from https://cordis.europa.eu/project/rcn/107425/factsheet/en

Council Directive. (2001). Council directive 2001/14/EC on the allocation of railway infrastructure capacity and the levying of charges for the use of railway infrastructure and safety certification.

Council Directive. (2008). Council directive 2008/114 / EC of 8 December 2008 on the Identification and design of European critical infrastructures and the assessment.

Crnčevič, D. (2015). Transport Risk Analysis. In Transportation Systems and Engineering, Concepts, Methodologies, Tools, and Applications. Hershey, PA: IGI Global. doi:10.4018/978-1-4666-8473-7.ch001

Dado, M., & Zahradnik, M. (2007). *Intelligent Transport Technologies and Services*. Zilina, Slovak Republic: University of Zilina.

DEMASST. (2019). Factsheet. Retrieved from https://cordis.europa.eu/project/rcn/91165/factsheet/en

Dolnak, I. (2008). *Security of information systems in railway transport* [Dissertation thesis]. University of Zilina.

Drdla, P. (2018). *Passenger Transport of Regional and Supra-regional Importance*. Pardubice, Czech Republic: University of Pardubice. (in Czech)

Dvorak, Z., Cizlak, M., Leitner, B., Sousek, R., & Sventekova, E. (2010). *Risk management in railway transport*. Pardubice, Czech Republic: Jan Perner Institute. (in Slovak)

Dvorak, Z., Sventekova, E., Rehak, D., & Cekerevac, Z. (2017). Assessment of Critical Infrastructure Elements in Transport. *Procedia Engineering*, *187*, 548–555.

E.C. transport Themes. (2011). Roadmap to a Single European Transport Area - Towards a competitive and resource efficient transport system.

ECTRI. (2019). Retrieved from https://www.ectri.org/

ERACHAIR. (2019). Retrieved from http://www.erachair.uniza.sk/

EU ERA. (2019). Retrieved from https://www.era.europa.eu/

Eur-Lex. (2019). Retrieved from https://eur-lex.europa.eu/LexUriServ/LexUriServ.do?uri=OJ:L:2008:345:0075:0082:EN:PDF

Evaluation study. (2019). European Union. Retrieved from https://ec.europa.eu/info/law/better-regulation/initiatives/ares-2018-1378074/public-consultation_en

FORTESS. (2019). Retrieved from https://cordis.europa.eu/project/rcn/185488/factsheet/en

Hofreiter, L. (2004). *Security and security risks and threats*. University of Zilina. (in Slovak)

Hruza, P., Sousek, R., & Szabo, S. (2014). Cyber-attacks and attack protection. In *World Multi-Conference on Systemics* (Vol. 18, pp. 170–174). Orlando, FL: Cybernetics, and Informatics.

IDEKO Project. (2019). Retrieved from http://oblast.cdv.cz/cz/O37/user/project/detail/2

INFRARISK. (2019). Retrieved from https://cordis.europa.eu/project/rcn/110820/factsheet/en

INTACT. (2019). Retrieved from https://cordis.europa.eu/project/rcn/185476/factsheet/en

Iromuanya, C., Hrgiss, K., & Howard C. (2015). *Critical risk path method*. Hershey, PA: IGI global. doi:10.4018/978-1-4666-8473-7.ch028

ISO. (2018). ISO 31000:2018 Risk management. Retrieved from https://www.iso.org/iso-31000-risk-management.html

Kadri, F., Babiga, B., & Châtelet, E. (2014). The impact of natural disasters on critical infrastructures. *Journal of Homeland Security and Emergency Management, 11*(2), 217–241.

Kelemen, M., & Jevcak, J. (2018). Security Management Education and Training of Critical Infrastructure Sectors' Experts. In *Proceedings of* NTAD 2018 - 13th International Scientific Conference - New Trends in Aviation Development (pp. 72–75).

KISDIS. (2019). Retrieved from https://starfos.tacr.cz/en/project/VG20122015070

Margaritis, D., Anagnostopoulou, A., Tromaras, A., & Boile, M. (2016). Electric commercial vehicles: Practical perspectives and future research directions. *Research in Transportation Business & Management, 18*, 4–10. doi:10.1016/j.rtbm.2016.01.005

MD CR. (2019). Retrieved from https://www.mdcr.cz/?lang=en-GB

MDV SR. (2019). Retrieved from https://www.mindop.sk/en

Michalik, D. (2010). *Safety culture. Methodological manual.* Prague, Czech Republic: Occupational Safety Research Institute. (in Czech)

MODSAFE. (2019). Retrieved from https://cordis.europa.eu/project/rcn/92875/factsheet/en

Moridpour, s., Pour, A.T., & Saghapour, T. (2019). *Big Data Analytics in Traffic and Transportation Engineering. Emerging Research and Opportunities*. Hershey, PA: IGI Global.

O'Connor, A., & Eichinger, E. M. (2007). Site-specific traffic load modeling for bridge assessment. *Proceedings of the Institution of Civil Engineers: Bridge Engineering, 160*(4), 185–194.

OSCE. (2012). Retrieved from https://www.osce.org/secretariat/99852?download=true

Ouyang, M. (2014). Review on the Modeling and Simulation of Interdependent Critical Infrastructure Systems. *Reliability Engineering & System Safety, 121*, 43–60. doi:10.1016/j.ress.2013.06.040

PLASA. (2019). Retrieved from https://cordis.europa.eu/project/rcn/207498/factsheet/en

Plos, V., Sousek, R., & Szabo, S. (2016). Risk-based indicators implementation and usage. In *Proceedings* the 20th World Multi-Conference on Systemics, Cybernetics and Informatics (Vol. *2*, pp. 235–237). Academic Press.

Postranecky, M., & Svitek, M. (2018). *Cities in future*. NADATUR. (in Czech)

Prentkovskis, O., Sokolovskij, E., & Bartulis, V. (2010). Investigating traffic accidents: A Collision of two motor vehicles. *Transport, 25*(2), 105–115. doi:10.3846/transport.2010.14

Quayle, M. (2006). Transport. In Purchasing and Supply Chain Management: Strategies and Realities. Hershey, PA: IGI Global.

RAIN. (2019). Retrieved from https://cordis.europa.eu/project/rcn/185513/factsheet/en

RASEN. (2019). Retrieved from https://cordis.europa.eu/project/rcn/105547/factsheet/en

Rehak, D., Markuci, J., Hromada, M., & Barcova, K. (2016). Quantitative Evaluation of the Synergistic Effects of Failures in a Critical Infrastructure System. *International Journal of Critical Infrastructure Protection*, *14*, 3–17. doi:10.1016/j.ijcip.2016.06.002

Rehak, D., Senovsky, P., Hromada, M., & Lovecek, T. (2019). Complex Approach to Assessing Resilience of Critical Infrastructure Elements. *International Journal of Critical Infrastructure Protection*, *25*, 125–138. doi:10.1016/j.ijcip.2019.03.003

Rehak, D., Senovsky, P., Hromada, M., Lovecek, T., & Novotny, P. (2018). Cascading Impact Assessment in a Critical Infrastructure System. *International Journal of Critical Infrastructure Protection*, *22*, 125–138. doi:10.1016/j.ijcip.2018.06.004

RESIST. (2019). Retrieved from https://cordis.europa.eu/project/rcn/215997/factsheet/en

Rinaldi, S., Peerenboom, J., & Kelly, T. (2001). Identifying, Understanding and Analyzing Critical Infrastructure Interdependencies. *IEEE Control Systems Magazine*, *21*(6), 11–25. doi:10.1109/37.969131

SAFE-10-T. (2019). Retrieved from https://cordis.europa.eu/project/rcn/209711/factsheet/en

SAFURE. (2019). Retrieved from https://cordis.europa.eu/project/rcn/194149/factsheet/en

SERON. (2019). Retrieved from https://cordis.europa.eu/project/rcn/92516/factsheet/en

SMART RAIL. (2019). Retrieved from https://cordis.europa.eu/project/rcn/100584/factsheet/en

STERSST. (2019). Retrieved from https://cordis.europa.eu/project/rcn/110339/factsheet/en

Transportgeography. (2019). Retrieved from https://transportgeography.org/?page_id=247

Trucco, P., Cagno, E., & De Ambroggi, M. (2012). Dynamic Functional Modeling of Vulnerability and Interoperability of Critical Infrastructures. *Reliability Engineering & System Safety*, *105*, 51–63. doi:10.1016/j.ress.2011.12.003

UIC. (2019). Retrieved from https://uic.org/

UITP. (2019). Retrieved from https://www.uitp.org/

UNECE. (2019). Retrieved from http://www.unece.org/trans/theme_infrastructure.html

Vidrikova, D., Boc, K., Dvorak, Z., & Rehak, D. (2017). *Critical Infrastructure and Integrated Protection*. The Association of Fire and Safety Engineering.

Young, W., Sobhani, A., Lenné, M., & Sarvi, M. (2014). Simulation of Safety: A Review of the State of the Art in Road Safety Simulation Modeling. *Accident; Analysis and Prevention*, *66*, 89–103. doi:10.1016/j.aap.2014.01.008 PMID:24531111

Zagorecki, A., Ristvej, J., & Klupa, K. (2015). Analytics for Protecting Critical Infrastructure. *Communications Scientific Letters of the University of Zilina*, *17*(1), 111–115.

ADDITIONAL READING

Biosca, S. A. O. (2018). *Transportation infrastructure: Assessment, management and challenges*. Nova Science Pub.

Dvorak, Z., Leitner, B., & Novak, L. (2012). National transport information system in Slovakia as a tool for security enhancing of critical accident locations. In *Transport Means - Proceedings of the International Conference 2012* (pp. 145-148). Academic Press.

Dvorak, Z., Sventekova, E., Rehak, D., & Cekerevac, Z. (2017). Assessment of Critical Infrastructure Elements in Transport. *Procedia Engineering*, *187*, 548–555. doi:10.1016/j.proeng.2017.04.413

Kekovic, Z., Caleta, D., Kesetovic, Z., & Jeftic, Z. (2013). *National critical infrastructure protection regional perspective*. Beograd: Faculty of Security Studies.

Leitner, B., Luskova, M., O'Connor, A., & Van Gelder, P. (2015). Quantification of impacts on the transport serviceability at the loss of functionality of significant road infrastructure objects. Communications - Scientific Letters of the University of Zilina, 17(1), 52-60.

Nævestad, T., Hesjevoll, I., & Phillips, R. (2018). How can we improve safety culture in transport organizations? A review of interventions, effects and influencing factors. *Transportation Research Part F: Traffic Psychology and Behaviour*, *54*, 28–46. doi:10.1016/j.trf.2018.01.002

Rehak, D., Radimsky, M., Hromada, M., & Dvorak, Z. (2019). Dynamic Impact Modeling as a Road Transport Crisis Management Support Tool. *Administrative sciences (Special Issue: Rational Decision Making in Risk Management)*, *9*(2).

Vidrikova, D., Boc, K., Dvorak, Z., & Rehak, D. (2017). Critical infrastructure and integrated protection. The Association of fire and safety engineering, Ostrava, Czech Republic.

Chapter 3
Cyber–Physical Security in Healthcare

Vasiliki Mantzana
Center for Security Studies (KEMEA), Greece

Eleni Darra
Center for Security Studies (KEMEA), Greece

Ilias Gkotsis
https://orcid.org/0000-0003-2228-1387
Center for Security Studies (KEMEA), Greece

ABSTRACT

The healthcare sector has been considered a part of critical infrastructure (CI) of society and has faced numerous physical-cyber threats that affect citizens' lives and habits, increase their fears, and influence hospital services provisions. The two most recent ransomware campaigns, WannaCry and Petya, have both managed to infect victims' systems by exploiting existing unpatched vulnerabilities. It is critical to develop an integrated approach in order to fight against combination of physical and cyber threats. In this chapter, key results of the SAFECARE project (H2020-GA787005), which aims is to provide solutions that will improve physical and cyber security, to prevent and detect complex attacks, to promote incident responses and mitigate impacts, will be presented. More specifically, healthcare critical asset vulnerabilities; cyber-physical threats that can affect them; architecture solutions, as well as, some indicative scenarios that will be validated during the project will be presented.

INTRODUCTION

Critical infrastructure (CI) is an asset or system which is essential for the maintenance of vital societal functions. The damage to a CI, its destruction or disruption by natural disasters, terrorism, criminal activity or malicious behavior, may have a significant negative impact for the security of the EU and the well-being of its citizens. Over the last decade, healthcare sector has been considered as a CI and

DOI: 10.4018/978-1-7998-3059-7.ch003

the most vulnerable one, facing numerous physical-cyber threats that affect citizens' lives and habits, increase their fears and influence hospital services provision. Healthcare sector as a CI is safeguarding people, services, systems and physical infrastructure providing vital operation to the health services.

Health sector is responsible for delivering services that improve, maintain or restore the health of individuals and their communities (World Health Organisation, 2019). These services are large and complex, affect and get affected by multiple interacting actors, such as doctors, nurses, patients, citizens, medical suppliers, health insurance providers etc., with different backgrounds, knowledge, organizational beliefs, interests and culture. Health systems could provide services that are personal and non-personal. Personal health services can be therapeutic or rehabilitative and are delivered to patients and citizens individually. Non-personal health services are actions applied to individuals or collectives and might refer to health education (Peters, Kandola, Elmendorf, & Chellaraj, 1999).

Health services are widely relying on information systems (IS) to optimize organization and costs, whereas ethics and privacy constraints severely restrict security controls and thus increase vulnerability.

Threats in the healthcare sector can result in economic damage, human casualties, property destruction, and hospital assets functionality disruption that can produce cascading effects and harm patients' and citizens' confidence. Today, healthcare landscape is facing major threats, which cannot be analyzed as physical or cyber independently, and therefore it is critical to develop an integrated approach in order to fight against such combination of threats. Cyber-physical attacks are the results of the exploitable vulnerabilities that need to be considered by hospitals.

In this chapter, we will focus on a framework related to the enhancement of Hospitals' security and resilience, and respective solutions that will improve physical and cyber security to prevent and detect complex attacks, to promote incident responses and mitigate the impacts, based on the work implemented and further foreseen under SAFECARE project (H2020-GA787005),. In these terms, we will initially analyze the normative literature on CI and healthcare sector security and safety challenges and protection. In doing this, we will present healthcare critical assets' vulnerabilities and describe cyber-physical threats that can affect them (as they have been identified in SAFECARE). In addition, SAFECARE architecture solution will be presented, including a cyber-physical security system and proposed strategies that can boost their protection; avoid threats' cascading effects; enhance internal and external stakeholders' communication and response; mitigate impacts; and maintain healthcare services secure and available to citizens. Finally, some representative scenarios that will be validated during the project will be analyzed.

BACKGROUND

Critical Infrastructures

Europe has a long-standing history of approaches to improve CI protection. Past terrorist attacks fostered the development and adoption of the European Programme on critical infrastructure protection (EPCIP). The EPCIP provides systematic, network-based guidelines for member states to identify Critical Infrastructure assets (Izuakor & White, 2016) The EPCIP comprises the following pillars (European Commission, 2006): (a) means for its implementation (e.g., EPCIP action plan, CIWIN), (b) support for member states concerning National CIP, (c) contingency planning, (d) external dimension (exchange of information with non-EU countries), (e) EU security research program on "prevention, preparedness and consequence management of terrorism and other security related risks" and (f) financial measures.

The Directive 2008/114/EC functions as the main instrument of the EPCIP. Firstly, it provides definitions of CIs and ECIs. According to the Directive, ECIs are: “Assets, systems or parts thereof located in EU member states, which are essential for vital societal functions [...] the disruption or destruction of which would have a significant impact on at least two EU member states” (European Commission, 2008). The directive provides concrete support for three phases of EPCIP. The phase of identification includes specific criteria to identify CIs: (a) sectoral criteria and (b) CI definition, (c) transboundary elements and (d) cross-cutting criteria. The phase of designation includes all steps to negotiate and to decide on the criticality of any specific infrastructure: (a) notification of affected member states, (b) bilateral discussions and agreements and (c) final decision by the ‘hosting country’. Finally, it provides two instruments that really contribute to the protection of infrastructures: (a) OSP (obligatory unless similar regulations are in place) and (b) liaison officer as contact point between the ECI owner/operator and relevant member state authorities.

CIs are complex and they are turning into cyber-physical infrastructures because information and communication technologies (ICT) are important in the context of infrastructure management. Today, most of organizations are susceptible to cyber threats because they are increasingly exposed to the internet and to the external world. Technological trends like Internet of Things (IoT), Industry 4.0 are driving this augmented connectivity. Nowadays, most CIs are controlled by industrial control systems (ICS) that need to be frequently updated during maintenance campaigns. Since the beginning of 21st century, Critical Infrastructures have faced multiple cyber-attacks.

Healthcare Sector Challenges, Incidents and Protection

According to WHO definition “Hospitals complement and amplify the effectiveness of many parts of the health system, providing continuous availability of services for acute and complex conditions” (World Health Organisation, 2019). They are an essential element to health systems as they support care coordination and integration and play a key role in supporting other health-care providers, such as primary health care, community outreach and home-based services. They also often provide a setting for education of doctors, nurses and other health-care professionals and are a critical base for clinical research.

Any physical or cyber incident that causes loss of infrastructure or massive patient surge, such as natural disasters, terrorist acts, or chemical, biological, radiological, nuclear, or explosive hazards could affect the health care services provision and could cause overwhelming pressure to the affected health systems. Hospitals not only provide care services; but they are also the last shelters for disaster victims seeking care and represent an icon of social security, connectivity and community trust (World Health Organization, 2015). Thus, in this context, it is fundamental for a hospital to remain resilient; maintain the level of provided care; and be able to scale up its service delivery in any given emergency situation.

Healthcare sector is one of the most targeted sectors; 81% of 223 organizations surveyed, and >110 million patients in the US had their data compromised in 2015 alone (KPMG, 2015), with only 50% of providers thinking that they could protect themselves from cyberattacks (KPMG, 2015). It has been reported that between 2009 and 2018 there have been 2.546 healthcare data breaches involving more than 500 records and resulting in theft/exposure of 189,945,874 records (HIPAA, 2018). In the healthcare sector, hacking and malware (including ransomware) are the leading attack type of health data breaches (HIPAA, 2018). These data breaches result in large financial losses, but also in loss of reputation and reduced patient safety. Some known data breaches in healthcare sector are the following:

- 2017 WannaCry attack infected more than 300,000 computers across the world demanding that users pay bitcoin ransoms. Despite this attack was not specifically directed at healthcare organizations, other ransomware has specifically targeted the healthcare sector and, according to US media, the Presbyterian Medical Centre shut down for 10 days until it paid a $17,000 ransom (Perlroth & Sanger, 2017).
- Medical Device Hijack (Medjack) is another known attack that injects malware into unprotected medical devices to move laterally across the hospital network (Hei, 2013); Between the first detection of Medjack in 2015 and now, there have been many variations of the attack with several hospitals' medical devices, including x-ray equipment, picture archive and communications systems (PACS) and blood gas analyzers (BGA), etc., having been attacked. The attacker establishes a backdoor within the medical device, and almost any form of manipulation of the unencrypted data stored and flowing through the device is possible.
- It was reported in the press that in January 2019 hackers performed a ransom attack in a heart specialist clinic in Melbourne, where the hackers hit patient files (Martin, 2019). As a result, staff was unable to access some patient files for more than three weeks. The Clinic could have mitigated the impact, if data was properly and fully backed and if they were investing consistently in IT security.
- A billing company based in USA, which operates the online payment system used by a network of 44 hospitals in USA, discovered that some of its databases that contained 2,652,537 patients' records, had been compromised in 2018. Upon discovery of the breach, access to data was terminated and forensic specialists were hired to review the incident, secure affected databases and improve security controls (HIPAA, 2018).
- 128,400 records were affected by a sophisticated phishing incident that happened at New York oncology and hematology clinic. More specifically, fourteen employee email accounts clicked on phishing emails, which exposed health information in the email accounts. The clinic hired forensic specialists to assess the breach and types of data affected. Moreover, improvements to data security following the incident included active monitoring of affected systems, regular password resets, additional employee training and new email protocols (HIPAA, 2018).
- A woman opened fire at a flat opposite a Catholic Hospital and then inside the hospital in the south-western town of Lorrach in Baden-Wuerttemberg, Germany, killing at least three, including one child, and wounding several patients before police shot her dead (The Times, 2010).
- A UK A&E registrar was held hostage by a patient brandishing a pair of steel surgical scissors in a cubicle, she did not panic. Moments earlier, she had gone in to check on the young patient, who was having a mental health episode after taking drugs, and had closed the door behind her. Unfortunately, the patient had managed to hide a pair of scissors, which she pulled out before backing the doctor into a corner. The police were eventually called, and restrained the patient (The Guardian, 2019).
- While a nurse was examining a female patient, the accompanying Roma (gypsies) group attacked her and slammed her on the face. The incident happened at the Salamina Island Health Center, Greece (POEDIN, 2018).
- General Hospital of Larnaka, Cyprus: A 27-year-old patient physically attacked a male and a female doctor examining his leg at the Emergency Department (OFFSITE, 2017).

It is prevalent that healthcare organizations have several assets that are essential for their operation and should be protected. ENISA identified assets and presented them in a report namely "Smart Hospitals: Security and Resilience for Smart Health Service and Infrastructures" (ENISA, 2016). Assets that can be attacked are the buildings and facilities, data, interconnected clinical Information Systems, mobile client devices, networking equipment, identification systems, networked medical devices and remote care systems.

The two most critical hospital's assets are the patients' health and their records (ENISA, 2016). The first one can be affected in many ways, for example, turning off a critical medical device can cause a serious injury to a patient. Patient records contains valuable information, as Personal Identifiable Information (PII) and Protected Health Information (PHI) that can be the most lucrative information for attackers. It has been reported that healthcare data is substantially more valuable than any other data, as the value for a full set of medical credentials can be over $1000 (Sulleyman, 2017). Data held within health organizations also has political value. For example, World Anti-Doping Agency was attacked and athletes' records had been made public (BBC, 2017).

It is prevalent, that healthcare organizations and their assets suffer from vulnerabilities that attackers can exploit to damage the environment and cause disruptions. These vulnerabilities can be either cyber (application & OS, control gaps and design flaws, unpatched devices, unprotected networks, weak credentials, lack of cyber threat prevention and detection, lack of smart sensors etc.) or physical (lack of access management, video monitoring, fire detection, smart sensors, security agents, policy, collaboration with police and firefighters, etc.).

Hackers have different goals, as they might wish to cause damage, obtain a ransom, cause the interruption of service, or collect data to prepare future impacting attacks. Therefore, the adversary' motivation should be considered while protecting a system. The attacks can be roughly divided in two categories:

- Untargeted attacks that do not have a specific target and the attacker chooses the targets that maximize their gain/cost ratio.
- Targeted attacks that have a specific target and the attacker knows what he/she wants and will do everything necessary to achieve his/her goal.

To develop an efficient and effective protection system it is important to consider both aforementioned types of attacks and understand the profile and sophistication of attackers. The table below, extracted from ISE's report, summarizes the profile and motivation of healthcare organizations attackers (Independent Security Evaluators, 2016). These assets can be also attacked in different ways, such as cyber (social engineering, spear phishing, malware, RATs, DDoS, vulnerability exploits), physically (intrusion, aggression, material destruction, bombing, manmade fire etc.) and/or from natural hazards (flood, earthquake, storm).

As such, health structures are pointed out as potential targets, which highlight the need to enhance the protection of these Critical Infrastructures. Healthcare organizations can take practical steps to protect themselves and reduce the effects of an attack, such as to strengthen resilience, as resilient organizations are less likely to be attacked and suffer less harm when attacks occur. For example, on Papworth Hospital in 2016, a ransomware attack took place just after the daily backup, so no data were lost. Moreover, Hospitals and healthcare organizations should not only invest on cyber perimeter security, but must also adopt technologies that support limiting damages in case of an attack (e.g. network segregation, data encryption, etc.) (INFOSEC, 2019). In addition to cyber protection measures, hospitals should also focus on physical protection and should ensure that the file cabinets and doors are properly locked when unattended; security cameras and other adequate physical security controls should be installed

Figure 1. Profile and motivation of healthcare organizations attacker (Independent Security Evaluators, 2016)

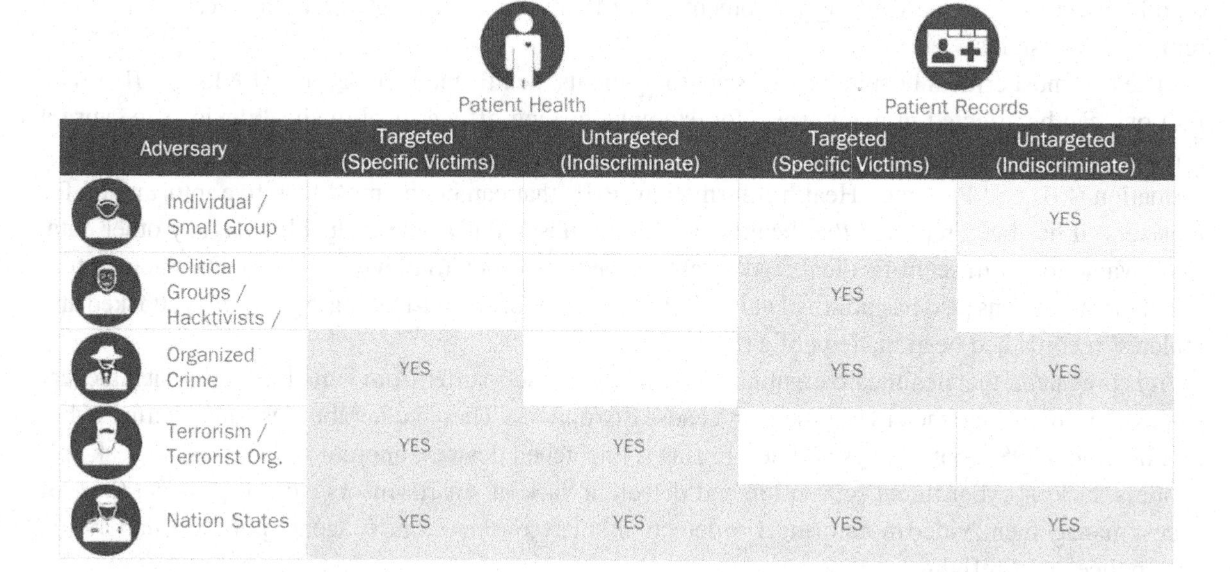

Adversary	Patient Health: Targeted (Specific Victims)	Patient Health: Untargeted (Indiscriminate)	Patient Records: Targeted (Specific Victims)	Patient Records: Untargeted (Indiscriminate)
Individual / Small Group				YES
Political Groups / Hacktivists /			YES	
Organized Crime	YES		YES	YES
Terrorism / Terrorist Org.	YES	YES		
Nation States	YES	YES	YES	YES

(INFOSEC, 2019). It is also crucial that healthcare staff (including researchers, administrators, front desk workers, medics, transcriptionists, handlers of medical claims to IT, and technical staffs) should be properly trained on physical and cyber security issues (Martin, 2017). Last but not least, healthcare organizations need to develop and adopt common healthcare security standards and adopt a clear security policy and response plan.

In tackling crisis management, which has been defined as "the developed capability of an organization to prepare for, anticipate, respond to and recover from crises" (British Standard Institute [BSI], 2014), the following steps have been identified and should be followed:

Preparedness: The aim is to prepare organizations and develop general capabilities that will enable them to deliver an appropriate response in any crisis. Preparedness refers to activities, programs, and systems developed before crisis that will enhance capabilities of individuals, businesses, communities, and governments to support the response to and recovery from future disasters.

Response: Response begins as soon as an event occurs and refers to the provision of search and rescue services, medical services, as well as repairing and to the restoration of communication and data systems during a crisis. A response plan can support the reduction of casualties, damage and recovery time.

Recovery: When crisis occurs, organizations must be able to carry on with their tasks in the midst of the crisis while simultaneously planning for how they will recover from the damage the crisis caused. Steps to return to normal operations and limit damage to organization and stakeholders continue after the incident or crisis (Deloitte, 2016).

Mitigation: Mitigation refers to the process of reducing or eliminating future loss of life and property and injuries resulting from hazards through short and long-term activities. Mitigation strategies may range in scope and size.

The efficient and timely identification of risks, threats and vulnerabilities of the healthcare sector infrastructures and services the disruption or destruction of which would have significant socio-economic and environmental impacts, unquestionably requires communication, coordination, and cooperation at national level as well, in order to deter, mitigate and neutralize any posed hazard and ensure the functionality, continuity and integrity of all affected assets and systems.

Critical Infrastructure Related EU Projects

The European Commission has put continuous effort, through several initiatives and research and innovation funding tools, in order to support the needs of CI operators for security enhancement. For example, there have been several funded projects under the topic CIP-01-2016-2017: Prevention, detection, response and mitigation of the combination of physical and cyber threats to the critical infrastructure of Europe, of Horizon 2020 programme, as depicted below:

1. DEFENDER (http://defender-project.eu/)

Defending the European Energy Infrastructures (DEFENDER) project is a project that will adapt, integrate, upscale, deploy and validate a number of different technologies and operational blueprints. The project has a vision to develop a new approach to safeguard existing and future European CEI operation over cyber-physical-social threats, based on novel protective concepts for lifecycle assessment, resilience and self-healing offering "security by design", and advanced intruder inspection and incident mitigation systems.

2. SAURON (https://www.sauronproject.eu/)

Scalable multidimensional situation awareness solution for protecting European ports (SAURON) project proposes a holistic situation awareness concept as an integrated, scalable and yet installation-specific solution for protecting EU ports and its surroundings. This solution combines the more advanced physical features with the newest techniques in prevention, detection and mitigation of cyber-threats, including the synthetic cyber space understanding using new visualization techniques (immersive interfaces, cyber 3D models and so on). In addition, a Hybrid Situation Awareness (HSA) application capable of determining the potential consequences of any threat will show the potential cascading effect of a detected threat in the two different domains (physical and cyber).

3. STOP-IT (https://stop-it-project.eu/)

Strategic, Tactical, Operational Protection of water Infrastructure against cyber-physical Threats (STOP-IT) assembles a team of major Water Utilities, industrial technology developers, high tech SMEs and top EU R&D providers, to find solutions to protect critical water infrastructure against physical and cyber threats. The main aims of STOP-IT are: a) Identification of current and future water infrastructure risks; b) co-development of an all-hazards risk management framework for the physical and cyber protection of critical water infrastructures.

4. RESISTO (http://www.resistoproject.eu/)

RESIlience enhancement and risk control platform for communication infraSTructure Operators (RESISTO) is an innovative solution for Communication Infrastructure providing holistic (cyber or physical) situation awareness and enhanced resilience. RESISTO will help Communications Infrastructures Operators to take the best countermeasures and reactive actions exploiting the combined use of risk and resilience preparatory analyses, detection and reaction technologies, applications and processes in the physical and cyber domains. RESISTO main objective is to improve risk control and resilience of modern Communication CIs, against a wide variety of cyber-physical threats, being those malicious attacks, natural disasters or even un-expected.

5. FINSEC (https://www.finsec-project.eu/)

Integrated Framework for Predictive and Collaborative Security of Financial Infrastructures (FINSEC) will develop, demonstrate and bring to market an integrated, intelligent, collaborative and predictive approach to the security of critical infrastructures in the financial sector. To this end, FINSEC will introduce, implement and validate a novel reference architecture for integrated physical and cyber security of critical infrastructures, which will enable handling of dynamic, advanced and asymmetric attacks, while at the same time boosting financial organizations' compliance to security standards and regulations. As a result, FINSEC will provide a blueprint for the next generation security systems for the critical infrastructures of the financial sector.

Independently of the thematic aim of the projects, there is a common focus on state-of-the-art solutions and frameworks that will assist CI operators, owners and security managers to address key emerging cyber-physical challenges in full of threats and uncertainties environment.

Physical and Cyber Security Solutions and Technologies

In the past decade, the security landscape has dramatically changed with the introduction of several new security technologies to deter, detect and react to more disparate attacks. Organizations are constantly introducing new technologies and upgrading existing ones in order to ensure the security of their most valuable assets such as people, infrastructure, and property. Typical systems include among others the following: (a) Fences/Walls, (b) Guards, (c) Building control, (d) Intrusion detection and access control, € Video surveillance, (f) Audio surveillance, (g) CBRNE sensors and (h) Physical Security Information Management (PSIM) systems.

As in physical protection, CI facilities may also be subject to Cyber-attacks, requiring additional Cyber-security measures. Most of them are meant to protect the CI against Information leakage/loss, prevent unauthorized access, and secure all communication in and out of the facilities. Typical systems include the following: (a) Data Protection, (b) Network Monitoring, (c) Intrusion Response Systems, (d) Endpoint Monitoring, (e) Authentication and Access Control, (f) Software Development based on privacy by design techniques, (g) IoT Sensors for Health and (h) Artificial Intelligence techniques.

SAFECARE PROJECT

In this direction, SAFECARE project (H2020-GA787005) (SAFECARE project, 2018) aims to provide solutions that will improve physical and cyber security in a seamless and cost-effective way. It will promote new technologies and novel approaches to enhance threat prevention, threat detection, incident response and mitigation of impacts. The SAFECARE solution will also participate in increasing the compliance between security tools and European regulations about ethics and privacy for health services. In doing this, SAFECARE will design, test, validate and demonstrate innovative elements, which will optimize the protection of Critical Infrastructures under operational conditions. These elements are interactive, cooperative and complementary (reinforcing in some cases), aiming at maximizing the potential utilization of the individual elements.

SAFECARE deals with the protection of Critical Infrastructures such as hospitals and public health national agencies. These organizations participate daily to the safety and security of European citizens. Health services breakdown can have tremendous negative effects on public healthcare and public order. Referring to the crisis events that affected over forty hospitals in UK by May 2017, cyber-attacks represent now a major threat for health organizations that mainly rely on IT systems and e-health devices.

The main aim of SAFECARE project is to build up a solution in order to better apprehend the combination of physical and cyber security threats. Furthermore, the project will focus on end users' needs (e.g. hospitals and a public health national agency), the risk assessment as well as and the study of applicable regulations. At the lens of these initial results, partners will be in position to develop innovative approaches and improve prevention and detection capacities on both physical and cyber security grounds.

Healthcare Organizations' Security Stakeholders

Healthcare organizations' security stakeholders are individuals or organizations that may contribute to be affected by or get involved in issues related to security planning, response or recovery from an emergency situation. Security stakeholders can be categorized according to their perceived proximity to the hospital into internal and external. Internal stakeholders are groups or individuals who work within the hospital, while external are individuals or groups outside the hospital, but who can affect or be affected by it. An indicative list of healthcare organizations' security stakeholders as well as their role are analyzed below.

Internal Stakeholders

- **Data Protection Officer (DPO):** The primary role of the Data Protection Officer is to ensure that organization processes the personal data of its staff, customers, providers or any other individuals (also referred to as data subjects) in compliance with the applicable data protection rules.
- **Physical Security Manager / Security Personnel:** The main role of a hospital security manager is to (a) develop and implement security policies, protocols and procedures, (b) manage budget and expenses for security operations, (c) manage training of security officers and guards, (d) plan and coordinate security operations and staff when responding to emergencies and alarm, (e) investigate and resolve issues and (f) review reports on incidents and breaches etc. Some hospitals might also subcontract security services to external security companies in order to support the permanent security personnel (and that is why hereafter is considered as internal).

- **IT Security Manager / Security Personnel:** IT Security manager is responsible for leading and managing the activities of the Information Security Risk Assessment and Security Operations team. In doing this, s/he is responsible to (a) manage security issues related to implementation, installation, monitoring and service/support of healthcare IT infrastructure (e.g. networks, platforms, applications, devices etc.), (b) manage (develop, assess, update and enforce) security plans and policies in accordance with IT policies, standards, and compliance requirements, (c) respond to cyberattacks and mitigate cyber risks, (d) provide reports on security issues/threats, (e) train the IT personnel.
- **Technical Manager/ Technical Staff:** The technical staff can identify the sensible technical components for a health structure, such as energy, elevators, technical gas/fluid, temperature, air control systems or building management). They are also responsible to manage access rights, hospital IT behavior system, threats and security events and security personnel related to healthcare organizations infrastructures and processes.
- Security and safety teams are responsible for safeguarding the Hospital against physical attacks: (a) technical assets (e.g. gas, electricity, water), (b) hazardous materials (e.g. radioactive, diagnostic or therapeutic materials), (c) personnel and patients, (d) against natural disasters and firefighting. These teams are continuously trained and participate in tabletop and field exercises and simulations with patients, staff, fire brigade, volunteers etc.
- Crisis Management Team (CMT) "focuses on detecting the early signs of a crisis; identifying the problem; preparation of a crisis management plan; encouraging the employees to face problems; and solving the crisis" (Mikušová, 2019).

External Stakeholders

- **Interconnected Critical Infrastructures and Related Organizations:** This category includes all types of CIs (as described identified in the EU Directive 114/2008 and the NIS Directive) and national policies), Member States NCP for the CIP and EU officials from different DGs related to CI resilience programs and regulatory work, and the scientific community. They also support incident management for physical and cyber threats and respond against respective security events
- **Law Enforcement Agencies (LEAs):** Their aim is to ensure peace and order as well as citizens' unhindered social development, a mission that includes general policing duties and to prevent and interdict crime.
- **Fire Service:** Its mission is to provide fire and rescue services for the citizens and their property. It operates during fires, forest fires, car accidents, other natural or man-made disasters (technological and other disasters, such as earthquakes, floods, chemical - biological - radiological - nuclear (CBRN) threats) and during rescue and assistance operations. It protects and safeguards property that has been destroyed or threatened by fires or other disasters, until handed over to police officers or its owners.
- Other healthcare control centers identified through the interviews conducted are the following: (a) Centre for Disease Control and Prevention; (b) National Health Operations Centre etc.
- General Secretariat for Civil Protection is the body responsible for promoting the country's civil protection relations with relevant international organizations and relevant civil protection agencies in other countries.

- Ministry of Health role in crisis management process is to support, coordinate and formulate crisis management process in healthcare organizations.

Finally, the combination of the aforementioned innovation elements will bring a significant improvement in protecting Critical Infrastructures. With the effective integration of relevant disciplines and project coordination, a successful global optimum solution will be developed to demonstrate improved systemic security and management of the combination of physical and cyber threats.

Healthcare Organizations' Critical Assets

Hospitals can be characterized as complex healthcare providing units that due to their dependency on interconnected internal and external support systems could potentially become highly vulnerable. Their continuing functionality depends on several factors varying from critical systems and vital equipment to essential clinical services and human recourses management as well. Within the scope of the project, a solid basis of understanding was created as to which aspects within a hospital determine its performance and therefore can be seen as critical assets.

An asset is a broad term that describes every entity of a system (Kersten & Klett, 2017). If an organization (like a hospital) can be described as a system (network) of processes (edges) and their inputs and outputs (nodes), an asset can be a process as well as an object as well as a bigger construct that is system emergent like organizational factors, e.g. culture. It appears that all aspects of a healthcare organization are assets.

Although all assets can be critical for a certain desired outcome, not every asset actually does. This means that it is necessary to define which outcomes are desired for a hospital in order to be able to evaluate the criticality of the assets (Bundesamt für Sicherheit in der Informationstechnik, 2013). Since these outcomes equal the objectives of a hospital, the assets are evaluated according to their impact on the following objectives (Bundesamt für Bevölkerungsschutz und Katastrophenhilfe, 2008):

- Preservation (= prevention of failure) of vital departments and restoration of functionality as quickly as possible respectively,
- Limitation of economic damage and restoration of efficiency and
- Securing (= prevention of endangerment) of human life.

The list below although not exhaustive, represents an attempt to identify all those elements that can be characterized as critical assets in health infrastructures:

- Specialist personnel: employees, persons with special functions, etc.
- Buildings and terrain: main and ancillary buildings, technical buildings, traffic areas, storage areas, open spaces, delivery, escape routes, etc.
- General technical plants and appliances: power supply, gas supply, district heating, water supply, wastewater disposal, kitchen, transport and traffic (including vehicles and supply of operating resources), access control, fire alarm system, building automation, etc.
- Special technical plants and special equipment: medical devices for diagnosis and treatment, laboratory apparatus, sterilizers, laundry inventory, etc.

- IT: networking equipment, telephone system, copy machines, software, software to monitor medical devices, firmware, entry points (wireless technologies, domain controller, old legacy protocols), etc.
- Data and records: patient data, billing data, company information, contracts, etc.
- Operating resources: medicinal products, medical consumables, laundry supply, sterile supply, food supply, etc.

Despite this categorization, to define the cyber-physical scenarios of threat (that will be validated during the project), it was necessary to consider a new set of categories, based in the assets categorization presented in the ENISA report (ENISA, 2016). The new assets categories are the following:

- **Specialist personnel:** employees, persons with special functions, etc.
- **Buildings and Facilities:** Main and ancillary buildings, technical buildings, power and climate regulation systems, temperature sensors, medical gas supply, room operation, automated door lock system, etc.
- **Identification Systems:** Tags, bracelets, badges, biometric scanners, CCTV (video surveillance) with recognition/authentication capabilities, RFID services, etc.
- **Networked Medical Devices:** Mobile devices (e.g. glucose measuring devices), wearable external devices (e.g. portable insulin pumps), implantable devices (e.g. cardiac pacemakers), stationary devices (e.g. computed tomography (CT) scanners), support devices (e.g. assistive robots), etc.
- **Networking Equipment:** Transmission media, network interface cards, network devices (e.g. hubs, switches, routers, etc.), telephone system, etc.
- **Interconnected Clinical Information Systems:** Hospital Information System (HIS), Laboratory Information System (LIS), Radiology information systems (RIS), Pharmacy Information System (PIS), Picture Archiving and Communication System (PACS), pathology information system blood bank system, etc.
- **Mobile Client Devices:** Mobile clients (e.g. laptops, tablets, smartphones), mobile applications for smartphone and tablets, alarm and emergency communication applications for mobile devices, etc.
- **Remote Care System Assets:** Medical equipment for tele-monitoring and tele-diagnosis, medical equipment for distribution of drugs and telehealth equipment (cameras, sensors, telehealth computer system for patients to register their physiological measurements themselves, etc.)
- **Data and Records**: Clinical and administrative patient data, financial, organizational and other hospital data, research data, staff data, vendor details, tracking logs, etc.
- **Operating Resources:** Medicinal products, medical consumables, laundry supply, sterile supply, food supply, etc.

In identifying an analytical list of hospitals critical assets, SAFECARE partners worked on focus groups, where participants were asked to name asset, justify their importance, and assign each asset to the business objectives over which it has an influence. Finally, the assets were assigned to the predefined categories.

Vulnerabilities and Threats

To design new solutions that will improve the physical and cyber security in healthcare sector, it is very important to be aware of the physical and cyber vulnerabilities in health infrastructures, how they facilitate malicious actions, and how they may increase the likelihood and impact of human errors and system failures. It is also crucial to have a clear understanding of actual security strategies and controls implemented at health targeted infrastructures (medical devices, ICT and network infrastructures, e-health services, Information Systems, etc.).

The most common vulnerabilities in Critical Infrastructures are:

- **Legacy Software**: Usually critical systems run on legacy software that lack sufficient user and system authentication, data authenticity verification, or data integrity checking features that allow attackers uncontrolled access to systems.
- **Default Configuration**: Out-of-box systems with default or simple passwords and baseline configurations make it easy for attackers to enumerate and compromise the systems.
- **Lack of Encryption**: Legacy controllers (e.g. SCADA controllers) and industrial protocols lack the ability to encrypt communication. Attackers use sniffing software to discover username and passwords.
- **Remote Access Policies**: Systems like SCADA are connected to unaudited dial-up lines or remote-access servers, giving attackers convenient backdoor access to the OT network as well as the corporate LAN.
- **Policies and Procedures**: Security gaps are created when IT and OT personnel differ in their approach to securing industrial controls. Different sides should work together to create a unified security policy that protects both IT and OT technology.
- **Lack of Employee Awareness and Education**: Employees are a weakest link in privacy and security; companies cannot rely on annual training to solve the problem. It is crucial to make continuous improvements in cybersecurity knowledge and behavior to ensure that employees know their responsibilities and also foster an organizational culture of information security compliance, improving the mitigation of cyber-physical attacks.

As can be observed in the previous list, vulnerabilities are not identified only within information systems, they can also be found in organizational governance. External relationships, namely, dependencies on energy sources, supply chains, information technologies, and telecommunications providers can also be a source of vulnerabilities.

In addition, ENISA identified and defined the following threat taxonomy for smart hospitals (ENISA, 2016):

- Malicious actions, usual correlated with the adversarial threats, are deliberate acts by a person or an organization. It is important to distinguish malicious actions from other deliberate actions that bypass policies and procedures without malicious intent. A person carrying out a malicious action may be an external or an internal from the perspective of the affected organization. All the critical threats typically recognized and performed by the attacker in this case are malware attacks, hijacking, medical device tampering, social engineering attacks, device and data theft, skimming and denial of service attacks. The protection from these attacks is of particular significance as they can lead to unavailability of the hospital's systems.

- Human errors occur during the configuration or operation of devices or information systems, or the execution of processes. Human errors are often related to inadequate processes or insufficient training. Major examples include the medical system configuration error, the absence of audit logs to allow for appropriate control, unauthorized access control or lack of processes, non-compliance especially in the Bring Your Own Device (BYOD) paradigm, physician and/or patient errors. All human errors are highly pertinent to smart hospitals particularly due to the patient's data involved in the process.
- System failures are highly relevant in the healthcare context, particularly due to the increasing complexity and dynamics of the systems. Some examples of system failures include the software failures, inadequate firmware, device failure or simply limited/reduced capability, network components failure, insufficient maintenance which may leave operational issues undetected and unresolved, overload can lead to unavailability of a system or service, communication between IoT and non-IoT.
- Supply chain failure is outside the direct control of the affected organization as it typically affects or falls under the responsibility of a third party. As hospitals are increasingly dependent on third parties, third-party failures may have far-reaching consequences for them. Examples of third parties a failure of which would have an adverse impact on smart hospital operation include cloud service providers, medical device manufacturer, network providers
- Natural phenomena may also be the cause of incidents, particularly due to their disruptive or destructive impact, particularly on healthcare facilities and ICT infrastructure. Moreover, natural phenomena may impact the provision of remote patient care services even if their impact is not targeted to or impacting the hospital itself (e.g. if the metro-level network infrastructure is disrupted due to an earthquake). Examples may include earthquakes, floods and fires.

It is worth mentioning that there are different types of threat actors as well as attack vectors in hospitals. Each of these threat actors have different attack surfaces available within smart hospitals. The threat actors could include insider threats (staff with malicious intent), malicious patients and guests, remote attackers and other environmental or accidental equipment/software failure causes. The attack vectors they can affect the hospitals could be physical interaction with IT assets, wireless communication with IT assets, wired communication with IT assets and the interaction with staff.

Solution Architecture

- Physical security solutions
- Cyber security solutions
- Integrated cyber-physical security solutions

The physical security solutions and the cyber security solutions consist of smart modules and efficient integrated technologies to respectively improve physical security and cyber security. More specifically, physical security solutions embed integrated intelligent video monitoring and interconnect building monitoring systems as well as management systems. While cyber security solutions correspond to cyber monitoring systems as well as threat detection systems related to IT, BMS and e-health systems.

Both physical security solutions and cyber security solutions are interconnected thanks to the integrated cyber-physical security solutions. The integrated cyber-physical security solutions consist of intelligent modules to integrate different data sources and better take into account the combination of physical and cyber security threats.

The following systems make up the physical security solutions (SAFECARE project, 2018):

- **Suspicious Behaviour Detection System:** The suspicious behavior detection system will capture video streams from surveillance cameras. It will perform near real-time analysis and it will trigger security alerts in case of crowding, loitering in restricted areas and suspicious activities on integrated devices.
- **Intrusion and Fire Detection System:** By interconnecting video monitoring system with existing access management and fire detection systems, it will be possible to notably extend threat detection capacities and reduce number of false positive incidents.
- **Data Collection System:** The system collects data from physical subsystems (such as ICS, SCADA, smart building sensors); sends fire detection events to the intrusion and fire detection system and to the Building Threat Monitoring System (BTMS).
- **Mobile alerting system.** The mobile alerting system will be used by local security agents providing the ability to quickly report specific categories of security threats or impacts correlated to a specific failure point. It will send notifications which consist of all the information needed to manage the security threat (e.g. location, emergency procedure, video etc.) The notifications will be sent to all the users according to their profile.
- **Building Threat Monitoring System:** The Building Threat Monitoring System is the basis for interaction of the physical security components with the rest of the architecture.

The following systems make up the cyber security solutions:

- **IT Threat Detection System:** The objective of the IT threat detection system is to detect the exploit of vulnerabilities in critical health infrastructure by monitoring the network traffic and analysing IT events (such as software logs) provided by the IT infrastructure.
- **BMS Threat Detection System:** A threat detection system will increase security protection and situational awareness by passively monitoring network traffic between the building management system and the building control sub-systems (e.g., lighting, power, medical devices, and lift). When threats are detected by the BMS threat detection system, the system generates security events.
- **Advanced File Analysis System:** The objective of the advanced file analysis system is to detect the malwares in critical health infrastructure by performing an in-depth analysis of files. The advanced file analysis system performs a static analysis to look for malicious code into the files and a dynamic analysis to check file behaviour in a sandboxing environment.
- **E-Health Devices Security Analytics:** E-Health device security analytics is a cybersecurity solution that is specialized to security monitoring, threat detection and reporting in healthcare critical infrastructures. The security analytics solution collects log data from medical devices, performs analytics to derive meaningful security data, and generates security insights, aggregated statistics and alerts upon detecting anomalous or suspicious security events.

Figure 2. SAFECARE architecture and interconnections (SAFECARE project, 2018)

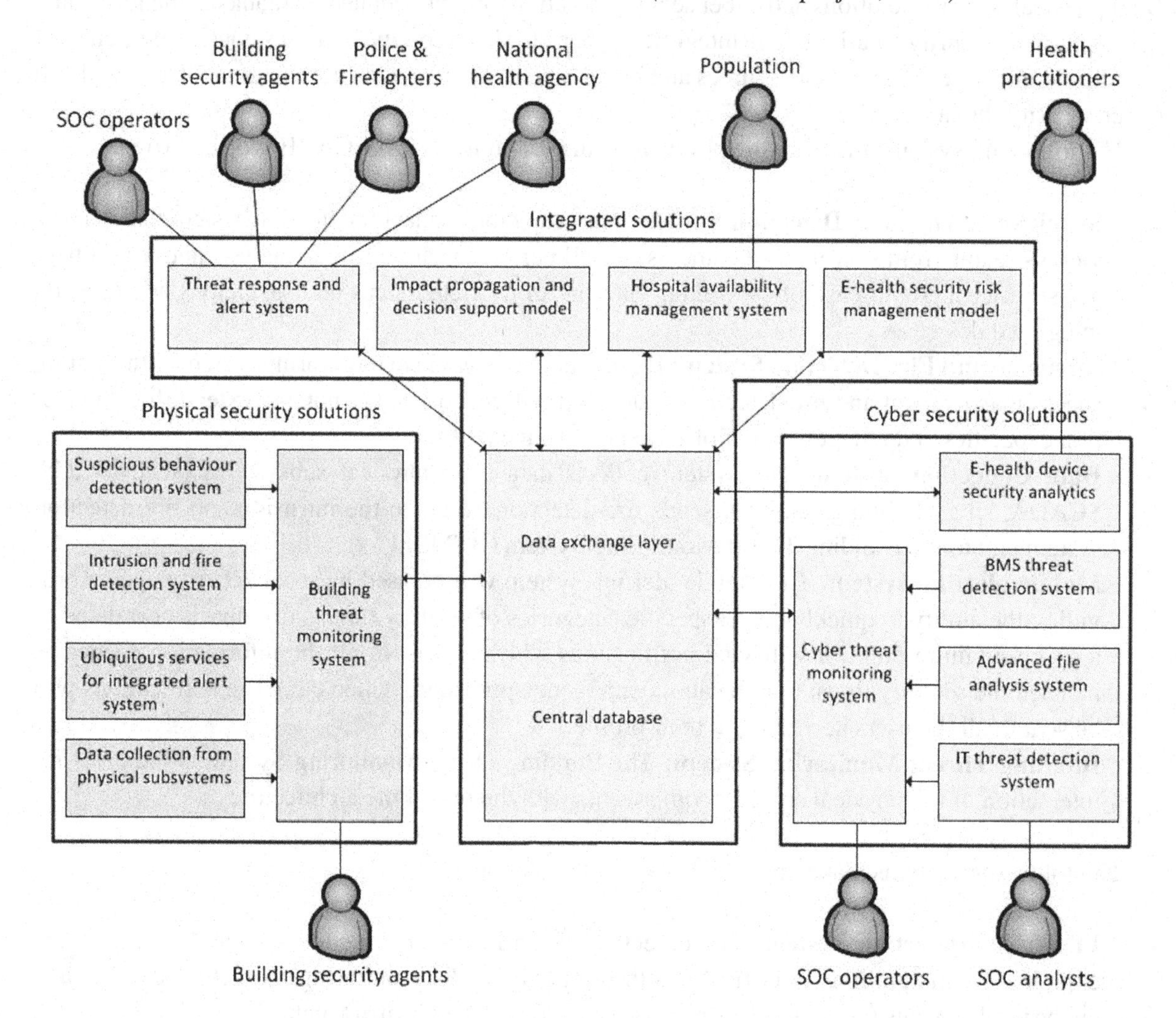

- **Cyber Threat Monitoring System:** The objective of the cyber threat monitoring system is to collect and centralize (a) security events and (b) potential impacts, organize the information and provide user-friendly interfaces to SOC operators so that they can visualize the threats and have an overview of the impacted assets.

The following systems make up the integrated cyber-physical security solutions:

- **Data Exchange Layer:** The data exchange layer triggers notifications to the other components when new physical and cyber incident (coming from the building threat monitoring system and the cyber threat monitoring system) or new impacts (coming from the impact propagation model) are sent.
- **Central Database**: Data centralization in a single database constitutes the pillar stone in order to build added-value indicators. Cross connecting data expands the capacity to create either more

consistent results or innovative results. The architecture of the database will take into account confidentiality, ethics and privacy constraints.

- **Impact Propagation and Decision Support Model:** The objectives of the Impact propagation model and decision support model are to (a) combine physical and cyber incidents that occur on assets, (b) infer cascading effects as impacts that could potentially affect the same or related assets and (c) alert other modules about the potential impacts and severity.
- **Threat Response and Alert System:** The threat response and alert system, once excited, will parse the "impact" data retrieved and run the corresponding response plan that will mainly consist in sending notification and alerts to internal and external practitioners. Notifications may include sending alerts to the Mobile alerting system.
- **Hospital Availability Management System (HAMS):** The Hospital Availability Management System will use the central database to get and store information about hospital assets and resources (e.g. department name and status, services, beds and staff availability, etc.).
- **E-health Security Risk Management:** E-health security risk management model goal is to effectively quantify the impact of security events in a uniform way for medical devices (e.g. visualize complex events by detailing how actors have obtained access to defined assets and what the potential risk outcome is for the related activity).

Scenarios Design and Description

To develop a secure and reliable protection system it is important to identify and formalize relevant use-cases and complex attack scenarios against Critical Infrastructures. These scenarios should exploit combined physical and cyber threats in the context of cascading attacks, and how they can impact and destabilize health services and will be simulated in the hospitals of Marseille, Turin and Amsterdam, involving security and health practitioners.

To define cyber-physical scenarios of threat, Expression of Needs and Identification of Security Objectives / Expression des Besoins et Identification des Objectifs de Sécurité (EBIOS) security risk assessment methodology was used. EBIOS was developed by the French Central Information Systems Security Division and is used to assess and treat risks related to Information Systems security (ISS). It can also be used to communicate this information within the organization and to partners, and therefore assists in the ISS risk management process since it is compliant with major IT security standards (National Cybersecurity Agency of France (ANSSI), 2018).

EBIOS uses a progressive risk management approach: it starts in the major missions of the object under study (highest level) and goes to the business functions and techniques (lowest level), studying possible risk scenarios. It aims to obtain a synergy between compliance and scenarios, positioning these two complementary concepts in the best way, i.e., where they bring the highest value. The compliance approach is used to determine the security base of the scenarios, particularly to develop targeted or sophisticated scenarios. This assumes that accidental and environmental risks are treated a priori by the compliance approach. Thus, scenario risk assessment focuses on intentional threats (National Cybersecurity Agency of France (ANSSI), 2018).

In the following paragraphs, some of the most representative scenarios of threat identified will be described:

Figure 3. Digital risk management pyramid (National Cybersecurity Agency of France (ANSSI), 2018)

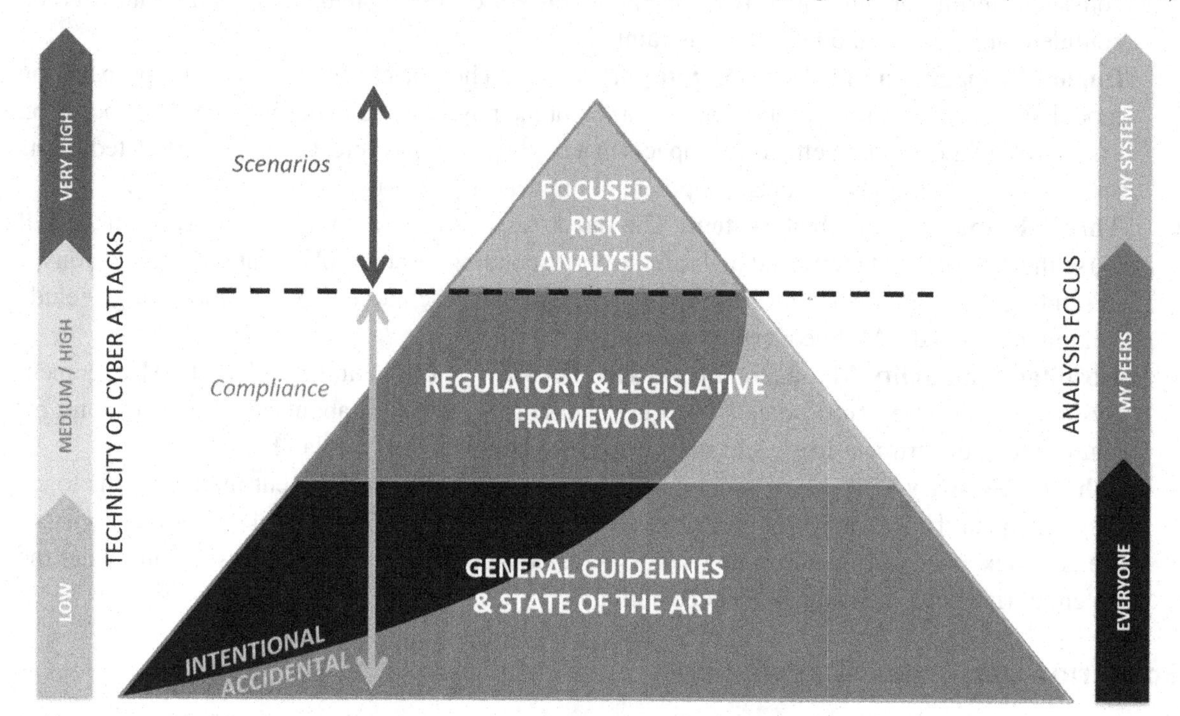

Scenario 1. Cyber-Physical Attack Targeting Power Supply of the Hospital

In the first scenario the aim of the attack is to cause damage in the power supply of the hospital to precipitate an energy breakdown. The attacker can be an external person who conducts the attack to have some gain (hacker) or an employee who has malicious intentions or makes a mistake. To get the necessary information to carry out the attack, the attacker can use social engineering, internet research, on-site visits, etc. Social engineering refers not only to person-to-person interaction but also to network interaction using, for example, social networks (Facebook, LinkedIn, etc.), or even phishing emails. All the information collected is used by the attacker for physical or cyber access to the PLC. This can be done through network scalation privileges or impersonating a PLC maintainer, for example.

Scenario 2. Cyber-Physical Attack to Steal Patient Data in the Hospital

Attackers are triggered by the strong possibility to acquire or even modify the patients' data. They can sell it with a big profit on the black market, ask for a ransom or even use it to damage the reputation of the hospital. The attacker can enter in the computer room stealing the data, damage the computer room or any other asset by triggering the fire alarm in order to take advantage of the confusion caused during the evacuation of the hospital

Scenario 3. Cyber-Physical Attack Targeting the Population, IT Systems and Medical Devices in the Hospital, and Patient Data Base

The aim of an attacker is to access, extort, sabotage or even intimidate valuable information from different sources including both service and clinical functionalities. The idea of the attack is to trigger the fire alarm in order to take advantage of the confusion caused during the evacuation of the hospital to enter in the computer room (a restricted area). In this way, the attacker can steal the data, damage the computer room or any other hospital asset, and take advantage of these actions. This type of intrusion in the hospital could be due to the fact that the IT defenses of the hospital were extremely difficult to overcome, and the data were of extreme value dictating a manual steal of hardware or connection with hardware from the inside.

Scenario 4. Cyber-Physical Attack Targeting the Air-Cooling System of the Hospital

As air cooling system is critical for the hospital environment, it can be attractive target for attackers. It is imperative to ensure good air quality and aseptic conditions to provide a safer environment for patients and staff. Using this attack vector, an attacker can not only disturb inpatients, but also cause serious damages to healthcare infrastructures, e.g. surgery rooms.

Scenario 5. Shooting, Explosive or Sabotage in Critical Places (Visible or Invisible)

Healthcare facilities, because of their importance in society, are a target of terrorist attacks harming hospital critical facilities planting a bomb. Usually, the main purpose of these attacks is to provoke a state of terror in people, but many other collateral problems arise. This is also applicable to other bad intention persons or groups like ideological activists, pathological evil-doer or even people looking for revenge.

Scenario 6: Theft at Hospital Equipment, Access to Hospital Network and IT Systems

It is known that many attacks are carried out from someone that works inside the healthcare institution. For example, for some reason an institution physician might want to erase some of the stored data referring to a medical operation. The physician may just want to hide some error, but the impact of his/her action on patient's safety can be very high.

FUTURE RESEARCH DIRECTIONS

Over a 36-month time frame, the SAFECARE consortium will design, test, validate and demonstrate innovative elements which will optimize the protection of Critical Infrastructures under operational conditions. These elements are interactive, cooperative and complementary (reinforcing in some cases), aiming at maximizing the potential utilization of the individual elements. The consortium will also engage with leading hospitals, national public health agencies and security forces across Europe to ensure that SAFECAREs global solution is flexible, scalable and adaptable to the operational needs of various hospitals across Europe, and meet the requirements of newly-emerging technologies and standards. Based on these, the SAFECARE project will produce additionally the following key results:

- A cyber threat detection system to improve Advanced Persistent Threat and zero-day attacks detection on Information Technology and Building Management Systems;
- A physical threat detection system using advanced video monitoring techniques and data cross-connection with other building monitoring sensors;
- An analytics and risk management tool to improve threat prevention on medical devices;
- A central database of cyber and physical incidents validated by security experts of the domain;
- A threat response system implementing cyber-physical defence strategies and automated notifications;
- A threat mitigation system to better manage hospitals availability and inform the population;
- A modular and scalable solution implementing standard communication protocols;
- An operational and cost-effective solution co-designed with health and security practitioners;
- Three demonstrators in Turin, Marseille and Amsterdam to test solutions in operational conditions;
- An analysis of the demonstrations to validate the approach and measure key performance indicators;
- Guidelines to enhance risk management plans and technological best practices in health services;
- Dissemination throughout health-user community to enhance awareness against multi-faced threats;
- Dissemination throughout scientific community and standardization bodies in order to support the establishment of certification mechanisms;
- A business plan to report economical perspectives of the industrial partners.

CONCLUSION

The SAFECARE project has as a notable ambition to contribute to Europe's strategy in reducing the vulnerabilities of Critical Infrastructure and increasing their resilience. Furthermore, it aims at contributing to some of the key initiatives undertaken by the Commission such as the European Programme for Critical Infrastructure Protection and the Critical Infrastructure Warning Information Network (CIWIN). The project will also participate in increasing the compliance between security tools and European regulations about ethics and privacy for health services. In terms of results, the project's main ambition is to bring a holistic approach that will consider both physical and cyber security. It will promote new technological uses and novel techniques to enhance prevention, detection, response and mitigation capacities against security threats. The project aims to demonstrate its results with a project pilot that will take place in Marseille with security practitioners, police and other local stakeholders. SAFECARE aims to contribute significantly towards the creation of new knowledge and the expansion of current knowledge and state-of-the-art in healthcare organizations' security and safety.

ACKNOWLEDGMENT

The work presented in this paper has been conducted in the framework of SAFECARE project, which has received funding from the European Union's H2020 research and innovation programme under grant agreement no. 787002.

REFERENCES

BBC. (2017). Who was hit by the NHS cyber-attack? Retrieved from https://www.bbc.com/news/health-39904851

BBC. (2017). Wiggins and Froome medical records released by Russian hackers. Retrieved from http://www.bbc.co.uk/news/world-37369705

British Standard Institute (BSI). (2014). *BS11200: Crisis Management – guidance and good practice*. BSI.

Bundesamt für Bevölkerungsschutz und Katastrophenhilfe. (2008). *Schutz Kritischer Infrastruktur: Risikomanagement im Krankenhaus.* Retrieved from https://www.google.com/url?sa=t&rct=j&q=&esrc=s&source=web&cd=1&cad=rja&uact=8&ved=2ahUKEwiC9oTthv3kAhVDb1AKHUiECG8QFjAAegQIAhAB&url=https%3A%2F%2Fwww.bbk.bund.de%2FDE%2FAufgabenundAusstattung%2FKritischeInfrastrukturen%2FPationenKritis%2FSchutz_KRITIS_Ri

Bundesamt für Sicherheit in der Informationstechnik. (2013). *Schutz Kritischer Infrastrukturen: Risikoanalyse Krankenhaus-IT. Bonn.* Retrieved from https://www.google.com/url?sa=t&rct=j&q=&esrc=s&source=web&cd=1&ved= 2ahUKEwjrxtyQhv3kAhUQJlAKHTkcC3wQFjAAegQIABAC&url=https% 3A%2F%2Fwww.bsi.bund.de%2FSharedDocs%2FDownloads%2FDE%2FBSI %2FPublikationen%2FBroschueren%2FRisikoanalyseKrankenhaus.pdf%3F__blob%3

Data Protection Authority. (2019). *Data Protection Authority*. Retrieved from https://www.dpa.gr/portal/page?_pageid=33,40911&_dad=portal&_schema=PORTAL

Deloitte. (2016). *Cyber crisis management: Readiness, response, and recovery.* Retrieved from https://www.google.com/url?sa=t&rct=j&q=&esrc=s&source=web&cd=16&cad=rja&uact=8&ved=2ahUKEwij0amRn_3lAhXISxUIHeu5AWAQFjAPegQICRAC&url= https%3A%2F%2Fwww2.deloitte.com%2Fcontent%2Fdam%2FDeloitte%2Fde%2FDocuments%2Frisk%2FDeloitte-Cyber-crisis-management-Rea

ENISA. (2016). *Good Practice Guide on Vulnerability Disclosure. From challenges to recommendations.* Retrieved from https://www.enisa.europa.eu/publications/vulnerability-disclosure

ENISA. (2016). *Securing Hospitals: A research study and blueprint. Independent Security Evaluators.* Retrieved from https://www.securityevaluators.com/wp- content/uploads/2017/07/securing_hospitals.pdf

ENISA. (2016). *Smart Hospitals: Security and Resilience for Smart Health Service and Infrastructures.* Retrieved from https://www.enisa.europa.eu/publications/cyber-security-and -resilience-for-smart-hospitals

ENISA. (2019). *Greek National Cyber Security Strategy*. Retrieved from https://www.enisa.europa.eu/topics/national-cyber-security-strategies/ncss-map/ national-cyber-security-strategies-interactive-map/strategies/national-cyber-security-strategy-greece/view

EU. (2008). *Council Directive 2008/114/EC.* Retrieved from https://eur-lex.europa.eu/legal-content/EN/TXT/?uri=uriserv%3AOJ.L_.2008.345.01.0075.01.ENG

EU. (2013). *Decision No 1082/2013/EU of The European Parliament and of the Council of 22 October 2013 on serious cross-border threats to health and repealing Decision No 2119/98/EC.* Retrieved from https://ec.europa.eu/health/sites/health/files/preparedness_response

EU. (2016). *Directive (EU) 2016/1148 of The European Parliament and of the Council of 6 July 2016 concerning measures for a high common level of security of network and information systems across the Union.* Retrieved from https://eur-lex.europa.eu/legal-content/EN/TXT

EU. (2016). *Regulation (EU) 2016/679 of the European Parliament and of The Council of 27 April 2016 on the protection of natural persons with regard to the processing of personal data and on the free movement of such data, and repealing Directive 95/46/EC (GDPR).* Retrieved from https://eur-lex.europa.eu/legal-content/EN/TXT/PDF/?uri=CELEX:32016R0679&from=EN

EU. (2017). *Cybersecurity Act.* Retrieved from https://eur-lex.europa.eu/legal-content/EN/TXT/?uri=COM:2017:0477:FIN

EU. (2017). *Regulation (EU) 2017/746.* Retrieved from https://eur-lex.europa.eu/legal-content/EN/TXT/?uri=CELEX:32017R0746

EU. (n.d.). REGULATION (EU) 2017/745. Retrieved from https://eur-lex.europa.eu/legal-content/EN/TXT/?uri=CELEX:32017R0745

European Commission. (2006). *Communication from the Commission on a European Programme for Critical Infrastructure Protection.* Brussels.

European Commission. (2008). *Council Directive 2008/114/EC on the Identification and Designation of European Critical Infrastructures and the Assessment of the Need to Improve their Protection.* Brussels.

Healthcare and Public Health Sector Coordinating Councils. (2017). *Health Industry Cybersecurity Practices: managing threats and protecting patients.* Retrieved from https://www.phe.gov/Preparedness/planning/405d/Documents/HICP-Main-508.pdf

Hei, X. D. X. (2013). Conclusion and Future Directions. In *Security for Wireless Implantable Medical Devices*. SpringerBriefs in Computer Science. doi:10.1007/978-1-4614-7153-0_5

Hellenic National Defence General Staff. (2019). *Hellenic National Defence General Staff.* Retrieved from http://www.geetha.mil.gr/en/hndgs-en/history-en.html

HIPAA. (2018). Healthcare Data Breach Statistics. Retrieved from https://www.hipaajournal.com/healthcare-data-breach-statistics/

HIPAA. (2018). Largest Healthcare Data Breaches of 2018. Retrieved from https://www.hipaajournal.com/largest-healthcare-data-breaches-of-2018/

Independent Security Evaluators. (2016). *Securing Hospitals - A research study and blueprinT.* Retrieved from https://www.securityevaluators.com/wp-content/uploads/2017/07/securing_hospitals.pdf

INFOSEC. (2019). Hospital Security. Retrieved from https://resources.infosecinstitute.com/category/healthcare-information-security/security-awareness-for-healthcare-professionals/hospital-security/

Izuakor, C., & White, R. (2016). Critical Infrastructure Asset Identification: Policy, Methodology and Gap Analysis. In *Proceedings of the 10th International Conference on Critical Infrastructure Protection (ICCIP)*, (pp. 27-41). Academic Press. 10.1007/978-3-319-48737-3_2

Kersten H., Klett G. (2017). Business Continuity und IT-Notfall management.

KPMG. (2015). *Health care and cyber security: increasing threats require increased capabilities*. Retrieved from https://assets.kpmg/content/dam/kpmg/pdf/2015/09/cyber-health-care-survey-kpmg-2015.pdf

Martin, G. M. P. (2017). Cybersecurity and healthcare: How safe are we? *BMJ (Clinical Research Ed.)*, *358*(j3179). PMID:28684400

Martin, L. (2019). Hackers scramble patient files in Melbourne heart clinic cyber attack. *The Guardian*. Retrieved from https://www.theguardian.com/technology/2019/feb/21/hackers-scramble-patient-files-in-melbourne-heart-clinic-cyber-attack

Mikušová, M., & Horváthová, P. (2019). Prepared for a crisis? Basic elements of crisis management in an organisation. *Economic Research-Ekonomska Istraživanja*, *32*(1), 1844–1868. doi:10.1080/1331677X.2019.1640625

National Cybersecurity Agency of France (ANSSI). (2018). *EBIOS Risk Manager – The method*. Retrieved from https://www.ssi.gouv.fr/en/guide/ebios-risk-manager-the-method/

NIS. (2019). *NIS*. Retrieved from http://www.nis.gr/portal/page/portal/NIS/

OFFSITE. (2017). 27χρονος επιτέθηκε σε γιατρούς στο Νοσοκομείο Λάρνακας. Retrieved from https://www.offsite.com.cy/articles/eidiseis/topika/231890-27hronos-epitethike-se-giatroys-sto-nosokomeio-larnakas

Perlroth, N., & Sanger, D. E. (2017). *The New York Times*. Retrieved from https://www.nytimes.com/2017/05/12/world/europe/uk-national-health-service-cyberattack.html

Peters, D., Kandola, K., Elmendorf, A. E., & Chellaraj, G. (1999). *Health expenditures, services, and outcomes in Africa: basic data and cross-national comparisons, 1990-1996 (English)*. Washington, D.C.: The World Bank. doi:10.1596/0-8213-4438-2

POEDIN. (2018). Κέντρα Υγείας σε Αποδιοργάνωση. *POEDIN*. Retrieved from https://www.poedhn.gr/deltia-typoy/item/3413-kentra-ygeias-se-apodiorganosi-viaiopragies-se-varos-iatrikoy-kai-nosileftikoy-prosopikoy--klopes-sto-kentro-ygeias-salaminas-epithesi-apo-omada-roma-se-giatro-tou-ky-lygouriou-pou-efimereve-sto-tep-tou-gnnafpl

SAFECARE project. (2018). *Grant Agreement Number 787005, European Commission H2020*. Retrieved from https://www.safecare-project.eu/

Solon, A. H. A. (2017). Petya ransomware cyber attack who what why how. *The Guardian*. Retrieved from https://www.theguardian.com/technology/2017/jun/27/petya-ransomware-cyber-attack-who-what-why-how

Sulleyman, A. (2017). *NHS cyber attack: Why stolen medical information is so much more valuable than financial data.* The Independent. Retrieved from https://www.independent.co.uk/life-style/gadgets-and-tech/news/nhs-cyber-attack-medical-data-records-stolen-why-so-valuable-to-sell-financial-a7733171.html

The Guardian. (2019). Violence in the NHS: staff face routine assault and intimidation. Retrieved from https://www.theguardian.com/society/2019/sep/04/violence-nhs-staff-face-routine-assault-intimidation

The Times. (2010). Woman kills 3 in hospital shooting spree. Retrieved from https://www.thetimes.co.uk/article/woman-kills-3-in-hospital-shooting-spree-n9q9nbhws9b

World Health Organisation. (2019). Hospitals. Retrieved from https://www.who.int/hospitals/en/

World Health Organisation. (2019). *Health Systems*. Retrieved from http://www.euro.who.int/en/health-topics/Health-systems/pages/health-systems

World Health Organization. (2015). *Hospital Safety Index: Guide for Evaluators.* Retrieved from https://www.google.com/url?sa=t&rct=j&q=&esrc=s&source=web&cd=1&cad=rja&uact=8&ved=2ahUKEwi_tdq7n__kAhVE2aQKHVZfBbkQFjAAegQIAxAC&url=https%3A%2F%2Fwww.who.int%2Fhac%2Ftechguidance%2Fhospital_safety_index_evaluators.pdf&usg=AOvVaw3Jb3x3xUgBh-IK84EtnKD8

ADDITIONAL READING

Abouzakhar, N. S., Jones, A., & Angelopoulou, O. (2018). Internet of Things Security: A Review of Risks and Threats to Healthcare Sector. In *Proceedings of the IEEE International Conference on Internet of Things (iThings) and IEEE Green Computing and Communications (GreenCom) and IEEE Cyber, Physical and Social Computing (CPSCom) and IEEE Smart Data (SmartData). IEEE Press.*

Ahmed, Y., Naqvi, S., & Josephs, M. (2019). Cybersecurity Metrics for Enhanced Protection of Healthcare IT Systems. In *Proceedings of the 13th International Symposium on Medical Information and Communication Technology (ISMICT).* Academic Press. 10.1109/ISMICT.2019.8744003

Alcaraz, C., & Zeadally, Z. (2015). Critical infrastructure protection: Requirements and challenges for the 21st century. *International Journal of Critical Infrastructure Protection*, *8*, 53–66. doi:10.1016/j.ijcip.2014.12.002

Baggett, R. K., & Simpkins, B. K. (2018). *Homeland Security and Critical Infrastructure Protection.* Praeger Security International.

ENISA. (2016). *Securing Hospitals: A research study and blueprint. Independent Security Evaluators.* Retrieved from https://www.securityevaluators.com/wp-content/uploads/2017/07/securing_hospitals.pdf

ENISA. (2016). *Smart Hospitals: Security and Resilience for Smart Health Service and Infrastructures.* Retrieved from https://www.enisa.europa.eu/publications/cyber-security-and-resilience-for-smart-hospitals

EU. (2016). *Directive (EU) 2016/1148 of The European Parliament and of the Council of 6 July 2016 concerning measures for a high common level of security of network and information systems across the Union.* Retrieved from https://eur-lex.europa.eu/legal-content/EN/TXT

EU. (2016). Regulation (EU) 2016/679 of the European Parliament and of The Council of 27 April 2016 on the protection of natural persons with regard to the processing of personal data and on the free movement of such data, and repealing Directive 95/46/EC (GDPR). Retrieved from https://eur-lex.europa.eu/legal-content/ EN/TXT/PDF/?uri=CELEX:32016R0679&from=EN

HIPAA. (2018). Largest Healthcare Data Breaches of 2018. Retrieved from https://www.hipaajournal.com/largest-healthcare-data-breaches-of-2018/

KEY TERMS AND DEFINITIONS

Asset: Something of either tangible or intangible value that is worth protecting, including people, information, infrastructure, finances and reputation.

Attack: Attempt to destroy, expose, alter, disable, steal or gain unauthorized access to or make unauthorized use of an asset

Cyber: Relating to, within, or through the medium of the interconnected information infrastructure of interactions among persons, processes, data, and information systems.

Cyber Incident: A cyber event that: a) jeopardizes the cyber security of an information system or the information the system processes, stores or transmits; or b) violates the security policies, security procedures or acceptable use policies, whether resulting from malicious activity or not.

Healthcare: Healthcare is defined as the prevention and treatment of diseases through medical professional services.

Hospital: An institution providing medical and surgical treatment and nursing care for sick or injured people.

Incident: Situation that can be, or could lead to, a disruption, loss, emergency or crisis

Prevention: Measures that enable an organization to avoid, preclude or limit the impact of an undesirable event or potential disruption

Threat: Potential cause of an unwanted incident, which may result in harm to individuals, assets, a system or organization, the environment or the community.

Vulnerability: Weakness of an asset or control that can be exploited by one or more threats. The existence of a weakness, design, or implementation error that can lead to an unexpected, undesirable event compromising the security of the computer system, network, application, or protocol involved.

Chapter 4
Security of Water Critical Infrastructure:
The Threat Footprint

David Birkett
Mettle Crisis Leaders, Australia

ABSTRACT

There is an identified and elevated threat level to water services by modern terrorists, in consideration of increasing levels of observed violence in recent terrorist attacks across Europe. This chapter raises significant aspects related to the security of the water critical infrastructure (water CI). Initially, dependencies and interdependencies of water CI, with other CIs, are highlighted as a potential incubating risk, which may well be hidden within the complexities of the modern water value chain of logistics and services. Threats to water CI including single points of failure are further described, followed by terrorist water attack planning methodologies and strategies. Finally, the water CI protection that may be considered to reduce any future threat levels from acts of terrorism is discussed.

INTRODUCTION

The current heightened risk of terrorist attacks in Western European countries (Europol, 2019), has elevated the threat level of water services to attack from adverse human intervention, such as terrorist groups. Various researchers consider that water services are more likely to be the critical infrastructure (referred to as 'CI' within this chapter) at the centre of human conflict for the indefinite future (USACHPPM, 2019; Kroll, 2010a; Maiolo & Pantusa, 2018; Meinhardt, 2006). This chapter provides detailed data and information related to the safety, security and protection of European urban water systems, which is considered by many researchers to be the most vulnerable and highest risk of all CIs (USACHPPM, 2019; Court-Young, 2003; Lee, 2009; Maiolo & Pantusa, 2018).

Areas of CI identified by various European Union (EU) countries have been designated as 'lifeline systems' (Wang, Hong, & Chen, 2012), which within our modern societies provide a reliable flow of the products and services essential to the defence and economic viability of modern society. Moreover,

DOI: 10.4018/978-1-7998-3059-7.ch004

the supply of water services certainly fit this definition, as it is essential for human existence, and is a key dependent CI for other CIs.

Why Water Critical Infrastructure?

The progression of modern terrorism has developed to a global phenomenon, and more specifically a potential threat to vulnerable and sensitive sectors of CI (such as water) in developed societies as a convenient alternate terrorist strategic action. A potential terrorist attack on water CI not only displays an immediate effect on society in the targeted area, but also across the rest of the world, engaging a receptive media and global community audience via the internet, news media and social media. Significantly, and more specifically, since the '9/11' terrorist attack in 2001 there is an increasing global awareness, in developed societies, that urban water systems are extremely vulnerable to forms of adverse human intervention, which may impact significantly upon large human populations.

Water Systems are vulnerable to a range of intentional threats, including damage or sabotage through physical destruction and cyber-attack (Maiolo & Pantusa, 2018).

The supply and distribution of potable drinking water, in most societies, is often a hidden, and unrecognised form of CI, despite being essential for life on our planet. Indeed, the average human is able to survive only for an average of three days without the ingestion of water to maintain the human body systems and functionality (Cohen, 2010; Lee, 2009). Deliberate contamination of a water distribution system serving a large European population centre potentially represents a mass casualty event, which may prove similar in outcome to the devastating and horrific attacks on New York, USA in 2001 (9/11) should this scenario occur.

As indicated, the supply of water is considered the most significant and critical, of the various identified sectors of CI due to linkages and interdependencies across all areas of CI.

Objectives

Initially, dependencies and interdependencies related to water CI in regard to other CI areas are defined. Due to potential various current elevated threat levels to water services, threats to water services, inclusive of single points of failure (SPOFs) in relation to any future terrorism based adverse human intervention, are described. Furthermore, the security of water CI is discussed detailing and addressing the methodologies and template strategies adopted by the modern terrorists, potentially planning, preparing and attacking water services. Finally, the protective strategies which may be adopted to anticipate, protect, deter and act in respect to any future terrorist activity or attack against water services are addressed.

HISTORICAL BACKGROUND

The application of deliberate adverse human contamination of water as a form of using water or wastewater as a weapon is not a new phenomenon, with recorded incidents of attacks on water systems extending from 4,500 years ago (Gleick, 2006), when Urlama, King of Lagash from 2450 to 2400 B.C., diverted water from Girsu, to boundary canals, drying up boundary ditches to deprive the Umma population of water.

A few selected global incidents where water was used as a threat or weapon between nations and external groups are detailed as follows:

- **1939:** The Japanese military poisoned Soviet water sources with intestinal typhoid bacteria in an area near the former Mongolian border (Kroll, 2010a).
- **1939 to 1942:** The Japanese military 'Unit 731' poisoned Chinese water sources, wells and bulk water storages with typhoid and other chemical and biological pathogens (Gleick & Heberger, 2014).
- **1944 (November):** Soviet troops flooded the area south of the Istra Reservoir near Moscow, Russia. This was a strategy to slow the German advance on Moscow. A few weeks later, German troops used the same tactic to create a water barrier to halt advances by the Soviet 16th Army (Gleick & Heberger, 2014).
- **1945:** The desperate and retreating German army poisoned a large reservoir in Bohemia with raw sewage (Kroll, 2010a).
- **1965:** Yasir Arafat's Fatah movement orchestrated several bombing attacks on Israeli water infrastructure. The Fatah movement was also responsible for additional attacks on Israeli water pipes (Kroll, 2010a).
- **1984:** In Australia, Geelong man jailed after threatening to poison Melbourne's water supply (Noble & Schrembi, 1984).
- **2000:** In Australia, intentional malicious attack on the supervisory control and data acquisition (SCADA) system in Maroochy, Queensland, causing 800,000 litres of raw sewage to be released into parks, rivers and the grounds of a luxury hotel (Abrams & Weiss, 2008).
- **2002 (July):** US Federal officials, in Denver, arrested two Al Qaeda operatives in possession of documents detailing how to poison the US water supplies (Kroll, 2010a & Locken, 2017).
- **2002:** In Rome, Italy, four men of the Salafist Group for Preaching and Combat (GSPC) were arrested in possession of cyanide based chemicals, false papers and detailed plans for the water supply network in the zone of the Embassy of the US (Balmer, 2004; Gleick & Heberger, 2014; Kroll, 2010b; Kroll, 2013; Locken, 2017).
- **2002:** Al Qaeda operatives arrested with plans to attack the water network surrounding the Eiffel Tower neighborhood in Paris, France (Kroll, 2008).
- **2003:** In Australia a man threatened to poison Melbourne's water supply unless certain shares on the Stock Market were to rise in value (ABC, 2003).
- **2004:** In the US 50 unwanted water system intrusion incidents were identified by 10 federal agencies involving SCADA computer systems that control water supply and wastewater systems (Gleick & Heberger, 2014; Kroll, 2006).
- **2004:** The FBI and the Department of Homeland Security issue a bulletin warning that terrorists are trying to recruit employees from water treatment plants as part of a drinking water poisoning project (Locken, 2017).
- **2006:** A water tank in Tring, England, is deliberately contaminated with herbicide. (Locken, 2017).
- **2006:** Strychnine (a pesticide) is intentionally released into a Danish artificial lake (Locken, 2017).
- **2007:** In China, 201 people die using water intentionally contaminated with fluoroacetamide (a pesticide) to prepared oatmeal (Locken, 2017).

- **2008:** In Varney, Virginia, a man was arrested in possession of 2 vials of cyanide to poison the water supply system (Locken, 2017).
- **2008:** The water supply system of a Burmese refugee camp in Thailand (with a population of 30,000) is intentionally poisoned with herbicide (Locken, 2017).
- **2009:** In the Philippines, the Moro Islamic Liberation Front (MILF) poisoned sources of water used by government soldiers and the population (Locken, 2017).
- **2010:** In the Kashmir region of India, Maoist rebels poison a pond used as a source of drinking water by the Central Reserve Police Force, a paramilitary group (Locken, 2017).
- **2010:** In England, two neo-Nazis, father and son, are convicted of several counts of terrorism, including castor-making and conspiracy with Serbian Nazis to poison water resources used by Muslims (Locken, 2017).
- **2011:** Documents seized during the US Military operation causing the death of Osama bin Laden revealed plans for poisoning water resources (Locken, 2017).
- **2011:** In Spain (La Línea de la Concepción, Cadiz), a conspiracy to poison water resources in response to the death of Osama bin Laden is thwarted (Locken, 2017; Silva, 2011).
- **2012:** In Australia, two 5,000-litre drinking water tanks are deliberately poisoned with Diuron (a toxic herbicide) (Locken, 2017).
- **2013:** Australia, a man charged with threats to poison Melbourne's water services faced 12 counts of collecting and possessing documents connected with a terrorist act. Items documented at this trial were associated with poisoning water supplies, among other forms of terrorism, with links to al Qaeda (Cooper, 2013).
- **2012:** In Afghanistan, hundreds of girls in a school fall ill after a deliberate poisoning of the water supply system (Locken, 2017).

Furthermore, Figure 1 clearly illustrates a significant rise in the attacks on water services, from minimal incidents indicated in the 1980's, with a defined rise from 1999, culminating in 30 attacks in 2012.

Additionally, a recent global analysis (Veilleux & Dinar, 2018) also indicates a significant rise in water related terrorist attacks on a global basis. Typical of these are the 'Shining Path' leftist Terrorist group who have targeted and destroyed precious water infrastructure in Peru, South America. Similarly the Islamic State of Iraq and the Levant (ISIL) also took control of the water infrastructure and dams at both Tabqa in 2013 and Mosul in 2014, indicating terrorist's understanding of the strategic nature of water as an essential resource and useful political form of CI.

In summation of earlier water related incidents, between 1970 and 1976, Veilleux & Dinar (2018) identified 675 attacks conducted by 124 terrorist organisations across 71 countries. These attacks resulted in approximately 3,400 dead or sickened people. The water infrastructure attacks mainly targeted the pipes, dams, weirs, levees and treatment plants associated with the storage, treatment and delivery of water. However, there is a new and more insidious perspective for the future with the potential terrorist application of 'cyber attacks' through SCADA and physical water distribution backflow incidents with toxins, chemicals or radioactive products, particularly in Western European large and more densely populated centres. For example, large population centres could be viewed as potentially vulnerable with the 2019 return of ISIL European Terrorists from Syria to Germany, Belgium, UK, France and Spain. Terrorist incidents attacking Water services in Western Europe are considered potentially more likely, due to the frequency of terrorist attacks over the past 10 years generally, in comparison with Eastern and Central European locations which are observed to being currently devoid of such attacks.

Figure 1. Recorded attacks on water CI 1980 – 2012 (adapted from (Abrams & Weiss, 2008; Peter H Gleick & Heberger, 2014; Dan Kroll, 2010a; Weiss, 2011))

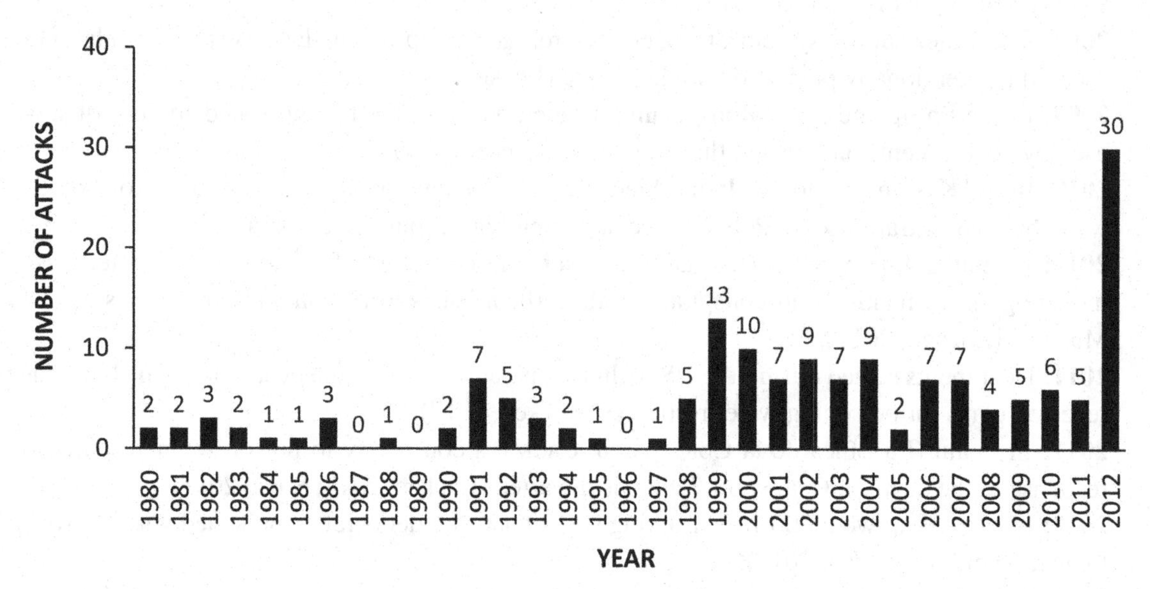

Presently, there are hundreds of threats against municipal water systems each year in the US (Beering, 2002). Furthermore, as indicated in the previous bulleted list water has been a target in modern times on a global basis.

SECURITY OF WATER CRITICAL INFRASTRUCTURE

Assessments of water CI security are extremely broad and complex tasks, with wide variations in risk management across EU water organisations. This section holistically addresses the threat footprint and protection of water services, commencing with dependency and interdependency of water infrastructure in relation to other CIs. Furthermore, the issue of systemic SPOFs is examined.

Defining Dependency and Interdependency

Modern society in the 21st century is increasingly dependent on CI which produces the essential componentry to provide society with the technology required to operate on a daily basis. Water and wastewater are considered to be amongst the most critical of CI, with the average person only able to survive only for three days without the intake of water to maintain human body system functionality. Additionally, modern society relies heavily on the effect and unseen removal and disposal of human waste through the wastewater disposal systems as a form of essential CI, which ensures the separation of treated drinking water and fecal contamination from human waste. This separation reduces the potential for community outbreaks of human diseases such as cholera and other forms of dysentery, as identified by Snow (1857).

Water Distribution Systems have grown exponentially and relative to the growth of the economy and population spread across geographical areas. However, over the recent decade the dependent and interdependent technological value chain has sustained a period of growth and extension to support the industry. Furthermore, Gillette, Fisher, Peerenboom, and Whitfield (2002) warn that:

Failure to understand how disruptions to one infrastructure could cascade to others, exacerbate response and recovery efforts, or result in common cause failures leaves planners, operators, and emergency response personnel unprepared to deal effectively with the impacts of such disruptions (Gillette et al., 2002).

It is this perception of the failure to adequately understand these often hidden dependence and interdependency nodes, which may impact the water sector suddenly, with minimal warning, and may well escalate to contribute to a series of cascading failures across other interdependent infrastructures.

Water Dependency

The continual progression of the computerisation of modern society, relative to CI, has created an expanding incremental dovetailing of complex and increasingly dependent linkages across the value chain of each CI, which is outlined as follows.

Mutual dependence and interconnectedness made possible by the information and communications infrastructure lead to the possibility that our infrastructures may be vulnerable in ways they never have been before (Marsh, 1997 as cited in Gillette et al., 2002).

As these rising levels and complexities of dependent linkages become increasingly dependent on each other, it is often difficult to identify the potential SPOFs that may occur should one of these supportive link nodes fail, with subsequent potential cascading impact on the operation of various CI, such as water and wastewater.

Water dependency may be considered as the most significant reliance factor of any other CI in continuing to effectively conduct normal business. As an example, this defined dependency on water services varies in a percentage from 67% - 99% reduction in critical manufacturing after 4 hours of water loss. Additionally, critical manufacturing, after a 6 hour period of wastewater loss, is predicted to experience a 34% - 66% reduction in production (Porod, Collins, & Petit, 2014).

A dependency analysis conducted in 2014 (Porod et al., 2014) across 12 identified CI areas, related the specific CIs of their dependency to water by percentage of dependency and estimated offline degradation without a water supply. Of specific interest to the issue of water vulnerability, as distinct from the actual mitigation of the water distribution system itself, is the aspect of interdependency, which is in the form of the mutual dependency of water with other CI. An example of this is electricity CI, as displaying a 92% dependency on water, and a 4 hour degradation period. The water is required for the generation plant's cooling systems, with the continued supply of electricity essential for water supply pumps, water and wastewater treatment processes, coupled with the SCADA communications network for the operation of the water and wastewater network.

Sullivan (2011) considers that most CI sectors are to a certain extent dependent on the water sector, with major incremental cascading effects across societies in the case of a sudden absence of treated water and wastewater services.

Figure 2. Dependency relationship of water CI on other CI's (Birkett, 2017)

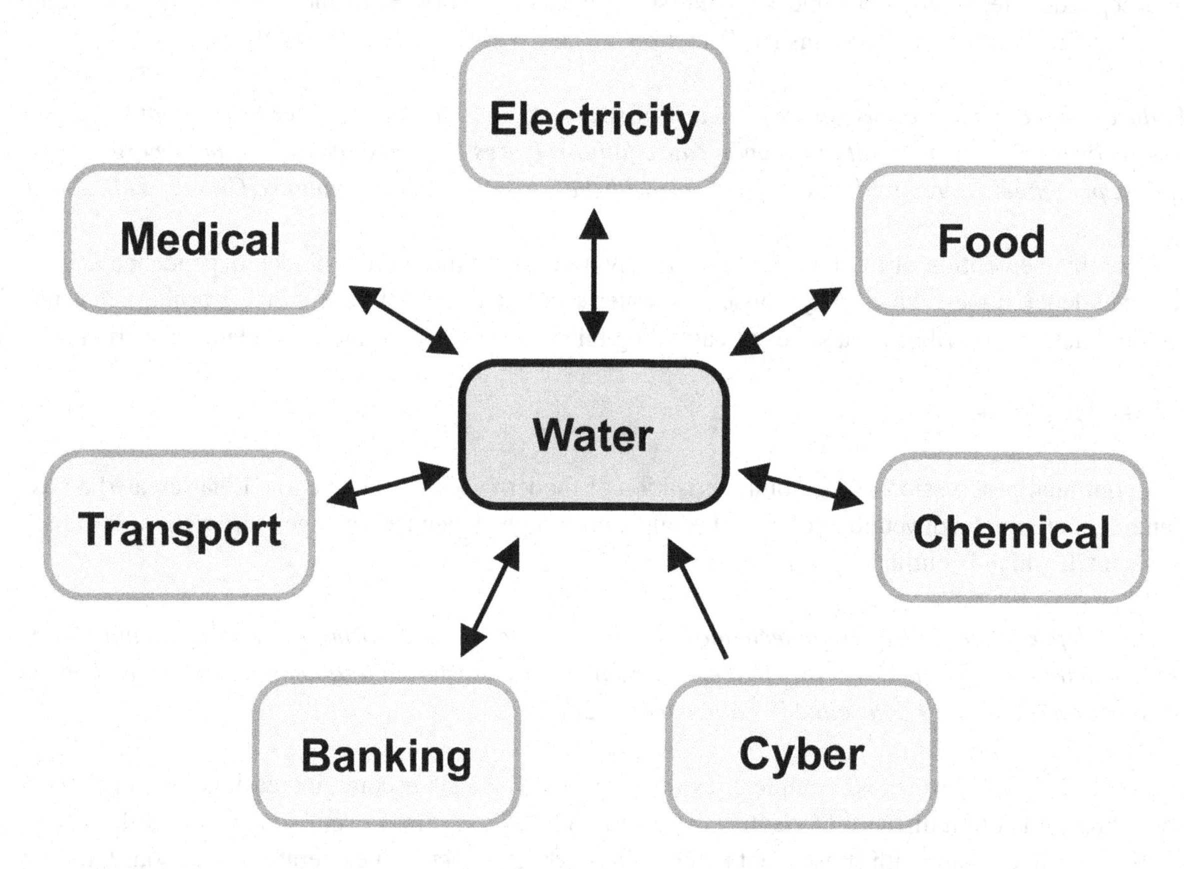

The Interdependency Supply Chain: The 'Soft Underbelly' of Water

As indicated in Figure 2, the water services CI is the most extensively dependent and interdependent CI with all other CI's, which contributes to its significance and potential for possible cascading failures in other CI's, subsequent to any future protracted water services failure. An interesting example of the cascading nature of a protracted water service failure is the recent London, UK extensive water supply failure which occurred on 13 June 2019 (Spear-Cole, 2019). Although this water failure was related to a life cycle asset failure associated with corroded water delivery pipes, it effectively illustrates the dependency flow on effect, with 100,000 properties without water supplies. Furthermore, schools, hospitals and other important services were closed, with bottled water at all supermarkets sold out within 1 hour. This example is of interest, as the water supply was reinstated incrementally within 2 days, but created significant stress and panic within a community of 100,000 people and had an economic impact closing businesses, education facilities and industry for a short period. Water outages or contaminated water incidents additionally tend to reduce the confidence of the community in the quality and delivery of potable water.

Furthermore, as the water industry progressively attains a greater dependency on computers, the supply chain vulnerabilities are often increasingly hidden within the context of an assumed product or service supply, and to a certain extent may be forgotten and neglected during water vulnerability assessments. Although water and wastewater are considered essential, to varying degrees, on a wide variety of other

CI's, water is also dependent, displaying an interdependency again of varying percentages. Electricity is considered an essential form of CI for water and wastewater, to supply energy to pumps, treatment and operational controls, as is the CI of chemicals to treat the water and wastewater. It is these often variable interdependencies and dependencies which are the 'soft underbelly' of water, displaying the often overlooked SPOFs, without which the water or wastewater service may not operate.

The challenge of analysing the risk or threat perception of interconnected interdependent systems of infrastructure is becoming increasingly complex as modern society and technology progresses, with computerised automated ordering systems reflecting variances in individual risks exposures and resistance to service failures (T. Brown, Beyeler, & Barton, 2004). However, some researchers have developed tools and processes to assess this arduous task, such as T. Brown, Beyeler, & Barton (2004), who describe the goal of assessing the "complex adaptive systems" as identifying the significant risks to critical systems and devising effective mechanisms for mitigating these risks.

Partnerships and communications linkages across the supply chain should be developed to discuss and increase awareness of the mutually dependent relationships. These improved relationships would tend to reduce the potential impact of any supply chain failure, and conversely to raise the awareness of the impact of possible water loss on other CI and supply chain owners.

Threats to Water Services

Although there are multiple valves, pipes, storages (treated & untreated) with the addition of reservoirs and pumping stations, it is of academic interest to observe and consider the potential generic water services targets that the 21st terrorist would consider if planning an attack.

Figure 3 illustrates the author's radial attack perspective with the most significant terrorist-initiated targets in the centre and the least likely on the perimeter of the circular diagram. The water distribution systems are considered to be the most vulnerable node of the water services to any external attack (Dan Kroll, 2006). Although this includes water system nodes of treated water storage tanks, this vulnerability is more significant in relation to a 'backflow' incident with the absence of water distribution non-return valves, or toxin/chemical detection systems in place, within the water distribution system. Furthermore, a potential attack on Wastewater systems would also cause significant disruption and health issues across large populations with the failure of waste disposal infrastructure in industry, education and domestic dwellings. The Tertiary targets would be directed against the assets, and human technical expertise within the water services to disrupt normal and repair operations.

The gradual global introduction of SCADA has some significant advantages and benefits with distant continuous monitoring, staff reductions and economic benefits. However, there are some potential challenges to be addressed such as many water treatment plants (WTPs) across Europe are now no longer staffed by water technicians, and are remotely controlled from a central control point some distance from the WTP. This increases the risk and vulnerability of adverse human intervention. Furthermore, there are some identified risks and weaknesses with nefarious and anonymous individuals hacking into the SCADA system to alter pump controls; chemical dosing, or to turn valves off and on.

Indeed, SCADA systems may well be a potential incubating weakness in water systems to an external cyber-attack. Between February 2000 and April 2000 an ex SCADA contractor transmitted illegal SCADA commands, from a vehicle, to the sewage system of Maroochy Shire Council in Queensland, Australia. The offender, an aggrieved ex- employee, carried out these illegal commands on 46 occasions, altering pumps, alarms and communications with the SCADA control centre. These activities released

Figure 3. Potential identified terrorist attack targets in water CI (Birkett & Mala-Jetmarova, 2014)

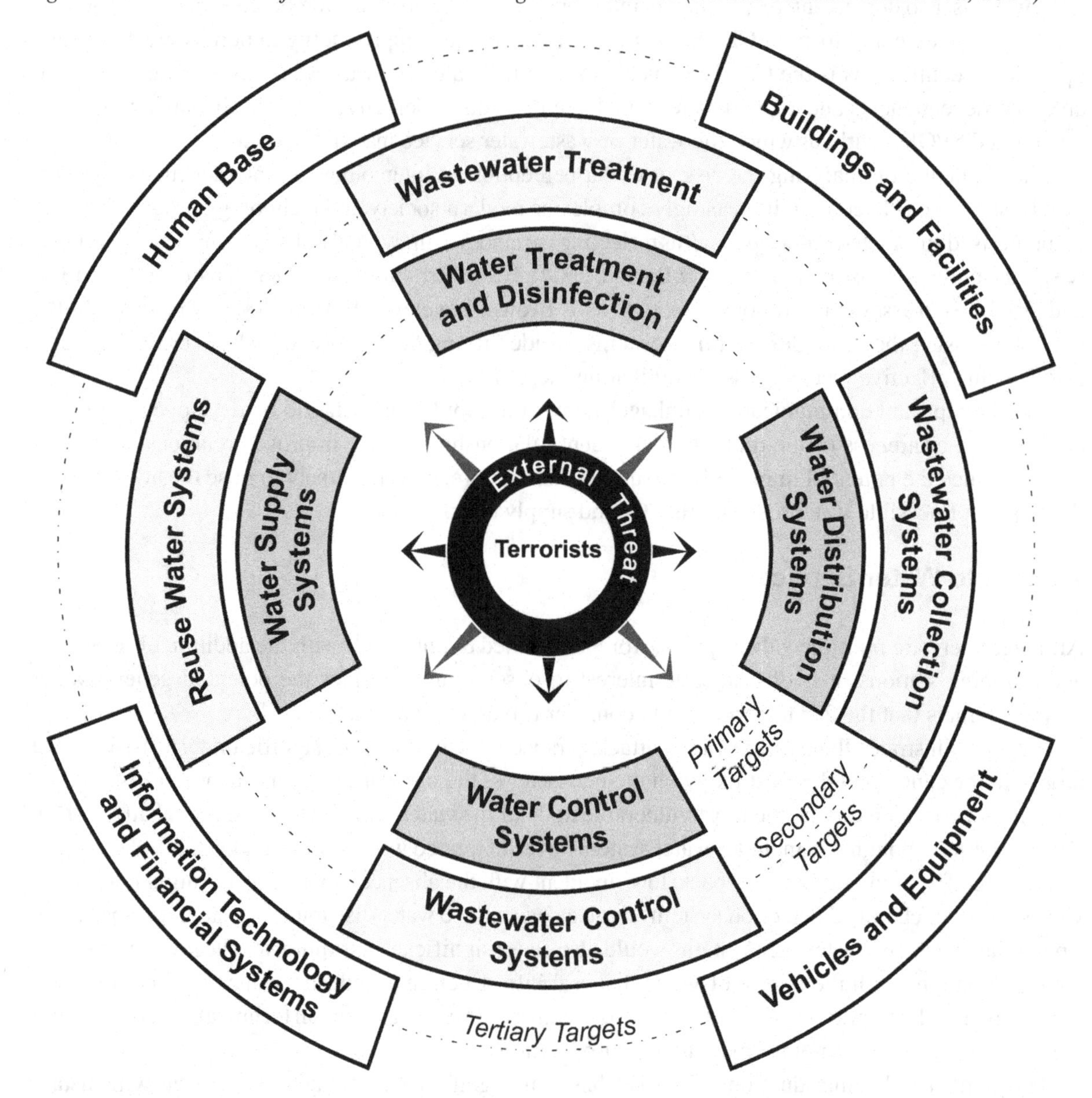

800,000 litres of raw sewage into parks, rivers and the grounds of a large luxury hotel (Abrams & Weiss, 2008). The offender was only apprehended by an observant Queensland Police Officer who stopped the offender's vehicle for a minor traffic offence.

As indicated in Figure 3, the terrorist targets are graded in suggested significance from the primary, secondary and tertiary levels, with the distribution system of treated water the most vulnerable.

Craig Stanners, the Director of IVL Flow Control in the UK issued a media statement in 2015 (Brockett, 2015) highlighting the potential vulnerability for deliberate water contamination in the UK:

The UK's drinking water is at far too great a risk from potential contamination of supply by terrorists, with current systems simply not quick enough to contain a chemical or biological attack (Brockett, 2015),

And, furthermore Stanners also stated (Brockett, 2015):

What's in place at the moment isn't anywhere near quick enough to cope, and those wishing to cause damage to our drinking water would laugh at our response that we are waiting five days for results to come back from the lab-by then it would be too late (Brockett, 2015).

The potential agents of toxins, biological agents and contaminants that may be used in a terrorist attack on water are extremely diverse in both literature and availability to the terrorist through the internet. However, there are many factors to be considered such as solubility in water, degradation by chlorine or ultraviolet light from the sun, danger of handling and detectability by the population served by the water supply (Dan Kroll, 2006). However, Peter H. Gleick (2006) suggests that it may not be necessary to actually contaminate drinking water with chemicals or toxins, as a plausible threat with harmless vegetable dye would be sufficient to create panic in the population from drinking water that may be purple, blue or green. This would potentially spread community fear, anxiety, and reduce confidence in drinking water supplies, which is one terrorist objective.

Furthermore, in July 2019, it was reported in a viral internet message on the 'WhatsApp' messaging application, that a woman claiming to be a nurse from a Paris Hospital advised people to not drink water from Paris drinking water outlets due to the alleged presence of radioactive titanium. From this social network message, rumours spread quickly throughout Paris that 'authorities' were requesting the population not to drink reticulated water. Subsequent to this rumour, hospitals and health centres were inundated with phone calls with people attempting to verify the validity of the news. In order to resolve the fast spreading panic, the water business released a public message that the drinking water posed no threat, with the Director of the Paris health body also releasing a statement that there was no problem with the water (Guay, 2019). The Paris prosecutor's office has opened an investigation into the crime of "publicising, spreading and reproducing false information intended to cause public disorder" (Guay, 2019), which risks a fine of up to 45,000 Euros to any person found guilty. This example clearly illustrates the panic and community fears of real or false drinking water contamination. Ironically on the day of the release of this messaging app of 'fake news' in Paris the temperature in Paris peaked at over 40 degrees Celsius, on a day of maximum demand and usage of drinking water.

An additional and appropriate example is to examine the May 2015 water contamination incident in Prague, Czech Republic (Lazarova, 2015; PDM, 2015b; Willoughby, 2015). Although this pathogen was alleged to be a 'natural occurring' pathogen identified as 'e-coli' or 'Norovirus', 250-300 people were admitted to hospital with severe symptoms as a consequence of this contamination. The 320,000 inhabitants of Prague 6 were advised not to drink the water with alternative water tanks supplied for drinking purposes (Lazarova, 2015; Willoughby, 2015).

Of specific interest was that the affected region of contaminated water was located in Prague 6, which is the area where most foreign missions and embassies are located within the Czech Republic. In this case, the identification of the drinking water contamination issue was notified by the Prague Health services, which may also reflect the reality that in the case of any future adverse human intervention into water services, prior to the contamination being recognised by scientific or technological assessment, the local population may be deceased or ill, with the primary alert being issued by the medical services (Meinhardt, 2005, 2006).

Moreover, Meinhardt (2006) considers that medical training, education and preparedness for these types of medical events may well be deficient, and this risk may only become visible in the case of any future event. As a consequence that most water managers have not encountered many adverse incidents of this nature, there appears to be a fundamental attitude that 'it probably will not occur' and so budgets are often alternatively allocated to the most common potential sources of asset failure and weather issues, as the most likely hazards & targets. The problem with this thinking is that threats and risks to water services are currently on the rise, due to an evolving threat environment (Van Leuven, 2011).

Historically, disastrous incidents have occurred by human error or negligence, of which the effects and outcomes may be considered in relation to deliberate damage to water CI. The wastewater collection system may act as a large explosive system, such as occurred in Guadalajara, Mexico on 22 April 1992 (Dugal, 1999). This incident involved fuel accidently leaking into the sewer system combined with the existing methane-based gas within the sewer, which subsequently caused a series of massive explosions, visible from a satellite in space. As a consequence, approximately 14 kilometers of streets including buildings and vehicles collapsed into a massive trench. Officially, 200 people were killed, 2,000 injured and 20,000 left homeless, with an estimated damage to buildings of $300 million, and 75 businesses failing to recover after this tragedy.

A medically reported example of contamination by human error in the UK in 1988 was the accidental discharge of 20 tonnes of aluminium sulphate into the treated water reservoir that supplied treated water to the Camelford area of Cornwall in the UK (Altman et al., 2006; Morris, 2012; Reid, 2007). The drinking water was heavily contaminated with aluminium sulphate, with the water, displaying a low ph. An extensive delay occurred in identifying the contamination from testing, with a significant community increase of rashes and gastrointestinal disturbances with some instances of green hair occurring within a few days, and later cases of musculoskeletal pains, malaise, impairment of concentration and memory. Subsequent to the medical alert of this incident, two years later approximately 400 persons were reported as still suffering from symptoms that they attribute to this incident (Altman et al., 2006; BBC & Rose, 2012; Morris, 2012; Reid, 2007).

As identified, and indicated by many researchers (Abrams & Weiss, 2008; Birkett & Mala-Jetmarova, 2012, 2014; Birkett, Truscott, Mala-Jetmarova, & Barton, 2011; Brockett, 2015; Cameron, 2002; Coleman, 2005; Copeland, 2010; Covert, 2008; Gleick, 1993; Kroll, 2006, 2010a), water and wastewater are recognised as vulnerable and soft targets related to deliberate, adverse human intervention under the categories of sabotage, terrorism or criminal incidents. Furthermore, Peter H. Gleick (1993) comments that as our civilization and development progresses forward in the twenty-first century, the likelihood of both military action and instruments of war is increasing to involve water and wastewater.

As we approach the twenty-first century, water and water-supply systems are increasingly likely to be both objectives of military action and instruments of war as human populations grow, as improving standards of living increase the demand for fresh water, and as global climatic changes make water supply and demand more problematic and uncertain (Gleick, 1993).

There are many historical examples of attacks and attempted attacks on the water sector. A documented example is the attempted terrorist contamination of drinking water in Rome in 2002. On 20th February 2002, a press release advised that four people had been arrested in Rome with cyanide and detailed maps of the Rome, Italy water supply network. Four Moroccan Terrorists of the 'Salafist Group for Call and Combat', an Algerian based terrorist group with links to the Al Qaeda terrorist group, possessed detail plans of the water network supplying the US Embassy in Rome. It was alleged that they planned to insert a cyanide derivative (Ferricyanide) which was a compound that was not considered poisonous. However, they were acquitted after a 2 year trial in the Italian court system (Kroll, 2006). Dan Kroll (2006) subsequently argues that the cyanide derivative was more likely to be a 'tracer chemical' to ascertain and identify which pipe actually fed the US Embassy site? This may well have been an attempted deliberate backflow attempt in the Rome water distribution system.

In the previously cited example it occurred in an Australian Melbourne court, where a man faced 12 counts of collecting and possessing documents connected with a terrorist act. The offender in this case possessed documents related to poisoning water. As indicated by Kroll, in 'Water as a Weapon', 28 recorded attempts and attacks on water are listed, as an applied instrument in conflict (Dan Kroll, 2010a). As previously mentioned, Meinhardt (2006) further advises that the primary and first indications of any deliberate water contamination would originate from the health sector, where training and education for this type of event may well be deficient. An interesting example supporting Meinhardt's medical hypothesis related to water contamination, is the 2015 water contamination incident which occurred in the 'Prague 6' region of the City of Prague in the Czech Republic. This incident was first reported by the Prague health services in lieu of the water company, as 250 - 300 people from Prague 6 were admitted to hospital with severe symptoms as a consequence of contamination from coliform bacteria and norovirus in the water distribution system (PDM, 2015a).

Should a major terrorist attack occur on a major water centre in the EU in the next 5 years, share prices will tumble; the chief executive officer (CEO) and possibly the board will come under increased pressure to explain; consumer confidence in the water systems will significantly reduce, and Governments may introduce a compulsory compliance issue to protect and reduce the threat to water services.

Single Points of Failure

Subsequent to 9/11, when government agencies particularly in the US, considered and analysed other potential targets of terrorist interest, water services were initially considered vulnerable at the upstream points of collection, such as water storage dams, rivers and water storages in general (Dan Kroll, 2006). Upstream water supplies or source water is located prior to treatment, which is now considered difficult to contaminate, due to the quantities of toxins or chemicals required by any deliberate offender. Furthermore, water treatment in many cases also may reduce the contamination prior to entering the consumer distribution system in the downstream locations.

There are many SPOFs across the entire water service systems. Water systems are so large and diverse there is some recognised difficulty for water companies to identify and protect all visible service nodes. However, when examined as to the most likely locations or nodes where terrorist attacks may occur in the future the following may be considered:

- Water services are vulnerable to physical and cyber-attacks on water treatment plants; mechanical and chemical components of water distribution systems such as pump stations, pipelines and treated water storages.

- The water distribution system after treatment is considered an 'at risk' area vulnerable to a back-flow attack by a terrorist, especially in consideration of the Rome attempt in 2002, and the number of documents seized from terrorists indicating this form of attack.
- Wastewater treatment plants are vulnerable to interference in the chemical, biological and physical treatment processes.

Both water and wastewater services are becoming increasingly computer automated, and subsequently vulnerable from the use of 'SCADA', which may also provide a potential future opportunity for a cyber-attack to alter chemical dosing, turn valves off and introduce 'water hammer' issues to damage the equipment (Clark, 2014; Maiolo & Pantusa, 2018).

Water Treatments Plants

There are many ways to inflict damage to a drinking water treatment plant. A terrorist may physically damage areas of the plant process, which may alter the flow and pressure of water. Furthermore, the chemical processes may be interfered with to alter the quality or disinfection of the drinking water. However, these are short term risks, which may be identified by the SCADA systems, alerting the water managers to an incident, with rectification mostly in a short time frame. The water treatment plant represents the final barrier between potential contamination and the end consumer. This point is also invariably the end point of continuous SCADA monitoring for chemical fluctuations and data.

Treated Water Holding Tanks

Treated water holding tanks are within the water system to provide a side demand management control facility and also to provide sources of water pressure in flat areas of supply. These tanks are vulnerable due to containing smaller sources of water, and usually, limited or displaying an absence of water treatment prior to delivery to the consumer.

Water Distribution System

The water distribution system is considered the most vulnerable to illegal water contamination through a deliberate back-flow attack by a terrorist or saboteur. A back-flow attack occurs when a pump is connected to a water system to overcome the water line pressure gradient in the system pipes to deliver a toxin, poison or other contaminants. Water line pressure is nominally 80 psi which is easily overcome using most commercial water pumps available from the local hardware store, or the internet. Subsequent to the pressure gradient overcome by the connection of a pump and the contaminant deliberately introduced, 'Bernoulli effects' produces a syphon, pulling the contaminant into the broader water system (Dan Kroll, 2006). The toxin/chemical then follows the natural flow of the water across the system delivering it to the consumer connection points.

Who would question a tanker truck connected to a fire hydrant within a city, with the operators in high visibility clothing and appearing on official business? It is not possible to understand whether the tanker truck is extracting water out, or pressurising water into the distribution system. In fact, the illegal process does not have to be so visible, as in consideration of the number of domestic and commercial connections to the water distribution system that may be accessed, devoid of the installation of 'non-return valves', a backflow event may occur, accidentally or deliberately. Backflow events do not entirely occur as deliberate activities, as they do occur by accident on a regular basis, mainly from commercial buildings devoid of backflow valve protection (Watts, 2010).

Backflow attacks have been recognised for a number of years, with a Wall Street Journal article in December 2001, indicating the potential danger from a terrorist attack using this methodology (Dreazen, 2001). Within the Wall Street Article, John Sullivan (Chief Engineer for the Boston Water & Sewer Commission in the USA) was quoted as:

There's no question that the distribution system is the most vulnerable spot we have ... ... Our Water Treatment Plants can be surrounded by cops and guards, but if there's an intentional attempt to create a backflow, there's no way to totally prevent it (Dreazen, 2001).

Water CI owners should consider initiating an assessment on the assets and supply chain related to criticality (how essential is the asset?); vulnerability (how susceptible are the range of supply assets to surveillance and attack); recoverability (how difficult will it be to recover from inflicted damage, in consideration of time, location, special repair equipment, and manpower to restore the supply chain assets to normal functional operation?) and, more importantly, what is the current threat level (how many water businesses regularly review the threat level from Terrorism?) (Brown, Carlyle, Salmeron, & Wood, 2005).

Security of Water Services

There are various concepts and understandings pertaining to the topic water security, and historically many organisations defined the security of water as the continuity of supply at a prescribed pressure and preservation of the quality of the water by virtue of the disinfection, and safety of the water to drink. In most water supply systems there is a new form of distant computer control monitoring or 'Supervisory Control and Data Acquisition' (SCADA) remote monitoring to assess and monitor water pressure; levels of disinfection in the system; ph.; and other mainly chemical variables in the system, ensuring that the water quality is delivered to the consumer.

Within an environment of the absence or extreme shortage of water, schools, hospitals, industry, and commercial sectors will be seriously affected and may shut-down for a temporary period until water supply is restored. Furthermore, extensive difficulty may be endured with private domestic dwellings being unable to cook, shower/bath, water gardens, wash the car or even to drink the nominal 0.2 to 15 litres of water per day to maintain healthy body systems (Court-Young, 2003). However, the 2013 United Nations water security definition is:

The capacity of a population to safeguard sustainable access to adequate quantities of acceptable quality water for sustaining livelihoods, human well-being, and socio-economic development, for ensuring protection against water-borne pollution and water-related disasters, and for preserving ecosystems in a climate of peace and political stability (UN-Water, 2013).

Moreover, the historical definitions were mainly created in a different climate of peace with minimal threats and perceptions of using water as a weapon of aggression, and also within a cultural and technological climate devoid of the internet and the rise of modern terrorist groups.

'Water Security' is now requiring an analysis and identification, relating to the protection of water from the perspective of potential external saboteurs, and more specifically the new modern terrorist. This new emergence of a form of terrorism which knows no boundaries of acceptable human behavior may now potentially threaten future water services in EU countries.

Nevertheless, to adequately understand current risk exposures within the spectrum of water and wastewater CI, the water systems require to be considered from the broader perspective. To be more concise, an analysis of the supply chain, in conjunction with the dependent services and industries requires to be undertaken in order to completely understand the vulnerabilities and threats to service functionality and integrity. Any future potential water and wastewater (often subliminal) failure and associated predicted industry degradation may be considered in relation to the complexities and number of reliant interdependent services. The main question is, is the future of water supply critical?

Water as an essential product for life on our planet is predicted to be in shorter supply by 2025, in consideration of the projected population increases, coupled with predicted increases in water consumption by individual countries (Gleick, 1993). Peter H. Gleick (1996) advises that hydrologists have identified that the minimum water requirement for each person residing in an efficient, modern, industrialised nation is 1,000 cubic metres of water per day, or 50 litres of water, per person, per day.

Peter H. Gleick (1993) further advises that there are many countries identified, in Table 1, to be deficient in this individual water requirement, requiring some future innovative water initiatives. The projected deficiencies elevates the profile of water as a more valuable resource, and potentially elevates the target value of water to a terrorist group. Table 1 indicates that only Malta is predicted to maintain the per capita water availability by year 2025. The reduced per capita water availability is based on population predictive increases and an anticipated increase in living standards within the individual countries resulting in an increase in water consumption (Pimentel et al., 1997).

Indeed, the future global availability of drinking water is considered at risk and a significant incubating international problem, with competition between countries and regions continuing to grow, placing political pressures on the future supply of water (Pimentel et al., 1997). The predictive reduction in availability potentially increases the likelihood of adverse human intervention to drinking water supplies and the interdependent supply linkages.

Adverse Human Interventions

Previous terrorist attacks in the European continent were occurring within the modes of guns, knives, explosives and deliberate vehicle ramming of pedestrians. These forms of attack are immediate with fear, shock and horror amongst the communities, with the offenders quite often apprehended after the event. However, as previously indicated, a water services attack, and more particularly water distribution contamination, opens a new and fearful threshold, with an estimated 3 to 5 days after any contamination, before identification by medical authorities and water businesses. The offenders in this case may well be on the other side of the world prior to the attack being identified, with a potential large number of casualties and affected people.

During past terrorist incidents in Europe populations may well choose to avoid train, bus or plane travel to reduce the risk of becoming a victim in a terrorist attack. They may also decide to shun more frequented bars, hotels and tourist destinations in order to avoid any potential attack. However, a future attack on water services strikes into homes where most populations have felt secure and safe in the past, with previous terrorist attacks on the news and media, occurring somewhere else and to someone else.

Table 1. Future Water Predictions in Selected Countries by 2025 adapted from (Gleick, 1993)

Country	Per capita Water availability 1990 (cubic meters/person/year)	Per capita Water availability 2025 (cubic meters/person/year)
Africa		
Algeria	750	380
Burundi	660	280
Cape Verde	500	220
Comoros	2,040	790
Djibouti	750	270
Egypt	1,070	620
Ethiopia	2,360	980
Kenya	590	190
Lesotho	2,220	930
Libya	160	60
Morocco	1,200	680
Nigeria	2,660	1,000
Rwanda	880	350
Somalia	1,510	610
South Africa	1,420	790
Tanzania	2,780	900
Tunisia	530	330
North & Central America		
Barbados	170	170
Haiti	1,690	960
South America		
Peru	1,790	980
Asia/Middle East		
Cyprus	1,290	1,000
Iran	2,080	960
Israel	470	310
Jordan	260	80
Kuwait	<10	<10
Lebanon	1,600	960
Oman	1,330	470
Qatar	15	20
Saudi Arabia	160	50
Singapore	220	190
United Arab Emirates	190	110
Yemen	240	80
Europe		
Malta	80	80

Global populations have frequently assumed the role of 'spectators' or 'viewers' of incidents on the internet or television in an age of 'Celebratory Terrorism' as described by (Howie, 2009; Dan Kroll, 2006), with a potential attack on water changing the psychological persona profile from a spectator, to that of a 'participant', imbibing fear and panic across large populations. The progress and development of social media has allowed terrorists to play to an increasing world audience.

Furthermore, terrorists thrive on the element of surprise and water managers should consider the 'unthinkable', in lieu of simply planning for what has occurred historically or more frequently. An example of this was the 2001 horrific attack on the World Trade Centre buildings in New York, USA, displaying a mass casualty event and alerting the world to terrorism's future creative directions. Contemplation of this form of adverse creativity should influence governments and water planners to think 'outside the box' to plan for and respond to 'unthinkable' acts of water terrorism.

Indeed, the level of threat from terrorism has not currently reduced across the EU nations. Furthermore, acts of terrorism have become increasingly complex with the use of the internet and social media networks creating new and more frightening methods in conducting terrorist acts (Europol, 2019). As Europol's Executive Director (Catherine De Bolle) stated in June 2019 (Europol, 2019):

Terrorists not only aim to kill but also to divide our societies and spread hatred. That feeling of insecurity that terrorists try to create must be of the greatest concern to us. Increasing polarization and the rise of extremist views is a concern for EU member States and Europol (Europol, 2019).

One of the main trends indicated in Europol's Report on Terrorism (2019) was that a general increase of Chemical, Biological, Radioactive and Nuclear (CBRN) terrorist propaganda, tutorials and threats have been observed, with the barriers for gaining this knowledge on the internet and other places decreasing (Europol, 2019).

Terrorist attacks in Europe over the past 20 years have displayed increasing forms and methodologies reflecting extreme brutality with an outcome of increased morbidity and injuries to the public and bystanders. This new form of terrorism appears to focus on the general public in lieu of previous government targets, and also appears to aim for a higher kill rate than previous attacks. Examples of these are: the 9/11 USA terrorist attacks using airplanes as a weapon, resulting in more than 3,000 deaths, and the simultaneous suicide bombings of American personnel in Nairobi and Dar-es-Salaam in 1998, resulting in a total of 224 deaths (Atwan, 2007). In consideration of the various methodologies and innovations applied by the modern terrorist it is feasible that they would consider attacking water services.

This is in sharp contrast to previous terrorist groups who historically have preferred: "Many people watching, but not a lot of people dead" (Jenkins, 1975).

Furthermore, most of the 1960's and 1970's terrorist groups generally had clear and defined objectives, such as agrarian reform, as in the case of the Moro Islamic Liberation Force in the Philippines, who desired a clearly defined area for the group in the Southern Philippines.

Although many have the view that terrorists are either insane, of low intellect or of a criminal nature, many, and in particular the planners and leaders, reflect patterns of normality, with many terrorists university graduates. Osama bin Laden was well educated, from a Wealthy Saudi family, and a qualified Civil Engineer (Coll, 2008). Additionally, the 4 pilots of the 9/11 attacks (Mohammed Atta; Ramzi bin al-Shibh; Marwan al Shehhi, and Ziad al-Jarrah) were educated at Hamburg University, Germany, in engineering (Miniter, 2011). Consequently, it is an error to fail to understand the intelligent terrorist processes, which may occur over a period of years to perfect, as was the 9/11 attack processes (Miniter, 2011).

Many recruits who travelled to al Qaeda bases for training were well educated and abandoned wealthy and established family structures (Atwan, 2007) indicating the ability to understand and design complex attack planning over many years.

Fortunately, many terrorist attacks across EU Countries have been foiled earlier, as plots, by efficient EU law enforcement. Consequently, many people fail to be aware of the currency, significance and frequency of planned attacks (LaFree, 2017; Europol, 2019).

As Enders & Sandler (1993) indicate, proactive policy measures and target hardening to ensure that potential terrorist attacks are more difficult, causes the terrorist to seek other targets that are 'softer' and easier to attack. This form of transference is well recognised and understood by researchers (Brandt, 2010), and when applied to the water industry could well shift to water and its supply chain, as a form of impact on a soft target to maximise the attack results and outcomes sought by the terrorist group. Indeed, as research (Kalil & Berns, 2004; Kroll, 2013; Kroll et al., 2010; Meinhardt, 2005, 2006) indicates, water is a soft target and some attack planning has been observed over the past 10 years. As Gleick suggests, it is not necessary to contaminate water supplies, with the terrorist objective achieved by a plausible threat to contaminate (Gleick, 2006). The mild ingress of a harmless vegetable dye would potentially be enough to create the sought objective of fear and chaos in the community.

A generic view of the 'Why, What, When and How' of terrorist planning is outlined in Figure 4, detailing the typical actions and thoughts that a terrorist group may consider when planning a typical CI attack. The illustration is based on previous Australian research conducted by the author in 2013, but is relevant on a global basis, inclusive of the EU Countries.

The modern terrorist utilises the internet to provide recruitment, training and templates for attacks on CI and other targets, contributing to a strain on law enforcement and intelligence agencies in trying to monitor any adverse behaviour leading to any future attack. Those fundamentalist groups with linkage to al Qaeda and other groups, follow a variable template for attacks adopting documentation circulated amongst the groups with information on the internet and within training camps conducted in various locations (Lee, 2009). ISIL & al Qaeda no longer exhibit a centralised and hierarchical chain of command (such as Hezbollah, a Lebanese umbrella organisation of radical Shiite groups and organisations), with methods and target inspiration filtered through the use of the internet, production of underground manuals and small clandestine 'sub rosé' (Mathams, 1982) group meetings in mosques, correctional centres and other places. The internet now provides an additional convenient and anonymous forum for recruitment, training, support, planning, target selection, attack methods and decision-making authority.

Especially significant is that there is an identified shift in targeting coupled with an expansion of tactics in relation to acts of violent extremism, with the modern terrorist continually seeking new and alternative methodologies to commit terrorist attacks (Wood & Van-Slyke, 2018). The 'ISIS' terrorist group in particular is predicted to possibly seek retaliation for the loss of its territorial 'caliphate' on European targets by the returning 'battle-hardened' fighters to the EU countries. Terrorists have been adopting low-technology, low-cost tactics, translating sequentially over time, applying vehicle ramming, knife attacks, incendiaries and suicide bombing. Similar to criminal activity, as law enforcement and governments take preventative action after an attack, the offenders have often adopted alternative tactics which are frequently more difficult to accurately predict, but indicating a distinct shift in targeting (Wood & Van-Slyke, 2018). By way of illustration, developed democratic societies are open societies

Figure 4. Terrorist Planning Considerations in Attacking CI (Birkett, 2017)

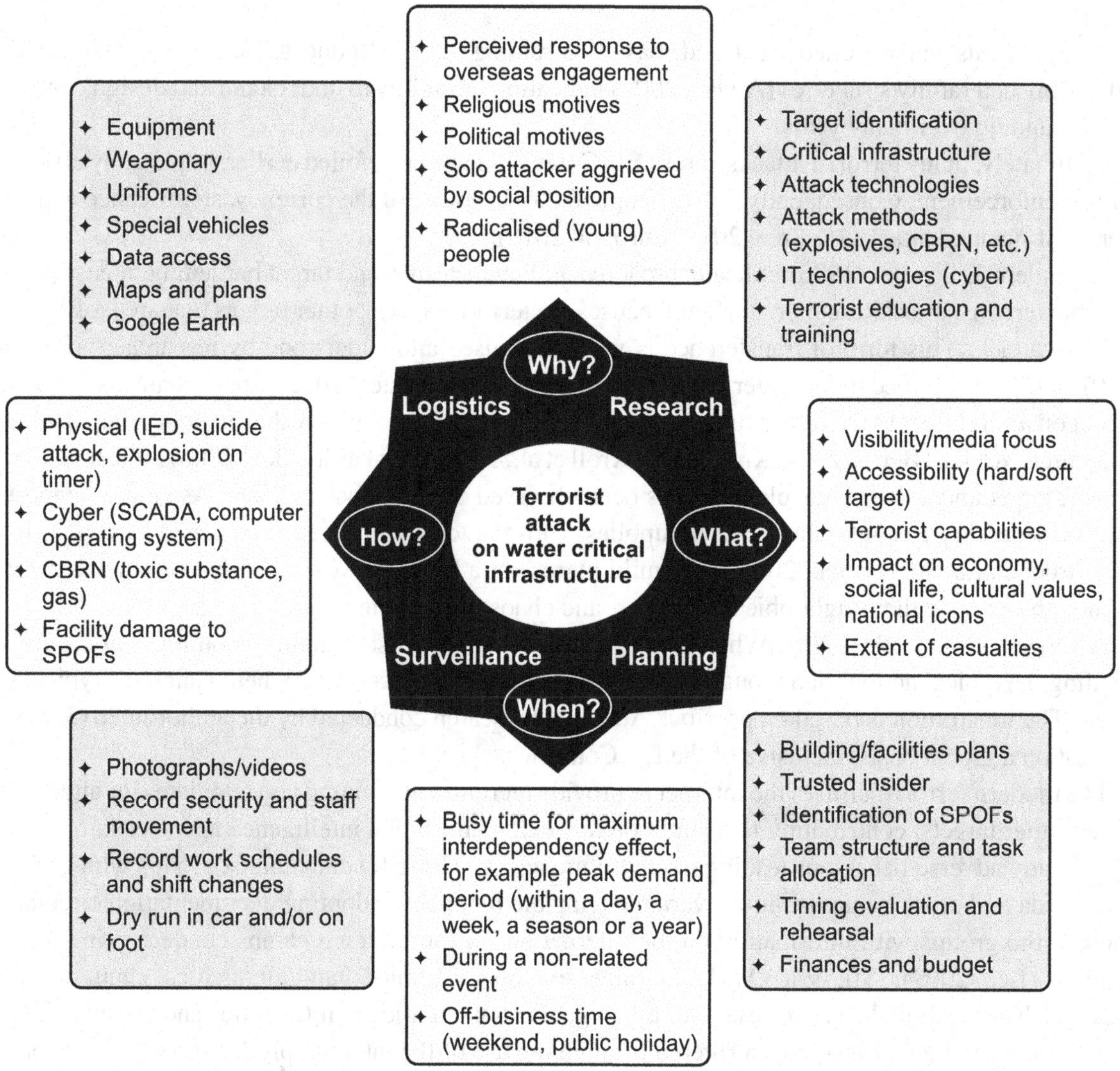

Terrorist attack on water critical infrastructure

Why?
- Perceived response to overseas engagement
- Religious motives
- Political motives
- Solo attacker aggrieved by social position
- Radicalised (young) people

Research
- Target identification
- Critical infrastructure
- Attack technologies
- Attack methods (explosives, CBRN, etc.)
- IT technologies (cyber)
- Terrorist education and training

What?
- Visibility/media focus
- Accessibility (hard/soft target)
- Terrorist capabilities
- Impact on economy, social life, cultural values, national icons
- Extent of casualties

Planning
- Building/facilities plans
- Trusted insider
- Identification of SPOFs
- Team structure and task allocation
- Timing evaluation and rehearsal
- Finances and budget

When?
- Busy time for maximum interdependency effect, for example peak demand period (within a day, a week, a season or a year)
- During a non-related event
- Off-business time (weekend, public holiday)

Surveillance
- Photographs/videos
- Record security and staff movement
- Record work schedules and shift changes
- Dry run in car and/or on foot

How?
- Physical (IED, suicide attack, explosion on timer)
- Cyber (SCADA, computer operating system)
- CBRN (toxic substance, gas)
- Facility damage to SPOFs

Logistics
- Equipment
- Weaponary
- Uniforms
- Special vehicles
- Data access
- Maps and plans
- Google Earth

LEGEND: CBRN = chemical, biological, radiological, nuclear; IED = improvised explosive device; IT = information technology; SCADA = supervisory control and data acquisition; SPOF = single point of failure

with extensive networks of CI, originally designed for ease of access for maintenance and inspection, and consequently not hardened to resist the 21st century threats of terrorism (Kroll, 2006).

Terrorists thrive on the element of surprise forcing Governments and water managers to consider the 'unthinkable' within water CI, and not simply plan for what has occurred historically in the past, or normal water related emergencies. Furthermore, water services CI are more specifically vulnerable, with potential cascading effects across most other CI in various levels with consideration of the human dependency on water within a 3 day average limit for actual survival.

A hypothetical terrorist CBRN water contamination incident may be identified as an 'incubating' or 'slow burn' weapon, as it may well be many hours or days prior to identification that any attack had occurred, allowing the offenders to escape to other global locations (Jackson, 2005). Such an attack in a major EU population centre may well kill and cause sickness in large clusters of the population, with the added effect of widespread panic and a loss of confidence in large urban water services. Dan Kroll (2006) advises that studies conducted utilising computer simulation by the US Airforce and Colorado University, replicating a toxic backflow incident indicated a 10 to 20% death rate in a population, dependent on the selection of chemical or toxic agent.

Water Critical Infrastructure Protection

Possibly due to a terrorist act on water being a 'low likelihood' incident in the past, and natural weather occurrences or accidents being more likely, the water companies have generally designated emergency budget money to the more likely events, such as flooding, life cycle maintenance and other more frequent occurring issues. However, it is the considered opinion of many researchers in this area of study, that it is not a matter of 'if' but more a realistic question of 'when' a major terrorist attack on water will occur in a large population centre, possibly across Western Europe. The author has discussed this issue with many water managers and CEO's with the most common response as: It has not happened yet…, but when it does, we will allocate increased budget to water security issues.

Most business continuity and risk matrixes tend to accentuate the priorities for 'high likelihood' events, whereas any terrorist attack in most areas is considered low likelihood but with high consequences related to the protection of life, property, the environment, with some reputational considerations. As Gilbert et al. (2003) highlight, the orderly historic process to mitigate natural or 'all-hazard' risks requires different levels of mitigation and protection than for deliberate, adverse human asset intervention.

The water CI sector is considered to represent a 'lifeline' sector to the majority of CI (Porod et al., 2014) and it is further suggested that

….it is essential to ensure both the protection and resilience of these systems in order for the successful treatment of water and wastewater to occur (Porod et al., 2014).

Preparedness

In order to maximise the preparedness for any form of adverse human intervention or even any adverse natural incident Water Service owners ought to have an emergency response plan (ERP) and a defined location that may be used as a Crisis or Emergency operations centre, which would facilitate the effective management of any future emergency, inclusive of consideration of a terrorist attack on water services. (USACHPPM, 2019). In addition to processes for resolving the issues, the ERP and emergency control centre (ECC) would firstly protect the health of consumers and then take defined steps to resolve the external attack problem in conjunction with relevant external agencies. An emergency & crisis plan should be prepared prior to any specific incident with exercise rehearsals conducted on an annual basis, involving field staff in addition to the water company management.

The exercises should incorporate all low likelihood events, (such as a terrorist act) as well as natural occurrences, (such as weather events and earthquakes). In addition to consideration of accidental events, as described at Camelford in the UK and the Prague incident in 2015 within the Czech Republic.

Four basic strategies are suggested as a primary line of defence against terrorist attacks on water services, in order to prevent the attack from occurring and to detect and confine any unauthorised insertions of toxins or pollutants prior to the consumer service point.

Anticipate

Water Managers in conjunction with Government Agencies, should conduct a 'Red Cell Review' to examine the total water services systems to analyse where the SPOFs exist (DSB, 2003). The systems, both drinking water supply and wastewater services, should be broken down into individual elements and examine the vulnerabilities and possibilities for an outsider with limited or specific knowledge to inflict damage which would be difficult to reinstate or significantly impact the health of the consumers.

Protect and Deter

Examine security across the system, inclusive of locking systems; CCTV coverage and quality. Forward looking water companies have installed computer-controlled locking systems due to the historical lack of control of Master Keys to buildings and infrastructure, coupled with the turnover of staff either not handing keys in or taking an unregistered copy of a Master Key. Water and Electricity organisations on a global basis have issued 'Master Keys' to staff for ease of maintenance in the past, prior to a heightened terrorist environment, as is currently the case. Intruder alarm systems, especially on water treatment plants and treated water storages, could be installed with either a connection to a SCADA control room, which many water companies now have, or to a telephone for an emergency manager's communication.

Due to reoccurring droughts and being the driest continent on the Earth, Australian houses, and more particularly in rural areas, have installed rainwater tanks. This action not only protects against drought, but can act as a secondary storage of boiled drinking water in the event of deliberate water contamination of treated water.

Detection

Apart from physical security and an in-depth review, there are now developed computer-based systems available, with sensors designed to be deployed within the water systems, to detect the unauthorised insertion of any toxic material or chemical into the drinking water. Dan Kroll from the US has published extensively on this topic and has many academic papers to explain the process and effectiveness of these new processes (Kroll, 2008; Kroll et al., 2010; Kroll & King, 2006; Kroll, King, & Klein, 2009).

Act

The role of the Water Corporation is to protect the safety of employees and the health of consumers from any unauthorised tampering or attack on water services. To do this effectively, water managers require an Emergency process, complete with a room that can be immediately transformed into an emergency or crisis centre. That is, complete with white boards; telephones; an emergency/ crisis plan; staff trained to occupy positions within the emergency centre, and to run a series of exercises to adequately train staff in their appointed positions. This is a very basic description, but again there is a plethora of emergency and crisis plans available on the internet with many crisis companies in all countries willing to assist and set up an operational emergency/ crisis process and plan.

The role of the governments in the EU is to liaise with the water corporations to ensure that (a) processes and an emergency/crisis capability are evident and exercised on an annual or a 6 month basis. (b) Government institutions should ensure that current intelligence support is shared with the designated person within the water corporation relative to any information concerning any heightened or elevated terrorist alert. Government agencies should also consider being involved with water companies in their annual exercise program.

Suggested Protective Strategies

Emergency management and crisis management strategies are steps often taken to reduce business damage, and restoring safety after an incident. However, some suggested protective mitigation strategies prior to any potential future incidents are as follows:

- The SCADA system should be separated from the company website utilising a strong 'firewall'. Furthermore, access codes and passwords should be regularly changed and controlled. This is particularly significant when internal staff resign or transfer to other areas.
- The effect of any deliberate or accidental backflow incident may be reduced by 'ring fencing' portions of the distribution system with strategically placed control valves. Additionally, there are commercially available toxin detection equipment that can link into the SCADA system to close valves and reduce the contamination spread across the system
- Identify facilities' physical security, security forces, security management, information sharing, protective measures, and dependencies related to preparedness, mitigation, response, resilience, and recovery. All nodes should be regularly reviewed on high risk areas.
- Identify security gaps.
- Create facility protective and resilience measures indices that can be compared to similar facilities across the EU.

Water services require increased protection as an obligation to the health of the large customer groups serviced by the water company. Protection can assume many guises and forms but the essential elements in addition to the points above are a comprehensive crisis/emergency management plan, an allocated location to be used in the potential incident occurring and regular practicing of the plan with a scripted exercise.

CONCLUSIONS AND RECOMMENDATIONS

The CI area of water services is extremely complex and has usually grown exponentially relative to individual European economies with spurts of growth associated with economic expansion, such as residential, industrial, agricultural and leisure facilities. The original design of water services was created in a relatively peaceful environment devoid of terrorism aspects that we all face in today's cultural climate. These aspects contribute to some difficulties in protecting water services.

Moreover, water and wastewater CI may be impacted by multiple and variable dependencies and interdependencies as an indicator of the progressive complexities of modern supply chains and interconnected services. These dependencies and interdependencies are not entirely understood and identified by industry, leading to potential significant points of failure in times of disaster or emergency situations. Frequently, these interdependent services and products are not considered in their entirety by their owners or in water resilience strategies, and can lead to an increase in the vulnerability of water and wastewater to sudden and protracted failure. The improved relationships across the supply chain would tend to reduce the potential impact of any failure, and conversely raise the awareness of the impact of possible water loss on other CIs.

As indicated in this chapter, terrorists have a long history of specific interest in mounting attacks on water CI, with water services individual nodes and outlets being considered an easy and soft target for the 21st terrorist. Recent contamination incidents from natural and accidental origins provide some evidence and awareness of the threat vulnerability. There should also be some consideration for increased awareness by EU governments and water service owners in relation to the SPOFs of water systems.

Water services require increased protection as an obligation to the health of the large customer groups serviced by the water company. The concept of water protection requires some consideration and hardening to increase the security of the water and wastewater product in particular from any potential external adverse human intervention, whether it originates as a terrorist act, saboteur or disgruntled internal employee. Protection can assume many guises and forms but the essential elements are a comprehensive crisis/emergency management plan, an allocated location to be used in the potential incident occurring and regular practicing of the plan with a scripted exercise. Water companies should implement these strategies to protect the health of consumers prior to any potential future incident (Truscott, 2015).

Moreover, water companies should be assessed by European Governments as to the compliance and actions relative to the protection of water services.

REFERENCES

ABC. (2003, February 21). Man Who Threatened Melbourne Water Supply Bailed. Retrieved from http://www.abc.net.au/news/2003-02-21/man-who-threatened-melbourne-water-supply-bailed/2689584

Abrams, M., & Weiss, J. (2008). *Malicious Control System Cyber Security Attack Case Study - Maroochy Water Services, Australia*. Retrieved from http://csrc.nist.gov/groups/SMA/fisma/ics/documents/Maroochy-Water-Services-Case-Study_report.pdf

Altman, P., Cunningham, J., Dahnesha, U., Ballard, M., Thompson, J., & Marsh, F. (2006). *Disturbance of Cerebral Function in People Exposed to Drinking Water Contaminated With Aluminium Sulphate: Retrospective Study of the Camelford Water Incident*. UK: Retrieved from Oxford. Retrieved from http://discovery.ucl.ac.uk/2010/1/807.pdf

Atwan, A. B. (2007). *The Secret History of Al-Qaida*. London: Saqi Books.

Balmer, C. (2004, April 29). Italian Court Acquits Nine in Alleged Plot Against US Embassy. Boston. Retrieved from http://www.boston.com/news/world/europe/articles/2004/04/29/italian_court_acquits_nine_in_alleged_plot_against_us_embassy/

Beering, P. S. (2002). Threats on Tap: Understanding the Terrorist Threat to Water. *Journal of Water Resources Planning and Management*, *128*(3), 163–167. doi:10.1061/(ASCE)0733-9496(2002)128:3(163)

Birkett, D. (2017). Water Critical Infrastructure Security and Its Dependencies. *Journal of Terrorism Research*, *8*(2), 1–21. doi:10.15664/jtr.1289

Birkett, D., & Mala-Jetmarova, H. (2012). Are Risk Mitigation Strategies of Water Critical Infrastructure Adequate within a European Environment of 21st Century. *Paper presented at the 10th International Conference on Hydroinformatics HIC2012*. Academic Press.

Birkett, D., & Mala-Jetmarova, H. (2014). Plan, Prepare and Safeguard: Water Critical Infrastructure Protection in Australia. In R. M. Clark & S. Hakim (Eds.), *Securing Water and Wastewater Systems, Protecting Critical Infrastructure 2*. Switzerland: Springer International Publishing. doi:10.1007/978-3-319-01092-2_14

Birkett, D., Truscott, J., Mala-Jetmarova, H., & Barton, A. F. (2011). Vulnerability of Water and Wastewater Infrastructure and its Protection from Acts of Terrorism: A Business Perspective. In R. M. Clark, S. Hakim, & A. Ostfeld (Eds.), *Handbook of Water and Wastewater Systems Protection, Series Protecting Critical Infrastructure* (pp. 457–483). New York, USA: Springer. doi:10.1007/978-1-4614-0189-6_23

Brandt, P. T., & Sandler, T. (2010). What do Transnational Terrorists Target? Has It Changed? Are We Safer? *The Journal of Conflict Resolution, 54*(2), 214–236. doi:10.1177/0022002709355437

Brockett, J. (2015, July 13). UK Water Networks Vulnerable to Terrorist Attack. Utility Week. Retrieved from http://utilityweek.co.uk/news/uk-water-networks-%E2%80%98vulnerable-to-terrorist-attack/1150512#.VkC-c8tOcyU

Brown, G., Carlyle, M., Salmeron, J., & Wood, K. (2005). Analyzing the Vulnerability of Critical Infrastructure to Attack and Planning Defenses. *Paper presented at the INFORMS Tutorials in Operations Research, Institute for Operations Research and the Management Sciences*. Academic Press. 10.1287/educ.1053.0018

Brown, T., Beyeler, W., & Barton, D. (2004). Assessing Infrastructure Interdependencies: The Challenge of Risk Analysis for Complex Adaptive Systems. *International Journal of Critical Infrastructures, 1*(1), 108–117. doi:10.1504/IJCIS.2004.003800

Cameron, C. (2002, July 30). Feds Arrest al-Queda Suspects with Plans to Poison Water Supplies. Fox News. Retrieved from http://www.foxnews.com/story/2002/07/30/feds-arrest-al-qaeda-suspects-with-plans-to-poison-water-supplies/

Clark, R. & Hakim, S. (2014). Securing Water and Wastewater Systems. Springer.

Cohen, F. (2010). What Makes Critical Infrastructures Critical? *International Journal of Critical Infrastructure Protection, 3*(2), 53–54. doi:10.1016/j.ijcip.2010.06.002

Coleman, K. (2005). *Protecting the Water Supply From Terrorism*. Directions Magazine. Retrieved from http://www.directionsmag.com/entry/protecting-the-water-supply-from-terrorism/123563

Coll, S. (2008). *The Bin Ladens - The Story of a Family and its Fortune*. London: Penguin books.

Cooper, A. (2013, April 9). Man's al-Qaeda Link, Court Reporting. The Melbourne Age, Australia.

Copeland, C. (2010). *Terrorism and Security Issues Facing the Water Infrastructure Sector. CRS Report for Congress*. Washington, USA: Congressional Research Service.

Court-Young, H. C. (2003). *Understanding Water and Terrorism* (September 2003 ed.). Denver, CO: Burg Young Publishing LLC.

Covert, A. J. (2008). Water: Vital to Life, Vulnerable to Terrorism [PhD] Air University.

Dreazen, Y. J. (2001, December 28). 'Backflow' Water-Line Attack Feared - Terrorists Could Reverse Flow in System to Introduce Toxins, Water Security. *The Wall Street Journal*. Retrieved from https://cryptome.org/backflow-panic.htm

DSB. (2003). *The role and Status of DoD Red Teaming Activities*. Retrieved from http://www.fas.org/irp/agency/dod/dsb/redteam.pdf

Dugal, J. (1999). Guadalajara Gas Explosion Disaster. *Disaster Recovery Journal, 5*(3). Retrieved from http://www.drj.com/drworld/content/w2_028.htm

Enders, W., & Sandler, T. (1993). The Effectiveness of Anti-Terrorism Policies: A Vector-Autoregression-Intervention Analysis. *The American Political Science Review*, *87*(4), 829–844. doi:10.2307/2938817

Europol. (2019). *Terrorism Situation and Trend Report 2019*.

Gilbert, P. H., Isenberg, J., Baecher, G. B., Papay, L. T., Spielvogel, L. G., Woodard, J. B., & Badolato, E. V. (2003). Infrastructure Issues for Cities - Countering Terrorist Threat. *Journal of Infrastructure Systems*, *9*(1), 44–54. doi:10.1061/(ASCE)1076-0342(2003)9:1(44)

Gillette, J., Fisher, R., Peerenboom, J., & Whitfield, R. (2002). Analysing Water/Wastewater Infrastructure Interdependencies. *Paper presented at the 6th International Conference on Probabilistic Safety Assessment and Management (PSAM6)*. Retrieved from http://www.ipd.anl.gov/anlpubs/2002/03/42598.pdf

Gleick, P. H. (1993). Water and Conflict: Fresh Water Resources and International Security. *International Security*, *18*(1), 79–112. doi:10.2307/2539033

Gleick, P. H. (1996). Basic Water Requirements for Human Activities: Meeting Basic Needs. *Water International*, *21*(2), 83–92. doi:10.1080/02508069608686494

Gleick, P. H. (2006). Water and Terrorism. *Water Policy,* (8), 481-503.

Gleick, P. H. (2006). *Water Conflict Chronology*. Retrieved from http://citeseerx.ist.psu.edu/viewdoc/download?rep=rep1&type=pdf&doi=10.1.1.204.9178

Gleick, P. H., & Heberger, M. (2014). *Water Conflict Chronology* (1597264210). World water. Retrieved from http://www.worldwater.org/wp-content/uploads/sites/22/2013/07/ww8-red- water-conflict-chronology-2014.pdf

Guay, B. (2019, July 23). Probe Opened in France Over Radioactive Water Rumours, Water contamination. *AFP*. Retrieved from https://www.afp.com/en/news/826/probe-opened-france-over- radioactive-water-rumours-doc-1j12mc2

Howie, L. (Ed.). (2009). *Terrorism, the Worker and the City - Simulations and Security in a Time of Terror*. Farnham, UK: Gower.

Jackson, J.H., Mountcastle, R., & Charles, E. (2005). *The Counter-Terrorist Handbook*. London: Michael O'Mara Books Ltd.

Jenkins, B. M. (1975). *Will Terrorists Go Nuclear*. Retrieved from http://www.defence.org.cn/aspnet/vip-usa/UploadFiles/2008-10/200810132327510156.pdf

Kalil, J. M., & Berns, D. (2004). *Drinking Supply: Terrorist Had Eyes on Water*. Defendyourh2o. Retrieved from http://www.defendyourh2o.com/pdfs/DRINKING%20SUPPLY.pdf

Kroll, D. (2006). *Securing Our Water Supply: Protecting a Vulnerable Resource* (S. Hill, Ed.). Tulsa, OK: PennWell Corporation.

Kroll, D. (2010a). *Aqua ut a Telum "Water as a Weapon."* Retrieved from http://hachhst.com/wp-content/uploads/2010/07/White-Paper_Water-as-a-weapon.pdf

Kroll, D. (2010b). *A Reinterpretation of the 2002 Attempted Water Terror Attack on the U.S. Embassy - Don't Underestimate the Enemy*. Hach. Retrieved from http://hachhst.com/wp- content/uploads/2010/07/White-Paper_-WATER-TERROR-ATTACK1.pdf

Kroll, D. (2008, September 22). Testing the Waters: Improving Water Quality and Security Via On-Line Monitoring. *Paper presented at the ICMA*. Academic Press. Retrieved from http://hachhst.com/wp-content/uploads/2010/07/Presentation_Water-Security-and-Quality.pdf

Kroll, D., & King, K. (2006). Laboratory and Flow Loop Validation and Testing of the Operational Effectiveness of an On-Line Security Platform for the Water Distribution System. *Paper presented at the 8th Annual Water Distribution Systems Analysis Symposium*. Academic Press.

Kroll, D., King, K., & Klein, G. (2009). Real World Experiences With Real-Time On-Line Monitoring for Security and Quality, Detecting and Responding to Events. *Paper presented at the International Workshop on Water and Wastewater Security Incidents*. Hach. Retrieved from http://hachhst.com/wp-content/uploads/2010/07/Presentation_Real-World-events.pdf

Kroll, D. J. (2013). The Terrorist Threat to Water and Technology's Role in Safeguarding Supplies. In *Proceedings of the 45th Session of the International Seminars on Nuclear War and Planetary Emergencies*. Academic Press. 10.1142/9789814531788_0030

Kroll, D. J., King, K., Engelhardt, T., Gibson, M., Craig, K., & Securities, H. H. (2010). *Terrorism Vulnerabilities to the Water Supply and the Role of the Consumer: A Water Security White Paper*. Waterworld. Retrieved from http://www.waterworld.com/articles/2010/03/terrorism- vulnerabilities-to-the-water-supply-and-the-role-of-the-consumer.html

LaFree, G. (2017, May 22). 6 Reasons Why Stopping Worldwide Terrorism is so Challenging. The Conversation. Retrieved from https://theconversation.com/6-reasons-why- stopping-worldwide-terrorism-is-so-challenging-70626

Lazarova, D. (2015, May 27). Problem With Contaminated Tap Water in Prague 6 Still Unresolved. Radio Prague International. Retrieved from http://www.radio.cz/en/section/news/problem-with-contaminated-tap-water-in-prague-6-still-unresolved

Lee, E. (2009). *Homeland Security and Private Sector Business: Corporations' Role in Critical Infrastructure Protection*. Boca Raton, FL: Auerbach Publications.

Lee, J.-J. M. (2009). A War for Water. *Paper presented at the World Environmental and Water Resources Congress 2009*. Academic Press. Retrieved from http://link.aip.org/link/?ASC/342/393/1

Locken. (2017, January 9). Understanding the Terrorist Threat to our Water Sector. Security Newsdesk. Retrieved from https://securitynewsdesk.com/locken-understanding-terrorist-threat-water-sector/

Maiolo, M., & Pantusa, D. (2018). Infrastructure Vulnerability Index of Drinking Water Systems to Terrorist Attacks. Academic Press. doi:10.1080/23311916.2018.1456710

Marsh, R. T. (1997). *Critical Foundations: Protecting America's Infrastructure*. President's Commission on Critical Infrastructure Protection.

Mathams, R. H. (1982). *Sub Rosa - Memoirs of an Australian Intelligence Analyst*. Sydney, Australia: George Allen & Unwin.

Meinhardt, P. L. (2005). Water and Bioterrorism: Preparing for the Potential Threat to U.S. Water Supplies and Public Health. *Annual Review of Public Health*, *24*(26), 213–237. doi:10.1146/annurev.publhealth.24.100901.140910 PMID:15760287

Meinhardt, P. L. (2006). Medical Preparedness for Acts of Water Terrorism. An On-Line Readiness Guide. American College of Preventative Medicine.

Miniter, R. (2011). *Mastermind, The many Faces of the 9/11 Architect, Khalid Shaikh Mohammad*. New York: Penguin Group.

Morris, S. (2012). The Camelford Poisoning: Blackwater, A Driver's Mistake. Retrieved from https://encrypted-tbn0.gstatic.com/images?q=tbn:ANd9GcQwx- FrraixjDahj7ftjHqWCBRBVF1WH0Aq-l0fE-qlEjIIZMnD

Noble, T., & Schrembi, J. (1984, September 17). Geelong Man Charged over Poisoning Threat. *The Melbourne Age*. Retrieved from https://news.google.com/newspapers?nid=1300&dat=19840917&id=7lVVAAAAIBAJ&sjid=n5UD AAAAIBAJ&pg=4756,7124&hl=en

Pimentel, D., Houser, J., Preiss, E., White, O., Fang, H., Mesnick, L., ... Alpert, S. (1997). Water Resources: Agriculture, the Environment, and Society. *Bioscience*, *47*(2), 97–106. doi:10.2307/1313020

Porod, C., Collins, M., & Petit, F. (2014). Water Treatment Dependencies. *The CIP Report, 13*(2), 10-13. Retrieved from http://cip.gmu.edu

Prague Daily Monitor (PDM). (2015a, June 8). Prague to Vaccinate Children From Districts With Contaminated Water. Retrieved from http://praguemonitor.com/2015/06/08/prague-vaccinate- children-districts-contaminated-water

Prague Daily Monitor (PDM). (2015b, May 28). Prague Town Hall Files Complaint Over Contaminated Drinking Water. Retrieved from http://www.praguemonitor.com/2015/05/28/prague- town-hall-files-complaint-over-contaminated-drinking-water

Reid, S. (2007, December 14). A Lethal Cover Up: Britain's Worst Water poisoning Scandal. Daily Mail. Retrieved from http://www.dailymail.co.uk/news/article-502442/A-lethal-cover- Britains-worst-water-poisoning-scandal.html

Rose, M. (2012, March 14). Camelford Water Poisoning: Authority 'Gambled With Lives.' BBC. Retrieved from http://www.bbc.co.uk/news/uk-england-cornwall-17367243

Silva, C. J. (2011, August 23). More on the Attempted Water-Poisoning in Spain. Gates of Vienna. Retrieved from http://gatesofvienna.blogspot.cz/2011/08/more-on-attempted-water- poisoning-in.html

Snow, J. (1857). Cholera, and the Water Supply in the South Districts of London. *British Medical Journal*, *4*(42), 864–865. doi:10.1136/bmj.s4-1.42.864

Spear-Cole, R. (2019, June 12). Thames Water: No Water Supply for 100,000 Properties and Schools Across Hampton, Twickenham and West London Due to Burst Pipes. *The Evening Standard*. Retrieved from https://www.standard.co.uk/news/london/100000-properties-left-without- water-in-south-and-west-london-after-pipe-bursts-in-hampton-a4165441.html#spark_wn=1

Sullivan, J. K. (2011). *Water Sector Interdependencies - Summary Report*. Retrieved from http://www.wef.org/uploadedFiles/Access_Water_Knowledge/Water_Security_and_Emergency_Res ponse/Final_WEF_Summary_WSI.pdf

Truscott, J. (2015). The Art of Crisis Leadership: Incident Management in the Digital Age. Mission Mode. Retrieved from http://www.missionmode.com/wp-content/uploads/2014/08/The-Art- of-Crisis-Leadership-Incident-Mgmnt-In-Digital-Age.pdf

UN Water. (2013, May 8). What is Water Security? Infographic. Retrieved from https://www.unwater.org/publications/water-security-infographic/

US Army Center for Health Promotion and Preventive Medicine (USACHPPM). (2019). *Countering Terrorism of Water Supplies.*

Van Leuven, L. J. (2011). Water/Wastewater Infrastructure Security: Threats and Vulnerabilities. In R.M. Clark, S. Hakim, & A. Ostfield (Ed.), Handbook of Water and Wastewater Protection (Vol. 1, pp. 27-46). New York, USA: Springer.

Veilleux, J., & Dinar, S. (2018, May 8). New Global Analysis Finds Water-Related Terrorism Is On the Rise. New security beat. Retrieved from https://www.newsecuritybeat.org/2018/05/global- analysis-finds-water-related-terrorism-rise/

Wang, S., Hong, L., & Chen, X. (2012). Vulnerability Analysis of Interdependent Infrastructure Systems: A Methodological Framework. *Physica A*, *391*(11), 3323–3335. doi:10.1016/j.physa.2011.12.043

Watts. (2010). Stop Backflow News! Retrieved from https://media.wattswater.com/F-SBN.pdf

Weiss, J. (2011, November 21). Hackers 'Hit' US Water Treatment Systems. *BBC News Technology*. Retrieved from http://www.bbc.co.uk/news/technology-15817335

Willoughby, I. (2015, May 26). Anger After Slow Announcement of Water Contamination Leaves Scores Sick in Prague. Retrieved from http://www.radio.cz/en/section/curraffrs/anger-after- slow-announcement-of-water-contamination-leaves-scores-sick-in-prague

Wood, J., & Van-Slyke, S. (2018). *Terrorism and New Ideologies*. Control Risks. Retrieved from https://www.controlrisks.com/-/media/3c6cfbc84397463bbe955f1c844cf78e.ashx

KEY TERMS AND DEFINITIONS

9/11: Terrorist attacks in New York, USA, on 9th September 2001.

Chemical, Biological, Radioactive and Nuclear (CBRN): Agents or products, which may be used to deliberately contaminate water.

Critical Infrastructure (CI): Necessary resources and structures for populations, which if impacted may cause major disruptions in society and the economy. Examples of CI are energy, water, transport and government services, with the selection of these categories differing from country to country.

Emergency Control Centre (ECC): A function activated during an incident, which acts as a communications conduit to the public and emergency personnel in the field.

Emergency Response Plan (ERP): A course of action developed to mitigate the damage of potential incidents, which could endanger an organisation's ability to function.

Islamic State of Iraq and the Levant (ISIL): A Sunni jihadist group with a particularly violent ideology, which calls itself a caliphate and claims religious authority over all Muslims.

Moro Islamic Liberation Front (MILF): A group formed in 1969 and based in Mindanao, which seeks land ownership and is in conflict with the Philippine Government.

Single Points of Failure (SPOFs): Elements in any system which if damaged may cause major operational disruptions of the system or in the worst-case scenario the entire system failure. Both of these may have a subsequent cascading impact on the operation of other CIs.

Supervisory Control and Data Acquisition (SCADA): A system generally referring to an industrial computer system, which distantly monitors and controls a process. In the case of water infrastructure, SCADA monitors and/or controls, for example, valves and water quality.

Water Treatment Plant (WTP): A facility producing potable water from raw water.

Chapter 5
Industry Process Safety: Major Accident Risk Assessment

Ales Bernatik
https://orcid.org/0000-0002-2625-0441
VSB – Technical University of Ostrava, Czech Republic

ABSTRACT

This chapter deals with the issue of process safety in industrial companies and major accident prevention. In the present-day technologically advanced world, industrial accidents appear ever more frequently, and the field of major accident prevention has become a dynamically developing discipline. With accelerating technical progress, risks of industrial accidents are to be reduced. In the first part, possible approaches to quantitative risk assessment are presented; and continuing it focuses on the system of risk management in industrial establishments. This chapter aims at providing experiences, knowledge, as well as new approaches to the prevention of major accidents caused by the implementation of the Seveso III Directive.

INTRODUCTION

In the past and also recently, major accidents have represented, owing to their consequences on human health, property and the environment, important events that should be avoided. The need to assess objectively the risk, to which employees or populations are exposed in connection with the production, processing, storage and transport of hazardous materials, appears. What will enable the regulation or management of risks to employees and the surrounding environment is only adequate evaluation.

The aim of major risk prevention is to operate process installations at an acceptable societal level. Risk acceptability limits can differ depending on the maturity of the state, the size of establishments and a number of other factors that must be, however, included into the analysis and assessment of risks.

The chapter describes benefits of carrying out the risk assessment. The results of risk assessment brings information on the identification of hazards to possible targets of impact; the targets can be employees and installations of the establishment, the surrounding human population and the environment. Risk

DOI: 10.4018/978-1-7998-3059-7.ch005

assessment presents information on possible preventive means and risk reduction priorities. Results of analysis assess preparedness for accidents and it is a source of information for emergency plan preparation.

The area of major accident prevention represents a relatively new branch of science. Nevertheless, transportation accident prevention requirements are an outgrowth of this branch. The first approach is to evaluate stationary risk sources with the largest amount of dangerous substances. However, dangerous substances limits have been decreased (amendments of Seveso Directives) and current attention was moved to so called "unclassified risk sources" and to mobile risk sources. Greater and greater amounts of dangerous substances are transported by road and rail. A growing interest in the subject area is also related to the threat of terrorist actions, possibly involving the stationary risk sources or the transport of dangerous substances.

The objectives of the chapter are to promote area of process safety for prevention of industrial accident.

BACKGROUND

Process safety focuses on preventing fires, explosions and accidental chemical releases in chemical process facilities or other facilities dealing with hazardous materials. Process safety considers a wide range of technical, management and operational disciplines coming together in an organized way. Main focus is on design and engineering of facilities, maintenance of equipment, effective alarms, effective control points, procedures and training. Process safety generally refers to the prevention of unintentional releases of chemicals, energy, or other potentially dangerous materials during the course of chemical processes. The consequences of so call major-accident can have a serious effect to the plant and environment.

By Seveso III Directive (Seveso, 2012) 'major accident' means an occurrence such as a major emission, fire, or explosion resulting from uncontrolled developments in the course of the operation of any establishment covered by this Directive, and leading to serious danger to human health or the environment, immediate or delayed, inside or outside the establishment, and involving one or more dangerous substances.

In the world, a number of major accidents have occurred; to the best known of them, FEYZIN (France – 1966), FLIXBOROUGH (Great Britain – 1974), SEVESO (Italy – 1976), BHÓPÁL (India – 1984), HOUSTON (USA – 1989), and others belong. The majority of significant accidents are described in the specialized literature in detail, see e.g. (Lees, 2005). Information on major accidents taking place in the countries of European Union is collected in the Joint Research Centre, MAHB (Major Accident Hazards Bureau) in Italian Ispra. The Major Accident Reporting System (eMARS) was established to handle the information on major accidents submitted by Member States of the European Union to the European Commission in accordance with the provisions of the Seveso Directive (eMars, 2019). Currently, eMARS holds data on more than 980 major accident events (by September 2019, see Figure 1).

In some countries, statistics on accidents involving the transport of dangerous substances are recorded. For example, in the U.S. Department of Transportation (DOT), PHMSA - the Pipeline and Hazardous Materials Safety Administration has public responsibility for the safe and secure movement of hazardous materials to industry and consumers by all transportation modes. Argonne National Laboratory (ANL, 2000) also reports on transport risk in the USA. The purpose of this National Transportation Risk Assessment (NTRA) study was to quantitatively characterize the risks associated with the transportation of selected hazardous materials on a national basis. Considering fatality risks on a per-ton-mile basis, the risks from LP gas substantially exceed the risks for gasoline, TIH materials (TIH - toxic by inhalation), and explosives (see Figure 2).

Figure 1. Numbers of EU major-accidents in database eMARS (eMARS, 2019)

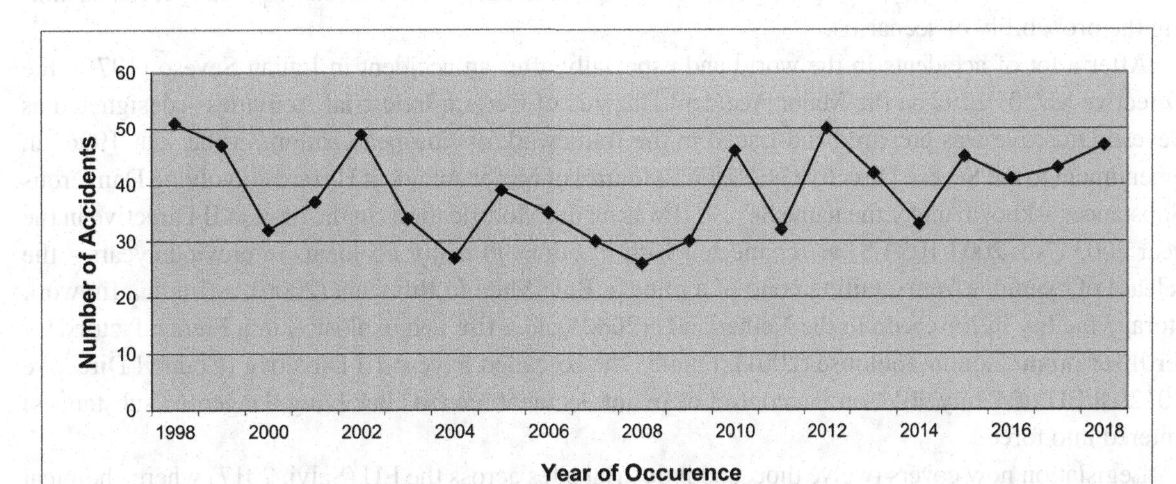

Figure 2. Fatalities normalized by commodity Flow (ANL, 2000)

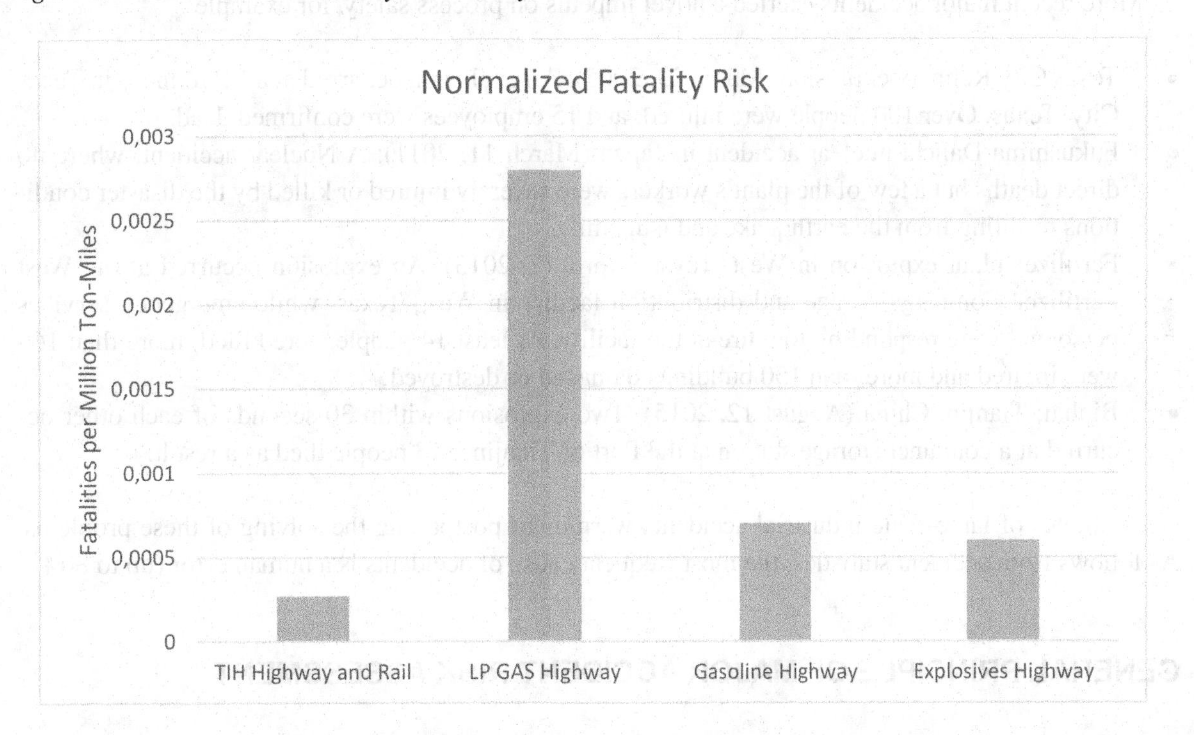

Well known database is ARIA - the Analysis, Research and Information on Accidents Database operated by the French Bureau for Analysis of Industrial Risks and Pollutions (BARPI). ARIA coverage geographically mainly France and Europe, but is well documented for accidents anywhere worldwide, covers more than 30 000 accidents. (Salvi, 2017)

Accident databases are an important source of information for risk analysis, especially for determining the probability of scenarios.

After a lot of accidents in the world and especially after an accident in Italian Seveso (1976), the Directive 82/501/EEC on the Major Accident Hazards of Certain Industrial Activities – designated as Seveso Directive was prepared and issued in the framework of European Union. In the year 1996, an amendment to the Seveso Directive 96/82/EC - Control of Major Accident Hazards Involving Dangerous Substances – known under the name Seveso II was made. Modifications in the Seveso II Directive in the year 2003 (No. 2003/105/ES) are connected with reactions to major accidents in previous years – the release of cyanides from a tailing pond of a mine at Baia Mare in Rumania (2000), a fire in a firework storage facility in Enschede in the Netherlands (2000) and a fire and explosion in a French factory for fertilizer production in Toulouse (2001). Finally, the so-called Seveso III Directive (Council Directive 2012/18/EU of 4 July 2012 on the control of major-accident hazards involving dangerous substances) entered into force.

Legislation now covers twelve thousand industrial sites across the EU (Salvi, 2017) where chemical substances are used or stored (upper / lower tier establishments). The latest modification of this Seveso Directive brought about the unification of the classification of dangerous chemical substances in the area of prevention of serious accidents in accordance with so called CLP (Regulation (EC) No 1272/2008).

More recent major accidents exerted a novel impetus on process safety, for example:

- Texas City Refinery explosion (March 23, 2005): An explosion occurred at a BP refinery in Texas City, Texas. Over 100 people were injured, and 15 employees were confirmed dead.
- Fukushima Daiichi nuclear accident in Japan (March 11, 2011): A Nuclear accident, where no direct deaths but a few of the plant's workers were severely injured or killed by the disaster conditions resulting from the earthquake and tsunami.
- Fertilizer plant explosion in West, Texas (April 17, 2013). An explosion occurred at the West Fertilizer Company storage and distribution facility in West, Texas, while emergency services personnel were responding to a fire at the facility. At least 14 people were killed, more than 160 were injured and more than 150 buildings damaged or destroyed.
- Binhai, Tianjin, China (August 12, 2015): Two explosions within 30 seconds of each other occurred at a container storage station at the Port of Tianjin, 173 people died as a result.

Examples of large-scale industrial accidents warn us of postponing the solving of these problems. As follows from accident statistics, the most frequent cause of accidents is a human error (up to 80%).

GENERAL PRINCIPLE OF MAJOR ACCIDENT RISK ASSESSMENT

Appropriate partial methods are recommended for individual risk management steps. An overview of the world-recognized techniques appropriate for each risk management step is given in the following diagram (see Figure 3). The scheme is based on the risk management process diagram in ISO 31010, some techniques can be used in multiple risk management steps.

These methods of Figure 3 in ISO 31010 are characterized in detail, it is clearly reported their utility, time-consuming and demands the specialists input data and the application. Methods are divided into qualitative and quantitative.

Figure 3. Methods / techniques in risk management process (ISO, 2019)

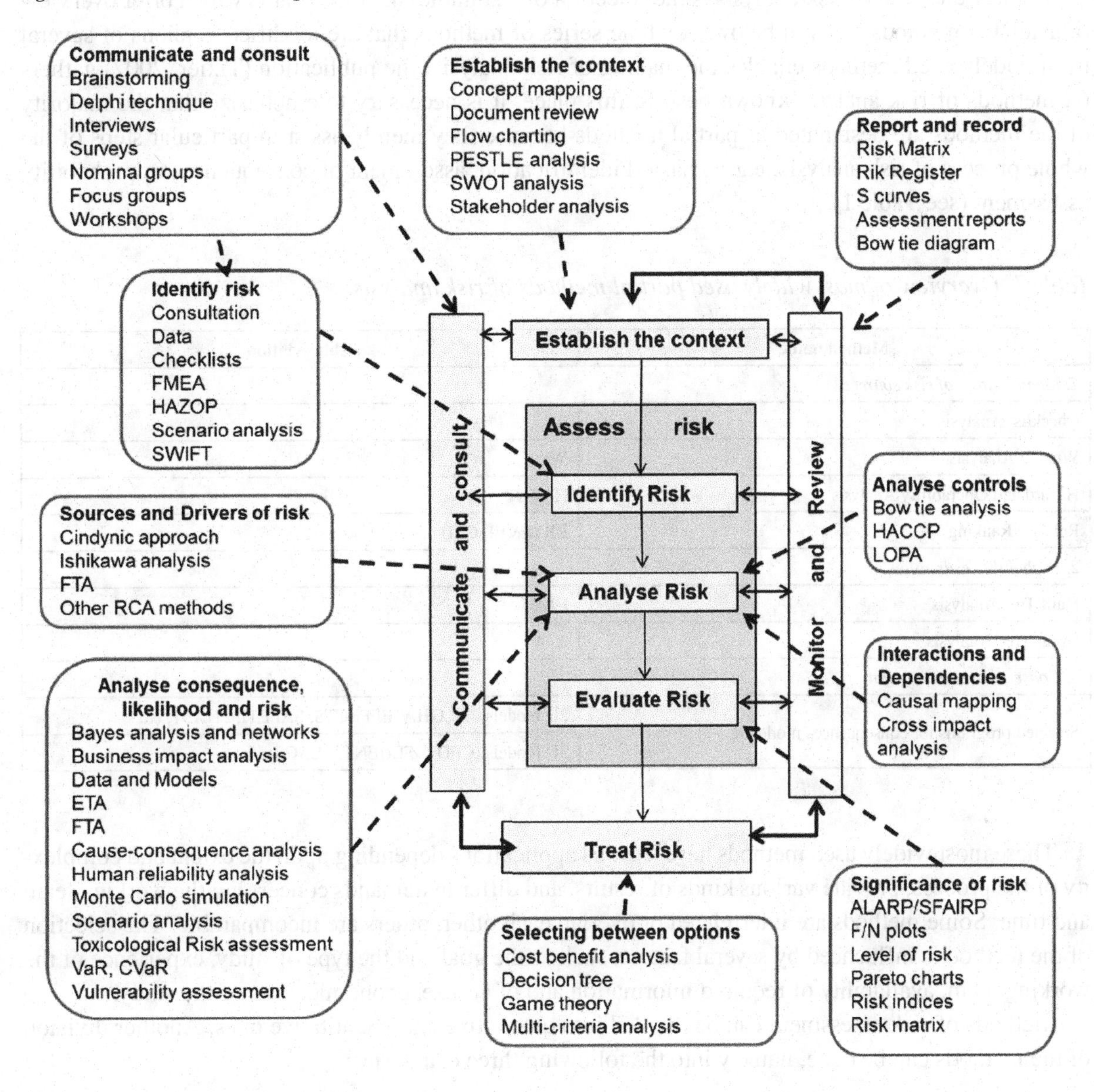

Generally, the following steps are to be taken in the major accident risk analysis and assessment:

Step 1: Identification of sources of risk (hazards) and formation of possible scenarios of events and their causes that may lead to major accidents.
Step 2: Estimation of probability of major accident scenarios.
Step 3: Estimation of effects of possible scenarios of major accidents on human health and lives, on the environment and property.
Step 4: Determination of risk level and assessment of acceptability of major accident risk.

The key question of risk analysis is the selection of a suitable method. That is why a brief overview of available methods is given below. A whole series of methods that are modified versions of several most widely used methods enables the making of risk analysis; the publication (Tixier, 2002) gathers 62 methods of risk analysis known best. In this place, it is necessary to emphasize that the majority of the methods are designated as partial methods because they merely assist in particular steps of the whole process of risk analysis, e.g. in hazard identification, assessment of consequences or probability assessment (see Table 1).

Table 1. Overview of most widely used partial methods of risk analysis

Method name	Abbreviation
1. Identification of risk sources	
Checklist Analysis	CL
What-If Analysis	WI
Hazard and Operability Analysis	HAZOP
Relative Ranking	RR (F&EI, CEI)
2. Probability estimation	
Fault Tree Analysis	FTA
Event Tree Analysis	ETA
3. Consequences estimation	
Selected programs for consequences modeling	2D models - ALOHA, EFFECTS, SAFETI, PHAST, etc.
	3D models (CFD) - FLUENT, FLACS, etc.

These most widely used methods have various applications depending upon the extent and complexity of the process, provide various kinds of results, and differ in demands concerning the working team and time. Some methods are interrelated or overlap each other, others are incomparable. The selection of the method is influenced by several factors, such as the goal and the type of study, experience of the working team, availability of required information and of course, economic costs of the study.

Methods of risk assessment can be divided into qualitative and quantitative ones. Another division of the methods can be done, namely into the following three categories:

- **deterministic** – based on the quantification of accident consequences;
- **probabilistic** – based on the probability or frequency of accident;
- combination of deterministic and probabilistic approaches.

Generally, it can be stated that the deterministic methods are used for the analysis of the whole industrial establishment, whereas the probabilistic methods for the analysis of a chosen part of the establishment requiring a more detailed and thus more demanding analysis. A trend in risk assessment is the hierarchization of results, especially in easily applicable methods; results are presented as hazard level indexes (so-called indexing or screening methods). For the risk sources that have the worst indexes it is then recommended to make a detailed analysis by using high precision methods. A similar new approach to the risk assessment of whole industrial establishments consists in the selection of major risk sources in the first phase and in the second phase, the detailed quantitative risk assessment (QRA) of most hazardous installations selected like that. Both these approaches aim at limiting the number of installations assessed in detail that are there in the industrial establishment, to simplify the whole risk analysis and to concentrate attention especially to the most major risk sources. It must be said that one method for making the whole risk analysis does not exist yet; in practice it is necessary to combine several methods.

The quantitative risk assessment (QRA) of major accidents has been published in various handbooks; with the most significant of them the following publications can be ranked: (Lees, 2005), (CCPS, 1989), (CCPS, 2001), (TNO, 1997) and (TNO, 2005). For the detailed assessment of major accident risks to the population, the below presented methodologies are used oftenest.

What enables an integrated risk analysis is the methodology CPQRA - Chemical Process Quantitative Risk Analysis (CCPS, 1989). This methodology was developed for the needs of chemical industry on the basis of experience of nuclear, aircraft and electronics industries, but its recommended procedure for the analysis is also applicable to other kinds of industry. CPQRA is a tool for risk quantification and reduction by means of partial methods and procedures.

Another acknowledged approach to the comprehensive risk assessment is a Dutch methodology "CPR 18E Guidelines for Quantitative Risk Assessment" known as Purple Book (TNO, 2005). This methodology consists of two parts, namely risk assessment for stationary installations and risk assessment for the transport of dangerous substances. In the further parts of this text, chosen chapters dealing with the assessment of stationary sources of risk will be presented briefly.

The methodology ARAMIS (Accidental Risk Assessment Methodology for Industries in the framework of the SEVESO II Directive) was invented as part of a project of the EU Fifth Framework Program in the years 2002 – 2004. The project ARAMIS proposed a harmonized methodology for risk assessment. The goal of it is above all to reduce uncertainties and variability in results and to include the assessment of risk management efficiency into the analysis. It is thus necessary to take ARAMIS as a global tool for carrying out effective risk identification and analysis with a whole series of pre-prepared and recommended steps (ARAMIS, 2004).

Industry needs a method to identify, assess and reduce the risk and demonstrate the risk reduction as required by the Art.9 of the SEVESO directive. This method and the demonstration have to be accepted by the competent authorities. The approach also has to bring useful information about the ways to reduce the risk and to manage it daily (ARAMIS, 2004).

ARAMIS is divided into the following major steps:

- Identification of major accident hazards (MIMAH).
- Identification of the safety barriers and assessment of their performances.
- Evaluation of safety management efficiency to barrier reliability.
- Identification of Reference Accident Scenarios (MIRAS).
- Assessment and mapping of the risk severity of reference scenarios.
- Evaluation and mapping of the vulnerability of the plant's surroundings.

The main idea of the ARAMIS methodology is based on the so-called bow-tie diagrams, the connection of the fault tree and the event tree (see Figure 4).

Figure 4: General scheme of the bowtie (ARAMIS, 2004)

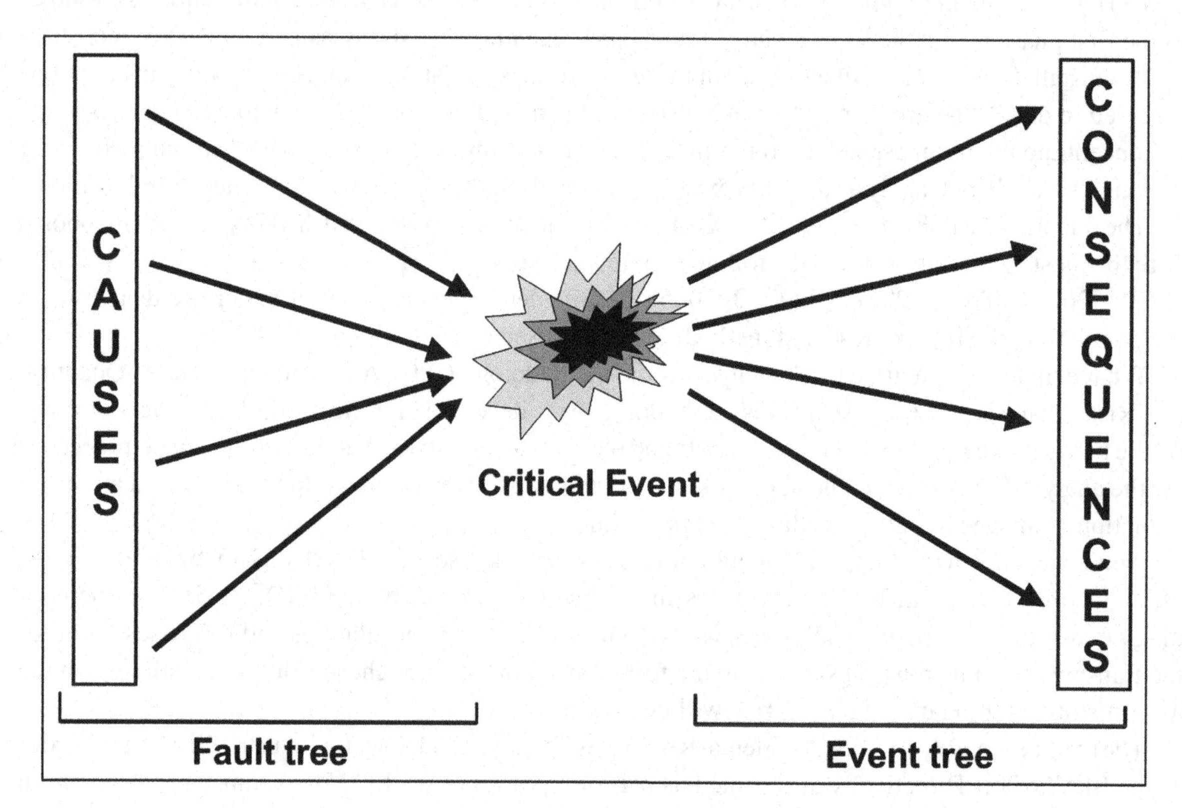

Step 1: Identify of Risk Sources and Scenario Creating

To identify hazards, it is appropriate to divide the hazard into several groups and subgroups, e.g.:

1. Mechanical hazards.
    a. Unprotected moving parts of machines.
    b. Dangerous surfaces (sharp edges, corners, protruding parts, etc.).
    c. Equipment under pressure.
    d. Equipment with necessary handling.
    e. Moving transport equipment.
    f. Falling, slipping, tripping.
    g. Falls from a height.
2. Electrical hazards.
    a. Electric shock.
    b. Electric arc.
    c. Electrostatic charge.
3. Chemical hazards.
    a. Chemical reactions.
    b. Use of flammable and explosive materials.
    c. Use of toxic substances.

    d. Irritants.
    e. Lack of oxygen.
4. Biological hazards.
    a. Infections caused by microorganisms.
    b. The presence of allergens.
5. Fire and explosion.
    a. Open fires.
    b. Hot surfaces.
    c. Sparks.
    d. Autoignition.
    e. Explosions of solids and liquids.
    f. Explosion of gases, vapors and dusts (explosive mixtures with air).
6. Other physical factors.
    a. Electromagnetic radiation (thermal, x-ray, ionizing radiation).
    b. Lasers.
    c. Noise and ultrasound.
    d. Vibration.
    e. Cold substances and environment.
    f. Media under pressure.
7. Factors of work environment and work organization.
    a. Inappropriate lighting.
    b. Inappropriate temperature, humidity, ventilation.
    c. Factors of work process (night work, physical exertion, rest, etc.).
    d. Monotonous work.
    e. Human factor errors.
    f. Insufficient work organization.
    g. Neglected maintenance, especially on safety equipment.
8. Other factors.
    a. Dangerous acts of other persons (terrorism, sabotage, crime).
    b. Working with animals.
    c. Unfavorable weather conditions.
    d. Natural disasters.
    e. Surrounding transport and neighboring industrial enterprises.

The list of hazards may not be complete, may be divided into other groups according to other criteria, but for the purposes of risk analysis it constitutes a useful default list for identifying sources of risk.

The most commonly used methods for hazard identification at the beginning of the risk analysis are briefly described below.

Selection Method

The selection method according to CPR 18E Purple Book (TNO, 2005) was developed for revealing such installations that contribute to a risk most. The installations selected in this way must be considered in the detailed quantitative risk analysis (QRA). Individual steps of the selection method are as follows:

1. An establishment is divided into independent installations (separated units).
2. A hazard related to each installation is determined on the basis of quantity of a substance, operational conditions and properties of dangerous substances. The indication number A expresses the level of real hazardousness of the installation.
3. A hazard related to the installation is determined for a set of points in the surroundings (on the boundary) of the establishment. The hazard related to the installation within a certain distance is determined on the basis of a known indication number and distance between the point being assessed and the installation. The level of hazard in the point being assessed is deduced from the value of selection number S.
4. For the analysis QRA, installations are selected on the basis of relative value of selection number S.

Dow's Fire and Explosion Index

The method "Fire and Explosion Index" (F&E Index) has been improved already for more than 30 years to get into the form of a complex index that is predicable to the relative level of risk of losses of the unit or installation being assessed from the point of view of a possible fire and explosion. In a broad conception, it represents a method for the relative classification of hazard of key units and installations. The stated index is used both by the Dow company and other users. It occupies the leading position among indexing methods in the chemical industry. The current version of F&E Index furnishes key data enabling the assessment of overall fire and explosion risk.

The goal of study using the method F&E Index is as follows (FEI, 1994):

1. To quantify expected damage as a consequence of a fire and explosion (not maximum loss, but maximum probable loss);
2. to identify the installations that could contribute to the occurrence and escalation of an accident;
3. To present obtained results of F&E Index to the management.

Each installation with the F&E Index higher than 128 requires another risk analysis. Table 2 is a listing of the F&EI values versus a description of the degree of hazard that gives some relative idea of the severity of the F&EI (FEI, 1994).

Table 2. F&E Index degree of hazard (FEI, 1994)

DEGREE OF HAZARD FOR F&EI	
F&E INDEX RANGE	**DEGREE OF HAZARD**
1 - 60	Light
61 - 96	Moderate
97 - 127	Intermediate
128 - 158	Heavy
159 - up	Severe

Dow's Chemical Exposure Index

The chemical exposure index (CEI) is a relatively simple method for the assessment of potential exposure of people in the vicinity of chemical plants, where a realistic possibility of release of dangerous chemical substance exists. It is very difficult to determine an absolute level of risk, the method CEI enables the relative comparison of various risk sources (CEI, 1994).

CEI can be used for installations intended for the storage and/or treatment of toxic substances both for new projects and existing installations. If the value of CEI is greater than 200, the unit requires another assessment of hazardousness.

In the text below, the methods HAZOP, FTA and ETA will be introduced because they oftenest follow the screening and indexing methods to specify possible causes of accidents or probabilities of accident occurrence.

Hazard and Operability Analysis (HAZOP)

This method was developed to identify and assess process hazards and to identify operational problems. It is used most frequently in the course and/or after the design stage of process; it is also used advantageously in existing processes. An interdisciplinary team (5 - 7 members) utilizes creative, systematic steps to reveal deviations from the project that may lead to undesirable consequences. For revealing, fix determined words are used (so-called key words – less, more, no, also, part, other, reverse, early, late) that are combined with process parameters. For instance, the key word "No" in conjunction with the parameter "Flow" gives a deviation "No flow". Results of team discussion are written into a table where individual columns represent causes, consequences and protective measures for process deviations. What is a disadvantage of this method is high time consumption as well as labour intensity (AIChE, 1992).

Creating of Scenarios

In case of accidents caused by dangerous substances, a lot of accident manifestations can occur. Basic forms of accident scenarios are summarized in the following diagram (see Figure 5) according to the methodical instruction of the Czech Ministry of Environment (MZP, 2017).

Three main categories of accident effects exist: toxic gases effects, thermal radiation effects, and pressure wave effects with possible flying debris. These effects can impact human lives, property damage, and environmental damage. From the synergetic effects (so called. domino effects) point of view, especially the effects of fires, explosions and flying debris are significant.

Step 2: Assessment of Scenario Probability

Failure and event trees are most commonly used to assess probability.

Fault Tree Analysis (FTA)

It is a deductive method that looks for individual accidents and system failures and determines causes of these events. FTA is a graphical model of various combinations of failures in installations and human errors that may result in a main system failure called "top event". It is also suitable for extensive systems,

Figure 5. Diagram of physical effects and possible accident expansion (MZP, 2007)

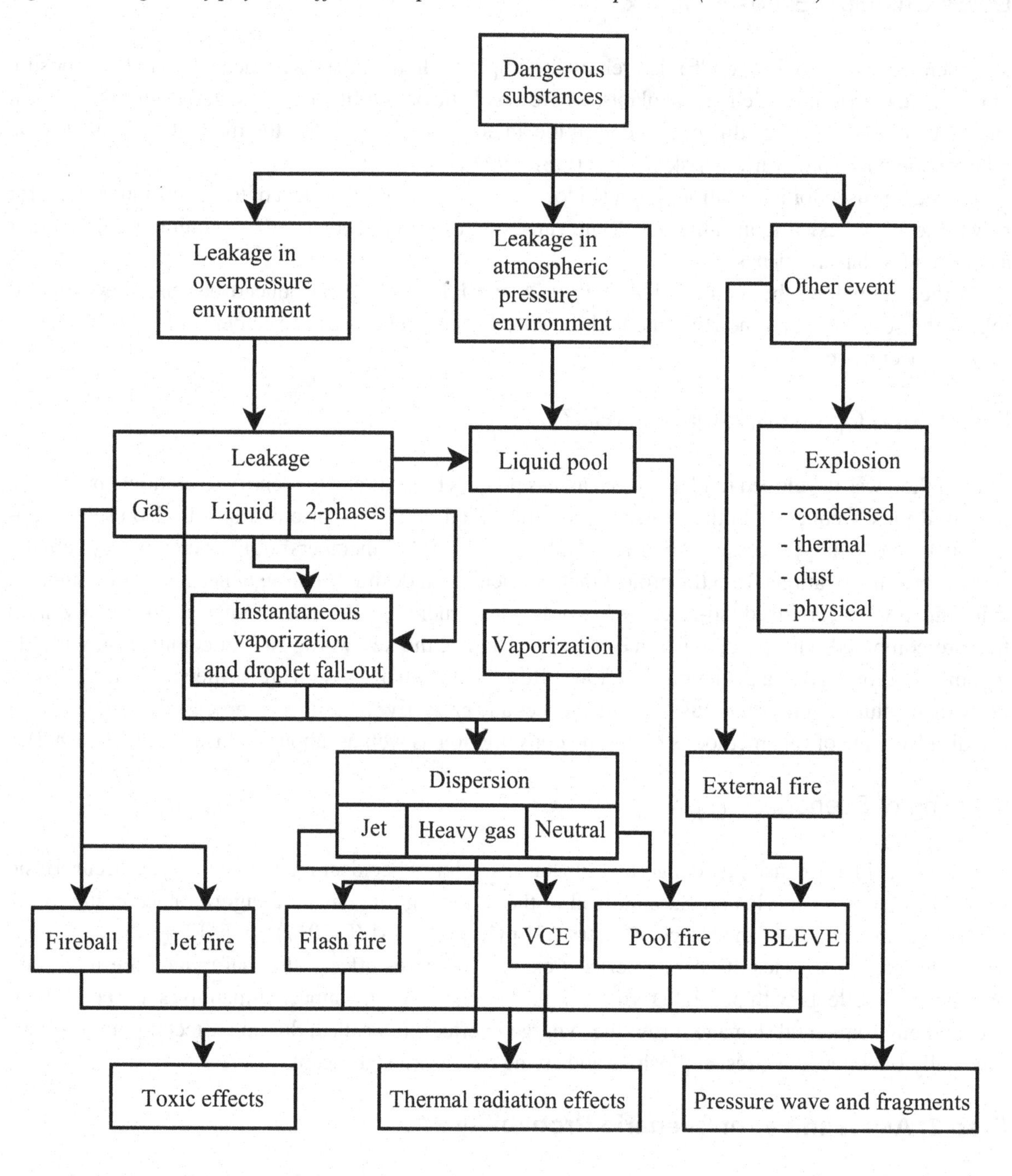

is able to determine a complete list of minimum failures. The model is based on Boolean algebra (gates "and", "or", and others) in looking for a minimum failure leading to a top event; outcomes are types of failures and quantitatively assigned probabilities of system failures, if we know probabilities of primary causes. The study can be carried out by one or more analysts who can recommend safety improvements in the process. The method is not suitable for early phases of design and is time-consuming; demands increase depending upon the complexity of the system (AIChE, 1992).

Event Tree Analysis (ETA)

The method graphically expresses possible consequences of an accident following from an initiation event. Outcomes are sequences of accidents, a series of failures and faults leading to an accident (system function success or failure is assessed). Sequences of accidents represent logical combinations of events, can be transferred to a fault tree model and further quantitatively assessed. The method is suitable for the analysis of comprehensive process that has several kinds of safety systems. The analysis can be made by one analyst, but 2 - 4 analysts are often preferred. Analysts can utilize outcomes for recommendations for reducing the probability and/or mitigating consequences of potential failures (AIChE, 1992).

Scenario Probability Assessment

To determine the probability of individual scenarios for major accidents, generic values stated in the special literature based on historical data are often used. One of the most frequently utilized publications is the Purple Book (TNO, 2005), where events in the course of which installation damage and a release of dangerous substance, so-called Loss of Containment (LOC), occur, are described. For these cases of failures, assumed frequencies of occurrence determined on the basis of accidents in the past are stated. The methodology thus presents the failures in installation that contribute to a societal risk and must be taken into account during quantitative risk assessment.

As examples of data on failure frequencies, values for atmospheric tanks are there in the following Table 3.

A single leak with immediate ignition may cause a BLEVE event and a fire ball. Example BLEVE type explosion effects (the probability of death of persons depending on the distance from the source) is shown below (Fig. 6) for storing 100 tons of propane.

Step 3: Assessment of Scenarios Consequences

1. Thermal radiation
   a. Continuous radiation
   b. Instantaneous radiation
2. Blast effects
3. Missiles
4. Toxic effects

All the effects represented in Table 4 make reference only to humans or structures but not to the environment.

For the detailed modeling of releases of dangerous substances and consequences of fires, explosions and spreading toxic clouds, a whole series of computer programs can be used (Bernatik, 2008). To the best known programs ALOHA, RMP Comp, SAFETI, PHAST, EFFECTS, CHARM, and others belong. Some programs are freely available on the Internet (e.g. ALOHA, RMP Comp), others are commercial products of significant companies concerned with risk analyses (e.g. company TNO / Gexcon– EFFECTS or company DNV – PHAST, SAFETI). In the further text, the program EFFECTS is described as an example.

Table 3. The failure frequencies for atmospheric tanks (TNO, 2005)

LOCs for atmospheric tanks						
G.1 Instantaneous release of the complete inventory a: directly to the atmosphere b: from the primary container into the unimpaired secondary container or outer shell G.2 Continuous release of the complete inventory in 10 min at a constant rate of release a: directly to the atmosphere b: from the primary container into the unimpaired secondary container or outer shell G.3 Continuous release from a hole with an effective diameter of 10 mm a: directly to the atmosphere b: from the primary container into the unimpaired secondary container or outer shell						
Installation (part)	**G.1a Instantan. release to atmosphere**	**G.1b Instantan. release to secondary container**	**G.2a Continuous 10 min release to atmosphere**	**G.2b Continuous 10 min release to secondary container**	**G.3a Continuous Ø10 mm release to atmosphere**	**G.3b Continuous Ø10 mm release to secondary container**
Single containment tank	5×10^{-6} y^{-1}		5×10^{-6} y^{-1}		1×10^{-4} y^{-1}	
tank with a protective outer shell	5×10^{-7} y^{-1}	5×10^{-7} y^{-1}	5×10^{-7} y^{-1}	5×10^{-7} y^{-1}		1×10^{-4} y^{-1}
Double containment tank	1.25×10^{-8} y^{-1}	5×10^{-8} y^{-1}	1.25×10^{-8} y^{-1}	5×10^{-8} y^{-1}		1×10^{-4} y^{-1}
Full containment tank	1×10^{-8} y^{-1}					
In-ground tank		1×10^{-8} y^{-1}				
mounded tank	1×10^{-8} y^{-1}					

Figure 6. The probability of death as a function of the distance for a BLEVE of a storage tank containing 100 tons of propane (TNO, 2005)

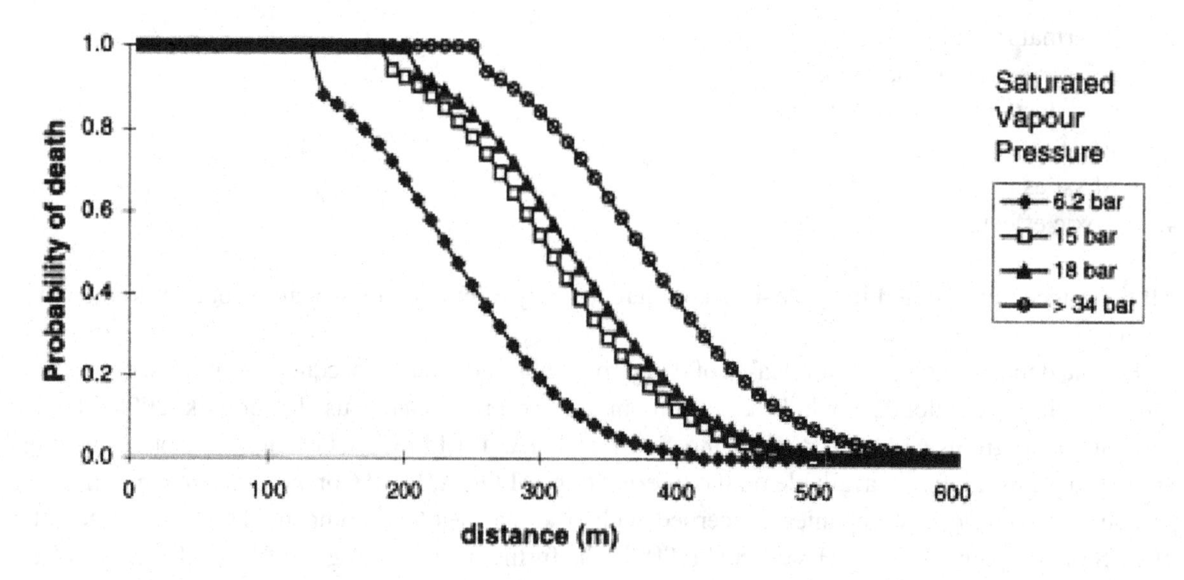

Table 4. Definition of the thresholds for the diverse levels of effects (ARAMIS, 2004)

Level of effects	Radiation (1) (kW/m²)	Instantaneous Radiation	Blast (mbar)	Missiles (2) (%)	Toxic effect (3)	Description
1	< 1,8	< 0.5 LFL	< 30	0	< TEEL-1	Small or non effects
2	1,8 - 3		30 - 50		TEEL-1 - TEEL-2	Reversible effects
3	3 - 5		50 - 140		TEEL-2 - TEEL-3	Irreversible effects
4	> 5	= 0.5 LFL	> 140	100	> TEEL-3	Start of lethality and/or domino effects

(1) For 60 s exposure
(2) Range distance of the indicated percentage of missiles
(3) TEEL values (Temporary Emergency Exposure Limits)

The EFFECTS model represents a high standard of modeling the consequences of major accidents. The program combines two accepted models for the calculation of physical effects after a release of dangerous substances EFFECTS and DAMAGE. The model EFFECTS make it possible to determine manifestations of accidents, such as pressure wave, heat radiation, gas concentration, the model DAMAGE makes it possible to determine consequences of accidents, e.g. human death rate, first-degree and second-degree burns, lung injury, eardrum damage, and others. The advantage of connection of these two models in one program is the inclusion of comprehensive calculations covering the range from initiation physical effects to accident consequences. Results are presented in text and graphical forms (EFFECTS, 2018). An example of monitor of the program EFFECTS is shown in the figure below (see Figure 7).

Figure 7. Example of monitor of software EFFECTS (Effects, 2018)

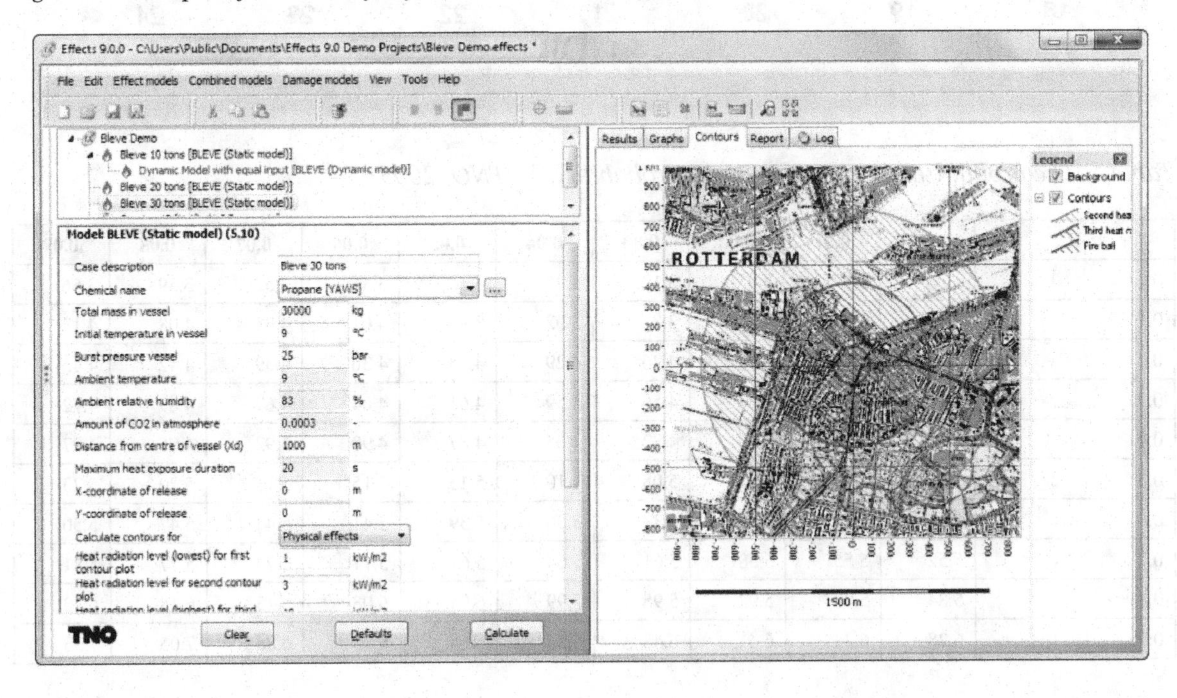

Step 4: Risk Evaluation - Risk Acceptability

The calculation of individual and societal risks includes the calculation of probability of human death for given exposure. The probability of death is calculated by using the probit function (kind of dose–response model dependence expressed by an equation). The relation between the probability of an effect and the exposure usually results in a sigmoid curve. The sigmoid curve is replaced with a straight line if the probit is used instead of the probability (see Figure 8 and Table 5).

Figure 8. The probability and the probit as a function of exposure to ammonia (TNO, 2005)

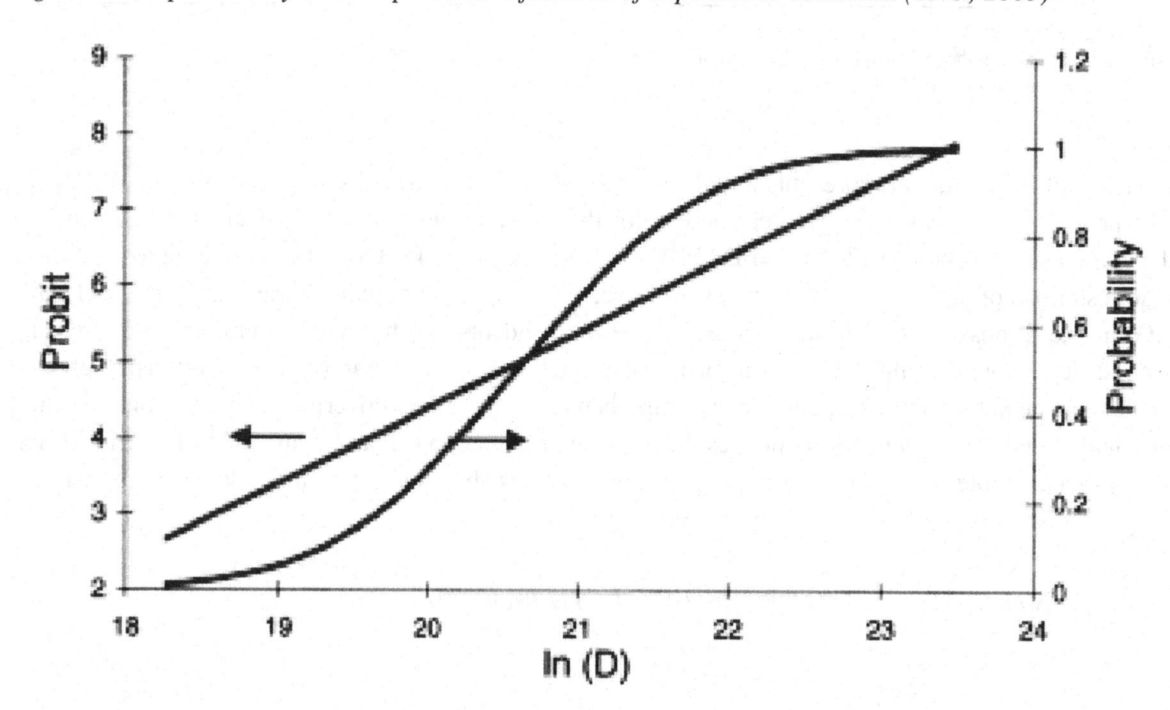

Table 5. The probit, Pr, as a function of the probability, P (TNO, 2005)

P	0	0.01	0.02	0.03	0.04	0.05	0.06	0.07	0.08	0.09
0	-	2.67	2.95	3.12	3.25	3.36	3.45	3.52	3.59	3.66
0.1	3.72	3.77	3.82	3.87	3.92	3.96	4.01	4.05	4.08	4.12
0.2	4.16	4.19	4.23	4.26	4.29	4.33	4.36	4.39	4.42	4.45
0.3	4.48	4.50	4.53	4.56	4.59	4.61	4.64	4.67	4.69	4.72
0.4	4.75	4.77	4.80	4.82	4.85	4.87	4.90	4.92	4.95	4.97
0.5	5.00	5.03	5.05	5.08	5.10	5.13	5.15	5.18	5.20	5.23
0.6	5.25	5.28	5.31	5.33	5.36	5.39	5.41	5.44	5.47	5.50
0.7	5.52	5.55	5.58	5.61	5.64	5.67	5.71	5.74	5.77	5.81
0.8	5.84	5.88	5.92	5.95	5.99	6.04	6.08	6.13	6.18	6.23
0.9	6.28	6.34	6.41	6.48	6.55	6.64	6.75	6.88	7.05	7.33

As example, the probability of death due to exposure to a toxic cloud, P_E, and the fraction of people indoors and outdoors dying, $F_{E,in}$ and $F_{E,out}$, is given in Figure 9.

The probability of death, P_E, is calculated with the use of a probit function and relation 1. The probit function for death due to toxic exposure is given by (TNO, 2005):

$$Pr = a + b \times \ln (C^n \times t) \tag{1}$$

where:

Pr probit function corresponding to the probability of death (-)
a, b, n constants describing the toxicity of a substance (-)
C concentration ($mg.m^{-3}$)
t exposure time (minutes)

Notes:

- The value of the constant, a, depends on the dimensions of the concentration, C, and the exposure time, t. The dimensions of the concentration and exposure time must correspond to the value of the constant, a, in the probit function.
- The probit is a function of the toxic dose to an individual. The toxic dose, D, is equal to $D = C^n \times t$ if the concentration, C, is constant during the time of exposure, t. If the concentration is not constant in time, the toxic dose is calculated as $D = \int C^n\, dt$; the probit, Pr, should be calculated correspondingly.
- The exposure time, t, is limited to a maximum of 30 minutes, starting from the arrival of the cloud. The arrival of the cloud can be defined as the moment when the probability of death, PE, exceeds 1%.
- Staying indoors reduces the toxic dose since the concentration indoors is lower than the concentration outdoors during cloud passage. This effect is accounted for by a generic factor 0.1 in the fraction of people dying indoors. (TNO, 2005)

Similarly, the probit approach is recommended for evaluation of consequence to fire and explosion. For example, the probability of death from heat radiation exposure is calculated by means of the probit function that is given by a relation presented bellow (TNO, 2005):

$Pr = -36.38 + 2.56 \times \ln (Q^{4/3} \times t)$ (2)

where:

Pr probit function corresponding to the probability of death (-)
Q heat radiation ($W.m^{-2}$)
t exposure time (s)
Calculation and Result Presentation

From the comprehensive point of view, a risk is understood to be a relation between the expected loss (health damage, loss of life, loss of property, and others) and the uncertainty of loss considered

Figure 9. Calculation of the probability of death, P_E, and the respective fractions of the population dying indoors and outdoors, $F_{E,in}$ and $F_{E,out}$, due to exposure to a toxic cloud. The function $f(a,b,n;C,t)$ is the probit function for exposure to toxic substances (TNO, 2005)

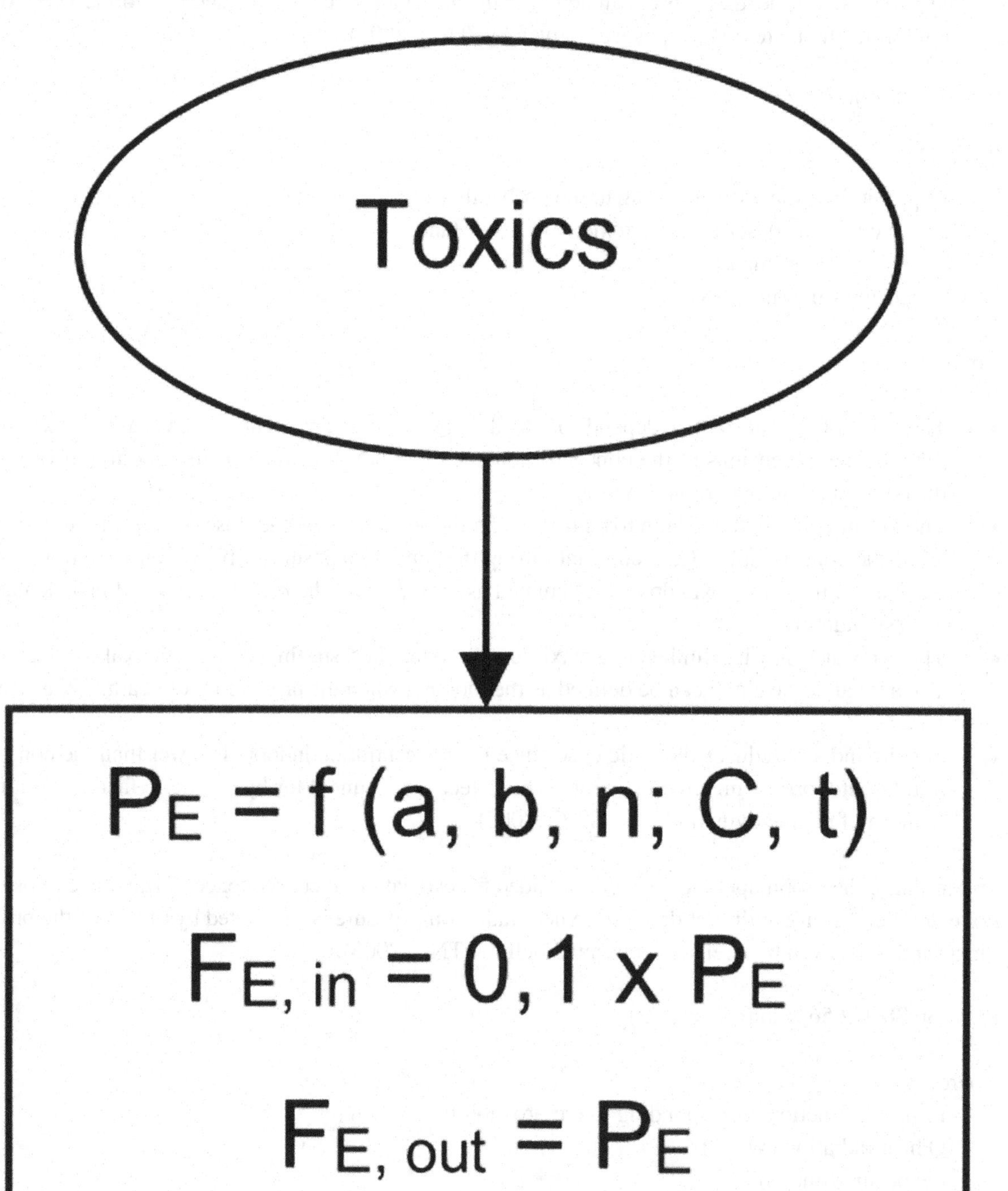

(usually expressed by the probability or frequency of occurrence of an unexpected event). The results of quantitative risk analysis (QRA) are the individual risk and the societal risk.

- The individual risk represents the frequency of death of an individual in connection with a case of installation failure (LOC). One can assume that the individual is not protected and is exposed to unfavorable conditions for the whole time of exposure. The individual risk is represented by contours in a topographic map.
- The societal risk represents the frequency of such events in the course of which more persons lose their lives together. The societal risk is represented by means of F-N curves, where N stands for the number of deaths and F is the cumulative frequency of events accompanied by N or more deaths.

Criteria for the determination of individual and societal risks are illustrated in Figure 10.

Figure 10. Theoretical examples of criteria for (a) individual and (b) societal risk (Christou, 2006) Note: ALARA - As Low as Reasonably Achievable

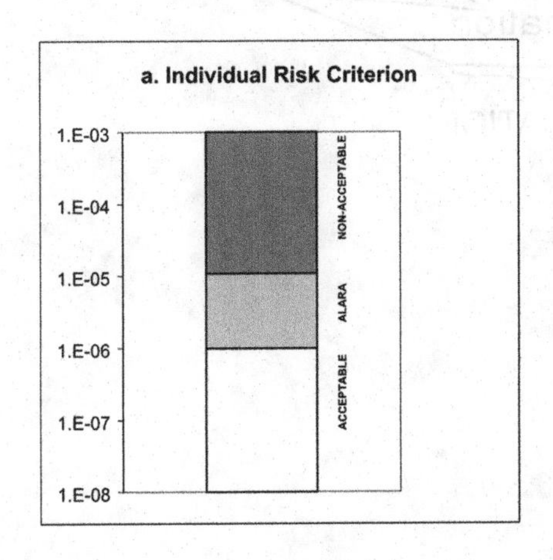

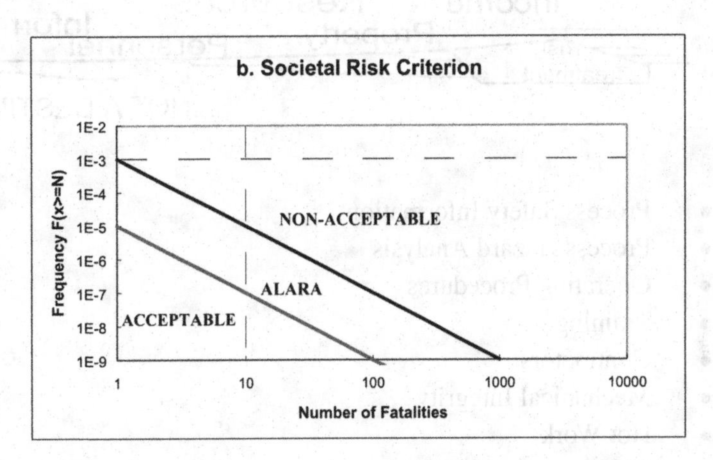

MAJOR ACCIDENT RISK MANAGEMENT

Risk management is the process of identifying, analyzing and responding to risk factors throughout the life of a project and in the best interests of its objectives. Proper risk management implies control of possible future events and is proactive rather than reactive (Stanleigh, 2016). The whole process of risk management is an endless journey ... (see Figure 11).

Risk management involves understanding, analyzing and addressing risk to make sure organizations achieve their objectives (IRM, 2016). Figure 12 presents the cycle of risk management processes as well as the basic idea of generally all methods.

The process safety management system program is divided into 14 elements. The U.S. Occupational Safety and Health Administration 1910.119 define all 14 elements of the process safety management system plan (OSHA, 2013).

Figure 11. Risk management (Knight, 2010)

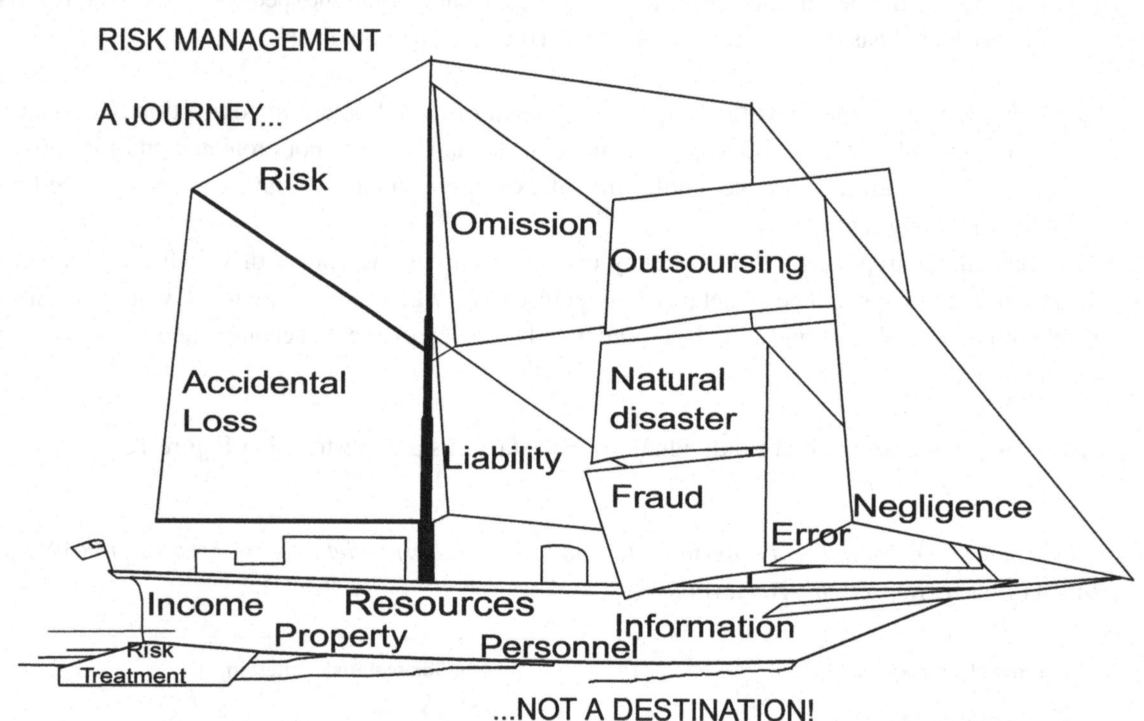

- Process Safety Information
- Process Hazard Analysis
- Operating Procedures
- Training
- Contractors
- Mechanical Integrity
- Hot Work
- Management of Change
- Incident Investigation
- Compliance Audits
- Trade Secrets
- Employee Participation
- Pre-startup Safety Review
- Emergency Planning and Response

All of those elements mentioned above are interlinked and interdependent (see Figure 13). There is a tremendous interdependency of the various elements of process safety management (PSM). All elements are related and are necessary to make up the entire PSM picture. Every element either contributes information to other elements for the completion or utilizes information from other elements in order to be completed.

Figure 12. Planning for risk management (Rahim, 2016)

The problematics of risk management is extensive, complex, with large number of individual elements and connections between them. Therefore, there is no way how to formulate a general method for all applications.

The need to assess and manage major accident risks follows from several factors, especially from a large number of accidents occurred in the past, leading to pressures to reduce risks associated with various process installations. Furthermore, this need is initiated by the necessity of prevention against accidents in land-use planning, i.e. approving the location of new establishments in relation to inhabited and protected areas, by the necessity of improving emergency preparedness, and others. Carrying out a risk assessment and following measures to reduce the risk can contribute to the prevention of accidents,

Figure 13. 14 Elements of Process Safety Management (Khamar, 2016)

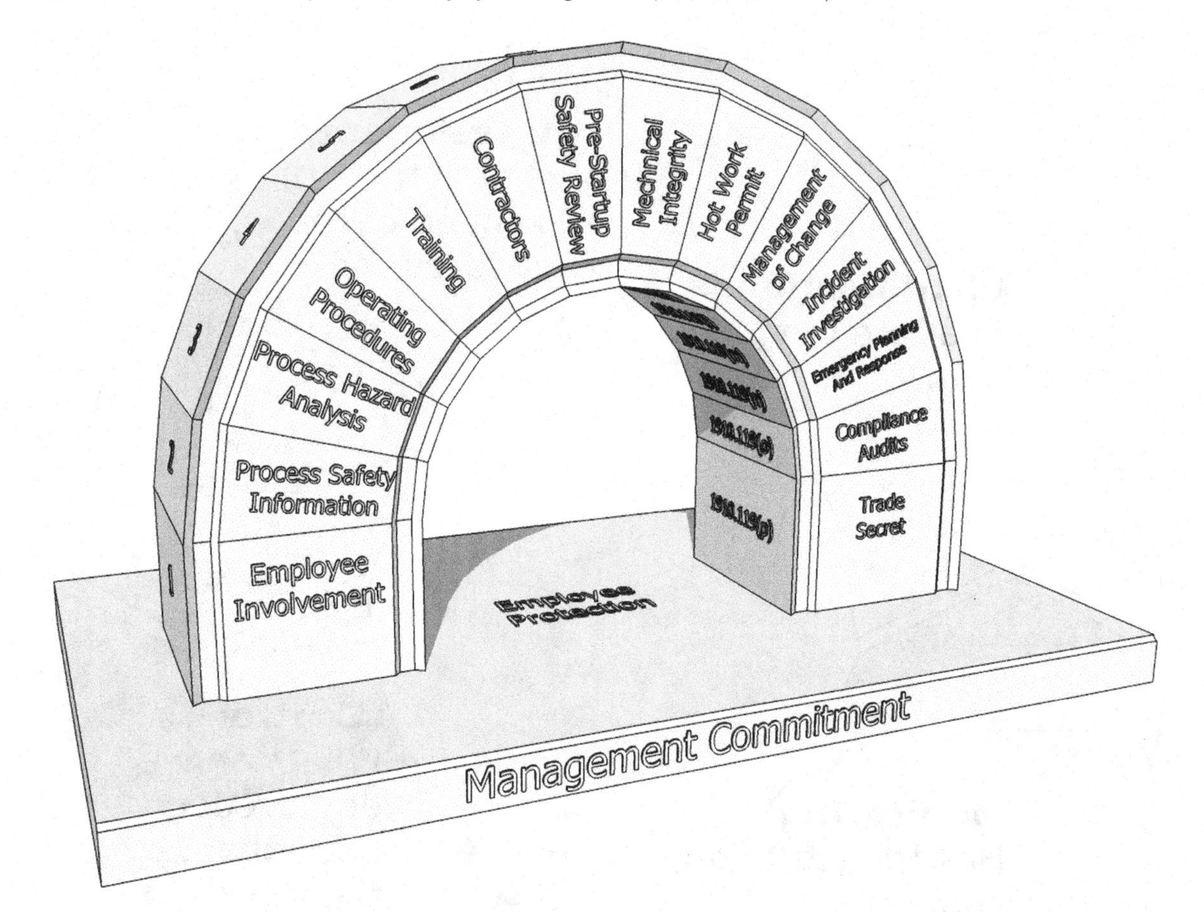

the mitigation of accident consequences on human lives, property and the environment, and also can prevent the unsuitable location of a new establishment in the vicinity of population centers and areas protected from the environmental point of view. Risk assessment should be done in the preparatory phase of building of a new establishment and during a major accident investigation to prevent accident re-occurrence, and in the phase of establishment operation when the risk assessment contributes to better awareness on risk sources, accident consequences and threatened target groups. Prepared accident scenarios are used for improvements in emergency plans and preparedness for effective interventions in case of accident.

SOLUTIONS AND RECOMMENDATION

On the basis of results of risk assessment, priorities in reducing the risks can be decided in the phase of risk management. This area of risk management is necessary for the sustainable development of a mature society. Basic approaches to the reducing of risks associated with process installations are as follows:

- The replacement of a dangerous substance by a substance less dangerous,

- A change in technology to become more advanced, with the presence of a smaller quantity of dangerous substances,
- Decreasing the stock of dangerous substances to a necessary minimum,
- The effective separation of the quantity of dangerous substances in the establishment (e.g. by remote-controlled valves) – decreasing the quantity of a releasing substance.

Safety approach based on inherent safety has developed since 1991 by Trevor Kletz. The four main methods for achieving inherently safer design are: (Khan, 2003)

- **Minimize**: Reducing the amount of hazardous material present at any one time, e.g. by using smaller batches.
- **Substitute**: Replacing one material with another of less hazard, e.g. cleaning with water and detergent rather than a flammable solvent.
- **Moderate**: Reducing the strength of an effect, e.g. having a cold liquid instead of a gas at high pressure or using material in a dilute rather than concentrated form.
- **Simplify**: Eliminating problems by design rather than adding additional equipment or features to deal with them. Only fitting options and using complex procedures if they are really necessary.

Two further principles are used by some: (Khan, 2003)

- **Error tolerance**: Equipment and processes can be designed to be capable of withstanding possible faults or deviations from design. A very simple example is making piping and joints capable of withstanding the maximum possible pressure, if outlets are closed.
- **Limit effects** by design, location or transportation of equipment so that the worst possible condition produces less danger, e.g. gravity will take a leak to a safe place, the use of bunds.

In terms of making plants more user-friendly Kletz added the following: (Kletz, 1991)

- Avoiding knock-on effects;
- Making incorrect assembly impossible;
- Making status clear;
- Ease of control;
- Software and management procedures.

Further, the recommended precautions and measures for risk reduction are given, where the field of risk management follows up with risk assessment. Basic division of safety precautions for risk reduction:

- **Technical Measures**: Such measures in equipment construction that lead to the increase of plant safety (examples: safety valves, sprinklers, automatic regulation, safety reservoirs, double-skin storage tanks…)
- **Organizational Measures**: Measures in work organization, regulations, technological procedures (including pertinent technical equipment) leading to the increase of plant safety.

Synergetic effects and domino effects can be also prevented using barriers limiting the possibility of initiation of consequent accidents and emergencies initiated by "covered" risks. Such barriers can be:

1. **Passive Barriers**: Functioning permanently without human intervention, energy sources, and without sources of information. Passive barriers can be physical (safety reservoirs, walls...), permanent barriers (corrosion prevention system) or internally safe design of equipment/facility.
2. **Active Barriers**: Must be activated to function. Active barriers always require the sequence: detection - diagnostics - action.
3. **Requiring Human Intervention**: Broadly interpreted, includes monitoring of all senses, communication, thinking, physical activities, emergency activities, etc. Human intervention can be part of the sequence: detection - diagnostics - action.

Automatic systems that eliminate a human factor are more suitable (e.g. detection and alarm systems, automatic systems like blocking and stop systems, fire and explosion protection systems, protections against releases of dangerous toxic substances, further safeguards against intrusion and unauthorized manipulation, central stations including the indication of protection system functionality). The majority of these technical safety measures are called preventive (precautions) because they contribute to a reduction in major accident occurrence.

Another group of safety measures consists of protective means and emergency response means that mitigate and limit the consequences of a major accident. Generally, e.g. fixed technical means (fixed extinguishing systems, ventilation systems), mobile technical means (pumps, fans, telescopic platforms, scum boards), transport means and special mechanisms (earth moving machines, truck tanker fuels, fire extinguishing tankers), emergency response and emergency materials, personal protective equipment, staffing (strength of staff ready to respond) can be stated. Into the safety measures, employee alarm and warning systems and methods and systems and methods of notification of entities concerned in case of emergency must be included.

Complete summary of safety measures from the point of view of effects risks minimization is given in the following list of barriers ranked into individual groups:

Safe process management

- Spare energy sources
- Cooling
- Emergency standstill
- Computer driven process
- Operational instructions /procedures
- Summary of reactive compounds
- Performs risk assessment
- Assembly of safety valve
- Safety valve against under pressure (vacuum)
- Regular maintenance and cleaning of equipment
- Protective work aids (for employees)

Separability of materials

- Remotely controlled armatures
- Reflux valves for the prevention of reverse flow
- Safety valves in pipelines
- Discharges / purge
- Drainage
- Blocking / Interlock
- Safe distances between equipment

Elements of explosion protection

- Control of potential explosion
- Explosion-proof instruments
- Flame fuses
- Usage of inert gas

Elements of fire protection

- Detection of gas leakage
- Construction steel / fire resistant finishing
- Fire water supply
- Special systems (CO_2, halons, smoke detectors, etc.)
- Water drench systems / sprinklers
- Automatic water curtains
- Foam spray
- Handheld fire extinguishers / fire annunciators
- Cable protection

Measures against dangerous liquids leakage

- Measurement of surface level in storage tank / equipment
- Double-skin storage tanks
- Place storage tanks with liquids into safety reservoirs or tanks
- Fitting the safety reservoirs by pump connected to plant sewage
- Sewage water control before leaving the plant

Measures against gaseous substances leakage

- Toxic gas detectors
- Devices detecting the oxygen concentration
- Fitting water drench systems to equipment (hand operated or automatic sprinklers)

Measures against external influences

- Flood and inundation protection

- Protection against natural external effects
- Monitoring cameras
- Registration of entry to the area
- Sensitive equipment must be locked and protected against unauthorized intrusion

Emergency planning

- Internal emergency plan for the evacuation and protection of employees
- Outer emergency plan for the protection of neighboring population

Below is a summary of the "Golden Rules" for the prevention of major accidents, as stated in the OECD publication (OECD, 2003).

ROLE OF ALL STAKEHOLDERS

- Make chemical risk reduction and accident prevention, as well as effective emergency preparedness and response, priorities in order to protect health, the environment and property.

While the risks of accidents are in the communities where hazardous installations are located, requiring efforts by stakeholders at the local level, there are also responsibilities for stakeholders at regional, national and international levels.

- Communicate and co-operate with other stakeholders on all aspects of accident prevention, preparedness and response.

Communication and co-operation should be based on a policy of openness, as well as the shared objective of reducing the likelihood of accidents and mitigating the adverse affects of any accidents that occur. One important aspect is that the potentially affected public should receive information needed to support prevention and preparedness objectives, and should have the opportunity to participate in decision-making related to hazardous installations, as appropriate.

ROLE OF INDUSTRY (INCLUDING MANAGEMENT AND LABOR)

Management

- Know the hazards and risks at installations where there are hazardous substances.

All enterprises that produce, use, store, or otherwise handle hazardous substances should undertake, in co-operation with other stakeholders, the hazard identification and risk assessment(s) needed for a complete understanding of the risks to employees, the public, the environment and property in the event of an accident. Hazard identification and risk assessments should be undertaken from the earliest stages of design and construction, throughout operation and maintenance, and should address the possibilities of human or technological failures, as well as releases resulting from natural disasters or deliberate acts (such as terrorism, sabotage, vandalism, or theft). Such assessments should be repeated periodically and whenever there are significant modifications to the installation.

- Promote a "safety culture" that is known and accepted throughout the enterprise.

The safety culture, reflected in an enterprise's Safety Policy, consists of both an attitude that safety is a priority (*e.g.,* accidents are preventable) and an appropriate infrastructure (*e.g.*, policies and procedures). To be effective, a safety culture requires visible top-level commitment to safety in the enterprise, and the support and participation of all employees[20] and their representatives.

- Establish safety management systems and monitor/review their implementation.

Safety management systems for hazardous installations include using appropriate technology and processes, as well as establishing an effective organizational structure (*e.g.,* operational procedures and practices, effective education and training programs, appropriate levels of well-trained staff, and allocation of necessary resources). These all contribute to the reduction of hazards and risks. In order to ensure the adequacy of safety management systems, it is critical to have appropriate and effective review schemes to monitor the systems (including policies, procedures and practices).

- Utilize "inherently safer technology" principles in designing and operating hazardous installations.

This should help reduce the likelihood of accidents and minimize the consequences of accidents that occur. For example, installations should take into account the following, to the extent that they would reduce risks: minimizing to the extent practicable the quantity of hazardous substances used; replacing hazardous substances with less hazardous ones; reducing operating pressures and/or temperatures; improving inventory control; and using simpler processes. This could be complemented by the use of back-up systems.

- Be especially diligent in managing change.

Any significant changes (including changes in process technology, staffing, and procedures), as well as maintenance/repairs, start-up and shut-down operations, increase the risk of an accident. It is therefore particularly important to be aware of this and to take appropriate safety measures when significant changes are planned - before they are implemented.

- Prepare for any accidents that might occur.

It is important to recognize that it is not possible to totally eliminate the risk of an accident. Therefore, it is critical to have appropriate preparedness planning in order to minimize the likelihood and extent of any adverse effects on health, the environment or property. This includes both on-site preparedness planning and contributing to off-site planning (including provision of information to the potentially affected public).

- Assist others to carry out their respective roles and responsibilities.

To this end, management should co-operate with all employees and their representatives, public authorities, local communities and other members of the public. In addition, management should strive to assist other enterprises (including suppliers and customers) to meet appropriate safety standards. For example, producers of hazardous substances should implement an effective Product Stewardship program.

- Seek continuous improvement.

Although it is not possible to eliminate all risks of accidents at hazardous installations, the goal should be to find improvements in technology, management systems, and staff skills in order to move closer toward the ultimate objective of zero accidents. In this regard, management should seek to learn from past experiences with accidents and near-misses, both within their own enterprises and at other enterprises.

Labor

- Act in accordance with the enterprise's safety culture, safety procedures, and training.

In the discharge of their responsibilities, labor should comply with all the procedures and practices relating to accident prevention, preparedness and response, in accordance with the training and instructions given by their employer. All employees (including contractors) should report to their supervisor any situation that they believe could present a significant risk.

- Make every effort to be informed, and to provide information and feedback to management.

It is important for all employees, including contractors, to understand the risks in the enterprise where they work, and to understand how to avoid creating or increasing the levels of risk. Labor should, to the extent possible, provide feedback to management concerning safety-related matters. In this regard, labor and their representatives should work together with management in the development and implementation of safety management systems, including procedures for ensuring adequate education and training/retraining of employees. Labor and their representatives should also have the opportunity to participate in monitoring and investigations by the employer, or by the competent authority, in connection with measures aimed at preventing, preparing for, and responding to chemical accidents.

- Be proactive in helping to inform and educate your community.

Fully informed and involved employees at a hazardous installation can act as important safety ambassadors within their community.

ROLE OF PUBLIC AUTHORITIES

- Seek to develop, enforce and continuously improve policies, regulations, and practices.

It is important for public authorities[21] to establish policies, regulations and practices, and have mechanisms in place to ensure their enforcement. Public authorities should also regularly review and update, as appropriate, policies, regulations, and practices. In this regard, public authorities should keep informed of, and take into account, relevant developments. These include changes in technology, business practices, and levels of risks in their communities, as well as experience in implementing existing laws and accident case histories. Public authorities should involve other stakeholders in the review and updating process.

- Provide leadership to motivate all stakeholders to fulfill their roles and responsibilities.

Within their own sphere of responsibility and influence, all relevant public authorities should seek to motivate other stakeholders to recognize the importance of accident prevention, preparedness and response, and to take the appropriate steps to minimize the risks of accidents and to mitigate the effects of any accidents that occur. In this regard, the authorities should establish and enforce appropriate regulatory regimes, promote voluntary initiatives, and establish mechanisms to facilitate education and information exchange.

- Monitor the industry to help ensure that risks are properly addressed.

Public authorities should establish mechanisms for monitoring hazardous installations to help ensure that all relevant laws and regulations are being followed, and that the elements of a safety management system are in place and are functioning properly, taking into account the nature of the risks at the installations (including the possibilities of deliberate releases). Public authorities can also take these opportunities to share experience with relevant employees of the installations.

- Help ensure that there is effective communication and co-operation among stakeholders.

Information is a critical component of safety programs. Public authorities have an important role in ensuring that appropriate information is provided to, and received by, all relevant stakeholders. Public authorities have a special role in facilitating education of the public concerning chemical risks in their community so that members of the public are reassured that safety measures are in place, that they understand what to do in the event of an accident, and that they can effectively participate in relevant decision-making processes. Public authorities are also in a position to facilitate the sharing of experience (within and across borders).

- Promote inter-agency co-ordination.

Chemical accident prevention, preparedness and response is, by nature, an inter-disciplinary activity involving authorities in different sectors and at different levels. To help ensure effective prevention, preparedness and response, and efficient use of resources, it is important that all relevant agencies coordinate their activities.

- Know the risks within your sphere of responsibility, and plan appropriately.

Public authorities are responsible for off-site emergency planning, taking into account the relevant onsite plans. This should be done in co-ordination with other stakeholders. In addition, public authorities should ensure that the resources necessary for response (*e.g.*, expertise, information, equipment, medical facilities, finances) are available.

- Mitigate the effects of accidents through appropriate response measures.

Public authorities (often at the local level) have primary responsibility for ensuring response to accidents that have off-site consequences, to help reduce deaths and injuries, and to protect the environment and property.

- Establish appropriate and coherent land-use planning policies and arrangements.

Land-use planning (*i.e.*, establishing and implementing both general zoning as well as specific siting of hazardous installations and other developments) can help to ensure that installations are appropriately located, with respect to protection of health, environment and property, in the event of an accident. Land-use planning policies and arrangements can also prevent the inappropriate placing of new developments near hazardous installations (*e.g.*, to avoid the construction of new residential, commercial or public buildings within certain distances of hazardous installations). Land-use planning policies and arrangements should also control inappropriate changes to existing installations (*e.g.*, new facilities or processes within the installation). They should also allow for the possibility of requiring changes to existing installations and buildings to meet current safety standards.

ROLE OF OTHER STAKEHOLDERS (E.G., COMMUNITIES/PUBLIC)

- Be aware of the risks in your community and know what to do in the event of an accident.

Members of communities near hazardous installations, and others that might be affected in the event of an accident, should make sure that they understand the risks they face and what to do in the event of an accident to mitigate possible adverse effects on health, the environment and property (*e.g.*, understand the warning signals, and what actions are appropriate). This involves reading and maintaining any information they receive, sharing this information with others in their household, and seeking additional information as appropriate.

- Participate in decision-making relating to hazardous installations.

The laws in many communities provide opportunities for members of the public to participate in decision-making related to hazardous installations, for example by commenting on proposed regulations or zoning decisions or providing input for procedures concerning licensing or sitting of specific installations. Members of the public should take advantage of these opportunities to present the perspective of the community. They should work towards ensuring that such opportunities exist, whenever appropriate, and that the public has the information necessary for effective participation.

- Co-operate with local authorities, and industry, in emergency planning and response.

Representatives of the community should take advantage of opportunities to provide input into the emergency planning process, both with respect to on-site and off-site plans. In addition, members of the public should co-operate with any tests or exercises of emergency plans, following directions and providing feedback, as appropriate. (OECD, 2003)

FUTURE RESEARCH DIRECTIONS

The need to solve the issues of major accident prevention follows from long-term development in this area, when at the beginning, stationary risk sources with the greatest content of dangerous substances were dealt with; at present, limits on dangerous substances covered by the Seveso III Directive drop and simultaneously, attention turns to the mobile risk sources, with which the number of accidents in the course of transport of dangerous substances grows. In the next phase, attention will be surely paid to risk sources unclassified in the context of Seveso III Directive that may, above all owing to their locations, represent significant societal risks.

Risk analysis is encountered in individual areas of security; from the point of view of the causes of emergencies, the whole area can be divided into two parts - Safety and Security.

In the area of "Safety", these are accidental phenomena such as accidents at work, major accidents, natural disasters, etc., and risk analysis can be used in the following areas:

- Work risks (occupational health and safety, hygiene, ergonomics)
- Process safety (prevention of major accidents)
- Fire protection
- Population protection (natural disasters)
- Environmental protection (environmental risks)

In the area of "Security" it is arbitrary abuse with the aim of harming, such as crime, sabotage, terrorism, etc. and risk analysis can be used in areas such as:

- Physical security of persons and property
- Information security (cyber security)

The risk analysis methods presented in this chapter can be used in the above areas of safety and related issues.

Future research direction could follow a new holistic approach and supporting tools according to the Process Safety Research Agenda 21st Century prioritized research topics and categorized to their main character. (De Rademaeker, 2014):

Technical safety topics:

- Hazardous phenomena, properties of substances
- Inherently safer design
- Safety technologies, protection layers, drilling
- Risk assessment, consequence analysis, NaTech
- Complex systems, resilience

Organizational safety topics:

- Process and occupational safety
- Human factors, safety management, safety culture
- Knowledge transfer, learning, standards, easy methods
- Risk management, decision making
- Complex systems, resilience

Current trends of process safety and risk related development show that research is mainly focusing on challenges such as data uncertainty, scarcity of information and complexity of process systems. It is clear that the current research trend has been in the area of inherent safety, dynamic and operational risk assessment, incorporation of human and organizational factors into risk assessment and integration of a safety protection layer (safety instrumented system) into risk assessment. Transition from traditional quantitative risk assessment (QRA) to dynamic quantitative risk assessment (DQRA) is a natural evolution. DQRA enables implementation of inherent safety principles, features most desired in hazardous processes. (Khan, 2015)

CONCLUSION

Risk assessment is defined as a comprehensive process of determining the severity and probability of an undesirable situations and deciding what safety measures will be taken to eliminate or reduce the risk to an acceptable level. A key question for risk analysis is the choice of an appropriate risk assessment method. Therefore, an overview of selected available risk analysis methods used worldwide has been included in this chapter. The aim of the text was to present basic information about the principles of risk analysis methods that can be used in various areas of safety and security, especially in process safety.

In the first part, possible approaches to quantitative risk assessment are presented; in the further part it focuses on the system of risk management in industrial establishments. The aim of major risk prevention is to operate process installations at an acceptable societal level. Risk acceptability limits can differ depending on the maturity of the state, the size of establishments and a number of other factors that must be, however, included into the analysis and assessment of risks.

That is why the benefits of carrying out the risk assessment are summarized here:

- Risk assessment brings information on the identification of hazards to possible targets of impact; the targets can be employees and installations of the establishment, the surrounding human population and the environment.

- Presents information on possible preventive means and risk reduction priorities.
- Assesses preparedness for accidents and is a source of information for emergency plan preparation.
- Contributes to the satisfaction of conditions of existing legislation.
- Can ensure advantages when making agreements with insurance companies.
- Contributes to prevention against accidents, and thus to lowering the costs of mitigating accident consequences, compensation to affected people, paying penalties for pollution of the environment.
- By suitable publishing the results of risk assessments, the awareness of employees and population will be increased, and thus also the image of the establishment will be improved.

REFERENCES

Center for Chemical Process Safety (AIChE). (1992). Guidelines for Hazard Evaluation Procedures.

ARAMIS. (2004). Accidental risk assessment methodology for industries in the framework of the SEVESO II directive, User Guide.

Bernatik, A., Zimmerman, W., Pitt, M., Strizik, M., Nevrly, V., & Zelinger, Z. (2008). Modelling accidental releases of dangerous gases into the lower troposphere from mobile sources. *Process Safety and Environmental Protection*, *86*(3B3), 198–207. doi:10.1016/j.psep.2007.12.002

Brown, D.F., Dunn, W.E., & Policastro, A.J. (2000). *A National Risk Assessment for Selected Hazardous Materials in Transportation*. Argonne National Laboratory.

CCPS. (1989). *Guidelines for Chemical Process Quantitative Risk Analysis - CPQRA*. New York: Center for Chemical Process Safety of the American Institute of Chemical Engineers.

CCPS. (2001). *Layer of Protection Analysis: Simplified Process Risk Assessment*. New York: American Institute of Chemical Engineers CCPS.

CEI. (1994). Manual – Dow's Chemical Exposure Index (1st ed.). New York: AIChE.

Christou, M.D., Struckl, M., & Biermann T. (2006). Land Use Planning Guidelines in the context of Article 12 of the Seveso II Directive 96/82/EC as amended by Directive 105/2003/EC.; JRC.

De Rademaeker, E., Suter, G., Pasman, H. J., & Fabiano, B. (2014). A review of the past, present and future of the European loss prevention and safety promotion in the process industries. *Process Safety and Environmental Protection*, *92*(4), 280–291. doi:10.1016/j.psep.2014.03.007

EFFECTS. (2018). Software EFFECTS, Gexcon. TNO. Retrieved from https://www.gexcon.com/products-services/EFFECTS/31/en

eMars. (2019). Major Accident Hazards Bureau (MAHB), eMARS Data. Retrieved from http://mahbsrv.jrc.it

FEI. (1994). Manual – Dow's Fire & Explosion Index, Hazard Classification Guide (7th ed.). Academic Press.

IRM. (2016). *About Risk Management*, Institute of risk management: Enterprise risk magazine. Retrieved from https://www.theirm.org/the-risk-profession/risk-management.aspx

ISO. (2019). ISO 31010:2019. Risk management - Risk assessment techniques.

Khamar. (2016). *CFR 1910.119 14 Elements of Process Safety Management*. Retrieved from https://commons.wikimedia.org/w/index.php?curid=49203116

Khan, F., Rathnayaka, S., & Ahmed, S. (2015). Methods and models in process safety and risk management: Past, present and future. *Process Safety and Environmental Protection*, *98*, 116–147. doi:10.1016/j.psep.2015.07.005

Khan, F. I., & Amoyette, P. R. (2003). How to make inherent safety practice a reality. *Canadian Journal of Chemical Engineering*, *81*(1), 2–16. doi:10.1002/cjce.5450810101

Kletz, T. A. (1991). *Plant Design for Safety – A User-Friendly Approach*. New York: Hemisphere.

Knight, A. M., & Kevin, W. (2010). *A Journey Not a Destination*. Moscow: Risk Management, Presentation to the RusRisk.

Lees, F. (2005). *Lees´ Loss Prevention in the Process Industries, Hazard identification, assessment and control* (S. Mannan, Ed.) (3rd ed.). Elsevier.

MZP. (2007). Methodological guideline of the Department of Environmental Risks of the Ministry of the Environment for the procedure of preparation of the document "Analysis and evaluation of major accident risks" pursuant to Act No. 59/2006 Coll., On prevention of major accidents. *Bulletin of the Ministry of the Environment*, *3*, 1–15.

OECD. (2003). OECD Guiding Principles for Chemical Accident Prevention, Preparedness and Response - Guidance for Industry (Including Management and Labor), Public Authorities, Communities, and other Stakeholders.

OSHA. (2013). *Occupational Safety and Health Standards 29 CFR 1910.119*. United States Department of Labor Occupational Safety and Health Administration. OSHA.

Rahim, E. (2016). The Benefits of Risk Management Planning. Retrieved from https://pmcenter.bellevue.edu/2016/06/19/the-benefits-of-risk-management-planning/

Salvi, O., Corden, C., Cherrier, V., Kreissig, J., Calero, J., & Mazri, Ch. (2017). Analysis and summary of Member States' reports on the implementation of Directive 96/82/EC on the control of major accident hazards involving dangerous substances, Amec Foster Wheeler, Directorate-General for Environment (European Commission), EU-VRI, Final report – Study. doi:10.2779/2037

Seveso. (2012). Council Directive 2012/18/EU of 4 July 2012 on the control of major-accident hazards involving dangerous substances, Official Journal of the European Communities (SEVESO Directive III).

Stanleigh, M. (2016). *Risk Management: The What, Why, and How, Business Improvement Architects*. Retrieved from https://bia.ca/risk-management-the-what-why-and-how/

Tixier, J., Dusserre, G., Salvi, O., & Gaston, D. (2002). Review of 62 risk analysis methodologies of industrial plants. *Journal of Loss Prevention in the Process Industries*, *15*(4), 291–303. doi:10.1016/S0950-4230(02)00008-6

TNO. (1997). Methods for the calculation of physical effects due to releases of hazardous materials (liquids and gases) (3rd ed.). Academic Press.

TNO. (2005). Guidelines for Quantitative Risk Assessment. Academic Press.

ADDITIONAL READING

Bernatik, A., & Libisova, M. (2004). Loss prevention in heavy industry: Risk assessment of large gasholders. *Journal of Loss Prevention in the Process Industries*, *17*(4), 271–278. doi:10.1016/j.jlp.2004.04.004

Sikorova, K., Bernatik, A., Lunghi, E., & Bruno, F. (2017). Lessons learned for environmental risk assessment in the framework of SEVESO Directive. *Journal of Loss Prevention in the Process Industries*, *49*, 47–60. doi:10.1016/j.jlp.2017.01.017

KEY TERMS AND DEFINITIONS

Hazard: The potential of a physical situation or a chemical to cause harm to human health, property or the environment.

Major Accident: An emergency event such as an explosion, fire or leakage of a toxic substance in an industrial plant that causes death or personal injury, property damage or environmental contamination.

Process Safety: It deals with the prevention of major accidents in technologies where dangerous chemical substances occur.

Risk: Combination of consequences and probabilities of undesirable event.

Risk Assessment: Hazard identification and consequence and probability assessment process that determines the level of risk.

Risk Management: An overall risk management process involving both risk assessment and risk reduction through technical and organizational barriers.

Safety Barriers: Risk reduction measures that may be a technical and organizational nature.

Chapter 6
Industrial Occupational Safety:
Industry 4.0 Upcoming Challenges

Susana Pinto da Costa
https://orcid.org/0000-0001-7440-8787
University of Minho, Portugal

Nélson Costa
https://orcid.org/0000-0002-9348-8038
University of Minho, Portugal

ABSTRACT

The new industrial revolution will encompass massive change. Manufacturing Companies are pursuing digitalization and trying to figure out how to implement collaborative robots, all the while trying to manage data safety and security. It is a big challenge to deal with all the needed infrastructures to handle the big data digitalization provides whilst having to account for the shielding of it. Even more so when one has to succeed at it while taking care of the workers, the sustainability of their jobs, the implementation of safe practices at work, based on the contributions of the whole, through efficient vertical communication, imbued with Safety Culture and aiming the sustainability of the Company itself. This chapter proposes to address the role of standardization in managing industry 4.0, where culture, Risk Management and Human Factors are key, and how the tools provided by these norms may contribute to nimbly balance each Company's needs.

INTRODUCTION

As the new industrial revolution sets in, competitive Companies find themselves overwhelmed with the Herculean task of diligently managing the traditional workers' safety and health, quality, environment, and the increasingly complex sustainability as they try to introduce into it the new challenges brought by industry 4.0. Starting with digitization, whereby all hard copies of product manuals, instructions, customer files and repair handbooks were progressively made available and accessible in a digital format, through digitalization, where the digitization of analog data was used for applications that simplify standard work

DOI: 10.4018/978-1-7998-3059-7.ch006

practices, all the way into digital transformation, made possible because of digitization and digitalization, which enable data to be easily accessible for use across several interfaces, platforms and devices. Digital transformation entails the devising new business applications that integrate all this digitized data and digitalized applications, and has brought artificially intelligent finite-state machines (FSM), predictive maintenance, crowdsourcing and augmented reality tools. Digital transformation business innovations are revolutionizing industry, and are aimed at saving companies' time and money. Like the demise of Blockbuster® and Kodak®, this new industrial revolution will take a toll on companies who are unable to keep pace with the digital transformation, as they are in serious danger of becoming obsolete.

The advent of digital transformation, along with automation and the development of the Internet of Things (IoT) are believed to be the catalysts of this fourth industrial revolution (Lampropoulos, Siakas, & Anastasiadis, 2019). Furthermore, their synergetic effect promotes their faster development. Nowadays, the IoT is ubiquitous; it is present in everyone's daily life in the most varied work and leisure activities, through the smart devices whose embedded systems for sensing, communicating, data collecting, storing and processing, allow us to always be connected to work, to each other, to a wide range of services, and so much more, by bridging the physical world and the digital world. *Transformation* really is one of the keywords to characterize this fourth revolution; not only because it applies to the creation of a virtual world from the transformation of the physical, but also for what it allows. As a matter of fact, these evolving technologies present great potential to companies, whereby these smart systems are able to remotely sense an array of physical dimensions, collecting and storing a bulk of data, processing these data, interpreting it and acting on it, by adapting, readjusting, delaying, stopping, accelerating, or otherwise performing according to its decision-making autonomous process. Industry 4.0 is, therefore, characterized by bringing together more traditional industrial and manufacturing practices and processes, and state-of-the-art innovative, disruptive technologies like IoT, large-scale machine-to-machine (M2M) communications and cyber-physical systems (CPSs). Undeniably, *smart* would be the second keyword that would best describe this revolution; by fostering "self-maintainability, self-optimization, self-cognition, and self -customization into the industry", Industry 4.0 envisions the transformation of the classic industries into intelligent industries (Lampropoulos et al., 2019). CPSs provide the environment for machines to process data via a wireless connected embedded system, posing as the bridge between the tangible, physical world and the intangible, digital world. So, basically, CPSs embody the duality *transformation* and *smart* manufacturing. According to Lampropoulos et al. (2019), CPSs differ from the traditional embedded systems in that they contain control algorithms and computational capacities that make up for "cybertwined services", and other physical assets and differentiated computational skills, establishing networked interactions and encompassing a bulk of methodologies that are transversal to several disciplines. These systems are projected to accept physical inputs and provide physical outputs while interacting with humans and support them on their tasks through innovative communication modalities. Hence, the safety issue related to CPSs is paramount, for the success of these systems in tied to how seamlessly and effectively this interaction with humans performs. The amount of data that will be available for analysts to decide upon through these technologies is overwhelming. But it will not necessarily mean having to spend a tremendous amount of physical space to store all the information that can be gathered simultaneously at various locations within the company, in fractions of seconds timeframes, and process it. The *Cloud* (short for Cloud computing) will manage all the storage and computing necessary in the digital world, through several computer servers and within-Cloud resources.

It is a big challenge to deal with all the needed infrastructures to handle the big data digitalization provides whilst having to account for the shielding of it. An even bigger challenge is to succeed at it while taking care of the workers, the sustainability of their jobs, the implementation of safe practices at work, based on the contributions of the whole, through efficient vertical communication, imbued with Safety Culture and aiming the sustainability of the Company itself. Standards have proven useful and efficient in the past, but it is the authors strong belief that these standards may be used to tackle transversely all the challenges that Industry 4.0 entails. Risk Management is the common way to perform risk monitoring, whereby installation, personnel, assets and outcomes are protected against the adverse consequences of risks and the severity and uncertainty of losses is diminished. It is composed of risk assessment (which, in turn, includes risk analysis) and risk control. It entails the identification of hazards and their extension (e.g., number of workers exposed), valuation of the risk and comparison with reference values and control of risks that endanger people, environment and heritage. It involves planning, coordination and control of activities, assets and resources to minimize the impact of uncertain events. Simply put, it consists of risk assessment and risk control and comprises a set of preventive measures and prevention policies that enable risk reduction or, better yet, elimination. This chapter proposes to address the role of standardization in managing industry 4.0, where Safety Culture, Risk Management and Human Factor are key, and how the tools provided by these norms may contribute to nimbly balance each Company's needs.

BACKGROUND

Occupational Safety and Hygiene are often perceived as a unique concept (or, at least, difficult to distinguish) when they are, in fact, very distinct terms. To this contributes the fact that, generally, one uses the "train" of words "Occupational Safety and Hygiene" to define a discipline which deals with the management of occupational hazards. Even though the areas of Hygiene and Safety intersect, they can be defined individually. Like this, whereas Occupational Hygiene is devoted to studying, assessing and controlling occupational environmental hazards (e.g., poor lighting, extreme temperatures, excessive noise, chemical contaminants, vibrations), aiming occupational disease prevention, Occupational Safety is focused on studying, assessing and controlling the risks of operation (e.g., obstacles, sharp edges, safety protection-lacking machinery, falling objects) in order to prevent the occurrence of accidents and other anomalous situations at work. Indeed, both disciplines may focus on the same hazard but will address it differently. For instance, considering the physical hazard *noise*, while Occupational Safety will address it as an element that may predispose workers to be more distracted or tired and hence more likely to suffer an accident (or which may prevent the hearing of relevant audible warnings thus compromising their physical integrity), Occupational Hygiene will address it as a factor that may contribute to the development of work-related hearing loss (or other occupational disease related to noise exposure).

The concern for the Health and Safety of workers had a scattered, almost inexpressive, beginning, having the first been attributed to Hippocrates, who defined saturnism as lead poisoning owing to metal extraction exposure, causing stomach contractions and hardening of the abdomen. Georgius Agricola and Georg Bauer, two German doctors who lived between the end of the sixteenth century and the beginning of the seventeenth century, left records of diseases afflicting the mineworkers, as did Paracelsus (Freitas, 2016; Rodrigues, 2006). Traditionally, Bernardo Ramazzini (1633-1714) is considered the founder of Occupational Safety and Health (OSH), as he was the first to systematically treat work-related diseases, having his work, "*De morbis artificium diatribe*" [Diseases of Workers], been published in 1700 (Freitas, 2016; Rodrigues, 2006). The foundation of the International Labour Organization (ILO) in 1919 marks a new era for the OSH, wherein concerns regarding this matter take a primary position (Rodgers, Lee, Swepston, & Van Daele, 2009). The advocacy of prevention measures (e.g., per trade, fields of activity, products manufactured and being used) begins (Miguel, 2014). It is estimated that, all over the world, and on a daily basis, 5,000 people die as a result of accidents or occupational diseases, accounting for two million workers' deaths each year. Worldwide, 270 million occupational accidents occur and 160 million occupational diseases are reported, per year (Freitas, 2016, citing the ILO). In the European Union alone, hundreds of millions of workdays are lost every year as a result of poor working conditions, which is disturbing, not only due to the human loss they represent, but also because of its average impact which is estimated of 4% on gross domestic product (Freitas, 2016; Leão, Costa, Costa, & Arezes, 2018).

MAIN FOCUS OF THE CHAPTER

According to Lima (2004), the transnationalization of economic and social relations has commanded profound changes in the organization of labour processes, imposed by the increasing productivity and costs reduction, which are not usually accompanied by improvements in working conditions. Indeed, in order to cope with the flexibility of markets and of labour itself imposed by globalization, Companies resort to atypical forms of labour, so as to adjust the quantity and availability of labour to market imperatives, which, to a certain extent, accentuate the workers' insecurities and lead to the loss of expectations, given their uncertainties in the performance of their functions, causing the decay of their ability to deal with the unexpected. Indeed, globalization has prompted the Industry 4.0 paradigm, whereby to Companies are compelled to evolve beyond mass production and thrive through more advanced Manufacturing, leading Companies to implement several technological solutions and process automation (Bragança, Costa, Castellucci, & Arezes, 2019; Colim, Costa, Cardoso, Arezes & Silva, 2019; Leão et al., 2018). This technological revolution aims to make industrial production more efficient, more flexible and of higher quality through the adoption of smart technology. These trends will predictively affect the work organization, the way it is carried out, and certainly the Health and Safety of workers (Colim et al., 2019; Rodrigues, 2006). The swift progress of information and communication technology (ICT) and internet of things (IoT) have laid the groundwork for making the adoption of new technologies by Manufacturing Companies (including automated Manufacturing systems) possible. However, for the Companies to thrive, they must keep up with technological development. For instance, even though technology has enabled Companies to gather a bulk of data regarding production and the workers themselves, the pace of the big data-supporting technologies is not keeping up with the Companies' needs, so Companies are, on the one hand, striving to make sense out of the amount of data they are collecting and, on the other hand, they are worried about keeping that data secured, by preventing cyber-attacks. One can only imagine the impact of a cyber-attack on a Company with advanced cyber-physical systems that closely work and cooperate with Human workers. And not all Companies can afford state-of-the-art technology.

Even though it is generally seen as a legal obligation, industrial Safety can actually have many benefits, including the assurance of regulatory compliance and increased productivity. There are several occupational Safety and Health systems, which ensure the correct identification of hazards and the occupational risks thorough assessment and management. It is important to systematize the process of Risk Management, by identifying the conditions of all machines and work tools, the nature of work and the production process so that any latent or manifest hazards, which may cause harm to workers, impact sustainability or otherwise compromise the integrity of the Company and its workers can be noticed and acted upon (Lima, 2004). Indeed, it has been proven that careful, rigorous and systematic risk assessment is one of the best prevention strategies to reduce the work-related accident occurrence (Colim et al., 2019; INE, 2019). The issue (which Companies have also been struggling with and sharing their concern) is that the work paradigm is undergoing a process of profound change at such a rapid pace that the efforts that have been made by Companies have not been fast enough nor efficient enough to for them cope. Automated solutions (robots) have been resorted to increase production and release the worker from more difficult and hazardous tasks, according to the ILO. Automation in Manufacturing context replaces, to some extent, cognitive and physical human labour. Since automated Manufacturing systems are perceived to be efficient, automation is often viewed as a tool that can potentially enhance Manufacturing competitiveness (Colim et al., 2019 citing Salim & Johansson, 2018). This provides many bottom-line benefits for manufacturers, but they are also inherently dangerous and it is the responsibility of manufacturers to ensure a safe production environment for workers (Colim et al., 2019).

The lack of resources available and poor education regarding the cost-benefit relationship of investing in Health and Safety prevention strategies which, in turn, cause constraints in complying with OSH regulations, have been associated with the occurrence of accidents (Leão et al., 2018, citing Giaccone, 2010).

Indeed, a study by Moktadir, Ali, Kusi-Sarpong and Shaikh (2018) focused on the Bangladeshi leather industry showed that the most pressing challenge that could hinder the implementation of Industry 4.0 was the "lack of technological infrastructure".

Sustainability may also profit greatly from Industry 4.0, as this progress may entail opportunities for manufacturers to protect and control environmental impacts using smart technology, which can be developed via ICT and IoT (Moktadir et al., 2018). But it would be reductive to think of sustainability constrained to as looking out for the environment. As Dyllick and Hockerts (2002) put it: "such a reduction misses several important criteria that firms have to satisfy if they want to become truly sustainable". In its full sense, true sustainability refers to a concomitant avocation of efficient ecological, economic and social causes in a determined time frame. In reality, corporate sustainability is developed by satisfying 6 criteria: eco-efficiency, socio-efficiency, eco-effectiveness, socio-effectiveness, sufficiency and ecological equity. So, sustainable Companies do not just manage financial capital, they also manage natural, human, and social capital. (Dyllick & Hockerts, 2002). Nowadays, at a global level and from the business point of view, sustainable development means adopting business strategies and activities that meet the needs of the enterprise and its stakeholders' while protecting, sustaining and enhancing the human and natural resources that will be needed in the future (IISD, 2019).

The influence of an Organization's culture on Safety outcomes can be as weighty as the Safety Management System. A positive culture includes mutual trust, shared perceptions and confidence. 'Safety Culture' is a subset of the overall Company culture. Even though many Companies refer to 'Safety Culture' as the tendency of their employees to comply or not with the rules or act (or not) safely, the culture and the style of management may be even more significant. Generalized, routine procedural violations, failure to comply with the Company's own Safety Management System and Management decisions that consistently prioritize production or cost before Safety are symptoms of poor cultural factors.

The key aspects of an effective culture are (HSE, 2019):

Management commitment, which produces higher levels of motivation and concern for Health and Safety throughout the organization, being the active involvement of senior management in the Health and Safety system considered as very important;

Visible management, whereby managers need to be seen and to lead by example when it comes to Health and Safety, being open to talking about Health and Safety and visibly demonstrating their commitment through their actions (a great example of commitment is to show a willingness to stop production to resolve issues). Employees will generally assume that they are expected to put commercial interests first, causing Safety initiatives or programmes to be undermined by cynicism if Management is perceived as not sincerely committed to Safety;

Good communication throughout all hierarchical levels of employees, with questions about Health and Safety being part of everyday work conversations, and having Management actively listening to what they are being told, and taking it seriously;

Active employee involvement in workshops, risk assessments, risk analyses and other, because participation in Safety is important for workers to build ownership of Safety at all levels of the hierarchy and because the Company can truly benefit from the unique knowledge that each worker has of their work;

Inspection, including non-threatening interviews with a suitable cross-section of the Company, in a sample size large enough to take account of differing views and experience.

In 2017, the ILO, the Finnish Ministry of Social Affairs and Health, the Finnish Institute of Occupational Health, the Workplace Safety and Health Institute of Singapore, the International Commission on Health at Work and the European Agency for Safety and Health at Work (EU-OSHA) undertook a major project regarding the estimation of costs and benefits of OSH. Findings showed that work-related injuries and illnesses amounted to a loss of 3.3% of the European gross domestic product (GDP), the equivalent of € 476 billion every year, while worldwide, the societal costs totalled 3.9% of GDP, which corresponded to an annual cost of about € 2,680 billion. The percentage of the societal costs were shown to vary widely between countries, particularly between western and non-western countries, depending on the characteristics of the industry, the legislative context and the incentives for prevention. The report also showed that work-related diseases accounted for 86% of all work-related deaths worldwide and 98% of deaths in the European Union (EU). Work-related injuries and illnesses caused a total of 123.3 million ADLs (disability-adjusted life years) to be lost worldwide, 7.1 million of which totalled solely in the EU. Of this, 67.8 million (3.4 million in the EU) were deaths and 55.5 million (3.7 million in the EU) were disabilities. In most European countries, work-related cancer was responsible for the largest share of the costs (€ 119.5 billion or 0.81% of EU GDP), with musculoskeletal problems taking second place (Heuvel et al., 2017).

Risk Management is part of the overall Management System. As a risk reduction and Risk Management tool, it always aims for optimization. There are several aspects that motivate a systematized Risk Management: a) the increased pressure on Companies (to perform better, to be more efficient, more sustainable); b) the need to preserve workers' health (healthier workers are more productive, miss less working days, cost less and the absence of accidents has a good impact on the Company's image; c) the need to increase protection of heritage; d) the reduction of insurance payment costs and; e) the assignment of Safety and trust to processes and procedures. According to EU-OSHA, poor occupational Safety and Health costs Companies' money, but a good level of OSH has advantages. Companies with higher occupational Safety and Health standards are more successful and more sustainable. Recent estimates show that for every euro invested in OSH there is an average return of 2.2 euro (ranging between 1.29 euro and 2.895 euro) and that the cost-benefit of enhancing occupational Safety and Health is favourable (European Commission, 2014, citing BenOSH).

Indeed, it may be an overwhelming task taking into account not only the variables that need to be considered but also the responsibility of the task, given the impact of the undesirable consequences. Hence the importance of standardized Management Systems.

The gold standard for Occupational Health and Safety Management is ISO 45001. ISO 45001:2018 specifies the OSH Management systems requirements for Companies to proactively improve their OSH performance. It is intended as a tool to help establish and improve the working environment in Health and Safety, to prevent accidents and, in many cases, to exceed current legislation, through continual improvement, systems' performance analysis and short term action plans and long term strategies for OSH. Developed by a committee of occupational Health and Safety experts, this standard follows other generic Management System approaches such as ISO 14001 and ISO 9001. It was based on earlier international standards in this area such as OHSAS 18001, the International Labour Organization's ILO-OSH Guidelines, various national standards and the ILO's international labour standards and conventions. As part of the ISO family, ISO 45001 will enable organizations to enjoy a more globally recognized technical standard (ISO, 2018).

The main changes introduced by ISO 45001 are, as follows:

- **Business Context:** Chapter 4.1, External and Internal Issues, introduces new clauses for the systematic determination and monitoring of the business context;
- **Workers and Other Stakeholders:** Chapter 4.2 introduces a focus on the needs and expectations of workers and other stakeholders and worker involvement. Systematically identifies and understands the factors that need to be managed within the Management System;
- **Risk and Opportunity Management:** Chapters 6.1.1, 6.1.2.3, 6.1.4, companies should determine, consider and, where necessary, take action to address any risks or opportunities that may impact (positively or negatively) the ability Management System to deliver desired outcomes, including improved workplace Health and Safety;
- **The commitment of leadership and Management:** Chapter 5.1, ISO 45001 emphasizes the need for the active involvement of senior management and the assumption of responsibility for the effectiveness of the Management System;
- Objectives and Performance: Enhanced focus on objectives as drivers for improvement (Chapters 6.2.1, 6.2.2) and benchmarking (Chapter 9.1.1);
- **Extended Requirements Related to:**
    - Participation and consultation (5.4);
    - Communication (7.4): More prescriptive about "mechanisms" of communication, including determining what, when and how to communicate;
    - Purchases, including outsourcing, and contractors (8.1.4).

Figure 1 shows all the ISO 45001:2018 requirements for establishing an occupational health and safety management system (OH&SMS).

Compared to its predecessors, ISO 45001 presents a new level of sophistication, raising workplace Safety to a more strategic position. More importantly, it strengthens worker participation, by focusing on the needs and expectations of workers and stakeholders, as well as their involvement in the continual process of improvement. It also shows an increased emphasis on risk-based thinking, as it is closely related to the active commitment of top management and context analysis. Organizations should determine and take appropriate action to address any risks or opportunities that could positively or negatively impact the Management System's ability to deliver intended outcomes, including improving on-site OSH. Leadership is paramount for success: top management needs to be committed and actively involved and is responsible for the effectiveness of the Management System. Top management should be a leader, not a

Figure 1. ISO 45001:2018 requirements for an occupational health and safety (OH&S) management system

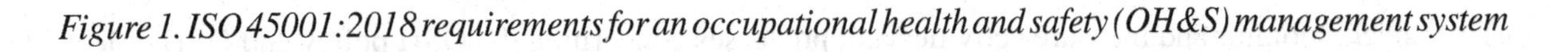

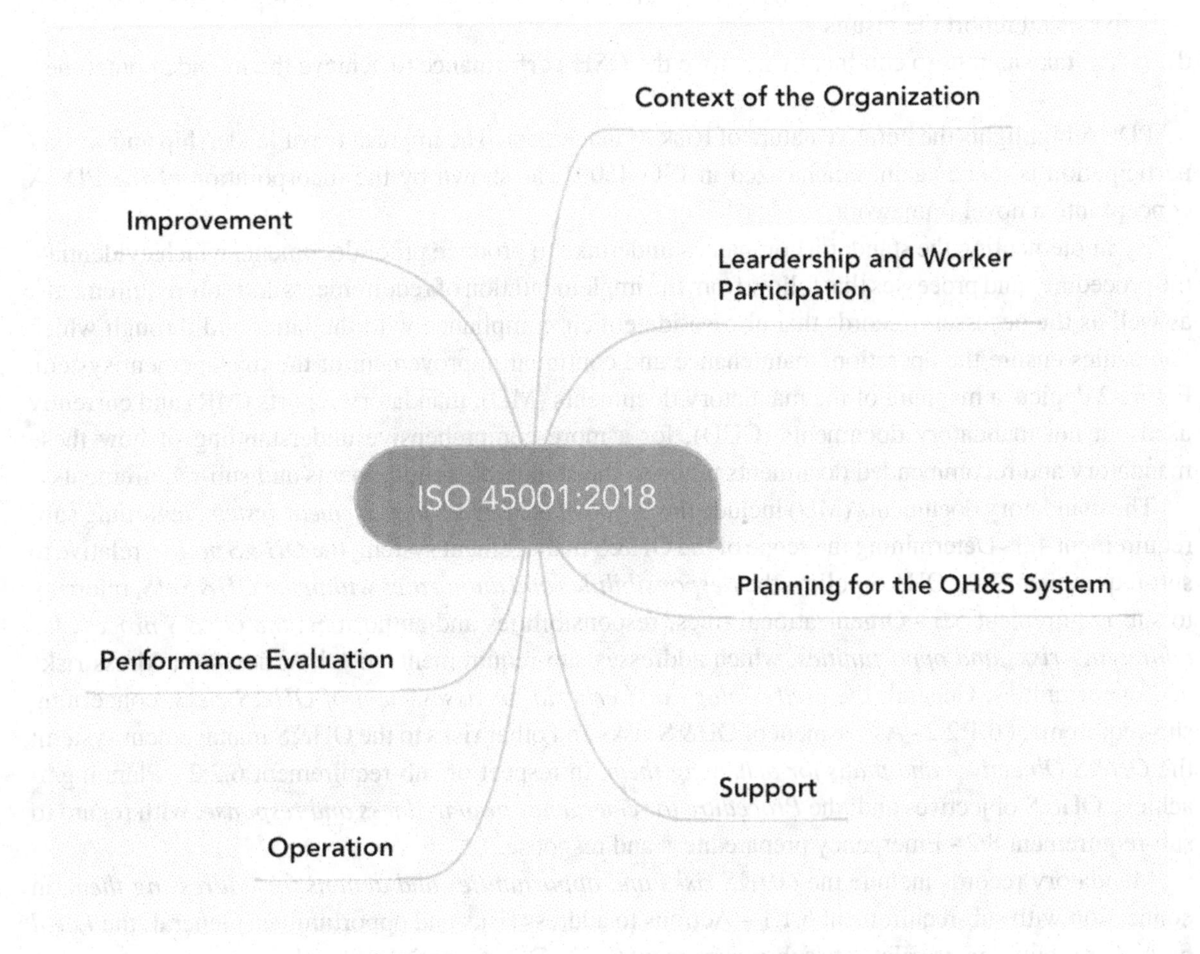

boss. There is also a special focus on monitoring outsourced activities and outside personnel concerning Safety and Health at work; they are considered part of the organization.

Surely, ISO 45001 allows users to systematically identify and understand the factors that need to be managed through the Management System. As part of the global Management System, OSH management systems provide a framework for managing OSH risks and opportunities, by setting management actions and procedures to be implemented by a Company, within a framework of definition of responsibilities, technical and financial means, aiming at the elimination or reduction of risks arising from the Company's activities, products or services. The OSH Management System approach applied in ISO 45001 is founded on the concept of Plan-Do-Check-Act (PDCA). It can be applied to the Management System and each of its elements, as follows (ISO, 2018):

a) Plan: determine and assess OSH risks, OSH opportunities and other risks and other opportunities, establish OSH objectives and processes necessary to deliver results following the organization's OSH policy;
b) Do: implement the processes as planned;

c) Check: monitor and measure activities and processes concerning the OSH policy and OSH objectives, and report the results
d) Act: take actions to continually improve the OSH performance to achieve the intended outcomes.

PDCA highlights the iterative nature of Risk Management. The importance of leadership and worker participation is, once again, emphasized in ISO 45001, as shown by the incorporation of the PDCA concept into a novel framework.

By implementing the standard, companies undertake to produce proof documents which evidentiate the procedures and processes that follow from the implementation of requirements and sub-requirements, as well as the necessary records that also validate their compliance with the latter and through which companies ensure the operation, maintenance and continual improvement of the management system. Figure 2 depicts a mapping of the mandatory documents (MD), mandatory reports (MR) and currently used but not mandatory documents (CUD), for a more comprehensive understanding of how these mandatory and recommended documents relate to the standard's requirements and sub-requirements.

The mandatory documents (MD) include the *Scope of the OH&S management system*, regarding sub-requirement 4.3 - Determining the scope of the OH&S management system; the *OH&S policy*, relative to sub-requirement 5.2 - OH&S policy; the *Responsibilities and authorities within the OH&SMS*, referring to sub-requirement 5.3 - Organizational roles, responsibilities and authorities; *the OH&S process for addressing risks and opportunities*, which addresses sub-requirement - 6.1.1 Actions to address risks and opportunities: General; the *Methodology and criteria for assessment of OH&S risks*, concerning sub-requirement 6.1.2.2 - Assessment of OH&S risks and other risks to the OH&S management system; the *OH&S Objectives and plans for achieving them*, in respect of sub-requirement 6.2.2 - Planning to achieve OH&S objectives and; the *Procedure for emergency preparedness and response*, with regard to sub-requirement 8.2 - Emergency preparedness and response.

Mandatory records include the *OH&S risks and opportunities and actions for addressing them*, in connection with sub-requirement 6.1.1 – Actions to address risks and opportunities: General; the *Legal and other requirements*, related to sub-requirement 6.1.3 - Determination of legal requirements and other requirements; the *Evidence of competence*, concerning sub-requirement 7.2 – Competence; the *Evidence of communications*, with regard to sub-requirement 7.4.1 – Communication – General; the *List of external documents*, referring to sub-requirement 7.5.3 - Control of documented information; the *Plans for responding to potential emergency situations*, relative to sub-requirement 8.2 - Emergency preparedness and response; the *Results on monitoring, measurements, analysis and performance* evaluation and the *Maintenance, calibration or verification of monitoring equipment*, both addressing sub-requirement 9.1.1 - Monitoring, measurement, analysis and performance evaluation: General; the *Compliance evaluation results*, in respect of sub-requirement 9.1.2 - Evaluation of compliance; the *Internal audit program* and *Internal audit results*, one and the other regarding sub-requirement 9.2.2 - Internal audit programme; the *Results of management review*, linked to sub-requirement 9.3 - Management review; the *Nature of incidents or nonconformities and any subsequent action taken* and the *Results of any action and corrective action, including their effectiveness*, the twain related to sub-requirement 10.2 - Incident, nonconformity and corrective action and; the *Evidence of the results of continual improvement*, which addresses sub-requirement 10.3 - Continual improvement.

In addition to the mandatory documentation, there are documents which refer to requirements that do not necessarily materialize in the documentation, but whose documentation may be useful for the company in establishing, implementing, and maintaining the OH&S management process. These documents

Figure 2. Mapping of the mandatory and common use documentation for the establishment, implementation and maintenance of an OH&S Management System according to ISO 45001:2018

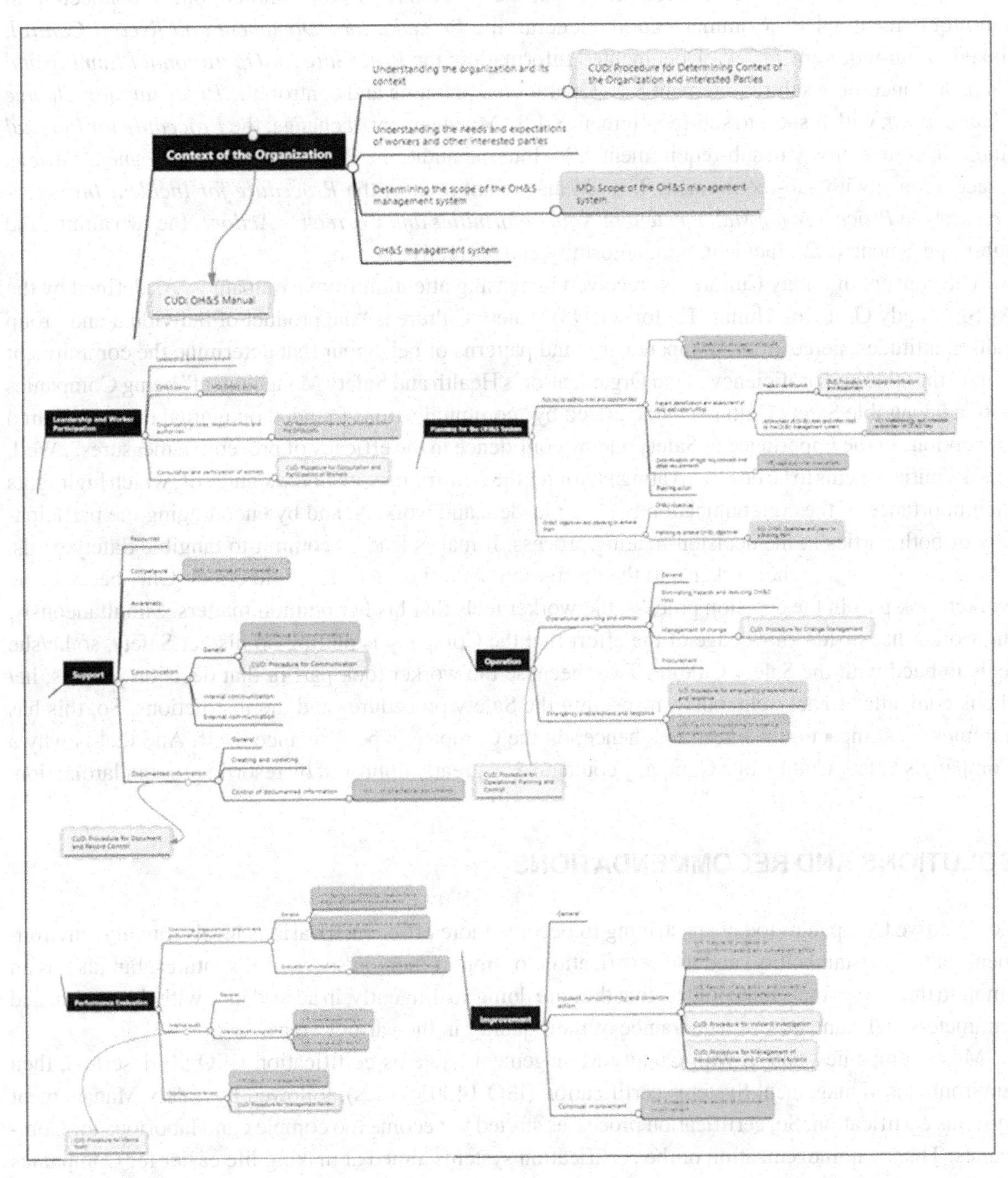

are commonly used but not mandatory (CUD) and include: The *OH&S Manual*, addressing requirement 4 – Context of the Organization; the *Procedure for Determining Context of the Organization and Interested Parties,* related to sub-requirement 4.1- Understanding the organization and its context; the *Procedure for Consultation and Participation of Workers*; referring to sub-requirement 5.4 - Consultation and participation of workers; the *Procedure for Hazard Identification and Assessment*, addressing sub-requirement 6.1.2.1 – Hazard Identification; the *Procedure for Communication*, in connection to sub-requirement 7.4.1 – Communication: General; the *Procedure for Document and Record Control*, linked to sub-requirement 7.5 – Documented Information; the *Procedure for Operational Planning and Control*, concerning sub-requirement 8.1 - Operational planning and control; the *Procedure for Change Management*, with respect to sub-requirement 8.1.3 - Management of change; the *Procedure for Internal Audit*, in conformity with sub-requirement 9.2 – Internal audit; the *Procedure for Management Review*, in accordance with sub-requirement 9.3 – Management review; the *Procedure for Incident Investigation* and the *Procedure for Management of Nonconformities and Corrective Actions,* the two aiming the sub-requirement 10.2 - Incident, nonconformity and corrective action.

The concept of Safety Culture has received increasing attention from Companies. As defined by the ACSNI Study Group on Human Factors (1993), Safety Culture is "the product of individual and group values, attitudes, perceptions, competencies, and patterns of behaviour that determine the commitment to, and the style and proficiency of, an Organization's Health and Safety Management" being Companies with a favourable Safety Culture characterized by "communications founded on mutual trust, by shared perceptions of the importance of Safety and by confidence in the efficacy of preventive measures". Well, this definition seems to be one motivating factor for the new framework of the standards, which highlights the importance of the communication between leaders and workers, and by encouraging the participation of both parties in the decision-making process. It makes leaders commit to tangible Safety goals, all the while bringing the worker into the discussion. And this has a two-fold effect. One, because the worker took part in the decision process, the worker feels that his/her opinion matters. Simultaneously, the worker has inside knowledge of the effort that the Company is putting on his/her Safety, so he/she feels imbued with the Safety Culture. Two, because the worker took part in that decision process, he/she is compelled to act righteously, respecting the Safety procedures and the instructions. So, this has a tremendous impact on workers and, hence, on the Company´s performance itself. And that is why a Company's Safety Culture of a Company could also be greatly improved by resorting to standardization.

SOLUTIONS AND RECOMMENDATIONS

Competitive Companies today are striving to become more efficient at various levels (quality, environment, Safety, sustainability) and use certification to support them in successful ventures, but also as an emblem that shows their stakeholders that they are doing so diligently, in accordance with the established parameters and standards, as a guarantee of their quality in the various areas.

Most Companies started with Quality Management Systems certification (ISO 9001 series), then Environment Management Systems certification (ISO 14001 series), followed by Safety Management Systems certification. So, certification processes started to become too complex and laborious for Companies. Then, a homogenization of the certification systems aimed at making life easier for Companies by simplifying the certification paperwork arose, with the Integrated Management System of Quality, Environment and Safety (IMS-QES).

The Integrated Management System of Quality, Environment and Safety (IMS-QES) benchmarks share the same dynamic cycle of continual improvement and the same fundamentals that contribute to the organization's competitive positioning, namely: customer orientation (whether internal or external); leadership commitment; knowledge and practical application of the legal requirements (and other);

prevention (of defects, incidents and accidents); involvement of all employees (including suppliers); ensuring proper training and education; performance regulation with procedures and work instructions; monitoring the systems to ensure their suitability; correction of deviations; considering suppliers in the internal processes; continuously improve performance. Furthermore, all three ISO 9001, ISO 14001 and ISO 45001 are based on a High-Level Structure, making them consistent with each other ISO Management System standards ISO 9001 (Quality) and ISO 14001 (Environment), both of which were updated in 2015. The high-level structure distributes clauses in 10 sections, aligned with the PDCA approach, to give logical sequence to management system requirements and proposes common text for very stable Management System requirements: scope, normative references, terms and definitions, context of the organization, leadership and worker participation, planning, support, operation, performance evaluation and improvement. The Plan-Do-Check-Act (PDCA) cycle, which is a cycle for improvement that was originated and made popular by two of modern quality control's gurus, Walter Shewhart and Edward Deming (which is why PDCA is also called Deming cycle), is used when implementing a change to improve a process. All three: Quality, Environment and Safety Management Systems approach follow the PDCA, which facilitates the process, not only of the individual implementation of the systems but also their integration. And it seems reasonable to think that PDCA will, as well, be proven useful for upcoming Industry 4.0 management challenges.

Quality, Environment and Safety Management System standards (ISO 9001, ISO 14001 and ISO 45001) are published following the standard of Annex SL, which facilitates the integration of the standards, reducing implementation costs and compliance.

The high-level structure, terms and common text set out in Annex SL of the ISO Directives were the means found to ease the application complexity, providing a foundation that facilitates the development and adoption of Management System standards, making them easier to read and interpret by the users and facilitating the integration of Management Systems in Organizations.

Even though the High-level structure cannot be changed, it begins with the generic requirements of Annex SL and becomes more specific as needed for applications in particular sectors. Thus, for different areas, it is possible to add sub-clauses and text specific to each area. And this is very advantageous! This is very valuable for companies that are striving to manage Industry 4.0 challenges because it provides them with an editable platform for creating an auditable reference, in line with one that has proven extremely useful and efficient and that fits the Company´s needs. Indeed, issues may differ (or rather evolve), but the way they are managed may be the same.

The gold standard for Risk Management is ISO 31000 (last revised in 2018), a document designed to support leaders who create and protect value in organizations by managing risks, making decisions, defining and achieving objectives and improving performance. Organizations of all types and sizes face external and internal factors and influences that make it difficult to achieve their goals. Despite ISO 31000 cannot be used for certification purposes, it provides direction for internal or external audit programmes. Organizations using it can compare their Risk Management practices with an internationally recognized benchmark, providing sound principles for effective Management and Corporate Governance.

Risk Management is, too, an iterative process, one that involves all activities associated with an Organization and includes interaction with stakeholders, and takes into account both the external and internal context of the Organization, including human behaviour and cultural factors, helping Organizations define strategies, reach goals and make informed decisions. So, Risk Management is part of governance and leadership and is critical to how the Organization is managed at all levels, for it contributes to improving all Management Systems. According to ISO 31000, Risk Management shall be based on the principles, structure, and process described in it, which may already exist (fully or partially) in the Company, however, they may need to be adapted or enhanced to manage risk in an efficient, effective and consistent manner.

The Risk Management process shall involve the systematic application of policies, procedures, practice in communication and consulting activities, setting the context and: assessing, treating, tracking, reviewing, registering and reporting risks. As principles, the standard establishes that the Risk Management structure and process are personalized and proportionate to the external and internal context of the organization, related to its objectives; and that risks may arise, change or disappear as changes in the external context and within an Organization occur; it also affirms that Risk Management must anticipate, detect, recognize and respond to these changes and events in an appropriate and timely manner and; reinforces that appropriate and timely stakeholder involvement allows their knowledge, viewpoints and perceptions to be considered, leading to better awareness and informed Risk Management itself. ISO 31000 recognizes the importance of human factors and Company Culture, stating that they significantly influence all aspects of Risk Management at every level and stage. Another important principle of this standard is that the inputs to Risk Management are based on historical and current information, but may also be based on future expectations. Risk Management is seen by the standard as transversal to all organizational activities and, because it establishes a structured and comprehensive approach to Risk Management, it contributes to consistent and comparable results. In short, ISO 31000 can (must!) be an integral part of management and decision making and integrate the structure, operations and processes of the Organization. Given its ability to be applied at several levels (strategic, operational, program or project levels) and its flexibility (there may be many applications of the Risk Management process within an Organization tailored to achieve objectives and to suit the external and internal context in which they are applied), this means that competitive Companies striving with the challenges that Industry 4.0 is bringing along should already be making use of this tool to manage Safety in an integrated manner, addressing the future challenges and dealing with the anticipated risks early on.

If Risk Management is the common way of performing risk monitoring, whereby the installation, personnel, assets and outcomes are protected against adverse consequences of risks and the severity and uncertainty of loss is decreased through planning, coordination and control of activities, assets and resources to minimize the impact of uncertain events, the new industrial revolution imposed challenges may be seen as sources of novel hazards that have to be managed. Thus, big data, cyber-physical systems and other industrial automation and all other disruptive technology that make up Industry 4.0 have to be seen in the perspective of Risk Management and dealt accordingly, by taking advantage of the proven efficient and advantageous inherited structure of standardization, thus tackling Industry 4.0 upcoming challenges by leveraging industrial Occupational Safety with standardization. This approach will enable Companies to smoothly adjust to the new work paradigm brought by Industry 4.0.

The document that assists in the implementation of ISO 31000:2009 by any public, private or community company, association, group or individual is ISO/TR 31004:2013. This standard provides a structure for organizations to follow and swiftly adapt their risk management framework to be consistent with ISO 31000 and coherent with the organization's needs, the disambiguation and clarification of the essential concepts of ISO 31000 and orientation on the risk management principles and framework particularities presented in ISO 31000:2009. Because it is not specific to any trade, industry or sector, nor is it specific to any type of risk in particular, ISO/TR 31004:2013 can be applied to all activities and to all parts of organizations. Public, private or community companies, associations, groups or individuals who are already consistent with ISO 31000:2009 but aim at the continual improvement of their framework and their management risk process as recommended in ISO 31000:2009, 4.6 and 5.6., may also benefit from the ISO 31004:2013 structured approach.

The effective tools for risk analysis can be retrieved from ISO/IEC 31010 (last revised in 2019). Although this standard does not deal specifically with Safety, it provides guidance on the selection and enforcement of risk assessment techniques in a broad array of contexts, designed to assist in the decision-making processes where there is uncertainty, to convey information about particular risks and as part of a process for managing risk. For such purpose, the standard provides summaries of those techniques and also refers to other documents where they can be found more fully described. There are several ways of classifying different risk assessment techniques. This classification is useful for nimbly grasping each technique's relative strengths and weaknesses, considering that the standard provides an overview of over 30 techniques. One shrewd classification of the risk assessment techniques is based on their applicability to each step of the risk assessment (strongly applicable, applicable and not applicable):

- Risk identification;
- Risk analysis - consequence analysis;
- Risk analysis - qualitative, semi-quantitative or quantitative probability estimation;
- Risk analysis - assessing the effectiveness of any existing controls and;
- Risk analysis - estimation the level of risk or risk assessment techniques;
- Risk evaluation.

Indeed, the standard provides a comparative table for the different techniques. It immediately informs the user of the level and type of risk analysis that is possible to entail with that particular tool. The numerous methodologies and techniques available present different levels of robustness and fragility, due to their specificity. They should not be used indiscriminately. The choice of the risk assessment technique will, hence, depend on the complexity of the problem and the methods required to analyze it, the nature and degree of uncertainty of the risk assessment (which will depend on the information at disposal and on what is necessary to satisfy the objectives), the quality of the resources (time available, level of expertise, data needs or cost), and whether or not the method can provide a quantitative output.

Considering that risk assessment is the global process of risk identification, risk analysis and risk evaluation, the methods and techniques used to perform risk assessment will, therefore, influence the risk assessment process, as will the context itself.

The "risk identification" concept refers to the process of finding, acknowledging and recording risks, with the purpose of identifying weaknesses within the organization that may hinder the accomplishment of their proposed objectives (what is already happening that should not be happening and what may happen). By acknowledging their existence, the organizations are able to develop tools to tackle them. Whenever possible, the found risks should be eliminated. Otherwise, they should be mitigated, preferably, or at least, controlled. The existing controls may include design features, people, processes and systems. The risk identification process involves finding the cause(s) and origin(s) of the risk, which may be an event, a situation or a circumstance. Examples of risk identification methods include check-lists (data evidence based method) and Hazard and Operability studies (HAZOP, which is an inductive reasoning technique). These methodologies may be aided by other supporting techniques such as Brainstorming and Delphi methodology, to improve the rigor of the risk identification process.

The "risk analysis" process is about understanding the risk. It consists of determining the consequences and their probabilities for the identified risks, considering the presence (or absence) of the existing controls and their effectiveness. The level of risk is given by the combination between the severity of the consequences and their probabilities. There are several methods which can be applied in the risk analysis, but whenever there are more complex situations, more than one technique may be required.

The "risk evaluation" concept stands for the understanding the significance of the estimated levels of risk (obtained during risk analysis) according to the risk criteria established. This evaluation, together with ethical, legal, financial and other considerations specific to the context will allow for deciding on future actions.

The different methods can still be classified as qualitative, quantitative and semi-quantitative. Qualitative methods consist of systematic examinations carried out in the workplace to identify situations that can cause harm to people. Quantitative methods involve the objective quantification of the different elements of risk, namely Probability and Severity of Consequences. When the assessment carried out by the qualitative methods becomes insufficient to achieve adequate risk valuation and the underlying complexity of quantitative methods does not justify the cost associated with its application, semi-quantitative methods are resorted to. For qualitative methods, an assessment of individual scenarios is generally used, estimating the different risks based on answering questions such as "What if…?", conducting qualitative assessments of Severity and Probability, without any associated numeric record. This type of methods is appropriate for assessing simple situations, whereby hazards can be easily identified by observation and compared with principles of good practice that exist for similar circumstances. Examples of qualitative risk assessment methods include HAZOP, Preliminary

Hazard Analysis (PHA), Failure Mode and Effect Analysis (FMEA), Check-lists, Cause-and-Effect Analysis and Delphi technique.

HAZOP is a general process of hazard identification where possible deviations from the expected or intended performance are identified. It should be noted that hazard is the term used for risk in the Safety context. Through a guideword based system, HAZOP allows for assessing the criticalities of the deviations. Originally developed as a more straightforward alternative to HAZOP, the Structured "What-if" Technique (SWIFT) is a systematic, teambased study, whereby a set of pre-made words or phrases is used to stimulate participants to identify risks. Resorting to standard 'What-if' phrase beginning, combined with the prompts, the participants investigate how a system, organization, process, worker (or any other element that makes up the specific working context) would be affected by deviations from normal operations and behaviour. The PHA is an inductive method of analysis aimed at identifying the hazards (and hazardous situations and circumstances) which may be prejudicial to an activity, facility or system. Failure modes and mechanisms, and their undesired effects may be identified by the FMEA technique. This technique derives some variants such as Design FMEA, System FMEA, Process FMEA, Service FMEA and Software FMEA. A Criticality Analysis may be performed after the FMEA, whereby the significance of each failure mode, qualitatively, semi-qualitatively, or quantitatively (FMECA) may be defined. Therefore, FMEA is technique that may, actually, provide a quantitative output. The Delphi technique is often mistaken by Brainstorming, but these are not synonyms. Brainstorming is a technique where well-informed participants discuss the identification of potential failure modes and related hazards, decision criteria and/or options for treatment, in a stimulating and encouraging free-flowing conversation environment with effective facilitation. Whereas the Delphi technique is a collaborative tool for building consensus among experts, having experts express their opinions in an individual and anonymous manner all the while having access to the other experts' considerations as the process evolves. Expert opinions may, hence, support the source and influence identification, probability and consequence estimation and risk evaluation.

For performing a Cause-and-Effect analysis, a tree structure or fishbone diagram is built based on the effect that needs to be studied (e.g., an undesired event). The effect (head of the fish) may have a several contributory categories (larger fish bones) of causal factors (smaller fish bone ramifications). Contributory factors are oftentimes identified through Brainstorming. The Check-lists are, probably, the best-known, most ubiquitous risk identification tool. Their success may be due to their simplicity and ease of use; given that they consist of lists of typical uncertainties which need to be considered. There are several pre-developed check-lists available, which may also contribute to their favoritism.

Quantitative methods assign a numerical value to the magnitude of the risk, using sophisticated calculation techniques that integrate data on the behaviour of the variables under analysis to determine the probability, by determining a pattern of regularity in the frequency of certain events and quantifying the severity, by resorting to mathematical models of consequences. These methods are particularly useful in cases of high risk or greater complexity (e.g., nuclear industry). Fault Tree Analysis (FTA), Event Tree Analysis (ETA) and Bow Tie are examples of quantitative methods.

The FTA is a technique which focuses an undesired event (top event) and, beginning with that top event, determines all the events that can lead to it, displaying the relationship between the several events graphically in a logical tree diagram. Once the fault tree is built, measures have to be considered to mitigate or eliminate potential causes and sources of those events. Resorting to inductive reasoning, ETA takes the eventual initiating events and transforms them into probabilities of possible outcomes. The Cause/Consequence Analysis is a result of the combination of Fault Tree Analysis and Event Tree Analysis, which includes the time delays. The Bow Tie analysis consists on the elaboration of a diagram that displays the pathways from the risk causes to its consequences. It can be thought of as a combination of the FTA – in what concerns to the cause of an event (which is represented by the knot of a bow tie) – and the ETA, regarding the consequences. Indeed, Bow Tie diagrams can be built based on pre-existing FTA and ETA. The resulting Bow Tie diagram presents not only the several causes of one risk and the several consequences of that same risk, but also the barriers between the causes and the risk, and the barriers between risk and consequences, which are the major object of analysis of this tool.

LOPA, which may also be called barrier analysis, is a semi-quantitative method for estimating the risks associated with an undesired event or scenario. It considers the existing controls and their effectiveness in the estimation, as it evaluates whether there are sufficient measures to control or mitigate the risk.

In this technique, a cause-consequence pair is selected. Afterwards, the layers of protection that stand between one and the other are identified. Finally, the computation of the order of magnitude is performed to assess whether the protection is adequate to reduce risk to a tolerable level.

For the application of several semi-quantitative methods, it is necessary to build hierarchy scales for Probability, Severity and Risk Index. The estimation of the numerical value of Occupational Risk Magnitude (R) is computed from the product between the estimation of the Probability of Risk (P) materialization and the expected Severity (S) of the injuries.

Risk matrices are semi-quantitative risk assessment methods par excellence (e.g., William Fine).

Check-lists and PHA are "Look-Up Methods", whereas Delphi, Brainstorming and SWIFT are "Supporting Methods". "Scenario Analysis" include FTA, ETA, Cause/Consequence Analysis and Cause-and-Effect Analysis. FMEA, FMECA and HAZOP are "Function Analysis" tools, and LOPA and Bow Tie analysis are "Controls Assessment" tools. More complex, "Statistical Methods" include Markov Analysis, Bayesian Analysis and Monte Carlo Analysis.

To prevent ambiguity and misuse of concepts (which would make it difficult to read and compare risk assessments from different sources), ISO/IEC Guide 73:2009 (last revised and confirmed in 2016) provides the definitions of generic terms related to Risk Management.

FUTURE RESEARCH DIRECTIONS

Industrial revolutions have been occurring at ever-smaller time steps. Although researchers are currently discussing all that Industry 4.0 brings and all the changes that it will encompass, it is necessary to think even further ahead, as changes are expected to be more and more frequent. It would be important in the future to create evolutionary tools that would adapt as needed, i.e., that keep up with the state-of-the-art industrial technologies, enabling managers to deal systemically with industry challenges in an agile manner. The authors are working closely with several manufacturing companies and gathering knowledge regarding their intervention needs in what relates to the strategies for companies to evolve so as to keep up with this industrial revolution. In a potential risks anticipation approach, authors will focus on pre-establishing measures to be implemented so that the risks are avoided or, if not possible, widely mitigated. A multidisciplinary team of specialists from several areas of knowledge (Ergonomics and Human Factors Engineering, Electronics and Computer Engineering, Systems Engineering and Psychologists) is to be established so that the most effective path is taken, while accounting for all the requirements (including technological, ergonomic, productive, corporate and sustainability requirements). To this end, the authors will rely on the standard references for risk management, resulting in a systematic and comprehensive approach that is believed to leave little to chance.

CONCLUSION

Industry 4.0 is an incipient concept, but major changes can be foreseen that will impact across the Industry and (also) services. From the way the work is executed to the way Companies will be managed at a macro level, the changes that this upcoming industrial revolution yields are significant. Thus, it makes sense that new ways of analyzing, assessing and managing the risks arising from this new paradigm be debated beforehand and implemented so that Companies are able, in the future, to maintain competitiveness in a sustained and sustainable manner, while benefiting from what the state-of-the-art technology brings to the environment. The Risk Management process shall involve the systematic application of policies, procedures, practice in communication and consulting activities, setting the context and: assessing, treating, tracking, reviewing, registering and reporting risks. In truth, Risk Management can be a daunting task, even in stabilized work paradigms. When, besides, one has to tackle an industrial revolution and remain competitive among one's peers, this task takes on a whole new magnified dimension. One good strategy to swiftly respond to all the upcoming demands is to rely on standardization, taking advantage of proven efficient and advantageous inherited structures, thoroughly managing each of the systems (areas) that compose the management of the Organization individually, according to a proven quality model, while a concomitant macro-management of the Company is carried out, owing to the ease of integration allowed by the inherent flexibility of the standards here presented, which allow for creating and editing an auditable reference that fits the Company´s needs.

ACKNOWLEDGEMENT

This work has been supported by FCT – Fundação para a Ciência e Tecnologia within the R&D Units Project Scope: UIDB/00319/2020.

REFERENCES

Advisory Committee on the Safety of Nuclear Installations (ACSNI). (1993). *ACSNI study group on human factors. United Kingdom.* London, United Kingdom: HM Stationery Office.

BenOSH. (n.d.). *Socio-economic costs of accidents at work and work-related ill health.* Retrieved from http://ec.europa.eu/social

Bragança, G., Siakas, K., & Anastasiadis, T. (2019). Internet of Things in the Context of Industry 4.0: An Overview. *International Journal of Entrepreneurial Knowledge, 7*(1), 4–19.

Bragança, S., Costa, E., Castellucci, I., & Arezes, P. M. (2019). A Brief Overview of the Use of Collaborative Robots in Industry 4.0: Human Role and Safety. In *Occupational and Environmental Safety and Health* (pp. 641–650). Cham: Springer. doi:10.1007/978-3-030-14730-3_68

Colim, A., Costa, S., Cardoso, A., Arezes, P., & Silva, C. (2019, July). Robots and Human Interaction in a Furniture Manufacturing Industry-Risk Assessment. In *Proceedings of the International Conference on Applied Human Factors and Ergonomics* (pp. 81-90). Springer.

Dyllick, T., & Hockerts, K. (2002). Beyond the Business Case for Corporate Sustainability. *Business Strategy and the Environment, 11*(2), 130–141. doi:10.1002/bse.323

European Commission. (2014). Communication from the Commission to the European Parliament, the Council, the European Economic and Social Committee and the Committee of the Regions on an EU Strategic Framework on Health and Safety at Work 2014–2020.

Freitas, L.C. (2016). Manual de segurança e saúde do trabalho. Lisboa, Portugal: Sílabo.

Giaccone, M. (2010). European Foundation for the Improvement of Living and Working Conditions. Health and safety at work in SMEs: Strategies for employee information and consultation. Retrieved from: https://www.eurofound.europa.eu/publications/report/2010/health-and-safety-at-work-in-smes-strategies-for-employee-information-and-consultation

Heuvel, S., Zwaan, L., Dam, L. V., Oude-Hengel, K. M., Eekhout, I., van Emmerik, M. L., . . . Wilhelm, C. (2017). *Estimating the costs of work-related accidents and ill-health: An analysis of European data sources.* European Agency for Safety and Health at Work (EU-OSHA). Retrieved from https://osha.europa.eu/pt/themes/good-osh-is-good-for-business

Health and Safety Executive (HSE). (2019). Common Topic 4: Safety Culture. Retrieved from http://www.hse.gov.uk/humanfactors/topics/common4.pdf

IISD. (2019). International Institute for Sustainable Development, Business Strategy for Sustainable Development. Retrieved from https://www.iisd.org/business/pdf/business_strategy.pdf

ILO. (2019). International Labour Organization. Retrieved from https://www.ilo.org/global/lang--en/index.htm

INE. (2019). Instituto Nacional de Estatística, Statistics Portugal. Retrieved from https://www.ine.pt/

ISO. (2009). ISO/IEC 31010:2009. Risk Assessment Techniques. International Organization for Standardization. Geneva-Switzerland.

ISO. (2009). ISO/IEC Guide 73:2009. Risk Management – Vocabulary. International Organization for Standardization. Geneva-Switzerland.

ISO. (2018). 31000:2018. Risk Management – Guidelines. ISO/TC, 262. International Organization for Standardization. Geneva, Switzerland.

ISO. (2018). 45001:2018. Occupational Health and Safety Management Systems – Requirements with Guidance for Use. International Organization for Standardization.

ISO. (2019). International Organization for Standardization. Retrieved from https://www.iso.org/iso-45001-occupational-health-and-safety.html

Lampropoulos, G., Siakas, K., & Anastasiadis, T. (2019). Internet of Things in the Context of Industry 4.0: An Overview. *International Journal of Entrepreneurial Knowledge*, *7*(1), 4–19.

Leão, C. P., Costa, S., Costa, N., & Arezes, P. (2018, October). Capturing the ups and downs of accidents' figures–the Portuguese case study. In *Proceedings of the International Conference on Human Systems Engineering and Design: Future Trends and Applications* (pp. 675-681). Springer.

Lima, T. M. (2004). Trabalho e Risco no Sector da Construção Civil em Portugal: Desafios a uma cultura de prevenção. *Oficina do CES*, *211*, 1–13.

Miguel, A. S. (2014). *Manual de Higiene e Segurança no Trabalho*. Porto, Portugal: Porto Editora.

Moktadir, M. A., Ali, S. M., Kusi-Sarpong, S., & Shaikh, M. A. A. (2018). Assessing challenges for implementing Industry 4.0: Implications for process safety and environmental protection. *Process Safety and Environmental Protection*, *117*, 730–741. doi:10.1016/j.psep.2018.04.020

Rodgers, G., Lee, E., Swepston, L., & Van Daele, J. (2009). The International Labour Organization and the quest for social justice, 1919-2009. *Book Samples*, 53. Retrieved from https://digitalcommons.ilr.cornell.edu/books/53

Rodrigues, C. (2006). *Higiene e Segurança do Trabalho – Manual Técnico do Formador*. Braga, Portugal: Nufec - Núcleo de Formação, Estudos e Consultoria.

Salim, R., & Johansson, J. (2018). Automation decisions in investment projects: A study in Swedish wood products industry. *Procedia Manufacturing*, *25*, 255–262. doi:10.1016/j.promfg.2018.06.081

ADDITIONAL READING

Aquilani, B., Piccarozzi, M., Silvestri, C., & Gatti, C. (2020). Achieving Environmental Sustainability Through Industry 4.0 Tools: The Case of the "Symbiosis" Digital Platform. In Customer Satisfaction and Sustainability Initiatives in the Fourth Industrial Revolution (pp. 37-62). Hershey, PA: IGI Global.

Avogaro, M. (2019). The Highest Skilled Workers of Industry 4.0: New Forms of Work Organization for New Professions. A Comparative Study. *E-Journal of International and Comparative Labour Studies, 8*(1).

Beard-Gunter, A., Ellis, D. G., & Found, P. A. (2019). TQM, games design and the implications of integration in Industry 4.0 systems. *International Journal of Quality and Service Sciences, 11*(2), 235–247. doi:10.1108/IJQSS-09-2018-0084

Chovancova, B., Dorocakova, M., & Malacka, V. (2018). Changes in industrial structure of GDP and stock indices also with regard to the Industry 4.0. *Business and Economic Horizons (BEH), 14*, 402-414.

Galati, F., & Bigliardi, B. (2019). Industry 4.0: Emerging themes and future research avenues using a text mining approach. *Computers in Industry, 109*, 100–113. doi:10.1016/j.compind.2019.04.018

Ivanov, D., Sethi, S., Dolgui, A., & Sokolov, B. (2018). A survey on control theory applications to operational systems, supply chain management, and Industry 4.0. *Annual Reviews in Control, 46*, 134–147. doi:10.1016/j.arcontrol.2018.10.014

Liao, Y., Deschamps, F., Loures, E. D. F. R., & Ramos, L. F. P. (2017). Past, present and future of Industry 4.0-a systematic literature review and research agenda proposal. *International Journal of Production Research, 55*(12), 3609–3629. doi:10.1080/00207543.2017.1308576

Liboni, L. B., Cezarino, L. O., Jabbour, C. J. C., Oliveira, B. G., & Stefanelli, N. O. (2019). Smart industry and the pathways to HRM 4.0: Implications for SCM. *Supply Chain Management, 24*(1), 124–146. doi:10.1108/SCM-03-2018-0150

Mishra, D., Roy, R. B., Dutta, S., Pal, S. K., & Chakravarty, D. (2018). A review on sensor based monitoring and control of friction stir welding process and a roadmap to Industry 4.0. *Journal of Manufacturing Processes, 36*, 373–397. doi:10.1016/j.jmapro.2018.10.016

Nascimento, D. L. M., Alencastro, V., Quelhas, O. L. G., Caiado, R. G. G., Garza-Reyes, J. A., Rocha-Lona, L., & Tortorella, G. (2019). Exploring Industry 4.0 technologies to enable circular economy practices in a manufacturing context: A business model proposal. *Journal of Manufacturing Technology Management, 30*(3), 607–627. doi:10.1108/JMTM-03-2018-0071

KEY TERMS AND DEFINITIONS

Human Factors: A science field that aims at understanding the interactions between humans and other elements of a system, not only by addressing the most current research challenges with a multidisciplinary approach, but also by applying theory, principles, data, modelling, and other methods to design, in order to optimize both human well-being and overall system performance.

Integrated Management Systems: An organization-wide system that integrates all the systems and processes of an Organization into a complete, efficient and effective structure, enabling the Organization to function as a single unit with unified objectives, optimizing resources for implementation, maintenance and audits to each Management system.

Manufacturing: The economic activity that uses a technique, generally dominated by the presence of machinery or machinery, to transform raw materials into production and consumption products.

Occupational Safety: The discipline that focuses on the study, assessment and control of the risks of operation (e.g., obstacles, sharp edges, safety protection-lacking machinery, falling objects) to prevent the occurrence of accidents and other anomalous situations at work.

Risk Management: The overall process of risk avoidance, mitigation and / or control, encompassing risk assessment, which in turn encompasses risk analysis.

Safety Culture: A subset of the overall Company culture, refers to "the product of individual and group values, attitudes, perceptions, competencies, and patterns of behaviour that determine the commitment to, and the style and proficiency of, an Organization's health and safety management" (ACSNI Study Group on Human Factors, 1993).

Security: The vigilance and protection against possible attacks or hackings on an institution or personality.

Sustainability: The quality of being able to exist over some time, enabled by the satisfaction of 6 criteria: eco-efficiency, socio-efficiency, eco-effectiveness, socio-effectiveness, sufficiency and ecological equity; concomitant avocation of efficient ecological, economic and social causes in a determined time frame.

Standards: Documents that provide requirements, specifications, guidelines or characteristics that can be used consistently to ensure that materials, products, processes and services are fit for their purpose (ISO).

Transnationalization (or transnationalism): explains how interconnected networks of social organization (individually or in society) can influence each other, either in the political or economic sphere or otherwise.

Chapter 7
Effects of Human Factors in Process Safety:
Analyses of Chemical Industrial Case Studies

Gargi Bhattacharjee
https://orcid.org/0000-0002-5592-0694
University of Calcutta, India & Prasanta Chandra Mahalanobis Mahavidyalaya, India

Sudip Kumar Das
https://orcid.org/0000-0002-9177-8381
University of Calcutta, India

ABSTRACT

Accidents and near-miss accidents in chemical industries are widespread. Most of the incidents occurred due to combinations of organizational and human factors. To identify the causes for an incident of an accident analysis is needed, because it reveals the possible causes behind the accidents. Accident analysis shows the human and organizational factors that support learning from the events. Literature review shows that human error plays an important role of accidents in process industries. The chapter discusses some case studies which are received very little media publicity and also no proper assessment. At first reports on the incidents were collected from newspapers and then the place was visited to conduct an interview with local people and present and past workers with the help of the PESO (M/S Petroleum and Explosive Safety Organization, Eastern Region, Govt. of India).

INTRODUCTION

Among industrial sectors chemical industry is one of the important and biggest industrial sectors. It is also an important source of employment sector. Small scale chemical industries manufacture products often run as a batch operation and large-scale production is mostly run as continuous operation. In both cases

DOI: 10.4018/978-1-7998-3059-7.ch007

there may be various kinds of technical systems like chemical reactors (which often have to withstand high or low temperatures and pressures), separation devices (distillation, filtration, etc.), and fluid systems for liquids and gases (pumps, valves, tanks), etc. Large or small scale both chemical industries handled harmful and hazardous chemical for large production. The complex nature of chemical plants increases the health risks of workers. As new industries develop with new technologies then automatically risk arises on human health. Each year all over the world a large number of workers from chemical industries are suffered from work related accidents and illness. Workers suffer from acute effects such as poisoning suffocation; long-term affects respiratory diseases, occupational cancers and health effects that can be both acute and long-term, such as skin diseases, allergies, reproductive problems and birth defects.

There are many types of chemical plants such as refineries, power plants, wastewater plants, biochemical plants etc. Recent years due to rapid development most of plant use more complex technology which create more unsafe process conditions. In general, as the chemical process plant handled dangerous substances and their plant complexity, they can be characterized by a high accident potential. The number of accidents occurred in chemical industries due to different reasons, but one of them is the worker ignorance, as they are not appropriately trained. Sometimes they have a lack of basic knowledge, and also skills. Plant operators are monitor and control the plant for proper functions, so they have responsibility to save equipment and workers health and safety. Due too much stress or pressure some time they play wrong performance. To identify the actual causes for an incident an accident analysis needed, because it informs the unsafe work procedure, analyze how human and organizations failures create unsafe situation which leads accidents that support learning from the events.

Accident analysis carried out to discover the real causes of an accident and also recommend the preventive measure. It also gives useful information to plant personnel in future how to prevent plant operations from accidents. For improvement of process plant safety, it is very important to analyses the past accidental case studies. Because it not only gives the useful information, it also increases knowledge of workers and inform the safe practices. In recent years many research works identified the causes of accidents and lessons learn. Accidental analysis showed that not only one single cause responsible for major accidents but most have multiple and interrelated factors. If accidents are properly reported or investigated then the causes are identified properly and recommended some preventive measures for control future similar accidents. At first collecting the facts, then analyses the gathering information, identify the main causes and then implement the plan of action for future prevention. Many researchers research on Process safety and have analyzed incident or accidents and find out new risk situations. This chapter focused on how to analyze human factors. According to Rassmussen et al. (1990) “Human Factor” plays a part of operators, designers or managerial role which leads error and also worked on the psychological factors and industrial analysis to develop human error taxonomies. Literature review shows that various works has been done on human factor analysis to control hazardous situations and also save life and property. PHA, QRA, HAZOP, HRA, WHAT-IF, FTA, FMEA, SWOT, JHA etc. analysis gain information about how and why situation become uncontrollable, wrong approaches, real unsafe situation, inadequate risk control management and also prevent similar adverse conditions, increase safety knowledge of workers.

PHA or Process Hazard Analysis used for finding, evaluating and controlling unsafe situations of process. Conducting PHA, also need a PHA team included safety specialist, engineers, supervisors, workers who have sufficient safety knowledge about safe plant operation. PHA is one of the most important elements of PSM (Process Safety Management). QRA or Quantitative Risk Assessment used for measuring total safety management of a chemical process industry. QRA identify the risk zone of storage area or handling process of chemical products. This process also recommends some preventable measures for save human live, environment and property. QRA widely used as for assessing oil and gas installations like refineries, tanks, pipelines, terminals, etc. HAZOP is a review process. HAZOP team finds out the possible causes and consequences of the deviation as well as existing safeguards protecting against the deviation. If any unsafe situations are detected, then they recommend some risk prevention actions. HRA or Human Reliability Analysis measures human reliability and the probability that workers perform the full task correctly in a fixed time. HRA is a group of many different methods like THERP, CREAM used as commonly. HRA is a method under of QRA. Swain and Guttmann (1983) present methods, models and estimated human error probabilities in nuclear power plants. According to Hollnagel (2005) HRA maximum used on its own both ways to assess the risk from human error and as a way to reduce system vulnerability. Kirwan (1994) stated that HRA has three principal functions as identifying what error can occur (Human Error Identification), deciding how likely the errors occur (Human Error Quantification) and enhancing human reliability by reducing error likelihood (Human Error Reduction). FTA is also known as Fault Tree Analysis is a software tool where analysis starts with an unwanted event and identifies the root causes. The result is represented as a logical diagram. FMEA or Failure Mode and Effect Analysis is used to eliminate faults in a system before they occur. Results are displayed in a table form like causes, effects, frequency, severity, probability and recommendations. WHAT-IF, SWOT and JHA analysis are briefly discussed in case study analysis section of this chapter.

The present chapter discusses the safety issues or risk factors associated with human factors. It informs about unsafe behavior of wrong task performance of workers and management. Three accident case studies of chemical industries are discussed in the present chapter where human factors play as a primary cause, but it is received very little media publicity and also no proper assessment. This chapter is useful for academician, practitioners, safety personnel, managers and workers who directly or indirectly associate with process safety management.

BACKGROUND

Process safety or workplace safety is not a new term. Research work started on workplace safety in the early 1980s. Zohar's (1980) study on safety climate received scant attention on organizational and psychological factors which are effects on safety climate. Nowadays, most companies/industries give priority to safety first. So, it is necessary for every process plant to include safety programme for their workers. For better safety performance, it is essential to investigate and analyze past accidents. It not only gives awareness about the causes behind accidents but also identifies, evaluate, and reduce occupational risks.

Most of the accidental factors related to:

1. **Task:** Ergonomics, process, materials, workers, safety work procedures, long hour of work
2. **Material:** Hazardous substances, faulty design of machine
3. **Environment**: Weather, temperature, improper light, personal protective equipment, noise, improper ventilation, workspace
4. **Human Factors:** Physical and mental capability like age, experience, training, health absenteeism, fatigue, frustration, alcohol addiction, carelessness, inattentive, ignorance
5. **Management Error:** Safety policies, safety knowledge, preventive measures

Industrial accidents directly effect on both public and environment. Accidents may damage the industrial economy and also results in injuries, pain, or even death of workers.

MAIN FOCUS OF THE CHAPTER

Industries are the backbone of economy to a nation. It produces goods from raw materials and engaged in making particular product or providing a particular service. There are many types of industries like agricultural industries, manufacturing industries, construction industries, craft industries, mining industries, chemical industries etc. Each industry has its own hazards. Among them chemical industries is one of the more accidents prone sectors.

Why Chemical Industries are More Unsafe

Chemical process plants handled and manufacture complex and hazardous chemicals in a large scale. Different chemical plants have different type of work procedure. All plant structure also different from designing. Some have fluid system and chemical reactor system, some have storage pressurized tank. Due to development of industrialization, the chemical plant structure change for continuous growth of complex materials with complex process equipment, technology in manufacturing process. Chemical process plant handled dangerous substances, so it is a high accident potential sector. Chemical plants need more attention because it stored large quantity of flammable and toxic materials. Hazardous chemicals have acute effects like poisoning, suffocation, long time effects like respiratory diseases and occupational cancer, skin injuries (skin burns, skin cancer), allergies, reproductive problem, birth defects etc. Chemical inhalation or irritant chemical substances may result asthma, chronic obstructive pulmonary diseases and chronic bronchitis. Exposure to diesel motor exhaust may cause of lung cancer. Certain substance from petro-chemical industries such as toluene and xylene effects on reproductive systems. Carcinogens that may be present in benzene, 1,3 butadiene, ethylene oxide etc.

Survey work done by Khan and Abbasi (1999) reveals that 1744 significant accidents have occurred in chemical industries during the period 1928 - 1997. Analysis of some major accidents and disasters in last few decades have given some valuable lessons like Texas city disaster in 1947, the worst ever accident in the chemical process industries involving toxic release occurred at Bhopal in 1984, offshore accident occurred in Piper Alpha in 1998, Flixborough disaster was an explosion at a chemical plant in 1974, Chernobyl disaster was a nuclear accident that occurred on 1986, Texaco refinery fire in 1994, Vishakhapatnam refinery in 1979, etc. A summary of major catastrophic accidents of chemical industries is presented in Table 1.

Indian industry benefited credibility from the past in the mind of the common man who saw in its hope for better tomorrow. However, the Bhopal tragedy shook this credibility, thereby set bay in motion a process of revolution and rethinking. Industries were established in India as early as in the 1850s. Large-scale industrialization, however, started after the country became independent in 1947. Today India is one among the top 10 most industrialized nations in the world. Planning commission (1997) of India revealed the information that Indian labour force had grown to 397.2 million, which is nearly 42% of the total population (951.2 million) of the country. Due to improvement of Industrialization in India the more and more complex industries established and work with hazardous materials for growth of production. So workers handled hazardous toxic and harmful materials which increase the number

Table 1. Major accidents in chemical industries

Year	Location	Chemical	Event	Deaths/Injured
1926	St. Auban, France	Chlorine	Toxic release	19/105
1928	Homburg, Germany	Phosgen	Toxic release	10/50
1939	Zarnesti, Romania	Chlorine	Toxic release	60
1943	Ludigshafen, Germany	Butadiene	Explosion	57/37
1944	Clevelland, OII	LNG	Fire and explosion	128/300
1947	Rauma, Findland	Chlorine	Toxic release	19/200
1947	Texas City, TX, USA	Ammonium nitrate	Explosion	552/3000
1948	Ludigshafen, Germany	Dimethyl ether	Explosion	245/2500
1950	Poza Rica, Mexico	Hydrogen sulphide	Toxic release	22/320
1954	Bitburg, Germany	Kerosine	Fire	32/16
1959	Meldrin, GA, USA	LPG	Explosion	23/78
1959	Ube, Japan	Ammonia plant	Explosion	11/40
1960	Kingsport, TN, USA	Aniline plane	Explosion	15/55
1962	New Berlin, TY, USA	LPG	Explosion	10/75
1965	Natchitoches, LA, USA	Natural gas	Explosion	17/56
1966	Freyzin, France	Propane	Fire and explosion	18/83
1968	Lievin, France	Ammonia	Toxic release	5/35
1969	Repcelak, Hungary	Carbon dioxide	Explosion	9/23
1970	Philadelphia, Panama	Catalytic cracker	Explosion	7/43
1972	Rio de Janerio, Brazil	Butane	Explosion	37/53
1972	Netherland	Hydrogen	Explosion	4/40
1972	West Virginia, USA	Gas	Explosion	21/20
1973	Kingman, AZ, USA	Propane	Fire	13/89
1974	Czechoslovakia	Ethylene	Explosion	14/79
1974	Flixborough, UK	Cyloheane	Explosion	28/76
1974	Madras, India	Potassium Sol.	Hot release	9/15
1975	Beek, Netherland	Propylene	Explosion	14/108
1976	Seveso, Italhy	TCDD	Toxic release	1/>300
1977	Gujarat, India	Hydrogen	Explosion	5/35
1978	Santa Cruz, Mexico	Propylene	Fire	52/88
1984	Sao Paulo, Brazil	Gasoline	Fire and Toxic release	508/221
1984	Mexico City, Mexico	LPG	Fire and explosion	550/23
1984	Bhopal, India	MIC	Toxic release	2500/100,000
1985	Brazil	Ammonia	Toxic release	>5000 evacuated
1985	Priola, Spain	Ethylene	Explosion	23/11
1985	Algerais, Spain	Naphtha	Explosion and Fire	18/56
1986	Basel, Switzerland	Fungicide	Toxic release	Severe damage to ecosystem
1988	Piper Alpha, UK	Hydrogen	Explosion	167/55
1988	Maharashtra, India	Naphtha	Fire	25/23
1989	Antwerp, Belgium	Aldehyde	Explosion	32/11
1989	USSR	Ammonia	Explosion and Fire	7/57
1989	Pasadena, TX, USA	Ethylene	Explosion	23/314
1989	Phillips, USA	Ethylene	Explosion	23/130
1990	Thane, India	Hydrocarbon	Fire and Explosion	35/10
1992	Sodegraura, Japan	Hydrogen	Explosion	10/7
1993	Panipat, India	Ammonia	Explosion and toxic release	3/25
1994	Dronka, Egypt	Fuel	Fire	410
1995	Gujarat, India	Natural gas	Fire	-
1996	Bombay, India	Hydrocarbon	Fire	2/45
1997	Chennai, India	LPG	Fire	3 /4
1997	Chennai, India	Molten metal	Explosion	2/5
1998	Hahnville, Louisiana, USA	Nitrogen	Asphyxiated by nitrogen	1/1
1998	New Jersey, USA	Yellow 96	Runaway reaction	0/7 2 severely burnt

continued on following page

Table 1. Continued

Year	Location	Chemical	Event	Deaths/Injured
1999	Martinez, California, USA	Naphtha	Fire	4
2001	Hangzhou, China P Rep	Aniline	Gas Leakage	0/700
2002	Missouri, USA	Chlorine	Toxic release	63 people seek medical evaluation
2002	Shenxian (Shandong province) China P Rep	Ammonia	Gas Leakage	13/11
2003	Kentueky, USA	Aquo ammonia	Explosion	1/26 evacuated
2003	Miami Township, Ohio, USA	Nitric oxide	Leakage & Explosion	Severe damage the plant
2004	Skikda, Algeria	LNG	Explosion	27/74
2004	Near Santiago, Chile	Acidic Chlorohydroquenone	Chemical spill	0/2 500 affected
2004	Songjia (Panjing, Liaoning), China P. Rep.	Hydrogen sulphide	Gas Leakage	0/120
2004	Oigihar (Heilongjiang) China P. Rep	Chlorine	Gas Leakage	0/134
2004	Fuzhou (Fujian province) China P. Rep.	Phosgen	Gas Leakage	1/300
2004	Houston, USA	Raw material for polyethylene wax	Explosion	2 firefighters injured during emergency response
2005	Perth Ambay, NJ, USA	Liquid waste in acetylene generator system	Explosion	3/11
2005	Juba, Sudan	Store ammunition depot	Explosion	31/150 4,000 homeless
2005	Hauinan (Jiangsu), China P. Rep.	Chlorine	Gas Leakage	27/285
2005	Texas City, USA	Hydrocarbon liquids and vapours	Sudden release of flammable liquids and vapours	15/170
2005	Lahore, Pakistan	Ammonia	Explosion	27/8

of occupational injuries and deaths. According to the International Labour Organisation informed that in India more than 403,000 workers died per year due to work related problems or due to unsafe working conditions. It also informed that every day more than 1,000 and per hour 46 workers died due to occupational cancers or diseases. There is very little awareness about workplace hazards due to lack of access to information, education, and training, working conditions effects on the body as well as the mind of workers. A various study indicates that in many factories and services in developing and developed countries, workers are exposed to a different physical and psychological problem. In present situation lack of safety awareness, safety education, safety training, implementation programmes are very poor.

Process Safety Information

Government has emerged several rules and regulations for reducing the impacts of fire and explosions occurred due to chemical releases in chemical process plant. These are developed based on previous incidents that occur in process industries and their impact on process industries. Hence, the incident records are necessary and essential to learn a lesson from the mistakes. Only few accident databases are available in developed nations like MARSH (Major Accidents Reporting system), MHIDAS (Major Hazard Incident Data Service), HSEES (Hazardous Substances Emergency Events Surveillance), ARIP (Accidental Release Information Program), ERNS (Emergency Response Notification System), ARIA

(Analysis, Research and Information on Accidents) etc. but unfortunately in India no such databases exist. Analysis of previous incidents as an essential aspect to reduce the frequency and severity of the future incident and also provides more accurate information which evaluates the industry's safety progress. Ministry of Labour and Employment of India tried to improve occupationally save the workers from mines, factories and ports. So, obtain to this principles incorporated in the constitution of India in the area of occupational health and safety in the above three categories implement laws and regulations which cover the areas of industrial safety, health and hygiene, industrial medicine, workers training and safe procedure of productivity, communication, primary accident hazards control, computer center.

Analysis of Incident

Investigation and analyzing any incident or accident are essential for process safety. Analyses are established both direct and indirect causes. Analyses results also measure the frequency and seriousness of accidents. Analysis method should be useful in

- Exposure of hazardous materials which will be susceptible to risk migration;
- Identify the high-risk operation;
- Analysis technique must be of use new safe process; at first, it analyzes new technology in a process and identifies hazards and changes and recommends safe operations.

So, analyses of incidents give more accurate and relevant information on the industry's safety progress in the future. For any general accident analysis, at first investigate the accidental data – gathering information, reconstructing events establishing what happens, then analysis the incident step by step. Various type of hazard analysis techniques used for analyzing accidents or incidents in recent years. Table 2 represents some accident analysis or accident investigation methods. Each methods has own strength and weaknesses and each has different application. Sometimes combination of methods used to analyses of complex accidents.

According to Bird and German (1992), Reason (1997) and Glendon et al. (2006) analysis of an accident is very important because it identify the possible unsafe working conditions like improper work planning of the organization, insufficient training, lack of attention, deficient maintenance of technical equipment, absent of job descriptions, wrong choice of materials or non-use of safety equipment and lack of attraction of employees. The qualitative, quantitative hazards, and risk techniques are commonly used are pointed below. Qualitative analysis is descriptive measurement and depends upon the decision making skills of organizational personnel to determine the impact of risk probability, like health and safety risk. Quantitative analysis is used to analyses accidents case studies in a more developed risk model, depending on the quality of the data inputted. This assessment can point out more useful information that can help safety risk assessment.

Qualitative Techniques

These methods are usually used to identify the potential accident scenarios and evaluate in detail to make a reasonable judgment of risk. These techniques are,

1. Preliminary Hazard Analysis (Pre HA/PrHA/PHA)

Table 2. Some incident/accident investigation/analysis methods

Sl. No.	Name of Incident/Accident Investigation/Analysis Methods
1.	ARCA – APOLLO Root Cause Analysis
2.	Black Bow Ties
3.	DORI – Defining Operational Readiness To Investigate
4.	ECFA – Events and Causal Analysis (Charting) and ECFA+ - Events and Conditional Factors Analysis
5.	Fishbone diagram
6.	HERA – Human Error Repository and Analysis System
7.	HERA-JANUS – Human Error Reduction in ATM (Air Traffic Management)
8.	HFACS – The Human Factors Analysis and Classification System
9.	HFAT – Human Factors Analysis Tools
10	HFIT – Human Factors Investigation Tool
11.	HSYS – Human System Interactions and allied industries (and others)
12.	ICAM – Incident Cause Analysis Method
13.	MEDA – Maintenance Error Decision Aid
14.	MORT – Management Oversight and Risk Tree
15.	PEAT – Procedural Event Analysis Tool
16.	PRISMA – Prevention and Recovery Information System for Monitoring and Analysis
17.	SCAT® – Systematic Cause Analysis Technique
18.	SOL – Safety through Organisational Learning
19.	SOURCE™ – Seeking Out the Underlying Root Causes of Events
20.	Storybuilder
21.	TapRooT®
22.	(Kelvin) Top-Set®
23.	STEP – Sequentially Timed Events Plotting
24.	TRACEr – Technique for Retrospective and Predictive Analysis of Cognitive Errors
25.	Tripod Beta
26.	WBA – Why-Because Analysis
27.	5 Whys
28.	Why Tree
29.	HRA – Human Reliability Analysis
30.	RTA – Root Cause Analysis
31.	CTA – Cognitive Task Analysis
32.	PHA – Process Hazard Analysis
33.	MCA – Maximum Credible Accident Analysis
34.	SCAP – Safety Credible Accident Probabilistic fault tree Analysis
35.	JHA – Job Hazard Analysis
36.	SWOT – Strength Weakness Opportunities Threats analysis
37.	FTA – Fault Tree Analysis
38.	FMEA – Failure Modes and Effects Analysis
39.	HAZOP - Hazard and Operability Analysis
40.	Cause-consequence analysis
41.	ETA - Event Tree Analysis
42.	LOPA - The Layer of Protection Analysis

2. Checklist
3. What-If Analysis
4. What-If/Checklist Analysis
5. Hazard and Operability (HAZOP) Analysis
6. Failure Modes and Effects Analysis (FMEA)

Quantitative Techniques

These methods judged the risk of judgment by provides more detailed and statistical evaluations of the risk of a specific accident scenario. These techniques are,

1. The Layer of Protection Analysis (LOPA)
2. Dow Fire and Explosion Index (F&EI)
3. Dow Chemical Exposure Index (DCEI)
4. Fault Tree Analysis (FTA)
5. Event Tree Analysis (ETA)
6. Cause-consequence analysis
7. Human Reliability Analysis (HRA)

The Human Factor in the Chemical Industry

In recent years the chemical process industries have achieved in optimization of plant design and also inherently safe plant. However, in chemical process plant small and sometimes major accidents do occur and recur, many of them have quite similar characteristics. Joschek (1981) and Bea (1996) showed that 80% accidents occurred in chemical and petrochemical have due to human failure as a primary cause, and these are approximately lost 30 million working days. Human error was mostly responsible in the 514 industrial incidents of potential major consequence in Canada as reported by Lees and Laundry (1989). Salminen and Tallberg (1996) reported that in Finland, human error is a prime factor of 84-94% of 300 hazardous material events in the workplace resulting in fatal and serious injuries. Chung and Jefferson (1998) observed that the chemical industry as a whole does not learn from past accidents. Most human errors are the result of either moment's forgetfulness/aberration, or the errors of judgment. Inadequate training or instruction or inadequate supervision are one of the main causes as reported by Kletz (2001). Löwe and Kariuki (2004) reported, based on a survey conducted by Technische Universität, Berlin, that 64% of the total in Germany, industrial incidents are due to human failure. Gangopadhyay and his subsequent publications (Gangopadhyay and Das, 2005, 2007 a, b, c, d, 2008 a, b, 2009 a, b) indicated that need of accidental analysis and also to find the root causes for the safety application in chemical process industries. Zhang and Zheng (2012) demonstrated that the chemical accidents in hazardous installation in China from the year 2006 – 2010 are due to human errors, i.e., exclusively 46.8% or partially (human and equipment) is 18.3% out of 1632 hazardous chemical accidents. Kidam and Hurme (2013) analyzed 549 accident cases from the Japanese Failure Knowledge Database (FKD, 2011). They observed about 66% (364 out of 549) of the cases are chemical process industry related. They observed that the accident contributors are the human and organizational failures (20%), contamination (12%), heat transfer (12%), flow-related problems (11%), reaction (9%), layout (7%), fabrication construction and installation (7%), corrosion (6%), construction materials (6%), static electricity (3%), mechanical failure (2%), external factor (2%), vibration (1%), erosion (1%), utility-related problems (1%). Qi et al. (2012) observed that the accidents occurred on a regular basis in the chemical process industries due to insufficient understanding to recognize to the best actions and lack of drive for safety improvement in the organization. They also reported the following issues for process industries,

- The organization has no memory – Similar incidents occurred in a different part of the world, and it is due to organizational failure as they failed to develop "organizational memory". Kletz (1993) and Mannan et al. (2009). The organizations failed to progress and execute effective learning from past incidents.
- Inadequate observation to main factor - Many factors contribute the safe operation of a plant, but the failure of maintenance of such factors like cost reduction, decrease training, the communication gap between management and workers, lack of supervision, workers fatigue, inadequate instrumentation, and improper equipment maintenance. Threat due to terrorism in chemical process industries has a new concern according to Bajpai and Gupta (2005), Reniers and Amyotte (2012).
- The growing complications of process operations and lack of communications – Modern chemical process industries are very complex and complicated. Poor management, lack of awareness, improper training may lead to an accident.
- Risk is also increased because of large population density in the off-site of the plant.

Human Factors and Human Errors

According to Health and Safety Executive (2005), the human factor is defined as "Environmental and organizational and job factors, system design, task attributes, and human characteristics that influence behaviour and affect health and safety".

The human error is defined as "Any human action or lack thereof that exceeds or fails to achieve some limit of acceptability, where limits of human performance are defined by the system" according to Lorenzo, (1990).

Bea et al., (1996) describe human error as a subset of the human factor, and the meaning of human error is unsafe acts made by worker in workplace. Human factors have received scant attention for various reasons in the process industries and are as follows:

- Lack of awareness
- Lack of understanding
- Lack of need
- Misunderstanding of human factors
- Lack of integration
- Lack of approaches to remediation of some human factors issues
- Lack of qualified analysts
- Lack of motivation

May and Deckker (2009) stated that the most of the human failures occurred due to lack of knowledge, underestimation of influence, unawareness, insufficient attention and negligence, poor memory or laps of memory, depending on others insufficient control, fault situation, unspecific definition of responsibilities etc. These are all cognitive factors of workers which reflect on the behavior of the workers. There are three related elements which determine the human factor and are

1. Job characteristics
2. The nature or behavior of the worker
3. Organizational systems and culture which may or may not include a safety management system.

Understanding Human Factors

Research shows that 50% – 90% of industrial incidents can be attributed to human error. So the attention must go to the unsafe acts by human as the most probable cause of frequent accidents happening in industries. Hence, human factors are an essential factor for accident prevention and are also necessary to control the occupational health of the workers. The traditional viewpoints of accidents are as follows

1. Accidents are caused by workers by inattention or ignorance of rules/ procedures;
2. Therefore accusation each other;
3. Prevention is then by following safeguard procedures, correct actions, safety training, strict supervision for the individual.

To increase process safety detect human factors, human abilities and barriers must first be understood at first,

- **Attention:** It is a cognitive process that means how human actively collect and save information and use them in a proper working conditions.
- **Perception:** It is a process where human collect information from their environment and used that information in order to interact with their environment safely. So, always correctly perceive. Work environments sometimes change human perception systems and information which not properly explained.
- **Memory:** It is the capacity to remembering information from the past experiences and in order to use this information in present situation often put undue pressure.
- **Logical Reasoning:** It is one of the fundamental skills of effecting thinking. In failures in reasoning and decision making can affect complex systems such as chemical plants, and tasks like maintenance and planning.

Classification of Human Error

Human error classifications facilitate the identification and analysis of human errors. There are various ways of classifying human errors. The simplest is the classification by mode or action according to Swain and Guttmann (1983) are as follows

- **Omission Error:** Proper working procedure is not followed.
- **Commission Error:** Wrong working procedure performed.

Generally, human error can be categorized into two steps, i.e., unintentional and intentional.

The unintentional step further categorized in more steps, as

- **Slips:** When the wrong action is performed.
- **Lapse:** When an action is omitted, or memory failures occurs.
- **Mistakes:** If a rule or work procedure has been forgotten or never fully understood.

The intentional step further categorized into the following step as,

- **Violation:** Operator breaking organizational rules or operates unsafe procedures intentionally, etc.

Meister (1962) categorized human errors as errors in operations, design, maintenance, fabrication, inspection, and contributory. Operating errors result from the wrong operation/handling of equipment by the human, and it may be due to motivational error, wrong action performance, failures to follow procedures etc. Design errors occur due to improper design practices. Maintenance errors are failure to carry out right procedure. Other maintenance errors are the error of calibration of equipment, repair and installation etc. Fabrication errors result from poor quality, use of incorrect component, use of incorrect material of construction, and departure from the drawing supplied. Inspection errors result from inadequate inspections that mean anybody breaking organizational safety rules or hide any faulty work process in the time of inspection.

Lincon (1960) indicates that operating error further classify five types of errors as

1. **Attention Errors:** The operator fails to concentrate that require attention.
2. **Memory Errors:** The operator does not remember the actual task procedure or fails memory to perform required task is call memory errors.
3. **Interpretation Errors:** When operator could not understand the right meaning of given information by plant personnel and perform the task incorrectly.
4. **Operation Errors:** The unsafe or incorrect work procedure followed by the worker.
5. **Identification Errors:** The items are not identified.

The combination of Attention errors and memory errors are the errors of omission whereas interpretation, operation, and identification errors are commission errors.

Managing Human Factors

According to HSE (Health and Safety Executive) the fundamental principles of managing Human Failure are

- Human failure is reasonable and predictable. Human factor can be identified by analyzing accidental case studies and it is also manageable. All process industries should include safety management system to reduce human factors.
- Implement good human factors in engineering. It is also called ergonomics, which are concerned with design jobs, operating process, mechanism of machines, and suitable working environment. Ergonomic factors like improper vision, improper motor controls, uncomfortable posture in work procedure and work overload in insufficient time schedule should be prevented.

- Human errors can be prevented by clear, accurate procedures and instructions which also improve workers skill and memory.
- Risk assessment and accidents investigation should identify where human failure can occur in safety critical tasks.
- Manage working areas conditions, e.g., extremes of heat, humidity, noise, vibration, poor lighting, and insufficient workspace.
- Manage workload.
- Social and organizational factors, e.g. insufficient staffing levels, long hour work schedules, unawareness about health and safety; should be manage.
- Psychological and social factors should be managing, like high levels of tiredness, family problems, health problems, addiction of alcohol and drugs.
- Providing proper supervision particularly for inexperienced staff, plant inspection and check proper maintenance in regular interval of time.
- Provide job relevant training and practices for managing abnormal and emergency situations.
- Develop new safer working methods through new procedures can reduce human factor and enhance safe environment in workplace.
- Identify unsafe behaviors associated with previous accidents.
- Inform the workers about how to use of personal protective equipment for individual safety and also give safety knowledge about general housekeeping, access to heights, lifting and bending, and contact with chemicals.

Three accidental case studies of chemical industrial accidents are discussed to find out the actual causes and human factors behind the incident.

CASE STUDY 1

Leakage of Chlorine from a Baby Cylinder

The Incident

A small unit manufactured calcium hypochloride solution by passing chlorine into lime solution. The occupier brought a cylinder of chlorine (baby cylinder – 60 kg. capacity) from the nearby dealer for the preparation of hypochloride solution. During the nighttime, the cylinder, which was kept lying in the ideal condition started leaking from its bottom all on a sudden. The worker brought some ice from the nearby shop and kept the cylinder in the ice pot with anticipation that the chlorine leak from the cylinder will stop if the cylinder is cooled down by ice. After a few minutes the leak increased instead of stopping.

Chlorine gases profusely spread from the cylinder and dispersed according to the direction of wind affecting the people on its path in the adjacent houses, 4 people died by the chlorine gas and other 87 people were injured. The calcium hypochloride unit was located in a heavily populated area in a metropolitan city, Kolkata and most of the buildings were multistoried, normally shops were in the ground floor and residential accommodation from first floor onward. As the chlorine gas is heavier (vapor density of 2.48 gm/cc) than air it spreads in very little quantities in the first floor level. 87 people were affected who needed only first aid treatment for the minor injury by inhalation of the chlorine gas.

Incident Analysis

The main event was identified as

1. The chlorine cylinder was badly corroded.
2. Chlorine dispersed according to the wind direction and caused the death and injury.
3. No emergency first aid response was present.

Major Contributing Conditions Were Also Identified

- The man engaged in the manufacturing process had no knowledge regarding the hazards of the chlorine.
- The management and the worker were ignorant. The management's reaction was to put the chlorine cylinder on the ice pot and the reaction of the metals of the chlorine cylinder with chlorine and water ultimately increased the diameter of the leaky portion of the cylinder.

Corrosion Mechanism

Carbon steel is most common material of construction of baby cylinder. The dry chlorine reacts with the iron present in carbon steel as

$$2Fe + Cl_2 = 2FeCl_3$$

The ferric chloride is formed a proactive layer on the steel surface, however it is very hygroscope, and absorbs moisture and water, which accelerates the rate of corrosion (The Chlorine Institute, 2000).

The chlorine reacts with water to form hypochlorous acid (HOCl) and hydrochloric acid (HCl) as

$$Cl_2 + H_2O = HOCl + HCl$$

The hypoclorous acid is unstable and dissociates as,

$$2HOCl = 2HCl + O_2$$

According to Updyke (1982) and Saroha (2006) in the presence of water the protective layer of ferric chloride is dissolved. The acids and the dissolved salt enhance the corrosion of the cylinder.

Lessons Learned From this Incident

1. Implement the safe maintenance and gain knowledge about risk management.
2. Always perform the job safety analysis under the guidance of trained personnel and also understand the outlines of total incidents. Methods statements should be prepared which clearly define roles, responsibilities and the controls to be applied.

3. Inform chlorine cylinder supplier's about the safe procedure of carrying chlorine cylinder.
4. Cylinder leakage should be addressed to the manufacturer or local task force to tackle this problem immediately and to provide PPE's.
5. Arrange awareness programme on chlorine related hazards and motivated to participation of all staff.

What-If Analysis

The What-If method is the least structured of the creative Process Hazard Analysis techniques. It is a brainstorming approach in which a group of experienced people familiar with the subject process ask questions or voice concerns about possible undesired events. This analysis starts a series of questions that begin with, "What if…?" What-If questions can be formulated around human errors, process upsets, and equipment failures and each question represents a potential failure or miss-operation of the process. Then the "Consequence" which is the effects of What-if condition traced all the way to the loss event. Then the operator's responses are evaluated to determine if there is a chance or possibility of potential hazard. If so, then the proper modifications of the system should be recommended.

This method quickly focuses on issues that are critical and the speed of What-If analysis also helps reduce the boredom. It is also good at analyzing global issues, such as loss of utilities or the impact of a major fire.

Due to its lack of structure, the success of a What-If analysis is highly dependent on the knowledge, creativity, experience and attitudes of the individual operators. The "What-if" table generated for the human factor analysis shown in Table 3.

Table 3. What-if analysis

What - if questions	Consequences	Recommendations
If the chlorine cylinder titled to facilitate the leak of chlorine gas	The amount of chlorine leak is minimum	Ideal solution for initial period of liquid chlorine leakage.
If the leak cylinder put on the ice-pot	Increase in leak due to corrosive action with water	Educating workers about the hazards of chlorine and also encourage to participate in training and monitoring programs in the work place and set-up emergency plan.
If chlorine gas come out from the chlorine cylinder	It may fume while on contact with moist air and create highly corrosive atmosphere	Establish an emergency response plan for responding to leaks and also arrange the proper training for workers.
If chlorine gas spread offsite	Severe irritant to the eyes and respiratory system and also burn the skin	Establish an emergency response plan for responding to leaks.
If the workers are illiterate or if the operating procedure is not properly followed	Accidents may occurr	General chemical and hazardous information should be displayed and provide contact no. for possible help. Proper information and training for proper handling chlorine should give all the workers.
1. If chlorine cylinder become corroded	There is a possibility of leakage	Cylinder should be checked in regular interval of time.

Human Errors

After investigation it was observed that the worker was illiterate and he had no knowledge about the hazardous nature of chlorine. The small establishment did not have a telephone for communication. The following human errors are identified,

- **Lack of Knowledge:** Unawareness of the hazards of chlorine.
- **Lack of Logical Thinking:** In any technological situation logical procedures are necessary, but illogical thinking or the behavior lead to accident. Liquid chlorine is leaking from the cylinder if the cylinder was titled to facilitate the leak of chlorine gas, then the amount of chlorine leak is minimum. One volume liquid chlorine spreads as 460 volume gas chlorine. But the worker put the leaked cylinder on the ice pot which increased the leak diameter and release liquid chlorine.
- **Mental Inefficiency:** A situation which does not have the mental capacity to deal with the information that must be processed in the activity. The worker is illiterate and his mental capacity is also very limited to process any information in acute conditions.

Accident Prevention and Recommendation as Preventive Measures

There is a need for major precautions while working with chlorine. Arrange safety workshops for safely handling of chlorine cylinders such as movement, appropriate storage facilities and maintenance procedures to be used. Consult the Safety Engineer/supplier's safety officers for safe operational procedures to be followed in an emergency situation stated by Spellman (2003) and Gangopadhyay et al. (2005b, 2007a).

Some general recommendations set up for handling chlorine cylinder, control human error, administrative control and safety work procedure control were followed.

Handling Chlorine Cylinder

Cylinder Movement

The appropriate technique for cylinder movement is the valve cap should be placed and dropping a cylinder or allowing an object to strike the container with extreme force must be prevented. Chlorine cylinders should not expose on heat and never lift a cylinder by its hood. Always use clamp support for cylinder movement. Lifting the cylinder by crane, rope sliding chain or magnetic device should not be use. Lifting the cylinder by holding the valve cap or its neck is prohibited.

Cylinder Storage

The chlorine cylinder storage area is to be a well-ventilated, secured and protected from weather away from heavily congested areas and emergency exits. Never cylinder stored near salted or other corrosive, combustible or flammable materials. Full or empty cylinder should be marked. To prevent full container from being stored for long periods of time use first-in first-out inventory system. Check the cylinders visually at last in every week to detect any indication of leakage or others problems. Provide leak detectors and high concentration audio visual alarm in storage area. To avoid liquid/water ingress in the cylinder while consuming the chlorine, outlet line from cylinder to consumption point should be provided with 32 ft. high barometric leg.

Control of Human Error

1. The workers need proper education and training about the hazards of chlorine.
2. They must have knowledge about the emergency response plan for responding to leaks or accidents.
3. Workers must wear PPE's and breathing apparatus for handling chlorine cylinders or compounds.
4. Follow the manufacturer's instructions during emergency and display the instructions from the material safety data sheet (MSDS).
5. Do not spray water on damaged or leaking containers; it can make the leak worse.
6. When handling chlorine compounds or cylinders the workers should not eat, drink or use tobacco products.
7. The hands and face should be washed before eating or drinking.

General Control

1. Dragging, rolling or dropping of cylinder should be avoided.
2. Fusible plug safety device on containers should not be tampered.
3. Chlorine supplier is to be immediately contacted if any damage is found.
4. Repair a container or its part like valve should not be done – immediately contact the supplier.
5. In any reason never place a container in hot water, or apply direct heat to increase the flow rate.
6. Once the cylinder has been connected to the process, cylinder is to be opened slowly and carefully.
7. If user experiences any difficulty in operating cylinder valve, disconnect use and contact supplier.
8. When tank valves are in closed do not perform any maintenance work.

Administrative Control

1. Arrange training on safe work procedure to handling chlorine for workers, inform them about the health hazards from exposure to chlorine.
2. Plan an emergency preparedness for handling leak or spills of chlorine cylinders at the work site.
3. Prepare emergency exit plans from the working areas and to move uphill and upwind.
4. Follow the rules for handling or maintaining hazardous materials.
5. Store the cylinders in well ventilated areas and training to the workers is necessary for the operation and maintenance of these systems.
6. In storage area where chlorine used as a raw material, always work with proper cloth and gloves and approved self-contained breathing apparatus in working areas.
7. The ventilation system in the chlorine storage room should be on and functional when workers are in the room.
8. Every plant must have chlorine alarm system of the outside area of the working zone.
9. Clean up chlorine spill as soon as possible and workers must use appropriate protective equipment and clothing.

Conclusion

The incident of chlorine leakage from a baby cylinder and the off-site consequences are described. The analysis of the incident has been carried out with respect to human errors and causes due technical faults. Lack of knowledge among workers about the hazardous nature of chlorine and also some maintenance ignorance on plant safety was the main cause of this incident.

CASE STUDY 2

Phenol Formaldehyde Runaway Reaction in a Resin Factory

The Incident

One incident occurred in a plant in Kalyani, Nadia District, West Bengal, India. The plant operator started a preparation of a trial process using ammonia catalyst instead of caustic soda under the supervision of an expert. The earlier operation was done by another catalyst caustic soda. When using ammonia catalyst, the plant does not maintain any safe operational procedure and the entire process was started using manual control under expert supervision. The plant stored sufficient amount of phenol formaldehyde and liquid ammonia.

At first 87% phenol in water was taken in the reactor then the 37% concentration of formalin was transferred to the reactor. Then the stirring was started slowly. After that the catalyst liquid ammonia was added manually. During the catalyst addition small quantity of liquid catalyst flashed on the hand of the operator and the operator felt some burning sensation. The temperature of the reaction mass was already raised by steam heating to nearly 100^0 C and after few minutes it was observed that the temperature was increasing abnormally. The supervisor felt some abnormality in the reactor and instructed to drain the reaction mass from the reactor in several 100/200 lit. drums as available in the plant. After filling the drums, the operators closed the lids of some drums and all those were taken away in the open space at a distance from the reaction zone. Till some reactants were kept in the reactor. After a few minutes the closed barrels exploded one after another and some were flying at a height of 30 – 40 ft. and struck the wall and roof of the shed of the nearby areas. The residual mass in the reactor drained to the other empty drums and lids of those drums were not closed and these drums did not explode.

Probable Causes of the Accident

It was observed that there is wrong catalyst used. The liquid which add as a catalyst was not liquid ammonia, but it was concentrated nitric acid. So, there is vigorous reaction occurred inside the reactor. Concentrated nitric acid converts the reaction as a runaway reaction, which increased temperature of the reactor mass. It is observed that abnormal temperature rises in the reactor and there is poor cooling arrangement to control the reactor temperature. So, the expert decided to transfer the reaction mass to 100/200 lit drums and in the few drums the lids were closed and all drums were placed away from the reactor zone. Due to strong reaction at high temperature (as rate of reaction doubled roughly at increase of every 10° C) and evolution of stream, the closed drums experienced high pressure. The high pressure led to the explosion of the drums. The drums without lids did not explode. Draining of the reaction mass saved the reactor. So, it was cleared that the operator added wrong catalyst, concentrated nitric acid instead of liquid ammonia to the process and also at a high rate. Investigation shows that the incident occurred due to human and maintenance error.

The incident analysis shows that operating and maintenance error are primary cause of this incident. The worker used the catalyst from unlabeled jar and which containing wrong chemical compound. Analysis also concluded that though the supervisor had safety knowledge, but the unsafe organizational management process was the leading cause of the accident. Job Hazard Analysis (JHA) and SWOT analysis used to identify the potential hazards in the works place.

Job Hazard Analysis

Job Hazard Analysis (JHA) is very useful procedure for detecting workplace hazards. It recognizes the possible unsafe working situations before they occur. Chao and Henshaw (2002) stated that the method used for identifying, evaluating and controlling risks in industrial procedures. It also identifies the connection between the operator, the working procedure, methods or tools and the work environment. JHA identified the proper job procedure after carefully studying and recording each step of the job, then identifying the existing or potential job hazards, safety and health to determine the best way to perform the job to reduce or completely eliminate the hazards potential. According to Holt, (2001) Job Hazard Analysis is a safest planning procedure of a work. The advantages this analysis is to

1. It prevents worker injuries and illnesses and enhance organizational health and safety of workers
2. It identifies more effectual safe work procedures
3. Increase plant production
4. Prevent harmful situation
5. Reduce workers compensation costs
6. It is a precious technique for safety training of new employees.

To avoid maintenance error the management of a plant must give a practical explanation to the worker on health and safety, working procedures such as how skillfully use of machines and perform other working procedures by taking the corrective measures as identified by JHA. JHA gain trust and reliability of the workers even under harmful situation stated by U.S. Department of Labor, OSHA (2002). According to Chao and Henshaw (2002) the JHA conducted in to three main steps as

1. **Identification:** Selecting a particular job, then divided into different sequence stages and identify probable accidental/ unsafe situation of working procedures.
2. **Assessment:** Assess the possible hazards situation may be occur during the work.
3. **Action:** Evaluate to decrease the risk.

Friend and Kohn (2003) and Ramsay et al. (2006) describe that JHA also define job procedures in a proper way and step by step that workers and trainers both can completely understand the working procedures and also gain knowledge about unsafe working procedures and follow the proper method of the job.

SWOT Analysis

The SWOT analysis (Strengths, Weakness, Opportunities and Threats) is a valuable and helpful method which recognize strengths, weaknesses and also to examine the opportunities and threats of an industry. Strength and weakness analysis focuses on the previous work procedures, existing strength, resources and capabilities. The opportunity analyses pay attention to the improvement of working procedures and working conditions. Threats indicate the barriers or difficulty of job procedures which influence the production.

The SWOT method is used to recognize a solution of a problem or establish a plan of work which evaluates the strengths and weakness within organization and the opportunities and threats outside the organization. The internal and external factors indicate the positive points, strengths and opportunities and negative points, weakness and threats of the organization. It recognizes the clear idea about the job and the working environment to the management and workers. Hence it decreases the accident/near misses situations and also raising safety awareness.

Table 4. Job Hazard Analysis

Basic Job Steps	Potential Hazards or Injuries	Cause	Required Safe Job Procedures
This process based on ammonia catalyst	Wrong catalysts causes the reaction to run out of control	Wrong catalyst, a concentration of nitric acid used	Proper catalyst should have been added.
Adequate cooling arrangement should be provided	Inadequate cooling increases temperature	The reaction rate is doubled approximate increase of 10°C and also evaluation of water vapor a reaction product is large amount	Adequate vent arrangement (proper sized vent) in the extreme condition should be provided, otherwise there was possibility of explosion due to over pressure, it causes complete destruction of the reactor vessel and possibilities of flying of metal pieces in an around the plant which is dangerous for the human in-site and out-site and also release of hazardous materials
In store material should keep in proper order, display MSDS in working areas, check delivery order etc. are required to avoid mixing the wrong chemical in the process	Proper reactants or catalyst should be identified and used for the reaction and also use in proper sequence	Wrong reactants or catalysts or wrong sequences in the input to the reactor can also lead the reaction runaway	Chemical should be labeled and the use of Personal Protective Equipment (PPE) for carrying the chemicals
The runaway reaction exothermic in nature; therefore, increases temperature, so the reaction mass transferred or draining away from the reactor zone in case of emergency	The reaction mass transferred in a container close lid conditions leads to the explosion of the container	Create high temperature (as rate of reaction doubled roughly at increases of every 10°C) inside the container and also release of steam – pressure rises	Always transferred the reaction mass in a container in open lid condition which does not creates high pressure inside the container
Any kind of operating negligence occurred	Accident, blast, explosion occurred	Reaction is very exothermic, so any negligence can lead to the accidental situation	Implement various protective measures; arrange operator training, gain knowledge about chemical reactions, showing them past accident histories

Table 5. SWOT analysis

	Positive	Negative
Internal	**STRENGTHS** 1. Standard process 2. Experience in the process technologies by the engineers, also the knowledge and experience in runaway reaction 3. Due to sufficient knowledge and experience dangerous situation is avoided, otherwise reactor may burst, create more adverse situation 4. Management is committed for the new product using ammonia as catalyst	**WEAKNESSES** 1. The chemicals are not labeled in the storage 2. Inadequate knowledge of the workers 3. Faulty management system 4. Management is not confident enough in the new process 5. No emergency planning for in-site and off-site 6. Lower levels staff are mostly illiterate not much knowledge of the process due to lack of training 7. Lack of communication skill of the workers 8. Lack of knowledge of the released chemicals and their amounts in the accident scenario
External	**OPPORTUNITIES** 1. Proper training – to impart knowledge about the process and the runaway condition of the process 2. Showing MSDS of the chemicals in the proper place 3. Adequate cooling arrangement in the reactor 4. Chemicals should be added in the reactor in proper sequence 5. Proper storage facilities 6. Separate racks for separate chemicals in the storage with proper labeling 7. Profit margin is going up due to high market demand particularly in locality and also export market	**THREATS** 1. Uncontrolled exothermic runaway reaction 2. Human error/maintenance error 3. Accident or explosion 4. Loss of life and productivity

Job Hazard Analysis (JHA) and SWOT analysis has been carried out to investigate the accident. Table 4 shows the basic step of JHA which represents the all possible factors of the incidents. It explains step by step the incident in a simple way. JHA indicates that operating error and maintenance error were the main causes of this incident. The maintenance error occurred due to management ignorance because the catalyst jar was not labeled so, the worker added wrong catalyst. This analysis also suggests the preventive measures shown in Table 4. Table 5 shows SWOT analysis which represents the strengths, weaknesses, opportunities and threats of this manufacturing plant which identified the reasons of this accident and also finds out the solution which prevent this type of accident in future. This analysis concluded that the supervisor of the plant had knowledge and experience about the process but faulty management was the main cause for accident.

Remedial Measures

All the chemical reactants should be kept in proper storage area with proper identification. The reactants are to be added standard measured quantity and maintaining a precise sequence as prescribed in the operating manual.

Lessons Learnt

1. Job Hazard Analysis and SWOT analysis conducted to analyze the incident. Both analyses recommend the following preventive measures
    a. For identifying the chemicals are labeled in the storage area.
    b. Chemicals should be stored in properly arranged manner.

  c. Chemicals should be added in the reactor in proper sequence.
  d. To avoid runaway reaction adequate cooling arrangement is to be provided to the reactor.
  e. Material Safety Data Sheet (MSDS) should be display in the working zone.
  f. Use Personal Protective Equipment (PPE) while carrying the chemicals.
  g. Safety training programme for the worker should be arranged at regular interval of time, at least once in a year to avoid operating error.
2. Administrative controls
  a. Management is to be prepared the SOPs (Standard Operating Procedures) of the process.
  b. Emergency preparedness programmes are to be prepared for onsite and offsite.
  c. SOPs are to be followed.
3. Proper sized emergency relief valve must be provided in the reactor.
4. Past accidental case studies give appropriate guidelines on how to control accidents in future. The workers also learnt safety lessons from it.

Steps to be Taken to Reduce Hazards

1. SOPs (Standard Operating Procedures) for the new ammonia catalyzed process are to be developed to improve the inherent process safety.
2. Implementation of various protective measures like temperature control, instrumentation and interlocks to eliminate the opportunities for minimize the human errors.
3. The operators must be understood the hazardous information about the chemical and the process hazards and the consequences are to be evaluated for the process hazards assessment.
4. Operators should be followed SOPs.
5. To prevent this type of incident in future, identify the root causes and lessons learnt and all recommendations are to be incorporated to prevent recurrence.
6. Management should arrange training programme for the operators at regular interval of time to improve workplace safety.

Conclusion

An accidental case study in the process of manufacturing of phenol-formaldehyde resin has been reported using ammonia as catalyst in the new trial run process. Operator and management ignorance allowed for concentrated nitric acid added to the reactor instead of ammonia and a runaway reaction was started. The probable causes and remedies discussed on the basis of The Job Hazard Analysis (JHA) and SWOT analysis.

CASE STUDY 3

Carbide Fire in an Acetylene Gas Plant

The Incident

The incident took place in an industrial gas plant in West Bengal, India in one night of March 1993. The shift officer was on routine visit in the plant at around 11p.m. It was a summer night and there was heavy downpour around for two hours from 8 to 10 p.m. There was water logging in the factory roads due to rains. The shift officer was on round by company car after the rain stopped. While passing beside the calcium carbide godown, heavy smoke was seen issuing from the entrance of the main godown. The officer then stopped the car on road and observed that smoke was issuing from a carbide drum. The drum was kept on floor outside the godown beside the covered walkway approaching acetylene gas generator room. No flame was found to come out of the drum. The drum lid was in open condition. The fire was extinguished by fire extinguisher immediately. There was no loss of life and no loss of property.

Process of Fire in the Plant

Water ingress to the carbide drum occurred during rain. Thus, a typical carbide fire occurred as soon as carbide came in contact with water followed by formation of acetylene gas as result of temperature raised to the flash point of acetylene and there was auto ignition of acetylene.

Probable Causes of the Incident

1. Carbide drum with dust and lump was kept for long time for necessary disposal
2. Carbide dust was partially exposed to atmospheric air
3. There was water ingress into exposed drum
4. Formation of acetylene in the drum
5. Auto ignition of acetylene inside the drum due to rise of temperature in drum
6. Safe procedure was not followed for disposal of carbide dust and lump
7. Strict supervision was lacking

Errors and the Accident Analysis

What-if analysis used to identify the possible errors, and the unsafe activities. The human error, unsafe act committed by the system operator which led to the accident. This unsafe act either involve doing something wrong or failing to do something. Table 6 shows what-if analysis. It reveals that the accident occurred due to maintenance and operating error. The maintenance failure is also a major cause of human errors. The management of the gas plant does not maintain the maintenance procedure of the carbide storage and the plant operators had insufficient knowledge as,

1. The carbide drum which was necessary for disposal was kept for long time outside the go down on floor beside the walkway approaching acetylene gas generator room.
2. The lid of the drum was also in open condition.

3. The maintenance of the roof of walkway was very poor, so the water ingress to the carbide drums during rain due to spillage of rainwater in the drum.

Thus, the plant was running poor supervision. Although there was no loss of life and no loss of property, but a big incident could occur.

Table 6. What-if analysis

What - if questions	Consequences	Recommendations
Are calcium carbide drums in contact with moisture?	Acetylene formations occur	Use local exhaust ventilation or workers should worn respirators and personal protective equipment (PPE).
If acetylene gas mixes with air?	Explosions occur	Establish an emergency response plan for responding in accidental situation.
If workers are illiterate or if operating procedures are not properly followed?	Accidents may occur	General chemical and hazardous information should be displayed to the workers and train them for proper handling calcium carbide.
If the calcium carbide rules, 1987 are not properly maintain?	Typical carbide fires occur	Follow the Calcium carbide Rules, 1987.
If acetylene gas spreads off-site?	Severely irritate and burn the eyes, skin, mouth, throat and shortness of breath	On skin contact immediately wash or shower to remove the chemical, use P.P.E and goggles.
If acetylene fire breaks down?	Explosion occurred	Establish an emergency response plan for responding in accidental situation and use explosion proof electrical equipment and fittings.

Lessons learnt

1. Calcium carbide dust, sweeping, calcium carbide lump etc. should be disposed as soon as possible
2. Carbide dust, lump, etc., be stored at dry place before disposal
3. Care must be taken for supervision of carbide handling and storage
4. Refresher training programme to be in place in close frequency for all concerned
5. Encourage good performance in maintenance work
6. Good communications in maintenance team, working conditions (enough light, not very hot or not very cold, well ventilated clean storage area) are maintained by the management.

Corrective Action

1. Training was conducted for all concerned responsible for acetylene plant operation, in regular interval of time
2. Intense care was taken for carbide dust and lump disposal in the regular frequency
3. Storing carbide outside carbide godown was bad practice, management should strictly follow the Calcium Carbide Rules, 1987, Chapter V for storage (Ministry of Industry, India).

Disaster Management

All the fire in acetylene plant hazards can lead to disaster within a short while in acetylene plant due to inflammable nature of acetylene. The acetylene plant must have a disaster management plan to meet emergency situation. A very short resume of disaster management plan is cited here. The salient features of the plan are as usual the following basic steps of a standard hazard management plan. They are as under,

1. Hazard identification, Safety audit, etc.
2. Hazard analysis / evaluation, HAZOP (Hazard and Operability Study) study, etc.
3. Mitigation of hazard.
4. Prepare emergency plan for working areas inside or outside of the plant.
5. Save workers from the worse situation and shift the entire staff form accidental zone.
6. Provide safety information about the plant operations.
7. Roles and responsibilities of managerial and worker.
8. Testing the plan.

Recommendations

1. Management must take the responsibility to train their personnel to be sure that they understand the hazards while handling flammable materials and training should be carried out in regular interval.
2. Carbide must always be stored and handled at dry and moisture free condition, as per calcium carbide rules, 1987.
3. Carbide must be used in the process on FIFO (First in First Out) principle.
4. Handle carbide/dust only with spark proof shovels. Disposal of carbide dust must be done in the open with copious supply of water as soon as possible.
5. System must be in place for periodic audit of operation and safety procedure of the plant by internal and external audit team.
6. At no condition acetylene release, leakage, etc., can be allowed in an acetylene producing plants. All leaks to be identified and stopped with immediate effect. No delay is permissible.
7. All maintenance jobs must be done with best care and under permit of work procedure.
8. All operating staff must use Personnel Protective Equipments (PPE's), e.g., air stream helmet, cotton suits, fireproof hand gloves, safety shoes while charging carbide to acetylene generator and other jobs to be performed with the recommended PPE's.
9. The alarm system should be provided in different work area to operate the plant safety.
10. All do's & don'ts must be displayed at the entrance and also important areas of the plant.
11. To detect human failure conduct HAZOP study and draw a plan for emergency disaster management.
12. The management is to manage available tools/equipment, reduce interruptions and distractions, housekeeping and tool control, manage fatigue-work schedules, manage boredom- by task assignment, appropriate rules procedures, reduce action slips, reduce memory lapses, improves task completion, improve attention and memory, reduces complacency and overconfidence, reduce risky decision making.

CONCLUSION

The analysis of the incident has been carried out with respect to human errors. The carbide drum was for disposal kept for long time outside the go down and water ingress into the drum due to rain. The management of the gas plant does not maintain safety rules. In future to prevent this type of incident must follow the suggestive and corrective actions and recommendations.

DISCUSSION

Chemical industrial accidents include fires, explosions, and leakages of toxic or hazardous materials are harmful for people and environment. The causes of accidents are poor design, inadequate maintenance, human error etc. The chapter discussed about human factors is one of the major causes for accidents. Human factors occurred due to lack of knowledge, lack of motivation, lack of attention, lack of perception, carelessness/negligence, lack of memory etc. So, operating error, maintenance error, inspection error, attention error, identification error, interpretation error etc. occur as human error. Three accidental case studies also find the probable human factors behind the incidents. Engineering students, safety personnel, workers and all other persons related with plant safety also benefited from this chapter.

REFERENCES

Bajpai, S., & Gupta, P. J. (2005). Site security for chemical process industries. *Journal of Loss Prevention in the Process Industries, 18*(4-6), 301-309.

Bea, R. G., Holdsworth, R. D., & Smith, C. (1996). Human and Organization Factors in the Safety of Offshore Platforms. *Paper presented at the International Workshop on Human Factors in Offshore Operations*. Academic Press.

Bird, F. E., & Germain, G. L. (Eds.). (1992). *Practical Loss Control Leadership*. Loganville, GA: International Loss Control Institute.

Chao, E. L., & Henshaw, J. L. (2002). *Job Hazard Analysis*. Occupational Safety and Health Administration, US Department of Labor. Retrieved from https://www.osha.gov/Publications/osha3071.pdf

Chung, P. W. H., & Jefferson, M. (1998). The integration of accident databases with computer tools in the chemical industry. *Computers & Chemical Engineering, 22*(Supplement 1), 729–732. doi:10.1016/S0098-1354(98)00135-5

Friend, M. A., & Kohn, J. P. (2003). Fundamentals of occupational safety and health. UK: The Scarecrow Press, Inc.

Gangopadhyay, R. K. (2005a). *Study of some accidents in chemical industries – Lessons learnt.* Unpublished doctoral dissertation, Birla Institute of Technology, Meshra, India.

Gangopadhyay, R. K., & Das, S. K. (2007a). Chlorine emission during the chlorination of water case study. *Environmental Monitoring and Assessment*, *125*(1-3), 197–200. doi:10.100710661-006-9252-3 PMID:16897510

Gangopadhyay, R. K., & Das, S. K. (2007b). Accident due to release of hydrogen sulphide in a manufacturing process of cobalt sulphide – case study. *Environmental Monitoring and Assessment*, *129*(1-3), 133–135. doi:10.100710661-006-9347-x PMID:17057980

Gangopadhyay, R. K., & Das, S. K. (2007c). Lesson learned – fire due to leakage of ethanol during transfer. *Process Safety Progress*, *26*(3), 235–236. doi:10.1002/prs.10197

Gangopadhyay, R. K., & Das, S. K. (2007d). Accidental emissions of sulphur oxides from the stack of a sulphuric acid plant – case study. *Process Safety Progress*, *26*(4), 283–288. doi:10.1002/prs.10199

Gangopadhyay, R. K., & Das, S. K. (2008a). Ammonia leakage from refrigeration plant and the management practice. *Process Safety Progress*, *27*(1), 15–20. doi:10.1002/prs.10208

Gangopadhyay, R. K., & Das, S. K. (2008b). Lessons learned from a fuming sulfuric acid tank overflow incident. *Journal of Chemical Health and Safety*, *15*(5), 13–15. doi:10.1016/j.jchas.2008.02.002

Gangopadhyay, R. K., & Das, S. K. (2009a). Fire during transfer of LPG from the road tanker to Horton sphere – rupture of the hose attached to the manifold. In I. Sogaard & H. Krogh (Ed.), Fire Safety (pp. 113-120). Nova Science Publishers, Inc.

Gangopadhyay, R. K., & Das, S. K. (2009b). Fire accident at a refinery in India: The causes and the lesson learnt. In I. Sogaard & H. Krogh (Ed.), Fire Safety (pp. 121-126). Nova Science Publishers, Inc.

Gangopadhyay, R. K., Das, S. K., & Mukherjee, M. (2005b). Chlorine leakage from bonnet of a valve in a bullet – a case study. *Journal of Loss Prevention in the Process Industries*, *18*(4-6), 526–530. doi:10.1016/j.jlp.2005.07.008

Glendon, A. I., Clarke, S. G., & McKenna, E. F. (Eds.). (2006). *Human Safety and Risk Managemen*. London, New york: Taylor & Francis Publishing.

Hollnagel, E. (2005). Human reliability assessment in context. *Nuclear Engineering and Technology*, *37*(2), 159–166.

Holt, A. S. J. (2001). *Principles of Construction Safety*. London: Blackwell Science Ltd. doi:10.1002/9780470690529

HSE. (2005). *Inspector tool kit. Human factors in the management of major accidents hazards – introduction to human factors.* Retrieved from http://www.hse.gov.uk/humanfactors/topics/humanfail.html

Joschek, H. I. (1981). Risk assessment in the chemical industry. In *Proceedings of the International topical meeting on probabilistic risk assessment* (pp. 122 – 131). New York: American Nuclear Society.

Khan, F. I., & Abbasi, S. A. (1999). Major accidents in process industries and analysis of causes and consequences. *Journal of Loss Prevention in the Process Industries*, *12*(5), 361–378. doi:10.1016/S0950-4230(98)00062-X

Kidam, K., & Hurme, M. (2013). Analysis of equipment failures as contributors to chemical process accidents. *Process Safety and Environmental Protection, 91*(1-2), 61–78. doi:10.1016/j.psep.2012.02.001

Kirwan, B. (1994). *A guide to practical human reliability assessment*. London: Taylor & Francis Publishing.

Kletz, T. A. (1993). *Lessons from disaster: How organizations have no memory and accidents recur.* U.K.: Gulf Professional Publishing.

Kletz, T. A. (2001). *Learning from accidents.* Oxford: Gulf Professional Publishing.

Lees, R. E., & Laudry, B. R. (1989). Increasing the understanding of industrial accidents: An analysis of potential major injury records. *Canadian Journal of Public Health, 80*(6), 423–426. PMID:2611739

Lincon, R. S. (1960). Human factors in the attainment of reliability. *Transactions on Reliability, 9*, 97–103.

Lorenzo, D. K. (1990). *A guide to reducing human errors, improving human performance in the chemical industry.* Washington, DC: The Chemical Manufacturers' Association, Inc.

Löwe, K., & Kariuki, S. G. (2004). Berücksichtigung des menschen beim design verfahrenstechnischer anlagen. Fortschritt-Berichte, 22(16), 88-103.

Mannan, M. S., O'Connor, T. M., & Keren, N. (2009). Patterns and trends in injuries due to chemical based on OSHA occupational injury and illness statistics. *Journal of Hazardous Materials, 163*(1), 349–356. doi:10.1016/j.jhazmat.2008.06.121 PMID:18718716

May, I., & Deckker, E. (2009). Reducing the risk of failure by better training and education. *Engineering Failure Analysis, 16*(4), 1153–1162. doi:10.1016/j.engfailanal.2008.07.006

Meister, D. (1962). The problem of human-initiated failures. In Proceedings 8th National System Reliability and Quality Control (pp. 234-239). Academic Press.

Ministry of Industry (India). Department of Industrial Development, (1987). Calcium Carbide Rules. Retrieved from https://dipp.gov.in/sites/default/files/Calcium_carbide_rules_1987_0.pdf

Pamphlet 64, 5th Edition. Emergency response plans for chlorine facilities. (2000). The Chlorine Institute, Inc.

Qi, R., Prem, K. P., Ng, D., Rana, M. A., Yun, G., & Mannan, M. S. (2012). Challenges and needs for process safety in the new millennium. *Process Safety and Environmental Protection, 90*(2), 91–100. doi:10.1016/j.psep.2011.08.002

Ramsay, J., Denny, F., Szirotnyak, K., Thomas, J., Corneliuson, E., & Paxton, K. L. (2006). Identifying nursing hazards in the emergency department: A new approach to nursing job hazard analysis. *Journal of Safety Research, 37*(1), 63–74. doi:10.1016/j.jsr.2005.10.018 PMID:16490215

Rasmussen, J., Nixon, P., & Warner, F. (1990). Human error and the problem of causality in analysis of accidents. *Philosophical Transactions of the Royal Society of London. Biological Sciences, (B),* 1241-1327 & *Human Factors in Hazardous Situations*, 449-462.

Reason, J. (1997). *Managing the Risks of Organizational Accidents.* Ashgate Publishing.

Reniers, G., & Amyotte, P. (2012). Prevention in the chemical and process industries: Future directions. *Journal of Loss Prevention in the Process Industries*, *25*(1), 227–231. doi:10.1016/j.jlp.2011.06.016

Salminen, S., & Tallberg, T. (1996). Human errors in fatal and serious occupational accidents in Findland. *Ergonomics*, *39*(7), 980–988. doi:10.1080/00140139608964518 PMID:8690011

Saroha, A. K. (2006). Safe handling of chlorine. *Journal of Chemical Health and Safety*, *13*(2), 5–11. doi:10.1016/j.chs.2005.02.005

Spellman, F. R. (2003). *Disinfection of wastewater. Handbook of water and wastewater treatment plant operations*. London: CRC Publication. doi:10.1201/9780203489833

Swain, A. D., & Guttmann, H. E. (1983). *Handbook of human reliability analysis with emphasis on nuclear power plant applications*. U.S. Nuclear Regulatory Commission. doi:10.2172/5752058

Updyke, L. J. (1982). Method for calculating water distribution in a chlorine condensing system. *Paper presented at the 25th Chlorine Plant Operations*. Academic Press.

Zhang, H. D., & Zheng, X. P. (2012). Characteristics of hazardous chemical accidents in China: A statistical investigation. *Journal of Loss Prevention in the Process Industries*, *25*(4), 686–693. doi:10.1016/j.jlp.2012.03.001

Zohar, D. (1980). Safety climate in industrial organization: Theoretical and applied implications. *The Journal of Applied Psychology*, *65*(1), 96–102. doi:10.1037/0021-9010.65.1.96 PMID:7364709

ADDITIONAL READING

Alper, S. J., & Karsh, B. T. (2009). A systematic review of safety violations in industry. *Accident; Analysis and Prevention*, *41*(4), 739–754. doi:10.1016/j.aap.2009.03.013 PMID:19540963

Foord, A. G., & Gulland, W. G. (2006). Can technology eliminate human error? *Process Safety and Environmental Protection*, *84*(3B3), 171–173. doi:10.1205/psep.05208

Hayes, B. E., Perander, J., Smecko, T., & Trask, J. (1998). Measuring perceptions of workplace safety: Development and validation of the work safety scale. *Journal of Safety Research*, *29*(3), 145–161. doi:10.1016/S0022-4375(98)00011-5

Prussia, G. E., Brown, K. A., & Willis, P. G. (2003). Mental models of safety: Do managers and employees see eye to eye? *Journal of Safety Research*, *34*(2), 143–156. doi:10.1016/S0022-4375(03)00011-2 PMID:12737953

Rao, S. (2007). Safety culture and accident analysis – A socio-management approach based on organizational safety social capital. *Journal of Hazardous Materials*, *142*(3), 730–740. doi:10.1016/j.jhazmat.2006.06.086 PMID:16911855

Rooney, J. J., Heuvel, L. N. V., & Lorenzo, D. K. (2002). Reduce human error. How to analyze near misses and sentinel events, determine root causes and implement corrective actions. *Quality progress,* 27-36.

Silber, K. H., & Foshay, W. R. (Eds.). (2010). *Handbook of improving performance in the workplace.* San Francisco, CA: Pfeiffer, A Willy Imprint publishing.

Ward, R. B. (2002). Analysing the past, planning the future, for the hazard of management. *Trans IChemE, 80*(Part B), 47-54.

Zheng, X., & Liu, M. (2009). An overview of accident forecasting methodologies. *Journal of Loss Prevention in the Process Industries, 22*(4), 484–491. doi:10.1016/j.jlp.2009.03.005

KEY TERMS AND DEFINITIONS

Accident: An unfortunate event which is unexpected, unavoidable, unintended and damage property and life is called accident.

Analysis: A systematic examination and evaluation of data or information, by breaking it into its component parts to uncover their interrelationships.

Chemical Industry: Chemical industry is one of important and biggest industrial sector which produce chemicals in a complex manufacturing process.

Errors: The primary meaning of error is mistake. Sometimes error defined as incorrect or wrong procedures.

Hazards: Any harmful component that can cause harm or damage to human, property or the environment.

Incident: An unpleasant or unusual event which make situations accidental.

Occupational Safety: Concerns with physical and psychological conditions of workers and employers in workplaces which primary focus on preventing hazards, risk factors, diseases, stress related disorders etc.

Process Safety: It means preventions from all accidental possibilities or situations in industries and also preventing unintentional releases of hazardous materials which can have a serious effect to the plant and environment.

Chapter 8
Explosion Process Safety: Basics and Application of Explosion Protection

Dieter Gabel
https://orcid.org/0000-0002-3814-0960
Otto von Guericke University, Germany

ABSTRACT

Explosions can be considered to be the most devastating events in industry. Reasons that lead to such an event are often very complex. Nonetheless, the basic phenomenon is generally simple. To understand what leads to an explosion and how this can be prevented, the underlying physical and chemical processes as well as the basic steps that lead to an explosion are clarified. Practically a system of standards and regulations ensures a framework to avoid unwanted events. This together with typical sequence of events will be given and lead to a general overview of explosion process safety.

INTRODUCTION

Explosions are the worst events to be expected in process industry. These lead to damages due to a pressure rise and the accompanied heat release. Often these are followed by secondary explosions, accidental releases and uncontrolled fires. In the field of process safety explosions can be defined as rapid exothermal oxidation reactions. Based on this fundamental process the systematic to avoid explosions is built, focusing on either the fuel, the oxidant or the source of ignition. Additionally, measures to limit the consequences have to be applied wherever an explosion cannot be avoided for sure. This system is reflected in the European regulations for explosions protection, as well as in other laws worldwide. An international market of cause makes it necessary to consider different regional regulatory approaches. In fact, regulations cannot cover all possible situations, as the complex world of process industry cannot be totally be reflected. For this reason, the limits and expected consequence when crossing these are theme. Altogether, the chapter will enable the reader to understand the principle of explosion safety in process industry, how it is applied and when situation occur where special precaution is needed.

DOI: 10.4018/978-1-7998-3059-7.ch008

THE FORMATION OF EXPLOSIONS

Underlying Physical and Chemical Processes

What an explosion is depends on the context in which the term is used. In general, they can be divided into physical, chemical and nuclear explosions. They have in common that a big amount of energy is released in a short period of time. The first discriminating factor is the source of energy. The nuclear part can be neglected for process industry and is only focus of a very specialized field, as there are nuclear power plants and nuclear weapons. In the same way explosives shall not be treated here, even if in chemical industry compounds and reactions exist that are not less critical. The other two, physical and chemical explosions, are to be considered separately. But as can be seen in Figure 1 the processes that are primarily independent, might be connected in the occurrence of events.

Figure 1. Differentiation of different kinds of explosions

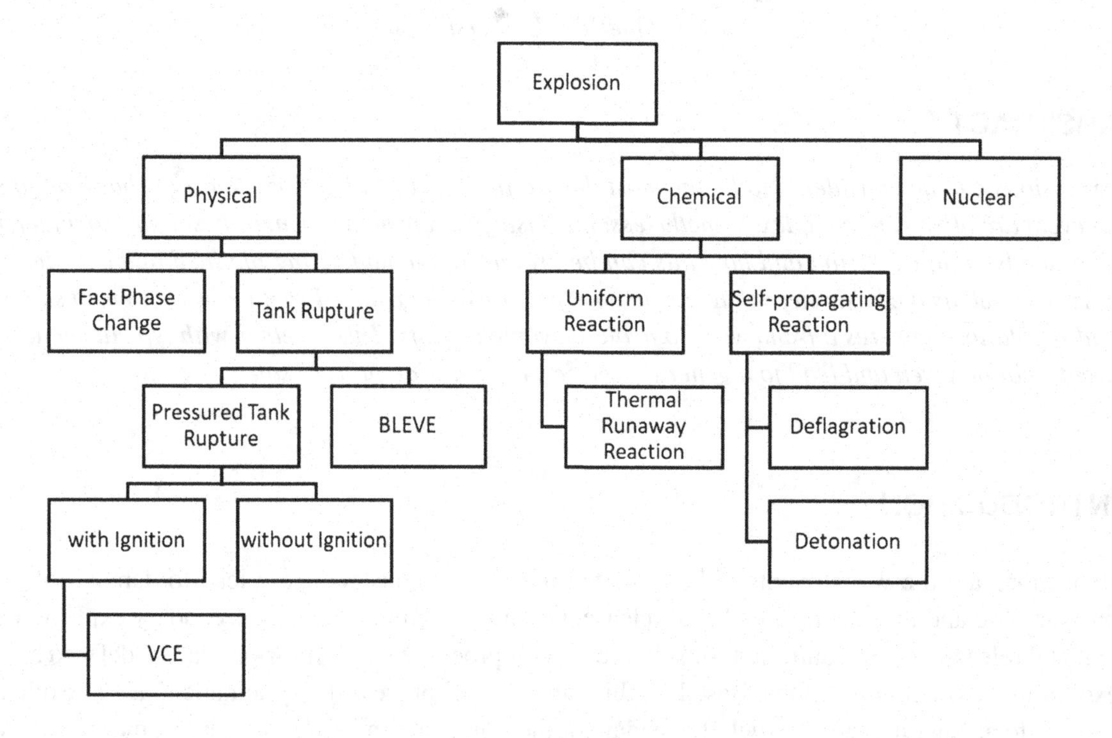

The differentiation given above is not unique and complete. Other authors like Abbasi (2010) have chosen a little bit different way in categorizing and went further in the level of detail. Here the categories of deflagration and detonation are put together and the following possibilities are listed:

- Vapor cloud explosion
- Aerosol or mist explosion
- Gas explosion
- Dust explosion

- Condensed phase explosion

That all kinds of explosions mentioned here can occur in a confined as well as an unconfined environment is true, whether all of them can really also lead to a detonation might be doubted. Abbasi (2010) does not only give "A scheme for the classification of explosions in the chemical process industry" but enhances the publication by a list of examples of real accidents with further literature sources.

Most common are the accidents that are connected to pressurized storage. They either lead to a Vapor Cloud Explosions (VCE) or a BLEVE (Boiling Liquid Expanding Vapor Explosion) that are explained in more detail later. In a narrowed way of looking at explosions in process safety, the definition is simplified to the occurrence of a chemical reaction. This specific reaction is in most cases an oxidation – the burning reaction.

By this, the second important differentiation is achieved. If the burning reaction is slow, it is called fire, if it is fast it is an explosion. How to differentiate between a fast fire, a flash fire, a deflagration, an explosion and a detonation needs to be discussed. The accompanied pressure rises and the reaction mechanism here play the crucial role. Further differentiation will be necessary, as boundaries are not always that clear.

As the focus is on burning reactions only, theses have to be explained in detail first. Whenever combustion plays a role, the fire triangle (Figure 2) is stated.

Symbolizing the three mayor parts that are needed to create a fire – Fuel, Oxygen, Heat – it overly simplifies a process that is quite complex. Nonetheless, it holds true, that if one part of the triangle is missing a fire or an explosion cannot occur. In real situations the three parts cannot always be seen that clearly.

For a combustion to take place it is necessary that a combustible substance (Fuel) gets in contact with an oxidizer (not only Oxygen is possible) and the self-sustaining reaction is started by supplying a sufficient amount of energy (Heat is only one form).

The energy needed to start the combustion reaction is called ignition energy and its manifestation defines the so-called sources of ignition that are listed later in this chapter. It has to be differentiated from the activation energy that in chemistry defines the amount of energy necessary to start any reaction. The main fundamental difference is that for an ignition it is necessary that the reaction is self-sustaining after the source of ignition is removed. This concept is not included in the activation energy. Consequently, the two cannot be compared, especially as the ignition energy is a purely experimental value determined under standardized procedures. In depth discussion on the ignition of burnable systems is given in Babrauskas (2003).

The oxidizing agents that takes part in the reaction can be any element or chemical compound that can be reduced. The strongest oxidizers are found in the chemical group of Halogens. However, as our planet earth has an atmosphere that consists of 21 vol-% Oxygen this, as part of the atmospheric conditions, is often take as set. Whilst this might hold true for many situations, the gas compositions in process plants might be totally different. This aspect plays a vital role in standardization and for Safety Characteristics. Consequently, tabulated values in data collections often cannot be directly applied for safety assessments.

Finally, the fuel that is chemical oxidized and, by the reaction, releases more energy than was introduce into the system to ignite it by the ignition source. Fuels can be all elements and chemical compound that can be oxidized. Practically, this includes all hydrocarbons, organic substances and metals. Answering the question, whether a given combination of materials can be ignited and lead to an explosion is more complicate, as it depends on various factors like there are:

Figure 2. Fire triangle

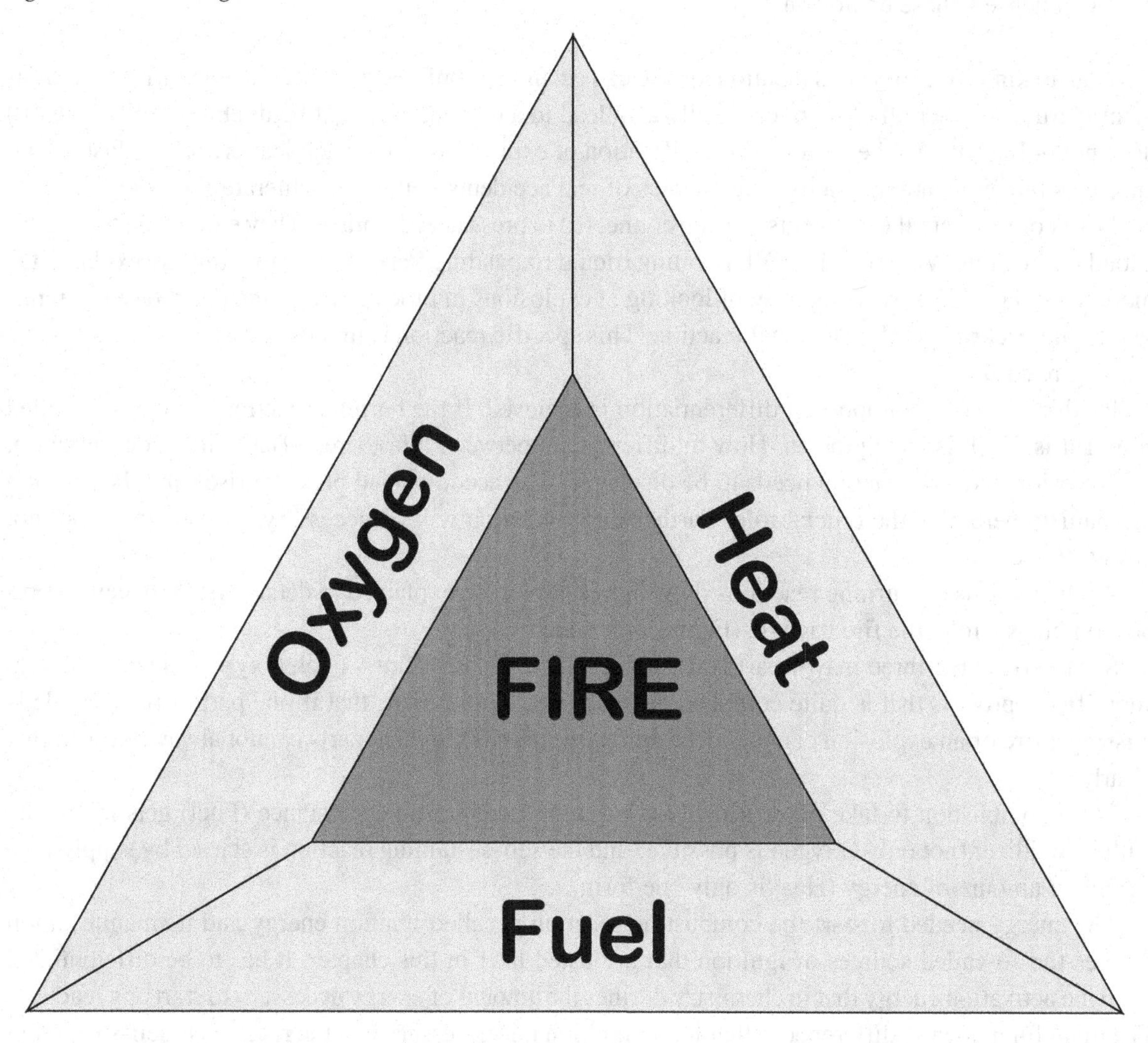

- Type and concentration of fuel
- Type and concentration of oxidant
- Physical state:
  - Gas/Vapor
  - Liquid
  - Solid
- Degree of processing:
  - Liquid droplet size distribution
  - Solid particle size distribution (dust)
- Mixture: ratio and uniformity
- Surrounding conditions:
  - Pressure
  - Temperature

    - Humidity
    - Windspeed
- Catalyst
- Source of ignition: Type and energy
- Enclosure / Damming

This listing is not complete and should only emphasis the necessity to take into account all relevant factors to describe this specific combustion process, the explosion. As these are usually unwanted events that occur accidentally the relevant situation needs to be predicted on basis of limited knowledge, mostly as part of a safety assessment. Here the most probable sequence of events needs to be chosen, as a worst-case scenario is generally inapplicable.

As outcome of all combustion processes we get heat released, as the reaction is exothermic. What distinguishes the explosion from a fire is the short time for the whole system to react. This directly leads to the other differentiation factor, the pressure that is accompanying the reaction. Three processes can lead to an increasing pressure:

- Temperature rise
- Phase transition
- Increasing amount of substance (number of moles)

Of course, these three leads to an increasing volume first, but even in an open environment a pressure increase is to be expected, as the reaction is very fast. The resulting pressure wave produces an over pressure the so-called free field overpressure. The numerical value is quite low and quickly decreases with the distance from the point of explosion. The consequence to be expected are uncertain and thus only can be expressed as probabilities. Values and calculated examples are presented in Hauptmanns (2015). For different situation different methods to estimate the expected overpressure exist and the impact to structures or humans often can only be predicted with the help of so-called Probit ("probability unit") -Equations.

In a confined or closed environment, the pressure rise is straightforward. The maximum pressure to be expected is a function of thermodynamic properties and generally does not exceed 10 bar. This value holds true for initially atmospheric conditions. The dependencies on the initial pressure, the turbulence state, the source of ignition and so on were investigated by different authors and are summarized in Mannan and Lees (2005). More important than the static overpressure is the time to reach the maximum, the pressure rise rate. This value in general depends on the explosion volume. This dependence will be discussed later, as this is only applicable in ideal conditions that are usually not given in an industrial environment.

To enable the user to predict the potential behavior of substances simplified and easy parameters are necessary. These are chosen in a way that in combination with regulation to apply these, a safe handling and processing in general is possible. Therefore, the safety characteristics are introduced.

Safety Characteristics

The determination of safety characteristics is strictly bound to procedures described in relevant standards. This is essential, as they are no physical or chemical characteristics but experimentally determined values.

Thus, they depend on the apparatus and procedures used, as well as on all parameters mentioned above for the combustion process in general.

The mayor differentiation is made between gases, vapors, liquids on one side and solids (bulk materials and dusts) on the other. Due to historic and practical reasons they are often treated in separate standards. The underlying principle can be explained together, referring to relevant different standards for details.

According to the general description above, only specific mixtures of substance can lead to self-sustaining combustion process. To describe these the following values are defined.

Flammability / Explosion Limits

Mixtures of burnable substance with air are only ignitable in a specific range of concentration. This range is limited by the lower and upper explosion or ignition or flammability limit as shown in Figure 3.

Figure 3. Explosion limits of fuel air mixtures

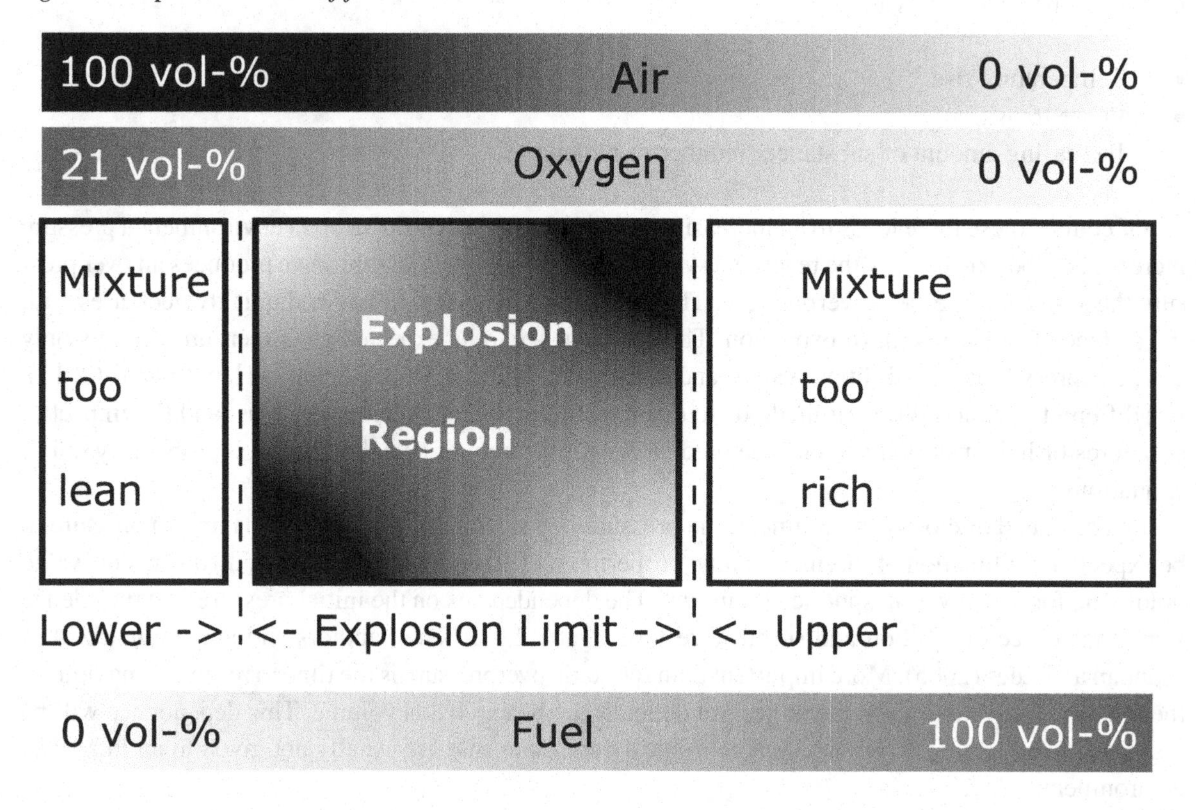

This scheme of cause simplifies the real situation in a user-friendly way. Below the lower flammability limit (LFL) the mixture is to lean for a self-sustaining reaction to go on. Locally – close to the ignition source - the burning reaction will occur. But the energy released is to low or the energy losses are to big for the reaction to be stable. Above the upper flammability limit (UFL) the mixture is to rich and the reaction rate is limited by the available oxygen. The standardized values revere to mixtures with air at environmental conditions (ranges can differ for different standards) only. Practically precaution is necessary at the lower end of the flammability range when liquids are evaporated or fuels are release and thus the concentration might increase. At the upper end of the flammability range a dilution of the fuel always can occur in open environments with contact to the atmosphere. It needs to be emphasized that the values for the ignition limits are determined under specific standardized conditions and all deviations will lead to difference of the real concentrations that are ignitable or not. For the determination, the mixture always is considered to be ideal, a state seldom really reach in real situations.

As for dusts an ideal mixture cannot be assured and the physical differences between the fuel and the oxidizer are much bigger than for gases or vapors, they are also referred to by minimum explosive concentration (MEC) given in grams per cubic meter and not in vol-%. The upper flammability limit is of no practical interest, as the dust concentration always can become lower due to the unavoidable settling process. Dust clouds always need an external force that distribute the dust in the air. Unfortunately, experience shows that this force might be provided by a first, usually small, local explosion. The resulting overpressure can create bigger dust clouds as dust deposits are often to be found were dust are processed. Consequently, "harmless" dust layers can become ignitable dust clouds and lead to secondary and ternary explosions and a cascading sequence of events.

Important is the fact, that the system referred to here is always air with an oxygen content of 21%. This fact needs to be emphasized, as in closed installation the atmosphere likely is different. If the oxygen content is changed, for example to enable an inerting process, another characteristic is to be applied.

Limiting Oxygen Concentration (LOC)

The LOC is the maximum oxygen concentration in the mixture of fuel and Nitrogen in which an explosion will not occur. This definition again is narrowed to the specific system of elements present in our atmosphere, O_2 and N_2. To extend this to other substance ternary flammability diagrams are used, that are presented in the next section.

Practically, the LOC is used for inerting processes in industry. In most cases Nitrogen or Carbon Dioxide is added to avoid an ignition of the system. Of cause, this is only possible in closed systems where the atmosphere can be controlled. To ensure safety the remaining Oxygen content always has to be monitored and an ingress of air from the surrounding atmosphere needs to be prevented.

Flammability Diagram

Triangle diagrams are used to represent the explosion region of mixtures of fuel, inert gas and oxidizer; Figure 4 gives an example. These need to be measured costly for all concentration of interest for each combination of three substances. An extensive collection of data for gases is published in Molnárné, Schendler, Schröder (2003).

Most common are the triangle diagrams for the system Fuel/Air/N_2 or CO_2 for inerting processes. For systems with Air the Lower Explosion Limits (LEL) and Upper Explosions Limits (UEL) could be found directly on the "left" axis (if the axes are arranged as in the example above). As presented in the example in Figure 4 the LEL and UEL lie on the so-called Air Line – all mixtures of fuel with N_2/O_2 with a ratio of 79/29. Generally, a line through one of the corners of the triangle gives all mixtures with the same ratio of two of the substances, like the Stoichiometric Line. Drawing a line parallel to the "right" Nitrogen axis allows to determine the LOC in the triangle diagram.

Figure 4. Triangle diagram to present the explosion region (Data from in Molnárné, Schendler, & Schröder (2003)).

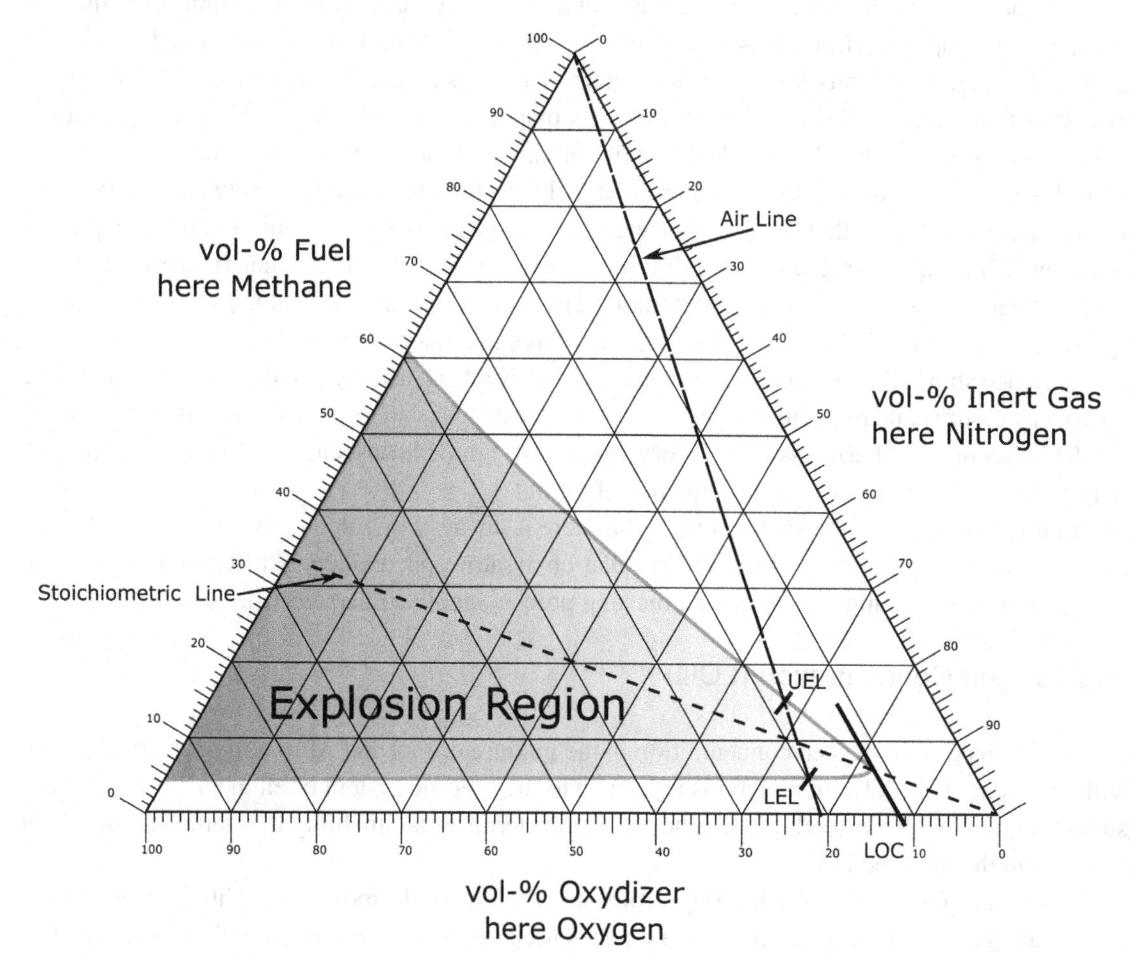

Other dependencies on the explosion region, like pressure or temperature, are presented and discussed in the above-mentioned literature.

Source of Ignition

The "Heat" in the fire triangle (Figure 2) symbolizes the energy needed to initiate the combustion. Indeed, the reaction needs a certain amount of energy to get started in a way, that it will continue by itself. This concept differentiates clearly from the activation energy which is a more theoretic value on molar basis. Thus, the ignition energy again depends on the systematic condition under which it is determined. It can be provided to the system by different so-called sources of ignition, as there are:

- Hot surfaces
- Flames and hot gases
- Mechanical sparks: mechanical impact or friction sparks
- Electrical systems: electric arcs or sparks in electrical equipment

- Electrical equalizing currents: potential differences
- Static electricity: electrostatic charging of non-grounded components or persons
- Lightning strike
- Electromagnetic fields: electromagnetically induced induction voltage in conductor loops and antennas
- Ionizing radiation
- Ultrasonic waves
- Adiabatic compression, shock waves
- Chemical reaction: flames and non-burning reactions

The two most important values or their expression as standardize safety characteristic will be explained here. Further possible characteristics are only mentioned as they often lead back to these two or are obviously always to be considered as strong enough to ignite a combustible mixture. Practically, a special focus needs to be drawn to static electricity and electromagnetic fields as they are invisible and a discharge can occur randomly. Upon other factors, these values depend on the concentration of fuel (in air) and need to be tested for a range of concentrations. In most cases, the lowest value is not to be found for the stoichiometric mixture.

Minimum Ignition Energy (MIE)

The MIE is the lowest amount of electrical energy stored in a capacitor that leads to an ignition when released in one spark discharge.

$$E = \frac{CU^2}{2}$$

With E the ignition energy [J], C the capacitance [F] and U the voltage [V].

In general, the ignition energy is significantly lower for vapors and gases than for dust. But there are exceptions on both sides, gases with high ignition energies and dust with very low MIEs. In both cases the standard procedures reach their limits of validity.

Minimum Ignition Temperature (MIT)

The MIT is the lowest temperature of a hot surface that leads to the ignition of a burnable substance. For gases and liquids this value is called Auto Ignition Temperature. A term, the author believes to be misleading, as it is a forced ignition by an external event and nothing that happens by itself (from Greek autos = self). It needs to be differentiated from the Self-Ignition Temperature of bulk materials that is determined in a hot storage test, too.

Self-Ignition

Summarized under chemical reaction in the list above the self-ignition process of bulk materials poses a challenge to the storage of usually solid Hydrocarbons. The same process can lead to an ignition of other burnable liquids on insulating media. The two have in common that slow exothermic processes in an isolating environment lead to an increase of the temperature. According to Arrhenius Law a higher temperature leads to a higher reaction rate and thus a self-reinforcing cycle can be created that heats the system to its ignition temperature. These situations are relevant for the storage of big amounts of organic material as well as for insulations contaminated with burnable liquids. Here an ignition source cannot be directly be assigned after an incident. The ignition temperature of dust layers (Glow temperature) is a safety relevant value closely related to this process It focuses on the minimum temperature of a surface to start an exothermic reaction. Here low layers of dust (5 mm) are tested that are expected to be deposited in a process environment where cleaning is not carried out properly. The self-heating process here could lead to glowing regions that could trigger a fire or an explosion. The fact that the hot surface itself directly is no ignition source practically "hides" the source of ignition.

Outcome of the Explosion

Contrary to normal burning processes the heat released in an explosion is of minor interest. The primary outcome is the pressure generated by the reacting system. For standardized procedures only closed experiments are applied. The maximum overpressure does not differ much for different substances and in most cases is below 10 bar. This is not the pressure to be measured in an open cloud explosion, where in a certain distance only a much lower pressure wave is to be expected. Contrarily, the Maximum Rate of Pressure Rise differs significantly for different substance. But this value on the other hand depends strongly on factors like the total volume and the turbulence. To be able to compare the severity of an explosion volume independently the cube root law is applied:

$$\left(\frac{dp}{dt}\right)_{max} * V^{\frac{1}{3}} = konst. = K_{G/St}$$

with V the Volume of the explosion vessel [m^3], p the pressure [bar] and t the time [s].

This leads to the $K_{G/St}$ value (G for gas and St for dust (Staub)) that does not vary with the volume and is used to design technical measures against explosions overpressure. For gases the requirements to apply the cube root law can theoretically be fulfilled:

- Ideal mixture
- Laminar state
- Point source ignition
- Infinitesimal thin flame front

For dusts, none of these factors can be fulfilled. (Hauptmanns, 2015).

Practically, it can be show that the cubic law is not valid, as demonstrated in Stahmer (2019) with an experimental study on alumina dust. The authors discuss the validity of the cubic laws for certain metal dust under the background that the K_{St} value is use in DIN EN 14491 for the specification of the pressure relief areas or the response times of the suppression systems for dust explosion venting protective systems. Therefore, the combustible dust a grouped in three categories (Table 1)

Table 1. Dust explosions classes (VDI 2263 (2018))

Dust explosion class	K_{St} value
St 1	> 0 bis 200
St 2	> 200 bis 300
St 3	> 300

Standards and Data Sources

A good compilation of definitions of Safety Characteristics for practitioner is published in Jedermann-Verlag (2007) and is available free of charge as downloadable file.

Basis to refer to are always the relevant standards as there are in the European community:

- Liquids
  - Flash Point EN ISO 1516, EN ISO 1523, EN ISO 2719, EN ISO 13736
  - Lower explosion point EN 15794
  - Upper explosion point EN 15794
- Gases, liquids and their vapors
  - Lower Explosion Limit (LEL) EN 1839
  - Upper Explosion Limit (UEL) EN 1839
  - Limiting Oxygen Concentration (LOC) EN 14756
  - Ignition Temperature EN 14522
  - Maximum Experimental Safe Gap EN 60079-20-1
  - Maximum explosion pressure EN 15967
  - Maximum rate of pressure rise EN 15967
  - Specific electrical conductivity EN 60247 and EN 60079-32-2
- Dust layers
  - Minimum Ignition Temperature (MIT) of a dust layer EN 50281-2-1
  - Auto Ignition Temperature (AIT) EN 15188
  - Specific electrical resistivity EN 61241-2-2 and EN 60079-32-2
- Dispersed dusts
  - Lower Explosion Limit (LEL) EN 14034-3
  - Limiting Oxygen Concentration (LOC) EN 14034-4
  - Minimum Ignition Temperature (MIT) of a dust cloud EN 50281-2-1
  - Minimum Ignition Energy (MIE) EN 13821
  - Maximum explosion pressure EN 14034-1
  - Maximum rate of pressure rise EN 14034-2

An also free data collection of Safety Characteristics is available in the CHEMSAFE database (https://www.chemsafe.ptb.de/en.html). Here only trustable and traceable results, mainly from the two German federal institution PTB (Physikalisch-Technische Bundesanstalt) and BAM (Bundesanstalt für Materialforschung und -prüfung) are published.

As dusts are not listed here the interested user can also refer to the GESTIS-STAUB-EX database (www.dguv.de/ifa/gestis-staub-ex), a project supported by the Commission of the European Union and produced and maintained by Institut für Arbeitsschutz der Deutschen Gesetzlichen Unfallversicherung (IFA, Institute for Occupational Safety and Health of the German Social Accident Insurance). Values for dust are always a to be treated carefully and are usually only a listed for comparison reason. Practically two dust are never equal in their explosion behavior as this depends on the:

- Chemical composition:
  - Often natural substance with varying mixtures and compounds are processed and treated as equal.
- Particle size distribution:
  - Generally, only the mean value is stated, ignoring the fact that the proportion of the finest particle play a paramount role for explosions.
- Moisture content:
  - A variation of the water content in the dust as well as in the surrounding air influence the different safety characteristics in unequally strong.
- Surface structure:
  - As the reactivity depends on the available surface a different structure might lead to different values even for comparable mean particle sizes.
- Degree of passivation:
  - For Metal dust an oxidized surface layer can significantly lower explosion properties.

For pure liquid or gaseous substances, the user can rely on published data, if the measurement methods are clearly defined and the values are determined by a certified test laboratory. Borders of the comparability are given by Krause (2019) in the program "Explosion Pressure". The authors can prove that their evaluation shows clearly that the current boundary conditions in the technical standards contribute to a diversification of results.

Safety Characteristics are determined by experiments and consequently "they are what they are". They are only valid in the given, usually standardized, experimental setup and under specific conditions. Due to their nature they can neither be calculated nor can they be used for further calculation. Even if they are in some cases the basis for the design of safety systems these are only engineering relation that are not based on purely physical, chemical or thermodynamic relations. The transfer to real applications is not straightforward.

It should also be mentioned that generally only pure or single substance are listed and tested. Mixtures, that frequently are processed in industry, are usually not in the focus. Up to know there are no indication that combinatory effects exist, that would make a mixture more reactive than the more reactive single substance. Consequently, taking the substance with the more critical values as basis can considered to be safe. This might be unwanted where very reactive substances are present in small amounts or traces only.

SYSTEMATIC EXPLOSION PROTECTION

General Structure: Primary, Secondary, Tertiary Explosion Protection

The general explosion protection principle that is introduced here lays down basic safety requirements that should be implemented. These include:

1. Prevent or reduce the formation of explosive atmospheres (primary explosion protection)
2. Avoidance of ignition sources (secondary explosion protection)
3. Limiting the impact of a possible explosion to an acceptable level (tertiary or constructive measures)

The Primary Explosion Protection comprises the two sides of the explosion triangle (Figure 1): Fuel and Oxygen. These two form the chemically reactive system that cannot be separated. The Secondary Explosion Protection clearly refers to the "Heat" that is needed to start the reaction. The Tertiary Explosion Protection focuses on the outcome if the explosions cannot be avoided.

These three measures must be followed in the given order.

Prevention of the Explosive Atmosphere - Primary Explosion Protection

First, a potentially explosible atmosphere has to be avoided or removed. Therefore, the following possibilities can be implemented:

- Substitution
    - To avoid the risk of an explosion or to minimize them first check whether the hazardous material can be replaced by other substances that pose no risk of explosion or where the probability of an explosion is reduced.
        - Typical example is a cleaning process, where an aqueous system replaces the solvent based system.
- Removal and dilution of explosive substances
    - All measures that keep the concentration of fuel below the LEL.
        - Where only small amount of gas or vapor are expected ventilation system can fulfill such purpose. Usually including concentration measurement systems with alarms.
        - In connection with dusts, cleaning of deposits is an effective measure to avoid the forming of ignitable clouds.
- Passivation of explosive substances
    - Some substance can be kept from a spontaneous exothermic reaction by allowing an oxidation in a controlled manner.
        - Metal dusts tend to build a passivated layer on their surface that makes them less reactive, without effective oxidation a high percentage of the mass.
- Inerting
    - The concentration of oxygen is kept below the Lower Oxygen Limit.
        - Inerting measures are usually accompanied by gas measurement system with alarms.
        - An inert gas cushion applied over a combustible flammable liquid makes an ignition impossible.

- Isolation
    - The explosive material should be stored or processed technically tightly, while avoiding the supply of air.
        - Inside of gas pipelines or tanks, the fuel concentration is kept above the UEL or the Oxygen concentration is below the LOC to avoid an ignition.

The application of Primary Explosion Protection measure is to be preferred above the other ones. But as substances are to be processed in industry this often is not possible. Mainly for existing systems they are often inapplicable or simply to expansive.

As soon as one of the Primary Explosion Protection measures ensures the absence of an ignitable and thus explosible mixtures no further steps need to be undertaken and the systems can be considered safe. If this is not possible the next possibility comes into the focus.

Avoidance of an Ignition Source - Secondary Explosion Protection

Therefore, the system has to be checked for potential sources of ignition as listed before. This firstly includes normal operation but the situation during expected possible failures need to be analyzed as well. For equipment to be used in potential explosive atmospheres the regulations provide the framework to determine reliable limits that are often connected to safety characteristics. An overview is given in Table 2. If this is not possible, a systematic checking of the system for potential ignition source needs to be undertaken. Here it is not sufficient to identify source of ignition, but it is necessary to judge whether the energy provided is able to ignite the system - to decide whether the ignition source is really an active one.

Table 2. Basic explosion protection principles and the belonging safety characteristics

Primary Explosion Protection	**Secondary Explosion Protection**	**Tertiary Explosion Protection**
Quantification of the mixture	Quantification of the ignition source	Quantification of the explosion effect
• Explosion Limits • Explosion Point • Flash Point • Burning Point • Limiting Oxygen Concentration • Oxygen Index	• Minimum Ignition Temperature • Minimum Ignition Energy • Maximum Experimental Safe Gap • Glow Temperature	• Flame Speed • Combustion Temperature • Maximum Explosion Pressure • $K_{G/St}$ Value • Calorific Value

Historically, only electrical systems were treated. Of cause also mechanical systems can provide the necessary energy to ignite a system.

Of special importance are situation where the source of ignition is not directly visible. One of these is the occurrence of electrostatic discharges. Whenever a substance is moving with a relative velocity to another one an electric charge can build up. This holds true for gases, liquids and solids. Grounding and the application of conductive materials can ensure a safe ground flow of unintended charges. Where isolating and ungrounded sections are present an accumulation of energy might occur, and random discharges can finally act as ignition sources. An important influencing factor in connection with electrostatic processes is the humidity of the surrounding atmosphere. The conductivity of the water steam can ensure a constant discharge and avoid unwanted charge build-ups.

The second typical "invisible" ignition source are self-ignition processes. These occur where the combustible material is placed in an isolated environment. Due to the random nature of chemical reactions a certain proportion a fuel will always undergo the exothermic chemical reaction. The energy released cannot be dissipated to the environment and thus lead to an increasing temperature of the system. This again leads to an increased reaction rate and thus can trigger a positive feedback loop. If the temperature reaches the ignition temperature of the substances an ignition of the total system occurs. One example for this process are dusts disposed on hot surfaces. Here the Ignition Temperature of Dust Layers (Glow Temperature) is the relevant Safety Characteristic. The same process leads to ignitions in bulk material storages. Additionally, organic reactions, e.g. due to bacteria, can provide the staring initial. As the processes in industrial scale have duration of several month, they are hard to monitor. Safety science tries to predict the behavior by hot storage experiments in lab size, extrapolating the results to industrial scales.

The European Commission provides a "Summary of references of harmonised standards published in the Official Journal – Directive 2014/34/EU 1 of the European Parliament and of the Council of 26 February 2014 on the harmonisation of the laws of the Member States relating to equipment and protective systems intended for use in potentially explosive atmospheres" that provides a reliable basis of applicable standards for dealing with potential ignition sources.

Summarizing a system systematically needs to be analyzed according to the presence of situations, apparatuses and material combination that are able to provide the necessary ignition energy to trigger a combustion.

Not in all cases it is possible to reduce the risk of igniting an explosible atmosphere to the required level only by the selection of appropriate equipment if the situation cannot be changed. Then additional protective measures are to be applied in order to control the effects of an explosion and to limit or to prevent a danger to people, the environment and the installation.

Limiting the Impact of an Explosion - Tertiary Explosion Protection

The tertiary explosion protection is to be applied when the measures of primary and secondary explosion protection cannot sufficiently assure safety. Additional explosion protection measures must then be applied in order to control and limit the effects of an explosion in order to exclude a personal hazard. When an explosion cannot be prevented for sure two main measures are applied. First, the explosion has to be limited in its spreading. Second, the effects need to be limited or controlled.

Some examples where Tertiary measures are applied are:

- Compressor for explosive gas mixtures, which cannot be excluded as ignition source,
- pneumatic conveying of explosive dusts,
- Dust filter systems
- Silos and bunkers for explosive dusts in which the formation (self-ignition) or introduction of smoldering nests is possible.

In general, additional action often need to be take in very vulnerable areas (zones 0 and 20 in Table 3). The following activities typically may be included:

- Pressure relief devices (pressure relief valves, bursting discs), which limit the explosion to a manageable level.
- Explosion proof or explosion pressure shock resistant design of equipment and structures that can withstand the explosion pressure (constructive measures).
- Flame arresters, which allow the gas flow but stop the reactive zone due to cooling.
- Water immersion in pipelines
- Automatic quick-closing devices in conjunction with appropriate detectors
- Autarkic closing valves in pipelines for gas, which barricade line sections in case of sudden pressure increase.
- Explosion suppression systems triggered by appropriate flame detectors or pressure sensors

Various measures exist that always need to be adopted to the present explosible situation. Special precaution needs to be given to surroundings installations, closed by traffic infrastructure, walkways or maintenance accesses, as they have to be included in the concept of explosions release.

Organizational Explosion Protection

Organizational measures extend this concept, enabling a flexible user-oriented explosion protection approach, where the classical measures fail or are otherwise inapplicable. But technical protective measures always take precedence over organizational protective measures. Additional organizational protection measures can significantly contribute to reducing the risk of explosion. Some examples are:

- Consistent spatial separation of activities in which the formation of explosive atmospheres (hazardous areas see Table 3) cannot be excluded from all other activities,
- Removal of all working equipment or devices and equipment from potentially explosive areas that are not necessarily required in these areas,
- Reduction of storage quantities of flammable liquids (order-related provision, disposal of old stocks),
- Regular cleaning,
- Regular maintenance and preventive maintenance.

These also can be applied to time varying processes where a potentially explosive atmosphere usually does not exist and only might form at predefined times. Here special measures are only to be taken during the special event and not during the normal operation. Examples include maintenance or cleaning of filters. Their tight housing limits the dangerous area to the interior when operating. If the enclosure is opened surrounding areas have to be considered endangered and time limited measures can be taken.

Further general aspects of the organizational explosion protection are for example:

- Introduction and monitoring of smoking bans
- Monitoring of the working environment
- Maintenance and service workings
- Create operating instructions and keeping them up to date
- Monitoring the operating parameters
- carry out tests and controls

- Clear and consistent marking of hazardous areas (Table 3)
- Coordination of responsibility
- Work release systems
- Sufficient qualification of employees

Regular safety training for all employees to make them aware of potential dangers and to avoid them to become blind to shortcoming in the companies processes is probably the most important aspect.

European Legislation

As one of the largest economic areas, the European Union has a unified legal system of explosion protection that is widely referred to as ATEX. The name ATEX derives from the French abbreviation for ATmosphères EXplosibles. It currently includes two directives in the field of explosion protection, namely the ATEX equipment directive 2014/34 /EU and the ATEX workplace directive 1999/92/EC. The main purpose of the directives is to protect persons working in potentially explosive atmospheres or who may be affected by explosions. The member states of the European Union are obliged to transpose at least the standards defined in these directives into national law. Deviations are possible but rather few. Hence it only makes sense to refer to the European original and not to national regulations.

The ATEX equipment directive 2014/34 /EU (also unofficially designated as ATEX 114, because of the relevant Article 114 of EC Treaty) for equipment and protective systems intended to be used in potentially explosive atmospheres sets the rules for placing on the market such equipment and systems. This directive is applicable for electrical and non-electrical devices, reflecting, that sources of ignition can be mechanical as well. It follows Directive 94/9/EC (ATEX 95) and must be transpose to national law since April 20, 2016. Manufacturers that bring an equipment on the European market need to proof conformity with harmonized standards. This is done by notified bodies (Nando, 2019) that are authorized to carry out third-party conformity assessment tasks under this Directive and

The 1999/92/EC (ATEX 137) defines the minimum requirements for improving the safety and health protection of workers potentially at risk from explosive atmospheres. It is based on the three explosion protection principles stated above. In Article 3 it is expressed as follows:

The prevention of the formation of explosive atmospheres, or where the nature of the activity does not allow that,

the avoidance of the ignition of explosive atmospheres, and

the mitigation of the detrimental effects of an explosion so as to ensure the health and safety of workers.

The employer is required to prepare an explosion protection document, keep it up to date. This includes regular reviewing in constant time intervals and whenever significant changes are made to the process. Also the employer has to classify areas where hazardous explosive atmosphere can occur (Table 3).

Besides the marking of potentially hazardous areas in appropriate plans, warning signs as illustrated in Figure 5 must be placed on site.

Table 3. Classification of hazardous areas according to ATEX Workplace Directive 1999/92/EC

Gases	**Zone 0** is an area in which an explosive atmosphere, consisting of a mixture of air and flammable gases, vapors or mists exists for long periods or frequently.	**Zone 1** is an area where at times during normal operation which an explosive atmosphere, consisting of a mixture of air and flammable gases, vapors or mists is formed.	**Zone 2** is an area in which under normal operation a dangerous explosive atmosphere as a mixture of air and flammable gases, vapors or mists normally does not occur or only briefly.
Dusts	**Zone 20** is an area in that the explosive atmosphere in form of a cloud of combustible dust in air exists constantly or for long periods or frequently.	**Zone 21** is an area in that the explosive atmosphere in form of a cloud of combustible dust in air exists at times during normal operation.	**Zone 22** is an area in that the explosive atmosphere in form of a cloud of combustible dust in air normally does not occurs or only briefly.

Figure 5. Warning sign for places where explosive atmospheres may occur [Henning (2006).]

According to the zones defined in the explosion protection document the requirement for the devices are defined according to Table 4. These are valid for normal process industry, where Equipment-group I is for mining only.

Table 4. Equipment-group II: Equipment for use in potentially explosive dust and gas atmospheres according to ATEX equipment directive 2014/34 /EU

	Category 1		Category 2		Category 3	
Danger	constantly, frequently or for long periods		occasional		rarely and briefly	
demand	very high safety		high safety		normal safety	
protection	even in the event of rare incidents relating to equipment, and is characterized by means of protection such that: • either, in the event of failure of one means of protection, at least an independent second means provides the requisite level of protection, • or the requisite level of protection is assured in the event of two faults occurring independently of each other.		in the event of frequently occurring disturbances or equipment faults which normally have to be taken into account.		during normal operation	
Zone	Zone 0	Zone 20	Zone 1	Zone 21	Zone 2	Zone 22
Atmosphere	Gas	Dust	Gas	Dust	Gas	Dust

Besides the equipment shall not be an active source of ignition to the predicted atmosphere. Therefore, the atmosphere has to be classified. Gases are divided in three explosion groups IIA, IIB and IIC according to their specific ignitability, which is determined by standardized procedures (Safety Characteristics). The decision is made following the parameters in Table 5

Table 5. Explosion groups for gases according to EN 60079-20-1 and IEC 60079-3

Explosion Group	Maximum Experimental Safe Gap (MESG)	Minimum Ignition Current ratio (MIC)
II A	> 0.9 mm	> 0.8
II B	$0.5 \text{ mm} \leq \text{MESG} \leq 0.9 \text{ mm}$	$0.45 \leq \text{MIC} \leq 0.8$
II C	< 0.5 mm	< 0.45

Table 6. Dust classification codes according to EN 60079

Code	Dust classification
III A	flammable fibers
III B	nonconductive dust
III C	conductive dust

The danger of the gas increases from explosion group II A to II C. Accordingly, the requirements for the equipment increase. Equipment approved for II C may also be used for all other explosion groups.

For dust three Groups are defined as well (Table 6). Here the ignitability does not play a role.

The Temperature Classes define a maximum allowed surface temperature according to the MIT of the gas (Table 7). For dust no classification is available and the maximum allowed surface temperature has to be given directly when labelling the equipment.

Examples for possible combinations of Explosion Groups and Temperature Classes are given in Table 8.

Besides these classes different protection principles are applied for electrical and non-electrical devices. They are classified in different types of protection according to EN 60079-0 (Table 9). The type of protection is not a quality feature but is a constructive solution chosen for the equipment for realizing the explosion protection.

Table 7. Temperature classes according to EN 60079-0 (example data from CHEMSAFE database)

Classes	max. surface temperature	Ignition temperature of some substances for comparison
T1	450 °C	Propane gas 510 °C, Natural gas 650 °C
T2	300 °C	Acetylene 305 °C, Diesel Fuel 310 °C
T3	200 °C	Gasoline 260-450 °C
T4	135 °C	Ether 170 °C
T5	100 °C	*No substance*
T6	85 °C	Carbon disulfide 95 °C

Table 8. Examples of Explosion Groups and Temperature Classes

Explosion Groups	Substances					
IIA	Ammonia Methane Ethane Propane	Ethanol Cyclohexene n-Butane	Petrol Diesel fuel Fuel oil n-Hexane	Acet- aldehyde		
IIB	City gas Acrylic nitrile	Ethylene Ethylene oxide	Ethyl glycol Hydrogen sulfide	Ethyl ether		
IIC	Hydrogen	Acetylene				Carbon disulfide
	T1 < 450°C					
	T2 < 300°C					
	T3 < 200°C					
	T4 < 135°C					
	T5 < 100°C					
	T6 < 85°C					
	Temperature Classes					

Table 9. Protection principles according to EN 60079-0

Code			Protection	Description	Regulation
Electrical equipment		Nonelectrical equipment			
in Gas	in Dust				
Ex i	Ex iD	Ex i	Intrinsic Safety	The supply of the electrical equipment is passed through a safety barrier, the power and voltage is limited so far that the minimum ignition energy and ignition temperature of an explosive mixture is not reached. The equipment is also subdivided into Ex ia for Zone 0 or 1 and Ex ib for Zone 1 or 2 respectively.	EN 60079-11
Ex d		Ex d	Flameproof enclosures	The components that can trigger an ignition are housed in a housing that withstands the explosion pressure. The openings of the housing are designed to prevent transmission of the explosion to the outside.	EN 60079-1
Ex e			Increased safety	The creation of sparks, arcs or impermissible temperatures, which could act as an ignition source, is prevented by additional measures and an increased degree of safety.	EN 60079-7
Ex p	Ex pD	Ex p	Pressurized enclosure	The housing of the devices is filled with a protective gas. An overpressure is maintained, so that an explosive gas mixture cannot reach the possible ignition sources inside the housing. Optionally, the housing is constantly flushed by a gas flow.	EN 60079-2
Ex o			Oil encapsulation (protection by liquid immersion)	The parts of the electrical equipment from which an ignition can be triggered are immersed in a protective oil.	EN 60079-6
Ex m	Ex mD		Encapsulation	The parts of the electrical equipment that can generate ignition sources are embedded in potting compound, so that an arc cannot pass through to an explosive mixture outside the enclosure.	EN 60079-18
Ex n			Ignition protection method n	As Ex q but only for Zone 2.	EN 60079-15
Ex q			Powder filling	Filling the housing of an electrical equipment with a fine-grained product ensures that when used as intended, a resulting arcing in the housing does not ignite a surrounding explosive atmosphere. There must be neither ignition by flame nor ignition by increased temperatures on the surface of the housing.	EN 60079-5
		Ex c	Design safety (constructional safety)	The devices are designed in such a way that they have no sources of ignition during normal operation. The risk of occurrence of mechanical faults, which can lead to the formation of ignition sources, is reduced to a very low level.	EN 80079-37
		Ex b	ignition source	Monitoring of potential sources of ignition such as vacuum pumps, etc. by sensors, in order to be able to detect dangerous conditions early.	EN 80079-37
		Ex k	Fluid encapsulation (protection by liquid immersion)	The parts of the electrical equipment from which an ignition can go out are immersed in a protective liquid.	EN 80079-37
	Ex tD		Protection by housing	The housing is securely sealed against the ingress of dust. This requires a housing seal according to IEC / EN 6052 from IP6x. In addition, dusty surfaces must not exceed a certain temperature.	EN 60079-31

Further regulations for equipment protection are listed in the "Commission communication in the framework of the implementation of Directive 2014/34/EU of the European Parliament" (European Commission, 2018).

Finally, the user has to choose the right combination of requirements for his specific application. Of cause a safer equipment always can be used, e.g. an equipment of Category 1 in Zone 1.

International Regulations

In North America in the field of Explosion protection of electrical equipment and installations techniques and systems that are significantly different. In the USA (NEC 500) and in Canada (CEC 18) define three categories:

- **Class I:** Combustible gases, vapors or mists
- **Class II:** Dusts
- **Class III:** Fibers and flying objects

By the frequency or duration of occurrence of these substances, the potentially explosive areas are divided in Divisions:

- **Division 1:** A location where the hazardous atmosphere is expected to be present during normal operations on a continuous, intermittent or periodic basis.
- **Division 2:** A location in which volatile flammable liquids or gases are handled, processed or used but in which they would normally be confined within closed containers or closed systems from which they can escape only in the event of an accidental rupture or breakdown of the containers or systems.

Accordingly, the National Electrical Manufacturers Association (NEMA) standardizes requirements for equipment for example for Explosion-Proof Enclosures. Although, a temperature classes system exists, that is comparable, but not completely equal to the European one.

The same holds true for the classification of gas atmospheres. The Maximum Experimental Safe Gap (MESG) is a standardized measurement of how easily a gas flame will pass through a narrow gap bordered by heat-absorbing metal. MESG is used to classify flammable gases for the design and/or selection of electrical equipment in hazardous areas, and flame arrestor devices. The National Electric Code classifies Class I hazardous locations into different groups depending on the respective MESG's of gases in the area. (Table 10). These slightly differ from the European grouping presented in Table 5.

Table 10. NEC Classification of gases (Class I)

NEC Class I Group	Gas MESG	Example Gas
Group A	0.25 mm	Acetylene
Group B	$\leq$0.45 mm (Except acetylene)	Hydrogen
Group C	0.45 mm $<$ MESG $\leq$ 0.75 mm	Ethylene
Group D	$>$ 0.75 mm	Propane

In the same way as European directives, transposed into national regulations, set the entry requirements for the European market these need to be fulfilled for the North American market.

As the basic principles of explosion protection are the same everywhere, it is obvious that the conditions for the approval of explosion-proof equipment should be standardized worldwide. The IECEx (International Electrotechnical Commission System for Certification to Standards Relating to Equipment for Use in Explosive Atmospheres; https://www.iecex.com/) System is a voluntary certification system with over 30 member states. Whilst now only few countries have legally recognized the certificate, there is a chance that the number will grow. An updated overview and comparison is given in "An Informative Guide comparing various elements of both IECEx and ATEX" (IECEx, 2018).

APPLICATION AND LIMITS

Typical Forms and Sequence of Events

Explosions in industry are unwanted events that are the consequence of failures, if we assume that prior the explosion protection document was made correctly. Following the basic scheme, one has to look for the explosible atmosphere first and ask how a significant amount of ignitable substance can be released. If the release is to the environment, the oxidizer is not limited, as our atmosphere provides the necessary oxygen. Additionally, the presences of an active source of ignition always has to be assumed.

One of the basic assumptions is the validity of the leak before break criterion. It assumes that a smaller leak, due to corrosion or a penetration is more likely than a complete rupture of the containment (Hauptmanns, 2015).

In such a case the amount that is accidentally released mainly depends on the factors

- Size of the leak
- Geometry of the leak
- Location of the leak

and different forms of release can be differentiated:

- Outflow of liquid from a container without pressure
- Outflow of a liquid from a pressurized container
- Discharge of gases below the speed of sound
- Discharge of gases at the speed of sound
- Discharge of pressure-liquefied gases
- Outflow of pressurized fluids above boiling point

After the release, further factors will influence the cloud formation like:

- Environmental temperature
- Windspeed
- Obstruction
- Soil temperature

These very different situations, their potential outcome and the likelihood are content of extensive publications, like Bartknecht and Zwahlen (2013), Bosch and Weterings (2005), Eckhoff (2008), Hattwig & Steen (2008), Hauptmanns (2015) and Mannan and Lees (2005). There the reader can find in depth information and further publications. In the following chapter only some major simplified events and remarks on the limits that restrict generalized descriptions are given.

Examples and Limits of Application

The first two major release situation are already mentioned in Figure 1; the VCE and the BLEVE.

Vapor Cloud Explosion

VCE occur after the release of a gas out of a containment or of a liquid that if further evaporated. In Bosch and Weterings (2005) the conditions for a VCE are explained in detail and can be summarized as follows:

1. The substance released must be flammable.
2. A cloud must form before the ignition.
3. Part of the cloud must be in the flammable concentration.
4. The overpressure produced is a function of the flame speed.

Even if the first condition seems to be trivial it needs to be considered that the condition of storage and that after the release differ and that in a process the location of the release might determine the composition of the fuel and is generally unknown.

The second point refers to the fact that an immediate ignition will lead to a pool fire or a jet fire instead of an explosion. It depends on the specific situation whether this might be less disastrous than a VCE. It has to be considered as well, that there by chance might be no ignition at all.

After the release, the flammable substance mixes with the surrounding air and for a partially ignitable cloud. The outer parts are usually to lean to be ignited and the mixture near the release or the pool will be to rich. How this ration between the three zones varies depends upon other factors on the wind speed and the time until ignition.

The major outcome of a VCE is the overpressure that is a function of the flame speed. Influencing factors are the specific material values of the fuel, the place of the ignition in the cloud and in general the turbulence created by the process and the environment (wind speed, congestion). In extreme situation this might lead to detonations as this process is self-reinforcing (Bosch and Weterings, 2005).

To judge what really is to be expected in an installation is obviously not an easy task; to calculate the potential outcome above all. The problematic is that to only focus on the worst-case scenario will lead to much too restrictive measures against accidents and thus to the situation that it can only be realized at too high cost or not at all. Contrarily, a belittlement of the situation might have catastrophic consequences.

Boiling Liquid Expanding Vapor Explosion

The BLEVE is a combination of physical and chemical explosion. Usually it occurs with pressurized liquefied gases that are stored in a containment. This can be small gas bottles, railroad tankers or big stationary storage facilities.

The first step is the accidental release of a pressurized liquefied gas. This may occur due to a failure of the containment directly or because of a fire that heats up the container. Usually such storage facilities are equipped with pressure relief facilities to avoid an unwanted pressure build up. In case of an accident they might fail or the pressure rise due to the heating might be that high, that a pressure is achieved where the containment fails completely. In such a case, the sudden depressurization leads to an evaporation due to the pressure drop and a rapid expansion of the vapor cloud. An intense mixing of the fuel with the surrounding air and high turbulent conditions are the consequence. These can lead to very strong (chemical) explosions. A BLEVE can have some or all the following outcomes (Hauptmanns, 2015):

- Pressure wave
- Missile flight
- Fireball
- Pool fire
- Vapor cloud explosion
- Flash fire

Dust Handling

Dusts pose a special problematic to process industry. They are processed in various branches, including such that are not considered to be that industrial as pharmaceutical or chemical industry. Typical examples deal with agricultural products for nutrition or heating. Besides the general lack of understanding for the problematic of explosions the materials there are often considered to be less harmful. Of cause, this is not true and lead to numerous accidental events as can be followed on the "Weekly Combustible Dust Update" (https://dustsafetyscience.com/).

Specific about dust explosions is the fact that solids usually are processed under environmental conditions. The logic consequence is that primary protection cannot be applied. Often secondary explosions protection is also not applicable, as for example the following processes cannot be avoided:

- Self-ignition in storages
- High energetic process steps (milling, grinding)
- Hot surface in connection with transport processes
- Electrostatic discharges.

Consequently, ternary explosion protection measures have to be applied. The problematic that arises here is the limited predictability and upscaling reliability of the current calculation methods. These base upon empiric models that are not valid for all materials. One of the relations that is basic for the sizing of ternary measures is the K_{St} value. However, for dusts none the physical foundation to apply this relation is fulfilled. Often operating experience or the knowledge of suppliers of protection equipment becomes the basis for decision making.

In general, the explosion protection measures are following a good engineering tradition. Limits of application are not always reflected in the regulations and therefore are focus of the current research activities.

FUTURE RESEARCH DIRECTIONS

Introducing standards to ensure safe handling and processing of burnable materials was an application-oriented approach. Throughout the years, the focus was more on understanding the underlying processes in detail. Nonetheless, it is yet not possible to foresee the reaction of complex systems in all cases. Typical fields of interest are listed here:

- Reactions mechanism of heterogeneous systems
- Ignition mechanism
- Deflagration and Detonation in very small geometries like micro reactors (Meye, 2015)
- Nonstandard conditions: behavior of systems and validity of Safety Characteristics (Tschirschwitz, 2015)
- Reaction of Mixtures: fuels with other fuels or inert materials of same or different aggregate state, e.g. hybrid mixtures (Addai, 2016)
- New materials: e.g. nano dusts (Krietsch, 2015)

The big aim in many research activities, besides increasing the safety in general and extending the field of application, is to provide better predictions or mathematic relations for simulation programs. These will find their way to standard CFD programs or are implemented in commercial (e.g. FLACS by Gexcon) or free software projects (Großhans, 2019).

CONCLUSION

The principle behind Explosion Protection in Process Industry can be understood by every engineer. The Fire Triangle as simplified as it is plays an important role as reminder of core basics. The underlying physical and chemical processes in their application lead to a variety of different situation that have the potential to become catastrophic. In fact, even the smallest accident that causes injuries, fatalities or losses should be avoided. Therefore, every situation must be looked upon individually.

The tool to avoid unwanted dangerous situation is the classification into the three categories of explosion protection. Following this schematic ensures a traceable and reliable explosion protection. To achieve this, the safety characteristics play a vital role. Users have to be careful, as they are limited in their direct application. The standardization behind the Safety Characteristics ensures the possibilities to compare substances but the transfer from the lab scale experiment to real industrial installation is only possible by means of engineering relations. Additionally, random processes have to be taken into account. Often user experience has to be used instead of profound scientific knowledge and relations.

The available framework of rules and regulation in general provides methods and measures to avoid explosions and ensure process safety. These prescriptions are not always straightforward and directly applicable. Country-specific difference and deviation have to be accounted for. Thus, the user might look for help to realize an implementation that fulfills all the regulations. Over the last decades, a lot of experience was collected and is now provided by associations, companies and specialized engineers. Whenever an installation becomes bigger or more complicate, external help could be a surplus of safety, also by its external point of view. This may lead to complex explosion risk assessments.

Apart from an installation that was profoundly designed with the help of safety engineers the employees play a key role to process safety. Many accidents could be avoided if the staff is trained on understanding potentially dangerous situation and a climate of safe operation is implemented and lived in the companies. Not all situations can be foreseen, as the work life is not static and prone to constant changes. Total safety is not achievable, but each explosion less, each fatality avoided and each staff less injured is worth the effort.

REFERENCES

Abbasi, T., Pasman, H. J., & Abbas, S. A. (2010). A scheme for the classification of explosions in the chemical process industry. *Journal of Hazardous Materials, 174*(1-3), 270–280. doi:10.1016/j.jhazmat.2009.09.047 PMID:19857922

Addai, E. A. (2016). *Investigation of explosion characteristics of multiphase fuel mixtures with air*. Powell, Wyoming: Western Engineering, Inc.

Babrauskas, V. (2003). *Ignition handbook principles and applications to fire safety engineering, fire investigation, risk management and forensic science*. Issaquah, WA: Fire Science Publ.

Bartknecht, W., & Zwahlen, G. (2013). *Explosionsschutz*. Berlin: Springer Berlin.

Bosch C., & van den Weterings R. (2005) *Methods for the calculation of the physical effects—due to releases of hazardous materials (liquids and gases)—yellow book*. CPR 14 E, The Hague.

Eckhoff, R. (2008). *Dust explosions in the process industries*. Amsterdam: Gulf Professional Pub.

European Commission (2018). Commission communication in the framework of the implementation of Directive 2014/34/EU of the European Parliament and of the Council on the harmonisation of the laws of the Member States relating to equipment and protective systems intended for use in potentially explosive atmospheres. *Official Journal of the European Union*, 61.

Ex Guide. (2018) IEC System for Certification to Standards relating to Equipment for use in Explosive Atmospheres - An Informative Guide comparing various elements of both IECEx and ATEX. Geneva: IEC.

Großhans, H. (2019). *Simulation of Explosion Hazards in Powder Flows*. Braunschweig: Physikalisch-Technische Bundesanstalt (PTB). doi:10.7795/210.20190521P

Hattwig, M., & Steen, H. (2008). *Handbook of Explosion Prevention and Protection*. Weinheim: Wiley-VCH.

Hauptmanns, U. (2015). *Process and Plant Safety*. Berlin, Heidelberg: Springer Berlin Heidelberg.

Henning, T. (2006). *Explosible region, warning sign*. Wikimedia. Retrieved from https://commons.wikimedia.org/wiki/File:D-W021_Warnung_vor_explosionsfaehiger_Atmosphaere_ty.svg

Jedermann-Verl. (2007). *Safety characteristics determination and rating*. Heidelberg.

Krause, T., Wu, J., & Markus, D. (2019). *Ringversuche im Bereich des Explosionsschutzes - Ergebnisse und Erkenntnisse aus dem Programm "Explosion Pressure"*. In Proceedings of the 15th BAM-PTB. Academic Press. doi:10.7795/210.20190521D

Krause, U. (2009). *Fires in silos*. Weinheim: Wiley-VCH. doi:10.1002/9783527623822

Krietsch, A. (2015). *Untersuchung der Brand- und Explosionsgefahren von Nanostäuben*. Magdeburg: Otto-von-Guericke-Universität.

Mannan, S., & Lees, F. (2005). *Lee's loss prevention in the process industries*. Burlington, MA: Elsevier Butterworth-Heinemann.

Meye, T. (2016). *Untersuchungen zu Gasdetonationen in Kapillaren für die Mikroreaktionstechnik*. Magdeburg: Otto-von-Guericke-Universität.

Molnárné, M., Schendler, T., & Schröder, V. (2003). *Explosionsbereiche von Gasgemischen*. Bremerhaven: Wirtschaftsverl. NW.

Nando (New Approach Notified and Designated Organisations) Information System. (2019). Retrieved from https://ec.europa.eu/growth/tools-databases/nando/index.cfm

Stahmer, K. W. (2019). Explosion parameters of aluminium dust in different volumes: the limits of the cube law. *Chemical Engineering Transactions, 75*.

Tschirschwitz, R. (2015). *Entwicklung von Bestimmungsverfahren für Explosionskenngrößen von Gasen und Dämpfen für nichtatmosphärische Bedingungen*. Magdeburg: Otto-von-Guericke-Universität.

VDI 2263. (2018). *Staubbrände und Staubexplosionen - Gefahren - Beurteilung – Schutzmaßnahmen*. VDI-Gesellschaft Energie und Umwelt.

ADDITIONAL READING

Abbas, Z., Zinke, R., Gabel, D., Addai, E. K., Darbanan, A. F., & Krause, U. (2019). Theoretical evaluation of lower explosion limit of hybrid mixtures. *Journal of Loss Prevention in the Process Industries, 60*, 296–302. doi:10.1016/j.jlp.2019.05.014

Crowl, D. A. (2019). *CHEMICAL PROCESS SAFETY : fundamentals with applications.*

Kletz, T., & Amyotte, P. (2019). *What went wrong?: case histories of process plant disasters and how they could have been avoided*. Amsterdam: Butterworth-Heinemann.

Ministries of Transport, Social Affairs and of the Interior. (2005). *Guidelines for quantitative risk assessment — Purple Book*. (CPR 18E). The Hague.

Ministries of Transport, Social Affairs and of the Interior. (2005). *Methods for determining and processing probabilities – Red Book*. (PGS 4). The Hague.

Nabert, K., Redeker, T., & Schön, G. (2004). *Sicherheitstechnische Kennzahlen brennbarer Gase und Dämpfe : Kenngrößen des Explosionsschutzes, zusammengestellt aus Schrifttum und eigenen Messungen* (Vol. 1). Hamburg: Dt. Eichverl.

Schröder, V. (2012). Flammable gases and vapors. In U. Hauptmanns (Ed.), Plant and process safety (Vol 2, 8th ed.). Wiley-VCH.

Siegfried Bussenius. (1996). *Wissenschaftliche Grundlagen des Brand- und Explosionsschutzes*. Stuttgart, Berlin, Köln: Verlag W. Kohlhammer, Cop.

KEY TERMS AND DEFINITIONS

Deflagration: A combustion that spreads at a speed which is lower than the speed of sound in the burning medium.

Detonation: An Explosion that is faster than a Deflagration and where the spread of the chemical reaction is coupled with a shock wave.

Explosion: In general process in which a very large amount of energy is released in a very short time; in process safety usually fast combustion process that liberates energy in form of heat and pressure.

Explosion Region: Concentration of fuel in air that is ignitable, limited by the lower and upper flammability limit.

Fuel: Any substance that can be oxidized so can be burned – a combustible substance

Ignition: The process to start a combustion in a way that it is self-sustaining.

Safety Characteristics: Standardized values to describe the potential danger of substance; used to define measures for save handling and processing.

Source of Ignition: Physical or chemical process that provides the necessary energy to ignite a substance.

Chapter 9
Simulation and Modelling in Fire Safety:
Virtual Reality for Smart Firefighting

Tomaz Hozjan
University of Ljubljana, Slovenia

Kamila Kempna
VSB – Technical University of Ostrava, Czech Republic

Jan Smolka
VSB – Technical University of Ostrava, Czech Republic

ABSTRACT

Actual and future concerns in fire safety in buildings and infrastructure are challenging. Modern technologies provide rapid development in area of fire safety, especially in education, training, and fire-engineering. Modelling as a tool in fire-engineering provides possibility to design specific fire scenarios and investigate fire spread, smoke movement or evacuation of occupants from buildings. Development of emerging technologies and software provides higher possibility to apply these models with interactions of augmented and virtual reality. Augmented reality and virtual reality expand effectivity of training and preparedness of first (fire wardens) and second (firefighters) responders. Limitations such as financial demands, scale and scenarios of practical training of first and second responders are much lower than in virtual reality. These technologies provide great opportunities in preparedness to crisis in a safety way with significantly limited budget. Some of these systems are already developed and applied in safety and security area e.g. XVR (firefighting, medical service).

DOI: 10.4018/978-1-7998-3059-7.ch009

INTRODUCTION

One of the main concerns of safety of infrastructure is fire safety. In case of fire occurrence, probability of domino effect is increased. For limitation of consequences, their impact and loses on life or property, effective reaction and suppression attack is the key. Effective suppression response and rescue operations require efficient and advanced training. This training is usually time consuming, expensive and time demanding. It also requires a place where some of the techniques need to be tested for possible future events. Modern technologies provide improving of this training and drill and decrease expenses and price of a firefighter's teaching program. One of the most emerging advanced technologies is Virtual or Augmented Reality.

It has been decades when firefighters do not only extinguish fire but also provide additional special service and duties in frame of rescue operations. It is also reason why fire brigades are commonly transform to Fire and Rescue Service. Activities of Fire and Rescue Services in frame of firefighting include structural firefighting, forest firefighting, extinguishing of vehicles, industrial firefighting, or any other types of fires where involving potentially injured or killed persons, animals or damaged property. Furthermore, activities of firefighters include services in HazMat leakages, intervention in traffic accidents related to collisions, search and rescue activities, rescue from waterscapes, rescue from heights and depths, technical assistance such as emergency opening of apartments, assistance in explosive findings and much more.

This is a wide list of activities that requires large amount of basic training and additional continual professional development which can deliver highly efficient persons able to react in the most difficult situations in modern technological and industrial world.

Intervention in mentioned activities are usually done under stress of the persons. Time pressure in some cases is life decisive. These tasks are usually realized in difficult conditions including high or low temperatures, when first responders hear people screaming, in darkness or low visibility caused by smoke, occurrence of toxic gases, products of combustion or any other danger substance. Every mistake can cause serious consequences on health of his own, persons in danger or property. In event of fire, firefighters provide suppression attack usually with breathing apparatus (SCBA), which increase load on firefighters' body very often exceeding 20 kilograms. With this equipment, a firefighter is able to operate in basements as well as in high-rise buildings and provide effective intervention in limited time defined by his own air consumption in SCBA, depending on health, psychological state, fitness and activity of firefighter, usually up to 40 minutes.

These conditions require physically and psychically health persons with ability of own decision making under pressure of surrounding circumstances and very well trained in his tasks. It requires a lot of time consumed by drills for specific activities, education and preparing for specific tasks, training cooperation, communication and coordination with other crew members.

Specifically, activities and specialization require additional training in facilities – for e.g. climbing in a polygon and diving in a smoke, try HazMat polygon. Firefighters learn a lot of theory about fire dynamics, fire behaviour, car systems, building structures, extinguishing methods and principles and much more.

There are several aspects and information needed to deliver to firefighters in so called basic course. Additional load of knowledge for firefighters necessary to understand and know can be learned during further continual professional development in their career. As training of firefighters requires to know theory, practical training and drill is equally important to deliver full preparedness for a practice or firefighters. In case of volunteer firefighters, it is even more complicated because of limited time of their training and preparedness due to training in their free time.

Virtual Reality and Augmented Reality are modern solutions and future possibilities to effectively deliver knowledge to firefighters and partly replace practical training. Use of these technologies dates back to early 90's when they have been used for training of astronauts, pilots of airplanes, first responders, military combat trainings and training of operators and drivers of vehicles or medical care service.

Since the first use of these technologies to near history, application of virtual environment has required high computing capacities for rendering, robust hardware and not very acceptable graphic cards and user-friendly environment. Development in the last years provide modern virtual reality systems with head-mounted displays, tracking movement or controllers allowing movement in virtual environment. Currently, modern gaming computers have enough computing power for operating virtual reality. When price of virtual reality system can be purchased from $400 and computer form $1000. Access of these technologies due relatively inexpensive price provides access to wider public. It also allows rapid development in research of this area. Additional factor helping to enrich development is open free accessible platform Unity 3D for advanced development in virtual reality software and virtual environment.

Current situation provides great opportunity for innovations and development of new educative technologies in various areas including industry, school educations or preparedness of first responders. Benefit of training in virtual environment is possibility to locate it anywhere in a room without needs of surrounding area. Firefighters have possibility to access virtual reality during duty whenever they have free time and interest in further personal skills and development and in a way suitable for them.

ADVANCED TECHNOLOGIES IN FIREFIGHTING TRAINING IN TECHNICAL INFRASTRUCTURE

Technical infrastructure has very wide meaning and similarly, also events where fire is involved can vary among industries and usually can have different scenarios. From designers of buildings or machines among operators to first responders, all of them need to think about effectivity and safety. In the 21st century we are at the point where novel technologies can help these professions to better and faster understand the problem and learn effectively. Considering safety and prevention, a man has usually only one chance to design or act wisely in a real life.

The only way how to learn right decision-making processes is a good training and own experience. For the first responders, real training exists but is not spread and affordable for everyone. Sometimes videos do not offer the firsthand experience to remember. In a reality created virtually, lots of scenarios and situations can be created and trainees can learn from good and wrong experiences with strong memorizing. Examples of practice where people can be given control to navigate through either the real or virtual environment and to interact with objects are augment reality and virtual reality.

Virtual Reality as a Training Tool

Virtual reality can be generally defined as "a computer-generated simulation of the real or imagined environment or world." (Gaddis 1998). Modern tools for virtual reality provide three-dimensional (3D) environment using on numerical data along with audio-visual hardware. In this way it provides experience for user to feel to be a part of the man-made environment.

Concept of virtual reality has been discussed early in 1990's when virtual reality was promising application (Milgram 1995). It was caused mostly by high performance requirements for graphical rendering, expensive technologies and computing power which resulted in uncomfortable, not realistic experience. However, bases for mixing elements from real life and virtual life were built and terms as augmented reality, augmented virtuality or virtual environment defined, Figure 1. The main application was usually for industry and military combat training, (Gaddis 1998).

Figure 1. Migram's reality-virtuality continuum as a tool in education for learning in the experience age. Adapted from (Milgram 1995)

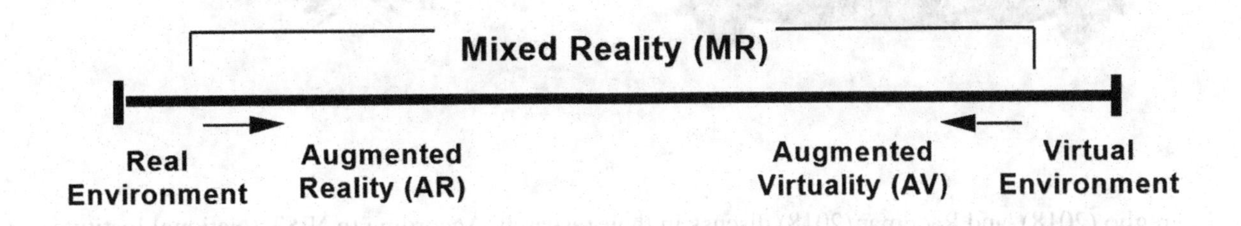

Current development in technologies provides novel equipment and technical possibilities allowing more realistic environment and experience for a user, (Royal 2019). Less expensive hardware such as equipment and gear for virtual reality starts at price that opens market for platforms development. For example, Unity provides this technology more accessible and allows further and rapid development of this area which nowadays allow researchers and companies to work together and create not only games but also usable training tools, (Nandavar 2018). More common augmented reality and virtual reality are further discussed.

Augmented Reality

Augmented reality (AR) as a modern technology is usually defined as a real-time direct or indirect view of real-world environment whose elements are augmented by adding virtual objects or components. Augmented reality primarily aims at enhancing real view in a real time and add virtual objects (Carmigniani, 2011), (Lovreglio 2018). The AR provides integration of digital information, simulation, artificial models and objects in real environment after reading specific stamps, marks or QR codes. AR can be delivered through smartphones or tables to view e.g. live text translation.

Virtual Reality

Virtual reality (VR) instead is created virtually completely to bring users out of the real world. It gives a trainee possibility to practice certain scenarios usually wearing e.g. head mounted display (HMDs), (Royal 2019).

Both, virtual reality and augmented reality have been used for various purposes related to disaster risk reduction including investigating of human behavior in evacuation as authors of Ronchi (2015),

Figure 2. First visible distinguishing between virtual and augmented reality is that augmented reality use computed modelled objects and put them into vision of real world (Adopted for comparison from [Martirosov 2017])

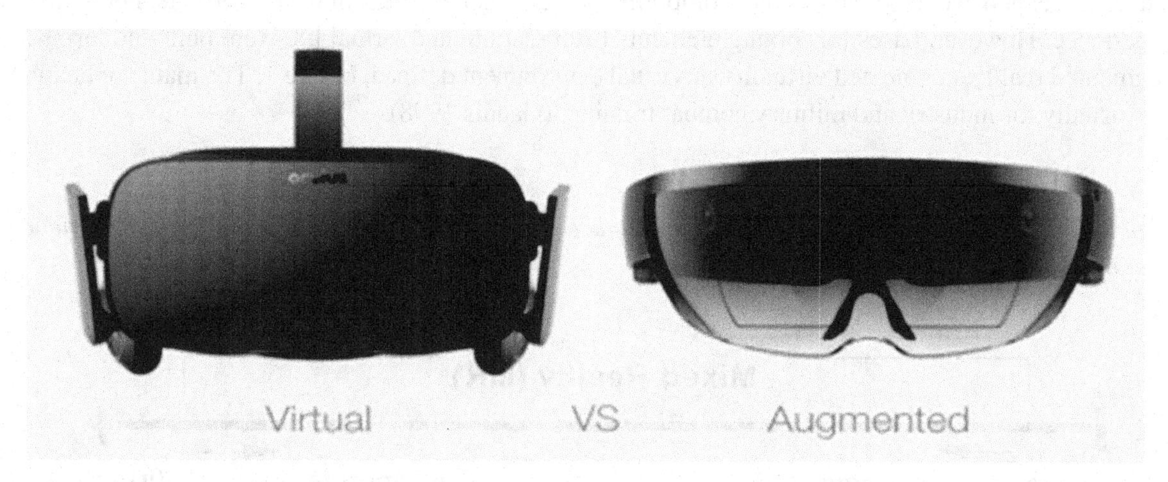

Lovreglio (2018), and Rådemar (2018) discuss in their research. According to NIST (National Institute of Standards and Technology) and The Fire Protection Research Foundation application and development of virtual and augmented reality need more research, innovation and development for Smart Fire Fighting as stated in their research road map (Jones 2015). Further systems used in infrastructure, environment and system of awareness such as BIM (Building Information Modeling) or IoT (Internet of Things) will together with right use and further work create safer future. For this challenge, several aspects need to be concerned.

Intellective Benefits of Virtual Reality Usage in Education and Training

For purposes of this book, chapter will be further focused mainly on application of virtual reality and main concern will be given to training and education of firefighters as an example.

Experimental and Active Learning

Environment of virtual reality provides experiences with use of new technologies. For the participants it requires active participation in education and assimilation of knowledge. Trainee obtains possibility to have freedom of movement and involvement in self-directed activities within the learning context.

It provides the way of education when "ongoing process of making sense out of new information- by creating their own version of reality instead of simply receiving the author's view" (McGuire 1996). Trainees also get better understanding of material used for manipulation and building of physical artefacts that can support knowledge of building construction. One of the examples of training with virtual reality for fire fighters is study of car rescuing, demounting car, active / passive protection, etc.

Visualization and Reification

The virtual reality provides novel methods of information and study material representation instead of classical presentations and only books or videos. By using virtual reality user can reach better understanding and visualization of the issue. In this way, virtual reality is a great tool for visualization of difficult information for abstract representation (e.g. physical or chemical concepts of extinguishing).

Learning Beyond the Comfort Zone

Presentation in virtual environment provides opportunity to show difficult or even potentially hazardous experiences for a practise of firefighters. It provides deeper understanding, information collection and their processing by trainee – e.g. movement in a burning and smoke-filled house, visualization of extreme fire behaviour, visualisation of temperature and smoke distribution, etc.

Motivation Enhancement

Using virtual reality is more playful, educational process together with stronger emotions increase interest of trainee in the subject. Interactive education motivates trainees in further and deeper learning, (Smith 2009), (Wells 2005).

Collaboration Fostering

Shared virtual reality in a group of people can encourage collaboration and foster learning process. Shared virtual environment provides opportunity to train and drill crew collaboration, teamwork and fulfilling even more complex tasks.

Adaptability

Another benefit provided by virtual reality is possibility to tailor scenarios to individual trainee's characteristics and needs. Trainees have opportunity to learn by themselves with their own comfortable way suitable for needs of the individuals, commanders, and group.

Evaluation and Assessment

Having virtual reality system with sensors and detectors for individuals provides evaluation system for each session. Overall behaviour of trainees can be monitored, evaluated and recorded. This provides more practical way of training, testing and final evaluation with verification of real understanding for individual trainees (Mantovani 2003).

The effectivity of people's perception during wearing virtual reality system is very dependent on how they feel during education or training, in the other words, level of immersion is crucial.

Possibilities and Categories of Virtual Reality

Virtual reality systems are categorized by their immersion. Main categories are non-immersive and immersive.

- **Non-immersive virtual reality system:** Non-immersive virtual reality system uses conventional graphics workstation with a monitor, keyboard and mouse.
- **Semi-immersive virtual reality system:** This system uses relatively high-performance graphic computing system connected with large surface to display the visual scene.
- **Fully immersive virtual reality system:** Projection is delivered through head mounted device or via large projection surface, which encases user and provides full feel of presence in the virtual reality.

Additionally, types of virtual reality can be recognized in three dimensions according to precepted presence of the user (Baus, 2014).

- **Personal presence**: Personal presence refers to the feeling of being in virtual environment (with e.g. physical room where the immersion is placed).
- **Environmental presence:** Environmental presence refers to the feeling that virtual reality seems to acknowledge the use by interaction with user.
- **Social presence:** Social presence refers to the feeling of not being alone in the virtual environment.

The strength of presence in the virtual reality moreover corresponds with ability of interaction between a user and virtual environment, objects, animated entities or other users.

Realistic feeling depends on possibility to isolate user from real environment surroundings, degree of convergence between the expectation of the user and actual experience in the Virtual Environment (VE) – e.g. smelling smoke and feeling heat. It means that high level of realism can be associated with virtual environment and advanced level of reality is delivered to user.

Virtual Reality Application

Virtual reality finds application in numerous of areas including healthcare, engineering, communication or entertainment.

Nowadays, virtual reality is more often used in training and drill of professionals. The application of virtual reality provides possibility to decrease costs and expenses of training, provides interactivity, deeper training with possibility to find out limitations of using specific equipment which would not be possible in real world due to high costs or protentional hazards.

Interactivity in virtual reality provides possibility to develop and prepare various very detailed scenarios at multiple levels and difficulty which deliver continual and more fluent learning for users. Several of applications of virtual reality in practical training can be highlighted. It widely used for visual inspection of aircrafts, operation of various vehicles, operation with cranes, rapid and effective decision making by medical services and solders in stressful situations, communication training for military deployments and emergency management, stress management or even fear and phobia management training (Baus, 2014).

Virtual Reality and Existing Hardware

In order to deliver very real experience, several types of hardware can be identified as useable for virtual reality on the market. The main equipment includes headsets, controllers, base station for movement monitoring. Advanced equipment used for virtual reality includes Cave virtual reality systems, Treadmills and special developed equipment, compartments, training equipment specially developed for specific purposes (airplane pilots, crane operators, etc.).

Headset is a head-mounted device with headphones provide virtual reality through two small displays. Head mounted displays place a screen in front of user's eyes during the use. A modern headset has integrated headphones and microphone, G-sensor, gyroscope and proximity providing function to track movement of user's head. Headset can be connected to pc wirelessly or through USB cables.

Another important part of hardware are controllers. There are usually several special controllers, including grip buttons, triggers, control buttons and multifunction trackpad allowing user to control virtual reality and interact with virtual objects.

For monitoring and tracking of user's movement in certain area apply base stations.

Treadmills is kind of equipment allowing user natural movement as well as monitoring his position and improvement regarding situation while being in virtual reality. These aspects are crucial for training of firefighters' drill. However nowadays treadmills are still in development, with limited application.

CAVE virtual reality system is virtual environment consisted of cube-shaped virtual reality - a room where surrounded walls, floor and ceiling are projection scenes. User usually wear headset and interact through input devices. The CAVE must be fully dark during its usage. User being synchronized with projectors can walk around and see projected reality from angles.

VIRTUAL REALITY AS A TRAINING TOOL FOR FIREFIGHTERS

The virtual reality becomes to be more common in systematic training and education for combat training in military, astronauts in NASA, airplane pilots or for first responder. The virtual reality provides new methods of education with higher efficiency even though this technology still requires further development and research. Work of a firefighter is very complex and requires preparations for several types of tasks. Main types of events where firefighters interevent on everyday bases worldwide follow.

Main Incidents and Activities of Firefighters

To identify the most common types of incidents involving intervention of firefighters, list of incidents and their types was taken as an example. This list of activities was taken from CFIF (CTIF, 2018) for international data and from Statistical Yearbook from Fire and Rescue Service of the Czech Republic (HZS, 2019) as for a particular country chosen based on the country's diversion.

Fires: fighting fires is intervention to any of undesirable combustion out of control which can cause harm to persons, animals, property or environment. According to CTIF, the highest number of firefighting interventions by are structural fires - 35.5%, other types - 20%, grass and brush - 18.8%, vehicles - 13.5%, rubbish – 8.9% and forests – 3.3%. Data comes from 2016 for illustration (CTIF, 2018).

Traffic Accidents: interventions related to collisions in transport means, where persons and animals were injured or killed or there is a property damage. Car accidents followed by fire are considered in fire in statistics.

HazMat Leakage: Intervention in emergencies associated with undesirable leakage of HazMat, including oil products during production, in transport or when handling and other substances. Intervention is aimed to limit or reduce the risk of uncontrolled release of flammable, explosive, corrosive, toxic, harmful, radioactive or any other hazardous substances, oil products or other substances into the environment (e.g. natural gas, acids and their salts, alkalis, ammonia), including serious accidents.

Leakage of Oil Products: Kind of intervention with goal mainly to prevent leakage and to limit its range of oil (gasoline, diesel, or oil).

Technical Accidents: intervention to eliminate hazards or hazardous conditions

Technical Assistance: intervention to eliminate hazards or dangerous conditions within smaller scales

- Rescue of persons from lifts.
- Emergency opening of apartments.
- Removing obstacles from roads and other areas.
- Disposal of fallen trees, electrical wires, etc.
- Ventilation.
- Rescue and evacuation of people and animals.
- Pumping water, water closing and water supply.
- Assistance in explosives findings.
- Provisional or other repairs.
- Extraction of objects and persons.
- Measurement of hazardous concentrations and radiation.

Technological Assistance: Intervention to eliminate hazards or hazardous condition in the technological operations of companies.

Other Assistance: Intervention, which cannot be defined as a technical accident, technical or technological assistance, such as transport of patients, searching missing persons, monitoring water streams, road accessibility control, etc.

Radiation Accident: Intervention in incidents related to the improper release of radioactive substances and ionizing radiation.

Other emergency Events: Intervention in other emergencies such as epidemics or infection, ensuring suspicious shipments and intervention for events that cannot be classified about mentioned types, (HZS, 2019).

There are additional firefighters' specialization requiring special competences and qualification for working in following professions.

Chemical Specialists: Chemical specialists are responsible for handling with chemical and radiation accidents, maintaining of chemical and SCBA equipment.

Transmitting Specialists: Transmitting specialists are responsible for ultra-high frequency (UHF) stations and communication in their unit including maintains and service.

Machine Specialists: Machine specialists are responsible for all maintains and operability of all engines and vehicles. These specialists are usually drivers of fire trucks with special training for driving cars with sirens.

Working at Height Specialists: Specialists of this area are trained and prepared to work on ropes at heights. Extensive qualification provides ability to provide rescue operation from helicopter.

Smoke Divers: Smoke divers provide search and rescue operations in burning houses and buildings, as they are trained and prepared for moving and orientation in smoke-filled areas and can provide rapid search activities.

Search and Rescue Specialists: In case of earthquakes, building can collapse, floods there are special units specialized for searching and rescuing people from debris.

Water Rescue Specialists: Trained specialists for using equipment for rescuing on waterscapes or during floods.

Fire Investigators: Another group of specialists are fire investigators. Creation of case studies from past fires can consider use of 3D models, 3D scanned area provides great opportunities for better training and preparedness for their work.

Emergency First Aid: Firefighters are responsible for emergency health care and transport to hospital in specific countries.

Number of activities has been increased (HZS, 2019). Tasks and duties included in everyday work of firefighters are very complex and people need to have overall overview, knowledge and experience for their accomplishing. The virtual reality is advanced technology useful for increasing efficiency of preparedness, training and drill of first responders. As there is number of different requirements on firefighters' skills and qualification in various areas, training of these competencies is time and financial consuming and demanding. One of the main tasks is structural firefighting where polygons are often used for training.

Training and Development of Firefighters' Skills

Process of firefighting training requires specific personal qualities and skills which need to be delivered on the highest possible level for approaching the most effective rescue and suppression operations. For development of virtual reality-based training equipment for firefighters it is needed to define those skills needed to be developed during the training tested further.

Working Under Stress with Limited Time

Psychical stress is very often part on the scene. Firefighters use SCBA equipment for ability to breathe and operate in areas with low level of oxygen and in events involving presence of toxic gases. It requires skills, training and psychological and physical preparedness of firefighters to operate in these spaces and fulfil tasks in the shortest time including search and rescue of persons, animals or subjects, orienting in the area or providing suppression actions.

Working with Low Level of Air Capacity

Operation in SCBA in smoke-filled area requires firefighters' overview about situation, monitoring of own level of air in SCBA and count with way back. When reserve of air is exceeding, the SCBA will start with signalization of low level of air in a tank. It increases physical pressure to firefighters to finish his activities and move out of the place as soon as possible. This situation can be really stressful.

Communication, Cooperation and Coordination

Crew cooperation and coordination by a commander is important part of leading firefighting attack where everyone has own role and know exactly what to do and when to do it. This skill can only be delivered by drill and training.

Orientation in Area with Low Visibility

Fighting fires involves smoke on the scene. Fire attacks are then happening in smoke-filled areas with low or even no visibility. Operations with low visibility require very good ability to keep orientation in area and also to find way out of the building where firefighting attack has been undertaken. It is important to be able to find origin of fire and lead effective suppression which decreases secondary loses and damages caused by water.

Fast Decision Making Without Discussion with Supervisor

Drill and training for firefighters is important to deliver real experience based on which a firefighter gets higher level of self-confidence. It is very important as it results in ability to avoid long unnecessary discussions and consultations with a supervisor or commander and safes time. This drill should provide self-confidence to act right and effectively in the shortest period of time as there is any second counted during rescuing and fire-fighting intervention. Another requirement on decision makings is ability to decide under difficult conditions where many times various sounds, screaming of people can occur in really short time. That can be delivered only by real experiences and training with good physical and psychological condition.

Suppression Methods and Techniques

Penciling, 3D extinguishing or other techniques such as application of water mist through structures are some other methods how to effectively suppress and extinguish fire with limited secondary damages caused by water. There are several types of suppression techniques and each should be tried by a firefighter to know their limitation and right use.

Predicting Fire Behaviour

Main part of training in this area is covered by compartment fire behaviour training. This requires large experience and understanding fire behaviour as well as basics of results of extreme fire behaviour (Smolka 2020) including Flashover, Backdraft, Smoke-explosion, Rollover and other ways of intensive fire spread.

Good understanding and having knowledge in compartment fire behaviour training is important part of skills of firefighters for providing the most efficient and safe suppression and rescue operations as well as self-rescue actions in case of fire danger phenomena occurrence. Compartment fire behaviour training also provides possible way how to extinguish burning objects and constructions with lowest possible water consumption and build culture of extinguishing by it.

Risk Awareness

Risk awareness is closely related to compartment fire behaviour training where understanding risks and hazards relating to firefighting attack and specific procedures are important to deliver effective and safety suppression and rescue actions. It is important to understand own limits and potential hazards, (NIST 2019).

Fighting Fires in Infrastructure and Training Facilities for Firefighters

Currently, various facilities exist for learning mentioned firefighters' skills and gathering experience. Firefighters are there tested, trained and prepared for real fires and other interventions. The main training facilities conclude training polygons (cages), flashover containers and fire houses, built structures equipped with smoke generators flame generators and old buildings before demolition which are provided to firefighters for their own burning and training.

Limited Possibilities to Testing Facilities and Possible Contamination

Training of effective and successful extinguishing with various extinguishing methods can result to high contamination of area, which can lead to avoiding further activities in this area. Usually firefighting training is not possible to be realized on fire stations located in urban areas. Under these conditions, it is difficult to train firefighters. A lot of scenarios for firefighters beneficial for training requires sophisticated facility. Specified facilities usually have long waiting list with limited time of training for each firefighting group. Such example is fighting industrial fires (e.g. extinguishing jet-fires) or airplanes fires, oil products storages, fires in mines. Additionally, HazMat crew training requires other specialized facility for intervention in case of HazMat material leakage in specific types of industrial properties (tubes, valves, etc.), radiation leakage. Another facility would be needed for first aid training.

Training Polygons - Cages

Training polygons are so called cages, polygons equipped with barriers such as tires, ropes, ladders, etc. where firefighters need to overcome. One of them is captured in Figure 3. Firefighters go fully equipped all way of this polygon with SCBA. All training is provided in a dark room with smoke and stroboscope. This training usually includes playing sounds as screaming people, sound of sirens, explosions, etc. In this cage firefighters fulfill specific tasks and activities e.g. removing SCBA in tight space, rescue people, searching for subjects, etc.

This training should provide identification of persons with claustrophobia, or other negative personal aspects which could limit firefighters work in their everyday work.

Real Structures

Another way of firefighters training is its undertaken in real compartments, are variously equipped with furniture, barricades, and searched subjects as Figure 3 represents. Conditions can vary every training, so persons won't remember previous scenarios. These compartments are fully smoke-filled, darkled and also equipped with stroboscope. Some of these compartments are equipped with fire generators and

Figure 3. An example of training cage in Europe. (Adapted from [HZS, 2017])

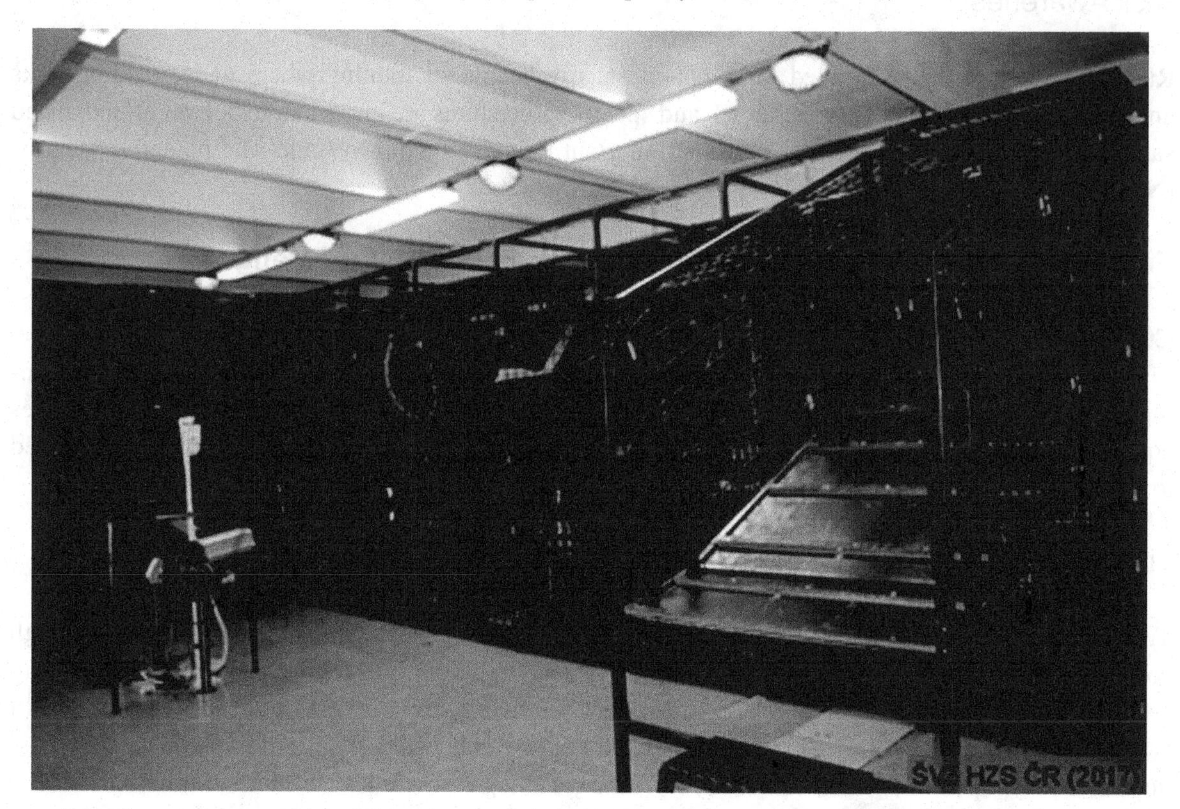

smoke generators with sound effects for providing training and experience as realistic as possible. As well as in polygons, in this compartment firefighters fulfil specific tasks, activities and train orientation in smoke-filled area. One of the examples in Europe Figure 4 captures.

Flashover Containers and Fire Houses

Flashover containers and Fire houses provide training with real fire. These facilities are part of compartment fire behaviour training, which usually include simulations and exhibition of extreme fire behaviour such as flashover, backdraft or rollover. Firefighters are prepared for intervention in these facilities and right behaviour during suppression enclosure fires are taught. Training includes effective methods of fire suppressing including 3D extinguishing, penciling, and other methods of effective extinguishing. It can also provide methods of self-protection in critical situations.

Buildings Before Demolition

Buildings meant for demolitions provide great opportunities to firefighters to test new methods of extinguishing, simulate various scenarios and methods of firefighting in real conditions. These possibilities are limited and not very common due limitation of buildings.

Figure 4. Example of training facility in Central Firefighting School – taken by authors in Bile Policany, the Czech Republic

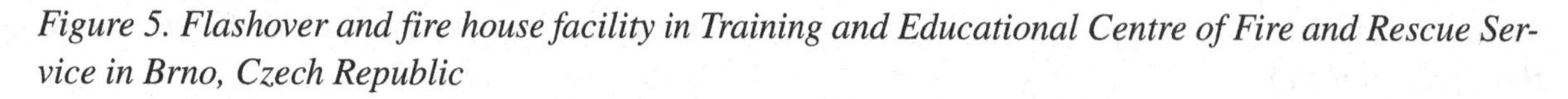

Figure 5. Flashover and fire house facility in Training and Educational Centre of Fire and Rescue Service in Brno, Czech Republic

Fighting Fires in Infrastructure and Firefighters' Equipment

Practice of firefighters is usually performed in dangerous conditions where high temperature, occurrence of toxic gases and smoke take place. Together with limited visibility, narrow spaces and another difficult conditions impact firefighters' comfort, movement and increase stress factor. For stress elimination and potential increase of firefighter's ability to withstand these conditions, there is a lot of tasks to drill and

Figure 6. Example of a training in a burning house before demolition in the Czech Republic

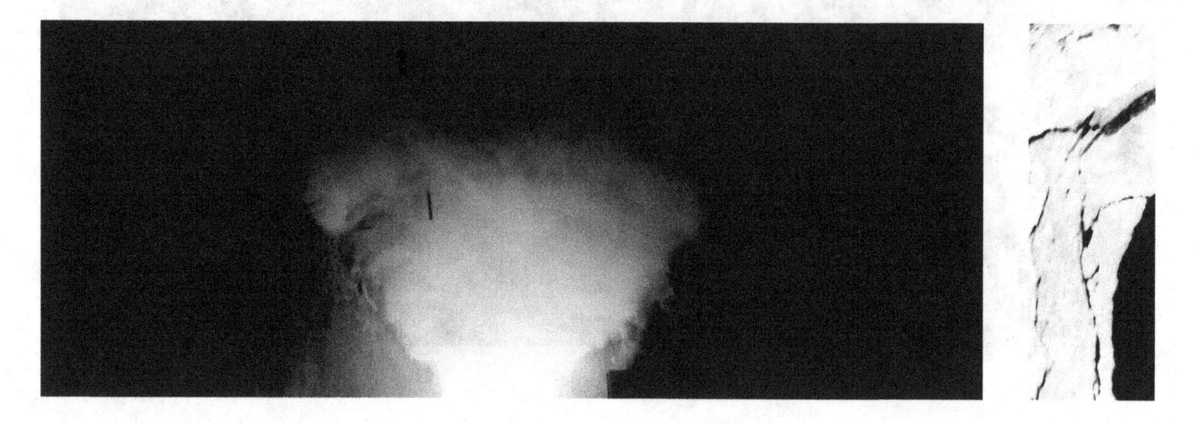

Figure 7. Firefighters equipment and its weight (Adapted from [Velazquez, 2016])

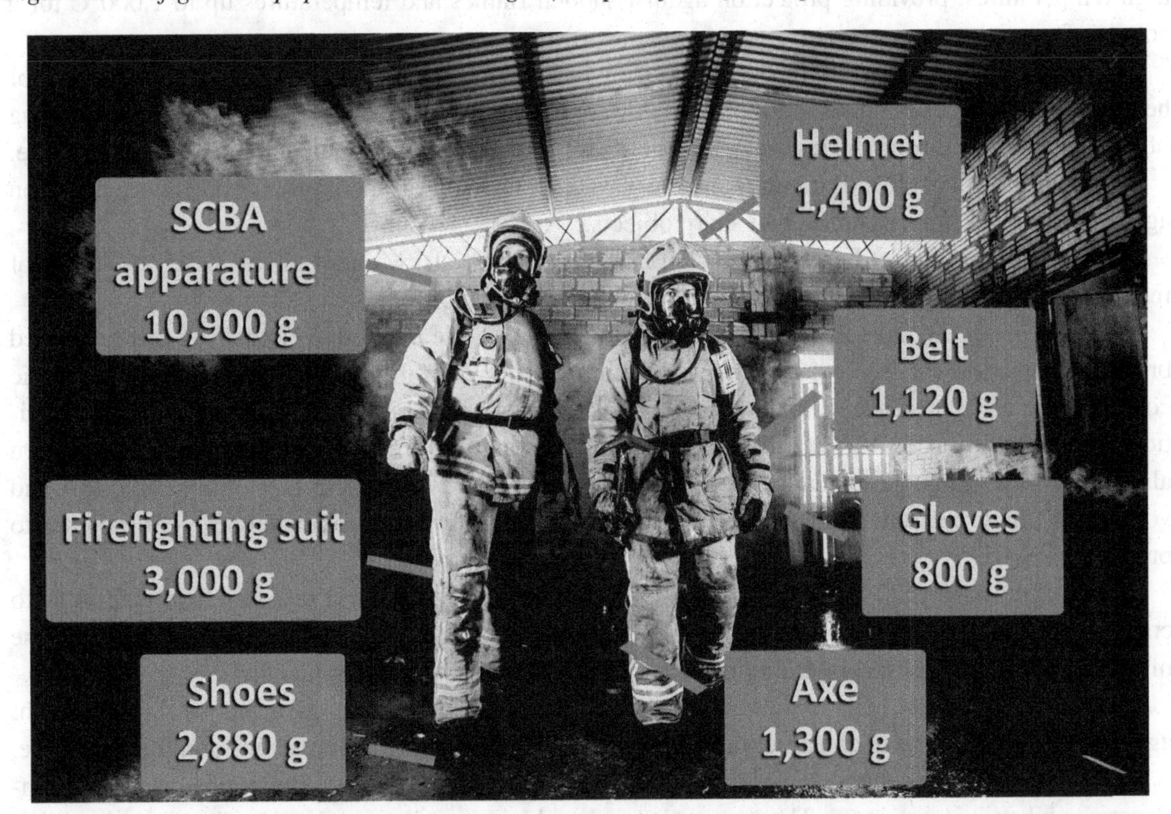

train with fully equipped gear. It is also the reason, why it is so important to elaborate virtual reality training considering firefighter's equipment (Williams-Bell, 2015; Fanfarova, 2017).

Possibility to connect equipment with a computer and monitor person's health condition together with tracking their movement can provide other important data could be useful not for only training of firefighting skill but also for further development of new firefighters methods and techniques, health care or psychological training.

Total weight of firefighters' equipment and protection can easily exceed 20 kg. Firefighters with this equipment are be able provide effective rescue and suppression operations, firefighting procedures and withstand high temperatures, low visibility etc. Training of fully equipped firefighters is important for further development and withstanding real conditions in action. That is also the reason why further virtual reality training should respect these requirements and the virtual reality should be implemented in this equipment. Following list of equipment discuss possibilities of application in virtual reality.

Helmet: Firefighting helmets provide protection against potentially falling materials and heat. Helmets usually include shield protection providing heat protection against radiative heat and protection to sparks and flying materials specific activities such as cutting, sawing, etc. Usually they have adapters for lights. Common weight of helmet is around 1,400 g.

Firefighters suit: Firefighters usually use two types of firefighting suits. Light suits are tended to be used in external fires such as forest / grass fires in open area. For firefighting in compartment fires there are three or more layered suits providing adequate protection for operations in environment with high temperatures, providing protection against sudden flames and temperatures up to 1,000°C for a couple of seconds.

Modern suits provide possibilities of measuring health conditions, body and ambient temperature of heat rate. These types of suites are more suitable for purposes of firefighter's training and monitoring of their health data during the training. Weight of suits is 3,000 – 4,000 g, depending on type and size.

Firefighting shoes: Beside comfort and every weather use, special shoes provide adequate protection against heat, sparks, water, acids, and sharp objects.

Gloves: Gloves protect hands against hot and sharp materials. Modern gloves have possibility of measuring temperature on holding objects. Weight of common gloves is ca. 800 g.

SCBA: One of the most important wearing equipment beside protective cloths is A self-contained breathing apparatus (SCBA) equipment. This equipment allows firefighters to enter areas with specific conditions to alive presence of hot and toxic smoke or gases and low level of oxygen. Health conditions and ability to work with SCBA is essential for estimation of time in action. The air consumption also affects activities undertaken by firefighters. Depending on type of activities, model of SCBA and volume of a high-pressure tank, an average time for a firefighter is approximately 30 – 50 minutes to breath with SCBA.

Usage of masks for SCBA limits view zone and increases psychological stress on firefighters due to constant control of breath for enduring in action as long as possible without need of a tank change. Use of this equipment needs special training and confident in use to be well prepared.

Belt: Firefighters use belts for holding firefighting axes and for protection during working at height against falls from ladders or constructions where suppression, demolition or rescuing activities are done.

Axe: Axe provides support to firefighters to remove construction, open doors or brake other constructions when it is necessary. The axe can be replaced by controllers or developed axes with motion tracking for implementation in virtual reality. In this way, users can train with axes activities such as movement in smoke-filled area, searching for lost persons, orientation in area, opening doors, and much more needed for training in firefighting practice.

Firefighting nozzle: Working with firefighting nozzle is one of the most important skill of firefighters. There is a number of various firefighting nozzle types. The most common is Turbo-Jet providing streams from solid to spray and water mist stream, for example Cobra Cold Cut System for delivering high-pressured water mist and Piercing Nozzle for low-pressure water mist extinguishing, (Smolka 2018), (Smolka, 2020).

Training of Fighting Fires with Virtual Reality

Development of right virtual reality training system can deliver great preparedness of firefighters for a practical training. The specific application and training program develop specialized skills needed for practical training in real conditions. Also, this training can provide more theoretical and practical preparedness. Ability to connect all, firefighter's equipment, virtual reality software and control panel of instructor for monitoring health conditions, status of activities and possibly improve scenarios of trainee can provide adequate training and psychologically prepare firefighters on practical training and real missions.

Training Skills in Virtual Reality

Identification of possible applications of virtual reality in training and monitoring firefighters' activities together with their psychological and health condition will help to find important directions of future steps.

Working Under Stress with Limited Time: Training and at the same time monitoring of firefighters' behaviour in stressed situations such as search and rescue in smoke-filled area, looking for origin of fire, looking at possible critical scenarios of an event can be really helpful for firefighters.

Working with Air Capacity Running Out: Tasks and activities when trainee suddenly starts to lose air and has limited time for escaping from place of a mission, or looking for escape routes in limited time while hearing sounds from SCBA informing about limited amount of air. These and similar scenarios which can be developed for training of firefighter's skills.

Communication, Cooperation and Coordination: Teamwork, cooperation and crew communication in difficult conditions are important way of training. Coordination of crew members by commanders is another important skill which is needed for effective rescue and extinguishing operation accomplishment. For purposes of decision making and firefighting strategies, software have been released and already used by first responders. This part of training is delivered by simulation systems, also being implemented and tested in Fire Brigades across various countries including North America or Europe.

Orientation in Areas with Low Visibility: Necessity to develop training program including operation in low or no visible conditions where a user could train own orientation, decision making is raising.

Economic Advantages of Virtual Reality Application

One of the most important aspects affecting development of virtual reality and implementation in educational system is price of the educational platform.

One of the most important economical aspect of training facilities is their location and its hinterland. For building training facilities for each specialist, especially in terms of compartment fire behaviour training area is required located in a distance from urban interface due generation of smoke. Instead of that, virtual reality requires a room which can be placed in any buildings without affecting surrounding area.

Virtual reality was for a long time inaccessible due to high price and requirements on computational power and graphic card performance and usually did not provide adequate quality. Development and emerging technologies allow now virtual reality to be more affordable and open to wider public. The starting price of computer supporting virtual reality is ca. $1000. Additionally, large amount of virtual reality hardware on today's market includes various additional hardware. Basic virtual reality system including headset, two handheld controllers, movement trackers starts at $500.

Virtual reality provides training in potentially danger situations and allows better exposure and experience of specific phenomena. Secondary benefit provides possibility to test and measure psychological conditions, health of trainees in potentially hazardous conditions in real conditions which can provide better psychological and health profile of a trainee. Generally, virtual reality provides safety of trainees and lectors

Virtual reality provides possibility to test and develop new equipment such as agronomical properties of firefighter's equipment during research and development in a non-destructive way.

Firefighting equipment is relatively expensive. By use of virtual reality can reduce damages on personal protective equipment. During the trainings, equipment is used more intensively with higher probability of damages. Additional expenses are needed for washing and cleaning. The application of virtual reality provides cleaner and safer training with limited damages on equipment due to limited exposure to water, smoke, flames and heat.

Virtual reality allows reduction of additional expenses related to training such as water, consumables – papers, books, or other material needed for training – fuel (gas, wood, gasoline), extinguishing agents, extinguishers etc.

Educational material such as software provides possibility of updates and continual development. Firefighters can be trained by themselves in their own free time and learn in their own specific way. More specific analysis of weak and strong aspects based on experience can be found in literature (Engelbrecht, Lindeman & Hoermann, 2019).

Compartment Fire Behaviour Training in Virtual Reality

The lack of modern firefighting systems is lack of development in fire dynamics which could improve development of firefighter's skills in Compartment Fire Behaviour Training (CFBT). compartment fire behaviour training is important part of firefighter's training and their reaction on specific fire behaviour and firefighter's reaction such as suppression methods.

Modelling of fire dynamics and fire behaviour requires high amount of computational power and time and currently real-time fire modelling systems which could be able precise fire modelling and react to suppression system do not exist.

Compartment Fire Behaviour Training (CFBT)

Very specific and large part of firefighters' training is compartment fire behaviour training. For firefighters, high amount of experience and knowledge is required to be able to "read" fire in time of their arrival. They need to decide fast what procedures for delivering the most effective fire suppression will be chosen. The indicators which affects fire behavior can be simplified for purposes of virtual reality system development in three categories – smoke, air track, heat, flame and building conditions. Figure 8 describes simplified fire behavior indicators which could be used in further development of training with virtual reality systems.

Specific conditions of fire development in specific scenarios affected by specific indicators in different ways results in extreme fire behavior which can be potentially be hazardous for firefighters. There is importance in further research of extreme fire behavior hand in hand with ability of numerical modelling and quantification of this behavior which could be implemented in educational systems of firefighters. Figure 8 provides basic list of protentional routes of fire behavior and resulting extreme fire behavior.

Figure 9 provides specific fire behavior scenarios occurring during firefighting attacks. Training firefighters in reading and forecasting of fire behavior increases their safety and efficiency of rescue and suppression operations.

The rescue work of firefighters includes various types of work. Firefighting attacks are nowadays only part of other rescue operations and activities including rescue in car accidents, rescue from waterscapes, rescue from heights, technical assistances, and much more. Also, firefighting attacks are different and various on external (e.g. forest fires, car fires) or internal fires – enclosure fires (e.g. small houses, high-rise buildings).

Enclosure fire behavior and scenarios of development, spread and intensity depends on many aspects and could be divided into two categories – external and internal fires. Both vary for each along types of

Figure 8. Complexity of fire behaviour in enclosures grows when a reader thinks about each aspect more deeply (Adapted from [Hartin 2019])

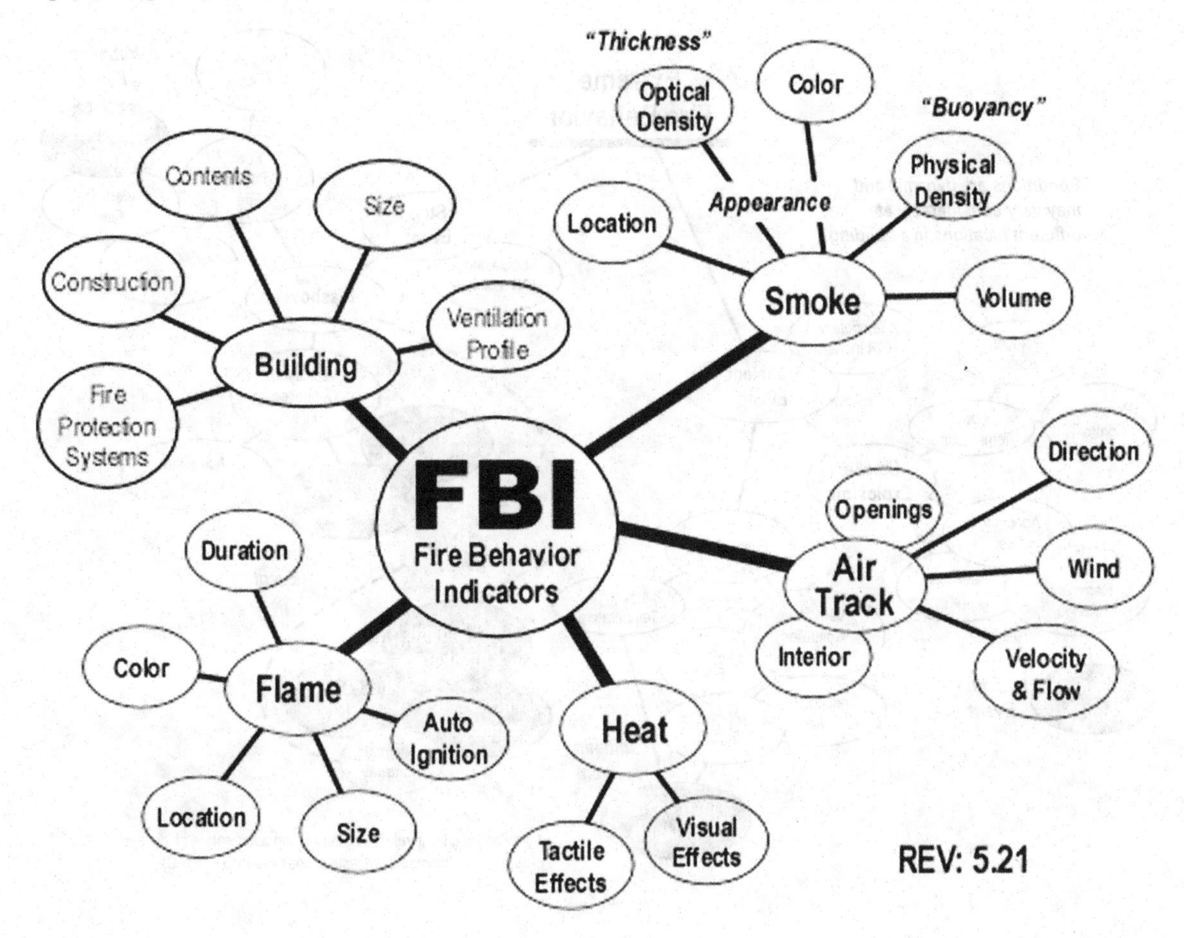

events and buildings. External conditions include height of building, airflow impact, aspect, etc. Internal conditions include type of construction, number of openings and their area and location, fuel load, flammability of structure, and much more. These aspects directly affect fire development scenarios and its spread and decide the state of fire when firefighters arrive on scene.

CFBT in Virtual Reality

Different scenarios have characteristic signs and so provide firefighters ability to read the situation, possibilities of fire attacks, how to lead the fire attack in the easiest and most effective way. These abilities require experience and knowledge about fire behavior, structures and their reaction to fire or effect of specific suppression methods under undefined conditions. It is also important to be able to predict protentional occurrence of sudden intensive fire development such as flashover, backdraft, smoke explosions, etc.

Collecting such experience, knowledge and skills requires long study of theory and practical training. As theoretical study is relatively cheap, practical training is much more expensive since it requires real fires with various scenarios. Firefighting techniques and procedures in enclosures are usually delivered

Figure 9. For hazard understanding and consequently to survive it is crucial for a firefighter to read signs of various scenarios (Adapted from [CTIF 2018])

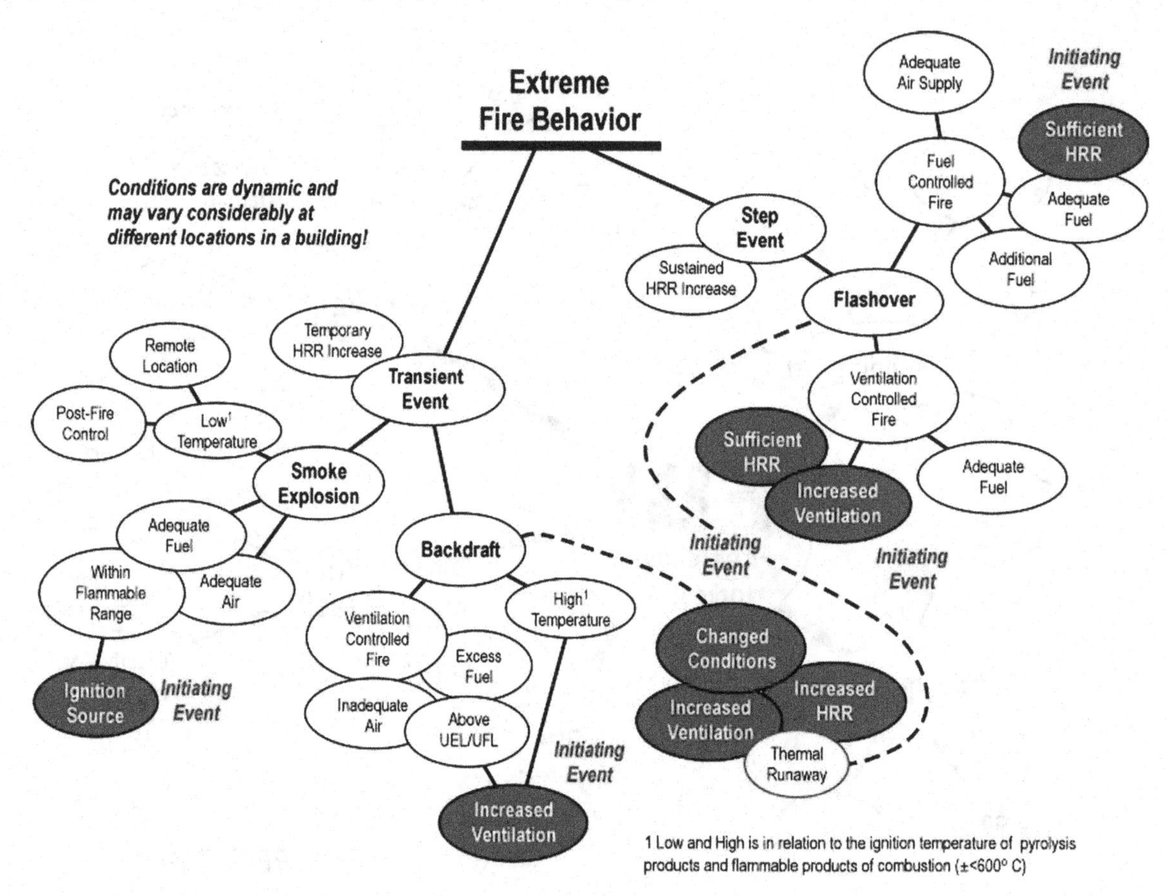

through the compartment fire behaviour training – Compartment Fire Behavior Training program (flash-over containers or burning houses). Number of these facilities are limited due to price, requirement on location for reason of producing certain amount of smoke and heat. Additionally, these facilities are not available in every country or area and require special training of instructors.

Even though the most valuable training of firefighters is indeed in a real situations, and obtained experience cannot be replaced by anything, the novel and advanced technologies such as virtual reality can provide new level of training for development of firefighters' skills and their preparedness for various fire scenarios and fire behavior. It is something that might be difficult to simulate in training facilities or would be complicated. Virtual reality can provide personal development and preparedness for compartment fire behaviour training to develop skills in decision making, coordination and cooperation including communication which could increase efficiency of common firefighters training. Additionally, this training can provide advanced compartment fire behaviour training and education, where firefighters can train and test specific equipment.

Existing Virtual Reality Training Systems and Supporting Software

There are several types of existing development platforms, game-based training systems as well as special training for virtual reality system such as XVR simulation or FLAIM Trainer.

Virtual Reality Development Platforms

The most common engines for development in virtual reality are based on game engines. Examples of mostly used engines are Unity 3D, Unreal Engine 4 or AppGameKit VR.

Unity 3D is commonly recognized and used. Free open access of this platform allows spread of knowledge and sharing good practice across the community. It also supports effective development of applications and spread to wider communities. It is supporting widely known and used virtual reality equipment including e.g. Oculus Rift, Seam virtual reality, HTC Vive, Playstation VR or openVR, operating in on common operational systems such as Windows, Linus or OSX. Customized environment can be designed in Unity 3D, including adjustable lighting, cameras, material, shaders and textures, particles, visual effects, etc. It provides possibility to develop dynamic environment from luminous day to the lighting signs at night. In this engine, amount of physical interaction including Rigidbody component enables physical behaviour and properties for objects as well as immediate interaction with gravity. Colliders components define shape of for the purposes of physical collisions, joints components are attached to rigids body or fixed points in environment to allow specific freedom in movement such as hinges. Final important component is for controlling character's (user's) interaction with environment movement in area providing control of irrelevant falls through floors or movement through walls.

Community using Unity 3D has developed number of online repositories with possibility to download specific behaviour of objects, animations, graphics, or other elements for virtual reality development engine. One of the most common and known is UnityList (Technologies, n.d.).

Supporting Software for Modelling the Fire and Smoke Spread and Evacuation

Currently, several advanced numerical models for fire spread, smoke movement and evacuation already exist. They are not yet merged with virtual reality for training systems. However, as such offer the capability in future to be merged with other virtual reality software and tools.

Fire and smoke modelling is hardware and software demanding and requires plenty of time depending on hardware to simulate specific fire conditions and behaviour. One example of commonly used software for this kind of simulations is Fire Dynamics Simulation (FDS) developed by NIST. However, none of current software provides real-time simulations and neither live interaction with decision of user such as firefighting attack, water delivering, and similarly.

Modelling of Fire and Smoke Spread – Pyrosim

Fire modelling is complex problematic depending on various indicators described in Chapter Predicting Fire Behaviour. FDS is currently used mostly for purposes of modelling in fire engineering for construction's fire resistance estimations, monitoring of fire spread and development and impact of specific factors on fire development. Another way of application can be found in evacuation modelling of occupants from place of fire.

Numerical modelling is still under research and need deeper investigation. Currently, the most known and used application is Fire Dynamic Simulation which provides complex calculation, numerical modelling and fire simulation. However, these simulations have high requirements on hardware for mathematical calculations and visualization. Furthermore, live software able to simulate fire behaviour with interaction of a user and to automatically re-calculate after user's interaction does not exist yet.

Company Thunderhead Engineering Consultants, Inc. has developed graphic user interface (GUI) platform Pyrosim for development in FDS engine. Implementation of the graphic user interface in FDS software provides advanced possibilities of application and usage. It allows professional application and development of more complex and sophisticated models. Moreover, it provides opportunities to import CAD models, use of property libraries for more uniform and easier designing of the scenarios etc. (Agent Based Evacuation Simulation, n.d.).

Modelling of Evacuation – Pathfinder

Another software of Thunderhead Engineering Consultants, Inc. with GUI is for development and modelling of human evacuation in buildings. The software is called Pathfinders and includes following features.

- Continuous movement mesh.
- Multiple simulation modes.
- Occupant movement to exits.
- 3D visualization of results.
- Contour plot of results.

Various ways of evacuation include assisted evacuation, group movement, different type of occupants, different types of routes with e.g. stairways, escalators, moving walkway or ramps.

As Pathfinder is developed by the same organization as Pyrosim, it is possible to import resulting fire scenarios and behaviour in Pathfinder and evacuation models. It provides realistic visualization of incidents and scenarios (Agent Based Evacuation Simulation, n.d.).

Previously mentioned PyroSim and Pathfinder give visualization of results for virtual reality environment and scenarios. Even though the software do not allow live interaction with user in virtual reality, they provide the most precise developed models of fire scenarios and occupants behaviour which can be effectively used for education and training of first responders, fire safety managers or engineers, etc.

Software for Training

One of the most emerging advanced technologies in area of fire safety research is virtual reality. Current software and hardware provide new possibilities of application, it is cheaper and more accessible to broader society. It provides further development and implementation in new areas of application including education and training. For actual training, various simulators can be used.

LUDUS Training Simulator

LUDUS is Italian training simulator targeting first responders for enhanced experiences. It is developed for tactical decision-making process during intervention for firefighters, police, military or other first responders. The simulator offers training of firefighters in several fire scenarios.

- Forest fires.
- Enclosure fires.
- Releases.
- Industrial fires.
- Ports.

Ludus training system varies in types of interventions including fire extinguishing, rescue, prevention squads, hazmat, ship fires or traffic accidents. During training it is even possible to dynamically change weather and environmental conditions.

Ludus provides statistical data of decision making and user's progress. Supervisor of the trainer can monitor simulation in real time and interact with the user by creating new conflicts and interventions (Virtual Reality for Firefighter Services, n.d.).

XVR Simulation Software

Another specific system for first responders' training with application in virtual reality on tactical and strategical leadership is XVR. It is the most advanced simulator these days. Either trainer can use model for a classroom, where a group of people is educated, or individual where a third person changes scenarios, or lastly, people can work in teams and train cooperation on various levels during an event.

This system provides possibility of interaction between students and an instructor and therefore is more focused on training of first responders in decision making on strategical or tactical level with direct disturbing of instructors.

FLAIM Trainer

Australian training system for firefighters incorporates a head-mounted virtual reality display, breathing apparatus kit, branch nozzle with standard hose-line and encapsulating hose, branch/nozzle and protective clothing with heat generation components (FLAIM, 2019) as Figure 10 shows.

RiVR

This system provides opportunity to scan a fire scene by a 3D scanner and replicate this fire scene to virtual reality environment. Therefore, firefighters and police can have advanced possibilities for further investigation of fire spreading and its origin and cause. This environment allows users to interact with virtual world and objects. Photorealistic system provides near-to-life immersive presence on scene which increases trustfulness of experience.

Figure 10. FLAIM Trainer solution for firefighters training in virtual reality (Adapted from [FLAIM 2019])

In this system, user can interact with environment i.e. look anywhere, open doors or closets or even pick up objects in the scene using controllers. This software includes forensic tools allowing evidence of observed objects and their record and analyses. RiVR allows also tracking and recording user's actions, use of scoring list to track actions and behaviour (Woodward, 2019).

FUTURE RESEARCH DIRECTIONS

Even though there is quantity of virtual reality equipment, virtual reality still needs development. As well as in military application, where special improved guns, helmets, headsets, and other equipment are developed for use in purposes of training for soldiers, it is necessary to develop equipment for firefighters' training purposes. Future research therefore can be focused on how various firefighting activities can be integrated in virtual reality and help to innovate training procedures.

Headset Improvement

There are several virtual reality headsets available but there is a lack of headset usable together with firefighters' helmet and breathing apparatus. The best solution could be development SCBA breathing mask with headset which could be used together with firefighting helmet.

Development of SCBA for Virtual Reality Purposes

Training firefighters with SCBA requires the same weight as would be used in real conditions. Breathing possibilities through mask with measuring air consumption with informing trainee about current level of air and time limitation would provide more realistic training. The SCBA should include ability to estimate volume of air, measuring air consumption regarding the activities and psychical stress.

Extinguishing Nozzles

For firefighting training and fire suppression training it is important to use adequate equipment for delivering drill as realistic as possible. The current controllers are not suitable and effective enough for firefighting training.

Heat Radiation Panels

For developing near realistic conditions, application of heat radiation panels could provide impact on firefighters' discomfort. It provides increase of stressed situation potentially affecting decision making and more realistic interaction firefighter with virtual reality.

Other

Furthermore, treadmills can be implemented for use of firefighters training. Monitors of temperatures could further be tested and developed for the training.

CONCLUSION

Services of first responders and firefighters, particularly, involve several tasks and require holistic approach to education and preparation. Fighting nowadays fires is becoming more complicated due to denser infrastructure, use of new technologies on buildings such as power storages, solar panels or electric cars. Additional aspects to consider are new engineering method allowing building more complex and sophisticated buildings with flammable materials involved. Impacts of climate change cause frequent natural disasters such as large forest fires, floods and flash floods, hailstorms, windstorms, tornados or snowstorms in some regions more than others. During these disasters, firefighters provide help to people, perform rescue operation and save property.

Another activity with firefighters involved is intervention in traffic accidents, HazMat and oil products with leakage intervention, assistance in explosive finds, working and rescue operations at height and depth, diving in water, rescuing on waterscapes, emergency opening of apartments, and much more.

This chapter helps to identify requirements on firefighters' physical state, skills and knowledge which allows them to operate in complicated stressful conditions and provide adequate activities to limit damages and rescue people, animals and property. Firefighters need also to be able to cooperate with other crew members, demanding on collaboration and coordination with other crews, and commanders. It requires certain time and energy for training, drill and studying. Every increase in efficiency and decrease in time consumption unneeded for training and learning is beneficial.

Virtual reality has potential to decrease expenses on training of firefighters and increase efficiency of their training. The use of virtual environment provides possibility to train firefighters in more suitable way according to preference. Education can be managed by trainees themselves and personal progress recorded. Additionally, game-based education increases interest of trainees and also efficiency of education. Virtual reality provides possibility to establish checkpoints and each level of education and training and provides possibility to monitor and record education and training of trainees. It allows lecturer to see if studied problematic was fully understood by a trainee. Virtual reality also present abstract and difficult to imagine information in a much understandable way, such as methods and mechanisms of extinguishing, extreme fire behaviour or location of active / passive protection in cars which can be virtually dissembled. Another example are complicated complexes of buildings or constructions. Virtual environment in training for firefighters allow to train in potentially dangerous conditions in real life, such as mechanisms of extreme fire behaviour that helps to understand issues without need of personal entering to a real fire room but with feeling of it. High potential of virtual reality in an application for firefighters training preparedness of specialists such as machine specialist and truck drivers, forensic investigators, smoke divers, search and rescue, etc.

Virtual reality cannot replace practical training but can efficiently help to prepare firefighters in theoretical aspects and prepare for practical training and real situations. Additionally, virtual reality provides great opportunity for continual professional development, which can be accessible anytime during their duty.

Moreover, application of virtual reality in training is a cost-effective solution reducing expenses on training including damages of personal protective equipment (PPE), consumption of extinguishing agents, fuel – wood, gasoline, gas, and maintaining of equipment and washing PPE. Additional benefits of virtual reality are lower requirements on placement of virtual reality system and its maintenance.

Currently, the most spread and advanced virtual reality system XVR simulation is used by first responders around globally. This system provides training of a single trainee with lecturer, group training and crew cooperative training. This system provides interacting between lecturer which has possibilities to improve scenarios of missions tasked to trainee during the testing.

Compartment fire behaviour training is well known for firefighting in technical infrastructures. This training is usually limited by time and number of people who can attend at a time. As variously frequent training is required, more complex fire models need to be developed, with real behaviour and advanced virtual reality equipment including heating system, firefighting equipment – SCBA with implemented virtual reality headset in breathing mask, advanced firefighting nozzles and piercing nozzles or standard nozzles. Several measurements can also be taken from the human during the training for better performance.

Currently, there is a lack of advanced computing technologies and software as well as fire numerical models which could provide real-time fire modelling and be able to interact with firefighter's suppression intervention. Well known nowadays is Fire Dynamics Simulation (FDS) software and software with implemented FDS – PyroSim, which provides advanced fire modelling. However, this system is not able to model fire in a real time. For development of virtual environment, additional platforms exist to allow creating virtual environment. Most of these environments are game-based such as Unity 3D or Unreal Engine. In fact, many development platforms are open accessed allowing broader community to spread knowledge between developers working on new virtual environments. The open access to platforms for development provides better education and training which potentially increase number of developers of VE enabling to develop virtual environment for firefighters' application in practice. For these aspects, there is necessity of further discussion between firefighters and developers and of providing them information they need such as preferred type of virtual environment, requirements on training software, their application and activities and which activities should be developed further.

Virtual reality is potentially great tool for training of firefighters and it still needs research, development and innovation especially in application of knowledge of fire behaviour and implementation of fire models. It could then enable real-time fire modelling and interaction with firefighter's suppression. This system will be useful also for training of compartment fire behaviour. It is necessary to develop special equipment suitable for firefighters' training such as breathing apparatus which could control and measure air consumption of firefighters, heating / cooling system of firefighters during training for approaching real conditions, sensors on firefighters which could monitor their health status and equipment related and suitable to practice of firefighters and connected to virtual reality such as firefighting nozzles, with possibility to provide more realistic firefighting training. Considering real requirements on persons from fire brigade, real conditions in fire or other types of intervention, modelling and virtual reality can be helpful in holistic approach to education and training with for fire fighters and fire engineers for a safer infrastructure.

REFERENCES

Agent Based Evacuation Simulation. (n.d.). Retrieved from https://www.thunderheadeng.com/pathfinder/

Carmigniani, J., Furht, B., Anisetti, M., Ceravolo, P., Damiani, E., & Ivkovic, M. (2011). Augmented reality technologies, systems and applications. *Multimedia Tools and Applications*, *51*(1), 341–377. doi:10.100711042-010-0660-6

Baus, O., & Bouchard, S. (2014). Moving from virtual reality exposure-based therapy to augmented reality exposure-based therapy: A review. *Frontiers in Human Neuroscience*, *8*(MAR). doi:10.3389/fnhum.2014.00112 PMID:24624073

Canadian Wood Council. (1996). Fire Safety Design in Buildings. A reference for applying the National Building Code of Canada fire safety requirements in building design Fire Safety Design in Buildings.

CTIF. (2018). *Extreme Fire Behavior*: Understanding the Hazard. Retrieved from https://www.ctif.org/news/extreme-fire-behavior-understanding-hazard

Engelbrecht, H., Lindeman, R. W., & Hoermann, S. (2019). A SWOT Analysis of the Field of Virtual Reality for Firefighter Training. *Frontiers in Robotics and AI*, *6*(October), 1–14. doi:10.3389/frobt.2019.00101

Fanfarová, A., & Mariš, L. (2017). Simulation tool for fire and rescue services. *Procedia Engineering*, *192*, 160–165.

FLAIM systems. (2019). *Virtual Reality for Firefighting*. Retrieved from https://www.flaimsystems.com/flaim-trainer/flaim-trainer-options/

Gaddis, T. (1998). *Virtual reality in the school. Virtual reality and Education Laboratory*. East Carolina University.

Hamins, A.P., Bryner, N.P., Jones, A.W., & Koepke, G.H. (2015). *Research Roadmap for Smart Fire Fighting Summary Report Research Roadmap for Smart Fire Fighting*. NIST and The Fire Protection Research Foundation. Retrieved from: www.nfpa.org/SmartFireFighting

Hartin, E. Compartment Fire Behaviour Training: Reading the Fire. Retrieved http://cfbt-us.com/wordpress/?p=38

HZS CR. (2017). *Polygon – Klecovy trenazer*. Retrieved from https://www.hzscr.cz/clanek/trenazery-polygon-klecovy-trenazer.aspx

HZS CR. (2019). *Statistical Yearbooks*. Retrieved from https://www.hzscr.cz/hasicien/article/statistical-yearbooks.aspx

Jiang, M., Zhou, G., & Zhang, Q. (2018). Fire-fighting Training System Based on Virtual reality. *IOP Conference Series: Earth and Environmental Science*, *170*(4). 10.1088/1755-1315/170/4/042113

Kahani, M. & Beadle, H. W. P. (1997). Comparing Immersive and Non-Immersive Virtual Reality User Interfaces for Management of Telecommunication for Management of Telecommunication Networks. Retrieved from https://www.researchgate.net/publication/228742087

Lovreglio, R. (2018). A Review of Augmented Reality Applications for Building Evacuation. Retrieved from http://arxiv.org/abs/1804.04186

Mantovani, F., Castelnuovo, G., Gaggioli, A., & Riva, G. (2003). Virtual reality training for health-care professionals. *Cyberpsychology & Behavior*, *6*(4), 389–395. doi:10.1089/109493103322278772 PMID:14511451

Martirosov, S. & Kopeček, P. (2017). *Virtual Reality and its Influence on Training and Education - Literature Review*. doi:. doi:10.2507/28th.daaam.proceedings

McGuire, E. G. (1996). Knowledge representation and construction in hypermedia and environments. *Telematics and Informatics*, *13*(4), 251–260. doi:10.1016/S0736-5853(96)00025-1

Milgram, P., Takemura, H., Utsumi, A., & Kishino, F. (1995). Augmented reality: a class of displays on the reality-virtuality continuum. In *Proceedings of Society of Photo-Optical Instrumentation Engineers: Telemanipulator and Telepresence Technologies*. Academic Press. doi:10.1117/12.197321

Nandavar, A., Petzold, F., Nassif, J., & Schubert, G. (2018). Interactive virtual reality tool for bim based on ifc: Development of openbim and game engine-based layout planning tool. In Proceedings of the *23rd International Conference on Computer-Aided Architectural Design Research in Asia: Learning, Prototyping and Adapting CAADRIA 2018* (Vol. 1, pp. 453–462). Academic Press.

NFPA. (2017). *FACT SHEET Smart Firefighting We're revolutionizing fire safety*. Retrieved from https://www.nfpa.org/-/media/Files/Code-or-topic-fact-sheets/SmartFFFactSheet.ashx

NIST-Blog. (2019). Making fighting fires virtual reality more real reality. Retrieved from https://www.nist.gov/blogs/taking-measure/making-fighting-fires-virtual-reality-more-real-reality

Rådemar, D., Blixt, D., Debrouwere, B., Melin, B. G., & Purchase, A. (2018). Practicalities and Limitations of Coupling FDS with Evacuation Software. In *SFPE 12th International Conference on Performance-Based Codes and Fire Safety Design Methods*. Academic Press. Retrieved from https://c.ymcdn.com/sites/www.sfpe.org/resource/resmgr/2018_Conference_&_Expo/PBD/Program/Hawaii_Program.pdf%0Ahttps://www.sfpe.org/events/EventDetails.aspx?id=901118&group=

Ronchi, E., Ph, D., & Bari, P. (2015). Evacuation modelling and Virtual Reality for fire safety engineering.

Royal, M. (2019). Heads-Up / Hands-Free Firefighting Solutions: Requirements, State of Technology Overview and Market Characterization. Defense research and development Canada.

Smith, S., & Ericson, E. (2009). Using immersive game-based virtual reality to teach fire-safety skills to children. *Virtual Reality*, *13*(2), 87–99. doi:10.100710055-009-0113-6

Smolka, J. (2020). Vliv proudění vzduchu na průběh požáru. Unpublished doctoral dissertation, Technical University of Ostrava, Czech Republic.

Smolka, J., Kempna, K., Hošťálková, M., & Hozjan, T. (n.d.). Guidance on Fire-fighting and Bio-Based Materials.

Unity Technologies. (n.d.). Unity Manual. Retrieved from https://docs.unity3d.com/540/Documentation/Manual/UnityManual.html

Velazquez, J. (2016) *Firefighters*. Retrieved from https://pixabay.com/photos/fire-volunteer-firefighters-rescue-2263406

Virtual Reality for Firefighter Services. (2019). Retrieved from https://www.ludus-vr.com/en/portfolio/firefighter-services/

Wang, B., Li, H., Rezgui, Y., Bradley, A., & Ong, H. N. (2014). BIM based virtual environment for fire emergency evacuation. *The Scientific World Journal*, (August). doi:10.1155/2014/589016 PMID:25197704

Wells, W. D. (2005). Naval Postgraduate School. *Security*, (September).

WoodwardB. (2019) RiVR. Retrieved from https://rivr.uk/

ADDITIONAL READING

Bengtsson Lars-Göran, & Hardestam, P. (2005). Enclosure fires. Karlstad: Swedish Rescue Services Agency.

Dillon Consulting. (2016). *FIRE-RESCUE STAFFING STUDY.*

Grimwood, P. (2008). Euro firefighter: global firefighting strategy and tactics. Retrieved from https://www.amazon.com/Euro-Firefighter-Firefighting-Strategy-Tactics/dp/1906600252

Izard, S. G., & Muñoz, S. V. (2017). Educational Virtual Reality Tool to Teach Students to Respect Environment. In EDULEARN17 Proceedings. Academic Press. doi:10.21125/edulearn.2017.1479

Svensson, S. (2005). *Fire ventilation.* Karlstad, Sweden: Swedish Rescue Services Agency.

Svensson, S. & Cedergårdh, E. (2009). *Tactics, command, leadership.* Karlstad: Swedish Civil Contingencies Agency.

KEY TERMS AND DEFINITIONS

Augment Reality: AR is a type of virtual reality technology that blends what the user sees in their real surroundings with digital content generated by computer software.

Building Information Modelling (BIM): It is an intelligent 3D model-based process that gives construction and other professionals the insight and tools to more efficiently plan, design, construct, and manage buildings and infrastructure.

Compartment Fire Behaviour Training (CFBT): CFB training integrates the topics of fire behavior, fire streams and ventilation within a structural firefighting context.

Fire Dynamics Simulator (FDS): is a computational fluid dynamics model of fire-driven fluid flow. FDS solves numerically a form of the Navier-Stokes equations appropriate for low-speed ($Ma < 0.3$), thermally driven flow with an emphasis on smoke and heat transport from fires.

Heat Release Rate: HRR is a measure of the rate at which heat energy is evolved by a material when burned. It is expressed in terms of power per unit area (kW/m^2).

NIST: NIST is abbreviation for National Institute of Standards and Technology, physical sciences laboratory.

Self-contained breathing apparatus (SCBA): is a device worn by rescue workers, firefighters, and others to provide breathable air in an immediately dangerous to life or health atmosphere.

Unity: Unity is development platform for virtual reality.

Virtual Environment (VE): Artificial environment that is created with software and presented to the user in such a way that the user suspends belief and accepts it as a real environment. It can be immersive, semi-immersive, non-immersive, or even augmented.

Virtual Reality: VR is a simulated experience that can be similar to or completely different from the real world, a mediated environment which creates the sensation in a user of being present in a (physical) surrounding.

Chapter 10
Current Safety Issues in Road Tunnel Construction

Miguel Vidueira
Spanish Fire Protection Association (Cepreven), Spain

Jiri Pokorny
https://orcid.org/0000-0002-1829-8437
VSB – Technical University of Ostrava, Czech Republic

Vladimir Vlcek
Fire Rescue Brigade of the Moravian-Silesian Region, Czech Republic

ABSTRACT

The construction of road tunnels is an important part of road infrastructure. The operation of road tunnels has historically been accompanied by a number of extraordinary events. Fires are among the most dangerous ones. Individual countries create their own safety standards that mutually differ to a large extent. Some of the differences of the requirements for safety devices, including the requirements for their functionality, are compared and commented on in this chapter. Moreover, attention is paid to the use of asphalt surfaces on roads and sidewalks in tunnels. This chapter also describes the approach to fire ventilation in tunnels, one of the most significant safety devices. Special attention is paid to the choice of the strategy of longitudinal ventilation, which has been the subject of many discussions. This chapter outlines the possible directions for a solution in the future.

INTRODUCTION

The construction and operation of road tunnels is nowadays an urgent topic in all countries all over the world. The reasons for the construction vary: it may be the need to shorten the length of a road, the need to get across mountainous areas, the need to reduce noise levels or the protection of the environment. The increasing number and lengths of road tunnels also bring about an increase in the emergency incidents that occur within these structures. The emergency incidents include fires, too.

DOI: 10.4018/978-1-7998-3059-7.ch010

The largest fires in road tunnels include the fire in the Salang Tunnel in Afghanistan in 1982, with 176 dead, the fire in the Mont Blanc Tunnel in France in 1999, with 39 dead, the fire in the Gotthard Tunnel in Switzerland in 2001, with 11 dead, or the fire in the Reigensdorf Tunnel in 2001 when a bus hit the tunnel's portal, with 24 injured. Other significant fires were the fire in the tunnel on the highway between Florence and Bologna in 1993, with 4 people killed, the fire in the Pfänder Tunnel near Bregenz in Austria in 1995, with 3 people killed, the fire in the road tunnel near Palermo in 1996, with 5 people killed, the fire in the Gleinalm Tunnel near Gratz in Austria in 2001, with 5 people dead, the fire in the Amberg Tunnel in Austria in 2001, with 3 people dead, the fire in the Guldborgsund Tunnel in Denmark in 2001, with 5 people dead, and the fire in the Fréjus Tunnel between France and Italy in 2005, during which 2 people were killed. The causes of these fires were traffic accidents or the ignition of vehicles. The reasons for such fires are traffic accidents or burning cars (PIARC, 2019).

In the course of 2013 to 2018 there were annually 2 to 5 fires in road tunnels in the Czech Republic.

This is a negligible number in relation to the overall number of fires in the Czech Republic, i.e. 20,000 fires a year approximately (HZS CR, 2019). Although the mentioned fires in the Czech Republic did not bring about any significant loss of life, material damage or damage to the environment, the historic events prove that the consequences of fires in tunnels may be catastrophic. For tunnel operation and solving emergency situations it is necessary to prepare the tunnel constructions from the structural, technical and operational viewpoint. The meeting of the aforementioned requirements is achieved by collaboration among the designer, the construction contractor, the tunnel operator and the rescue services. The work of researchers is of wide use in relation to the consequences of emergency incidents and securing the safety of tunnel constructions (Li & Ingason, 2018).

BACKGROUND

On the basis of the requirements of the Regulation of the European Parliament and Council (EU) no. 305/2011, which defines harmonized terms for the introduction of structural products to the market and cancels the Direction of the Council 89/106/EEC (Regulation of the EP and of the Council, 2011), the properties that the construction of road tunnels shall meet include requirements for fire safety. The mentioned requirements comprise the retention of the load capacity in the case of fire, the limitation of fire spreading inside and outside of buildings, providing for the evacuation and rescue of persons and securing the safety of fire and rescue units.

The minimum safety level requirements for tunnels in the European union is given by the European parliament directive 2004/54/ES, on minimum safety requirements in trans-European road networks. It stipulates the technical and maintenance requirements for tunnels longer than 500 m. (the EP and Council, 2004)

In individual countries, safety requirements for the construction of road tunnels are elaborated in national regulations. Examples of these are the documents Sicherheitsanforderungen an Tunnel im Nationalstrassennetz from Switzerland (ASTRA 74001, 2010), Richtlinien für die Ausstattung und den Betrieb von Straßentunneln RABT 2006 from Germany (RABT, 2006), Manual 021 Norwegian Public Roads Administration Standard Road Tunnels 03.04 from Norway (Manual 021, 2004), DMRB Volume 2 Section 2 Part 9 (BD 78/99) Highway structures: Design (substructures and special structures) material. Special structures. Design of road tunnels from Great Britain (DMRB, 1999) or NFPA 502 Standard for Road Tunnels, Bridges, and Other Limited Access Highways from the USA (NFPA, 2017).

In the Czech Republic and in the Slovak Republic these regulations are represented by government decrees (NV, 2006; NV, 2009), technical standards (ČSN 73 7507, 2013; STN 73 7507, 2008) and methodological instructions issued by state authorities (PPJK, 2018).

This enumeration of the regulations concerned is not complete. Many countries supplement the mentioned documents with others that focus on specific areas of safety.

Given the scope of the regulations setting safety requirements it is obvious that the requirements for safety in road tunnels vary from country to country. The context of the aforementioned facts leads to questions such as "What is the scope of the equipment of safety devices in tunnels in individual countries in relation to their characteristic parameters?", "For what reason are there differences in the scope of safety devices in individual countries?" or "How is the functionality of safety devices verified?"

One of the discussed questions that relate to the safety of road tunnels is the flammability of the surfaces of roads and sidewalks. The majority of countries, e.g. Austria, Spain, France, the Czech Republic and the Slovak Republic, prescribe certain restriction as to the use of asphalt surfaces.

The requirement for the limitation of the asphalt surfaces in tunnels is supported by research presented by the European Concrete Paving Association that has led to the conclusion that "bitumen surfaces can increase fire loading in tunnels and in the case of burning they release smoke and toxic substances" (Romero & Rueda, 2010).

The opposite of the stated arguments may be a recommendation of organizations that focus on road tunnels. The World Road Association PIARC has compared a number of fires in road tunnels and tried to determine the possible negative influence of flammable surfaces in relation to the dynamics of fire. The document states that "standard asphalt surfaces do not significantly decrease fire safety during the development of fire in the phase when people are evacuated and may be used in tunnels" (PIARC, 1999).

The research of the European Asphalt Pavement Association states that "in tunnels it is generally recommended to use inflammable materials for the ceiling and materials with limited flammability for walls, however there are no special recommendations for pavements". Furthermore, they state that "A large amount of heat is needed to ignite asphalt surfaces. Such heat exposures may only be reached in the immediate vicinity of a burning vehicle. Research shows that when bitumen strips ignite, only the surface layer participates in the fire. The heat released in this way is very low in comparison with the burning vehicle. The amount of gases produced and the temperature reached during fire are not sufficient to sufficiently worsen evacuation" (AEPA, 2008).

Thus it is obvious that opinions largely differ as to the use of asphalt pavement surfaces in road tunnels. There are other questions that are still topical nowadays, such as "As far as safety is concerned, are asphalt road surfaces suitable for the construction of road tunnels or do they increase their risk factor?", "Does the heat released from the material contribute in the case of fire?", "May the softening of asphalt surfaces caused by heat complicate the intervention of rescue corps?"

One of the most important safety devices in tunnels in the case of fire is ventilation. The type of ventilation selected and the strategy of its use depend on the characteristics of the tunnel. Longitudinal ventilation belongs to the types used relatively often. The objective of longitudinal ventilation is directing the movement of smoke and the prevention of reverse smoke flow (the "backlayering" effect). The mentioned goal is normally achieved by the so-called critical speed of flow, which is mostly within the interval 2.5 to 3.5 $m.s^{-1}$. In the case of tunnels with a higher probability of congestion or in tunnels with bidirectional traffic, lower speed of air flow are used for the reason of "preserving at least a partial stratification of smoke"; these however do not prevent the reverse flow of smoke. Two questions arise:

"To what extent does the lowering of airflow speed contribute to the preservation of its stratification?" and "Is this strategy an asset for the evacuation of people and the intervention of firemen?"

MAIN FOCUS OF THE CHAPTER

In the chapter concerned, the focus will be on the partial safety aspects of the construction of road tunnels that have been points of discussion among experts for quite some time. The aspects discussed will be divided into three main chapters:

- The scope of the equipment with safety devices in relation to the characteristic parameters of tunnels and the verification of their functionality,
- The influence of bitumen road surfaces on the increase of the amount of released heat during fires And the complication for the intervention of rescue units due to their softening,
- the strategy of the design of longitudinal ventilation in tunnels for the creation of conditions suitable for the evacuation of persons and the activities of rescue units.

SAFETY DEVICES IN TUNNELS AND THE VERIFICATION OF THEIR FUNCTIONALITY

The required level of safety is achieved in such situations when the risk of occurrence and consequences of emergency incidents is so low that it is considered acceptable. Safety measures may be described by the so-called "safety chain" which has a repetitive character depicted in figure 1.

Figure 1. Schematic depiction of the safety chain

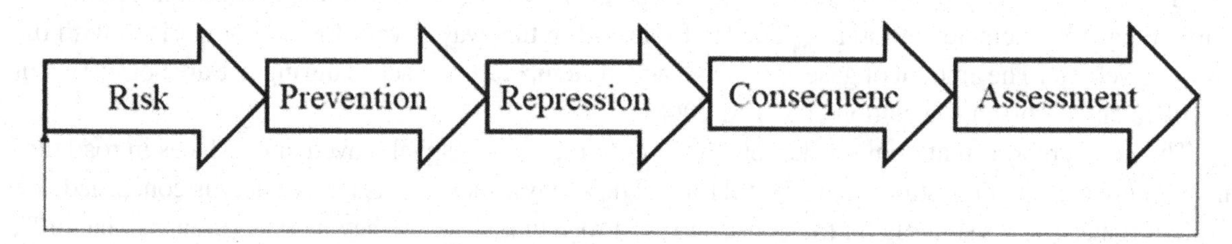

Safety Devices

Safety measures in road tunnels are based on the evaluation of all influencing aspects, namely the infrastructure, operation, tunnel users and vehicles. The following parameters are considered:

- The length of the tunnel,
- The number of tunnel pipes,
- The number of traffic lanes,
- The geometry of the cross section,

- Vertical and horizontal routing,
- The type of construction
- One-way or bidirectional traffic,
- The intensity of traffic,
- The risk of traffic congestion,
- The time needed to begin the intervention of rescue services,
- The presence and share of trucks,
- The presence, share and type of transport of hazardous load,
- The characteristics of access communications,
- The width of traffic lanes,
- Speed limits,
- The geographic and meteorological conditions (the EP and Council, 2004).

The equipment of road tunnels with safety devices is based on their categorization into "safety categories". Even though the method of classification of tunnels differs by country, in general the following influences are considered:

- The length of the tunnel,
- The intensity of traffic,
- The character of traffic.

As to length, road tunnels are usually as follows: up to 500 m (i.e. short tunnels), longer than 500 and up to 1,000 m (i.e. medium tunnels) and longer than 1,000 m (i.e. long tunnels) (the EP and Council, 2004). The intensity of traffic means the number of vehicles that passes through a single lane or tunnel tube within a time unit. As to the character of traffic, the tunnels are divided into one-way and bidirectional tunnels. The mentioned influences may also be designated as "the characteristic parameters of the tunnel" that have crucial impact on its structural design and the equipment with safety devices.

Beside the aforementioned influences other facts may be considered, e.g. the merging of lanes, the occurrence of crossroads, the permitted or forbidden transport of hazardous substances. The method of classification into safety categories differs by country.

For the sake of clarity the following paragraphs focus on the comparison of the requirements for the installation of some safety devices installed in road tunnels in the Czech Republic, Italy, France, Germany, Norway, Austria, the Slovak Republic, Switzerland and Great Britain.

Table 1 compares the following safety devices:

- Ventilation,
- Emergency lighting of the tunnel and escape routes,
- Points for emergency calls (emergency telephone),
- Automatic detection of events,
- Detection of fire,
- Sound system in the tunnel,
- Backup source of electricity,
- Portable fire extinguishers and internal hydrants,
- Stable extinguisher devices,

Table 1 states whether the selected countries have rules with a requirement concerning the compared safety devices in road tunnels (black points in the table mean that such a requirement exists). Table 1 proves that the scope of requirements for the installation of safety devices in road tunnels differs by country.

Table 1. Overview of requirements for the installation of safety devices in selected countries (Höj, 2005)

Safety devices	Czech Republic	France	Netherlands	Italy	Germany	Norway	Austria	Slovakia	Switzerland	Great Britain
Longitudinal ventilation	•	•	•		•	•	•	•		•
Transverse ventilation	•	•	•		•	•	•	•	•	•
Ventilation of emergency and rescue routes	•	•	•		•			•	•	•
Emergency lighting in tunnel	•	•	•		•		•	•		•
Emergency lighting of escape routes	•	•	•	•	•	•	•	•	•	•
Points for emergency calls (emergency telephone)	•	•	•	•	•	•	•	•		•
Automatic detection of events	•		•			•		•	•	
Detection of fire and smoke	•	•	•		•		•	•		•
Loudspeakers	•	•	•		•		•	•	•	•
Backup sources of energy	•	•	•		•	•	•	•		•
Powder fire extinguishers, internal hydrants	•	•	•		•	•	•	•	•	•
Stable extinguisher devices			•					•	•	•

The Functionality of Safety Devices

The differences in individual countries do not apply to the requirement for the installation of safety devices itself, the specifying conditions concerning the devices differ too. An example of this is given in table 2 for the conditions for the installation of fire detection devices, stable fire extinguisher devices and devices for initial fire extinguishing.

Table 2 shows that specifying conditions for the installation of safety devices also differ by country.

In order to secure the safety of road tunnels they must not only be equipped with safety devices, the error-free functions of these devices are also necessary. The correct function of safety devices is verified by tests. In relation to "the lifespan phase" of the devices these tests can be divided into tests in the stage of:

- Production devices,
- Design tunnels,
- Putting the construction into operation,
- The operation of the construction (MD, 2010).

In the production stage, the qualities of products are verified before they are introduced to the market. In the design phase, usually the design parameters of devices are verified by computer simulation. Before a safety device is put into operation functional tests are done. In the case of a combination of two or more fire safety devices that mutually influence one another, it is also necessary to do function coordination tests.

When tunnels are operated the tests of safety devices are done periodically. The period between tests is defined by national regulations.

Table 2. The overview of specifying conditions for the installation of fire detection devices, stable extinguisher devices and devices for initial fire extinguishing (Höj, 2004; STN 73 7507, 2008; TP 13, 2015)

Country	Safety devices		
	Detection of fire and smoke	Stable extinguisher devices	Initial fire extinguishing
Czech Republic	In tunnels longer than 300 m		2 pcs of extinguishers with the weight of the content of 6 kg placed at emergency call points located in the vicinity of portals and then after each 150 m
France	In tunnels without human supervisors		2 pcs of extinguishers with the weight of the content of 6 kg placed at each end of the tunnel and at emergency call points located after each 200 m
Netherlands	In tunnels with a controlling center, tunnels without such a center are equipped only if the operation of ventilation during fire is different from the usual operation	Limited use for the cooling of structures	2 pcs of extinguishers with the weight of the content of 6 kg placed at emergency call points located in the vicinity of portals and then after each 150 m in the case of new tunnels and after each 250 m in the case of already built tunnels
Italy			2 pcs of extinguishers with the weight of the content of 6 kg placed at emergency call points located in the vicinity of portals and then after each 150 m in the case of new tunnels and after each 250 m in the case of already built tunnels
Germany	In tunnels longer than 400 m with forced ventilation		2 pcs of extinguishers with the weight of the content of 6 kg in tunnels longer than 400 m placed after each 150 m
Norway			1 extinguisher with the weight of the content of 6 kg in boxes (compartments) at distances of 62.5 m to 250 m as per the tunnel category
Austria	If forced ventilation is installed		2 pcs of extinguishers with the weight of the content of 6 kg in tunnels longer than 500 m placed on both sides after each 250 m at emergency call points
Slovakia	In all spaces of the tunnel	Spaces with a backup energy source, electricity substations and collectors	1 pc of CO2 fire extinguishers with the weight of the content of 5 kg and 1 pc of powder fire extinguisher with the weight of the content of 6 kg at emergency call points placed at each 150 m
Switzerland	Automatic detection responding to temperature connected to the control room and traffic signs		2 pcs with the weight of the content of 6 kg at emergency call points placed in bidirectional tunnels at each 150m and in one-way tunnels at each 300 m along the outer side
Great Britain	In tunnels with a controlling center, tunnels without such a center are equipped only if the operation of ventilation during fire is different from the usual operation	Self-acting extinguisher systems are not suitable for traffic spaces, gas and foam extinguishers are not suitable for areas with the presence of people, water extinguisher systems may cool smoke	After each 100 m at emergency points

Besides the tests of safety devices, the preparedness of safety devices is also verified by the training of the tunnel operators and of the rescue corps. During this training, the safety devices are used and thus their correct function can be verified.

However, the extent (level) of the verification of the functions of safety devices when the construction is commissioned and during its operation may vary significantly. The functionality of devices may be verified by:

- Verifying the design parameters by physical measuring,
- Software simulations,
- Tests approaching the real state.

The efforts to bring the test closer to the real state are widely discussed and these tests are mostly designated as procedures "with the real energy source". The high level of getting close to the real situation of a fire provides the most relevant results; however, it also raises concerns as to the possible damage to the construction of the tunnel or the installed devices.

In recent years there have been efforts to develop "modified procedures" that would create the real situation of a fire "to a certain extent" without threatening any damage to the construction or devices. These are tests "with cold smoke" (smoke with temperatures significantly lower than those occurring during real fires). The disadvantage of these procedures is the very low temperature of smoke, which at lower temperatures behaves quite differently than smoke at temperatures achieved during real fires. Thus, the tests do not provide clear results from the viewpoint of the functionality of the devices in the case of real fires. For the utilization of the described modified procedures there is currently not sufficient unequivocal evidence that would assess the deviation caused by the low temperature of smoke. Therefore the use of the modified procedures is unconvincing.

INFLUENCE OF BITUMEN SURFACES ON SAFETY DURING FIRES IN TUNNELS

One of the significant elements of road tunnels are roads.

The Characteristics of the Road and the Road Surface

A land road is usually understood to be a transport road meant to be used by cars and other vehicles and pedestrians, including stable devices that are necessary to allow for this use and its safety. Land roads may be divided into the following categories:

- Highways,
- Roads,
- Local roads,
- Purpose-built roads (Act, 1997).

The road surface is the stabilized part of the road consisting in one or more layers with sufficient carrying capacity and even surface to allow for the safe passage of vehicles for the period of its life. The road surface, as a rule, consists of five layers: the abrasion layer and the bedding layer, the top and bottom sublayers and the protective layer (see figure 2).

The surface layers of the road may be formed by various materials, such as concrete or material glued together by asphalt binders (Kasa, 2012). Material glued by asphalt binders can also be called "asphalt

Figure 2. Schematic depiction of the composition of the road surface in a tunnel (Pokorny et al., 2018)

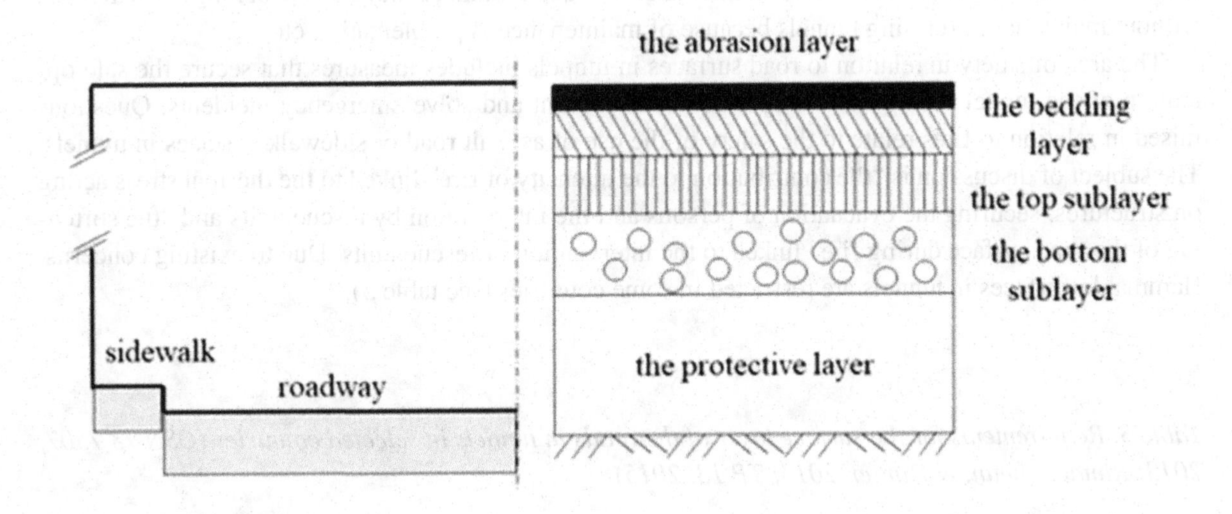

mix" that differs in the type and amount of binder, the content of individual components in the mix and the method of processing and laying. The asphalt mix is formed by two fundamental components, i.e. aggregate and binder. Additives may also be added to the asphalt mix.

Asphalt mixes may be divided into:

- Asphalt concrete,
- Asphalt concrete for very thin layers,
- Very soft asphalt mix,
- Asphalt mix compressed while hot,
- Mastic asphalt carpet,
- Cast asphalt,
- Drainage asphalt carpet (Kasa, 2012).

Both, concrete and asphalt road surfaces are used when road tunnels are constructed.

Requirements For Road Surfaces In Road Tunnels

Road surfaces in tunnels shall meet a number of structural, operational and safety requirements. These requirements may be divided into the following areas:

- Safety,
- Comfort,
- Sustainability,
- Cost effectiveness (AEPA, 2008).

The requirements for comfort, sustainability and cost effectiveness of road surfaces in tunnels are related to costs of the construction and maintenance of roads in tunnels, the securing of long-term lifespan without maintenance (closing tunnels because of maintenance is problematic), etc.

The area of safety in relation to road surfaces in tunnels includes measures that secure the safe operation of the tunnel during normal operation and prevent and solve emergency incidents. Questions raised in relation to fires relate to the safety of the use of asphalt road or sidewalk surfaces in tunnels. The subject of discussion is "the contribution to the intensity of fire" linked to the thermal stress acting on structures, securing the evacuation of persons and the intervention by rescue units and "the softening of the road surface during fire" linked to the intervention of rescue units. Due to existing concerns, flammable surfaces in tunnels are restricted in some countries (see table 3).

Table 3. Requirements for the surface layer of the roads in tunnels in selected countries (ČSN 73 7507, 2013; Rimac, Šimun, & Dimter, 2014; TP 13, 2015)

Country	Requirement for surface finish of road	Note
Bosnia and Herzegovina	asphalt surfaces permitted	
Croatia	no requirement defined	
Czech Republic	concrete surface in tunnels longer than 1 km	concrete surface is also recommended for tunnels longer than 500 m
Austria	concrete surface in tunnels longer than 1 km	
Slovenia	concrete surface in tunnels longer than 1 km	
Slovakia	concrete surfaces in new tunnels	required class of reaction to fire A1
Spain	concrete surfaces in tunnels	recommendation

Findings from Partial Surveys Executed in Relation to Asphalt Roads in Tunnels

In recent years, there have been a number of fire experiments concerning fires in tunnels. However, the flammability of the asphalt mix has been examined only marginally. Examples when asphalt mixes were evaluated include the experimental study of the behavior of asphalts during burning, executed by A. Noumowe from Cerge-Pentoise University in France or research the objective of which was to characterize the behavior of typical asphalt surfaces during fire in tunnels, which was done by the BRE Centre for Fire Safety Engineering, University of Edinburgh in Great Britain (Carvel & Torero, 2006).

The executed experiments have shown that the asphalt surface of the road can be ignited already by the action of a heat flow density of 40 $kW.m^{-2}$. With typical asphalt samples heated at the standard-based heat curve the ignition occurred at temperatures 480 to 530 °C (Carvel & Torero, 2006). However, the stated values of temperatures can be considered as relevant only with caution. The structures in tunnels are normally assessed by other curves than the standard-based heat curve and these curves are based on the assumption of a very intense development of fire. Heat curves for tunnels are closer to the carbon-hydrogen curve than to the standard-based heat curve (EN 1991-1-2, 2002).

Other studies done on samples of small dimensions with heat flow densities 40, 50 and 60 $kW.m^{-2}$ prove that asphalt mixes may not only catch fire but, moreover, they may also release significant amounts of heat. The amount of heat released can be divided into two phases, i.e. the phase with a lower amount of heat released when only the surface layer burns and the phase with a higher amount of heat released when the binder thaws and evaporates. In some experiments the heat flow density reached as much as 100 $kW.m^{-2}$. Even though the possible area of burning asphalt is mostly limited to the immediate vicinity of the focus of the fire, when the semitrailer of the truck catches fire, this area may be up to 50 m^2 and the amount of heat released may be up to 5 MW. Thus, the contribution to the total heat output of the fire may be significant (Carvel & Torero, 2006).

Other important experiments include large-scale tests performed in the Runehamar tunnel (Brekelmans & Bosh, 2003). The heat outputs in this case were 75 kW, 150 kW and 200 kW. The results of the tests show that the significance of the softening of asphalt mixes related to the possible "firemen getting bogged down" during the intervention shall be evaluated in the context of the distance from which firefighters will be realistically able to intervene. This distance is influenced by the conditions in the vicinity of the fire, technical options and the tactical procedures of firefighters. In the course of experiments, the heat flow density at a distance of 10 m from the centre of the fire was approximately 4 to 12 $kW.m^{-2}$. The stated heat flow densities may cause the softening of asphalt. The temperatures of the softening of common asphalt mixes used for roads are usually lower than 100°C (EN 12591, 2009). However, the temperatures at distance of 10 m exceeded 400°C, which is quite problematic from the viewpoint of the execution of an intervention (Kasa, 2012).

In the Czech Republic, asphalt mixes were examined for their safety between 2014 and 2017 within the project "Technological agencies of the Czech Republic TA04031642 Asphalts in road tunnels". The goal of the project was the development of "a new type of asphalt mix" that would show good operational characteristics and at the same time would not reduce the fire safety of tunnels. Another objective of the project was the proposal of a certified testing methodology to determine classes of the reactions of materials to the fire of asphalt surfaces in tunnels. The new type of asphalt mix is formed by a modified variant of asphalt PA 8 with added rubber granulate and meets the operational and fire requirements as the research has shown. Nonetheless, detailed results of the performed tests have not been published (Kudrna, Bebčák, Bebčáková, Šperka, & Stoklásek, 2018).

Recently, asphalt surfaces in tunnels have also been tested in relation to fire retardants. Based on the comparison of the progress of the burning of asphalt surfaces without retardants and with retardants, the asphalt surfaces of roads with the retardant added have been recommended as a possible alternative from the viewpoint of safety. The research also examined the possible use of nanomaterials in retardants (Li, Zhou, Deng & You, 2017; Qiu, Yang, Wang, Wang & Zhang, 2019).

THE STRATEGY OF THE DESIGN OF LONGITUDINAL VENTILATION IN TUNNELS

Fire ventilation in tunnels provides support for the evacuation and rescue of persons and also creates conditions for the effective intervention of rescue units. The design of fire ventilation in tunnels is mainly based on the characteristic parameters of the tunnel, which were described in the preceding part of this chapter (ASTRA, 2008; NFPA, 2017).

Based on the characteristic parameters of the tunnel it is possible, for the needs of ventilation, to divide tunnels into the following categories (ASTRA, 2008; MP, 2013; NFPA, 2017; RABT, 2006):

- Tunnels with one-direction traffic and low probability of congestion (usual highway tunnels),
- Tunnels with one-direction traffic and high probability of congestion (usual city tunnels),
- Tunnels with bidirectional traffic.

The following types of fire ventilation are used for the design of fire ventilation (ASTRA, 2008; NFPA, 2017; RABT, 2006):

- Natural (longitudinal) ventilation,
- Natural longitudinal ventilation with fixed setting,
- Forced longitudinal ventilation with flow regulation to a defined value,
- Transverse ventilation.

The Strategy of Longitudinal Fire Ventilation in Tunnels

The movement of smoke that occurs as one of the products of burning during fires in tunnels is, similar to other structures, influenced by a set of factors of major or minor importance. Basic agents acting on the movement of smoke from the viewpoint of tunnel structures include the following (Pokorny & Gondek, 2016):

- The geometry of the tunnel
- The stack effect
- The influence of standing vehicles
- Wind
- The upward pressure effect created by fire
- The increase in the volumes of gases
- Air conditioning devices

When designing fire ventilation in tunnels it is necessary to take the described influences into consideration.

Suitable strategies are used for the longitudinal ventilation of road tunnels. On the basis of the selected strategy the fire ventilation shall meet the defined objective.

The goal of the longitudinal ventilation in one-way tunnels with a low probability of congestion is to direct (force out) smoke in the direction of passing cars and completely prevent its spread against standing cars. The critical air speed is usually between 2.5 to 3.5 $m.s^{-1}$, however, it is influenced by a number of factors (ASTRA, 2008; Li, Li, Cheng & Chow, 2019; Yang et al., 2019).

The objective of longitudinal ventilation in tunnels with one-way traffic with a high probability of congestion and in bidirectional tunnels is to direct smoke and limit its spread, or at least reduce the speed of its flowing and create for a certain time conditions for smoke stratification. Usually the air speed of 1 to 2 $m.s^{-1}$is used (ASTRA, 2008; ČSN 73 7507, 2013).

In the case of longitudinal ventilation in tunnels, the air flow speed is a significant quantity. The requirements for air speed flow vary largely by country and they are presented for the sake of clarity in table 4.

Table 4 shows that in one-way tunnels with a higher occurrence of congestion and in bidirectional tunnels, as a rule, a lower air flow rate is required (under critical speed). The principle of reducing the air (smoke) flow rate is based on the general assumption that lower flow rates will cause a smaller overall smoking of the tunnel (smoke stratification will be preserved to a certain extent), while on the other hand higher flow rates will lead to a more intensive smoking of the tunnel (the smoke stratification will be disrupted very fast and the smoke will be flowing in one direction).

Verification of Assumptions with a Model

The assumption of preserving the stratification of smoke at lower air flow rates has been verified with a fire model. A case study was done using the Fire Dynamics Simulator (FDS) mathematical model of fire. The FDS model is a computational fluid dynamics model developed by the National Institute of Standards and Technology in Maryland in the USA, which enables setting a number of parameters accompanying the development of fires, and that may be done for partial volumes of the space. The model uses the Navier-Stokes equations for the solution since these equations are suitable for the assessment of flows, taking into consideration the transport of smoke and heat. The aforementioned fire model uses Smokeview software as a tool for the visualization of numeric calculations. (FDS-SMV, 2019)

The Determination of the Subject of the Study and of the Limit Criteria

A case study has been created for a real road tunnel in Klimkovice in the Moravia-Silesian Region near the city of Ostrava in the Czech Republic The tunnel is constructed as a one-way tunnel with two tunnel pipes. The length of the tunnel is about 1,000 m. The width of the roadway is 9.5 m, the width of two-sided sidewalks is 1 and 1.2 m, the height of the passage cross-section is 4.8 m. The tunnel pipes have a longitudinal incline of 0.6% and are interconnected by 5 shafts.

The tunnel is equipped with longitudinal ventilation composed of 8 pairs of ventilators. The ventilators are put into operation after every 5 seconds. The operation of the tunnel started in 2008.

Within the study the tunnel was assessed as an isolated system, i.e. without considering the external influences on the tunnel's portal (such as wind). The focus of the fire was located at a distance of 256 m from the higher portal in the direction of the traffic (approximately in one third of the tunnel's length). Thus, the main draft of longitudinal ventilation is expected to go against the tunnel's inclination. The heat output of the fire was simulated for the output of 30 MW. The airflow rate in the tunnel was estimated for the situation when the ventilation was not operating and the air flow rate was 0 $m.s^{-1}$ and for situations when the ventilation was working and the airflow rate was 0.5 $m.s^{-1}$, 1 $m.s^{-1}$, 1.5 $m.s^{-1}$, 2 $m.s^{-1}$, 2.5 $m.s^{-1}$, 3 $m.s^{-1}$, 3.5 $m.s^{-1}$ and 5 $m.s^{-1}$. The geometry of the tunnel and the location of the focus of the fire are depicted in figure 3.

Within the study the influence of airflow rate on the preservation of air stratification was evaluated. The visibility of the environment was selected as a limit criterion.

The visibility was evaluated at 20 evenly distributed measuring points in the tunnel's axis (always one measuring point at each 50 m of the tunnel's length) at the height of 2.5 m above the level of the road

Table 4. The required airflow speed values during longitudinal ventilation

Country	Characteristics of the tunnel	Required flow speed during longitudinal ventilation ($m.s^{-1}$)	Note
France (CI, 2000)	One-way out-of-town tunnels	3	
	One-way city tunnels	3	Recommended for tunnels up to 500 m long
		1 – 2 (1st phase) 3 (2nd phase)	1st phase – evacuation of persons 2nd phase – support of rescue units
	Bidirectional	3	
Czech Republic (MP, 2013)	One-way tunnels with a lower occurrence of congestion (T1)	Critical airflow speed of up to 10	Critical speed is usually about 3 $m.s^{-1}$
	One-way tunnels with a higher occurrence of congestion (T2)	1.2	
	Bidirectional tunnels (T3)	1.2	
Germany (RABT, 2006)	One-way tunnels with a lower occurrence of congestion	2.3 – 3.6	The value depends on the incline of the tunnel, the shape of the pipe (rectangular, hipped) and the output of fire
	One-way tunnels with a higher occurrence of congestion	1.5	
	Bidirectional tunnels	1.5	
Netherlands (Huijben, Fournier, & Rigter, 2006)	No distinctions in the nature of the tunnel	2.5	Minimal flow speeds of ventilation are recommended for various fire outputs
Norway (Manual 021, 2004)	Tunnels longer than 500 m and with the incline ≥ 2°	min. 2	Fire ventilation is specified by calculation
	Other tunnels with incline < 2°	2	Fire output 5 MW
		3.5	Fire output 20 MW
Austria (RVS, 2008)	No distinctions in the nature of the tunnel	2	Possible volume flow rate of air in the tunnel 120 $m^3.s^{-1}$
Switzerland (ASTRA, 2008)	One-way with a lower occurrence of congestion (RV 1)	3	
	One-way with a higher occurrence of congestion (RV 2)	1.5 – 3	Depending on the incline of the tunnel and direction of ventilation
	Bidirectional (GV)	1.5	
Slovakia (TP 049, 2018)	One-way traffic with a low probability of congestion (normal highway tunnels) (A)	1.5 – 2 (1 – 1.5 when bidirectional traffic is exceptional)	1st phase – evacuation of persons If smoke is exhausted at longitudinal ventilation 1.5 – 2 $m.s^{-1}$ from both sides to the point of exhaustion
	One-way traffic with a high probability of congestion (normal city tunnels) (B)	1-1.5	
	Tunnels with bidirectional traffic (C)	1-1.5	
	All variants	Critical air flow rate	2nd phase – support of rescue units Starting at the instruction of rescue units
USA (NFPA, 2017)	No distinctions in the nature of the tunnel	Critical flow rate	
Great Britain (DMRB, 1999)	No distinctions in the nature of the tunnel	1.3	Fire output 3 MW
		3	Fire output 25 MW
		7	Fire output 50 to 100 MW

Figure 3. The depiction of the geometry of the tunnel and the placement of the focus of the fire

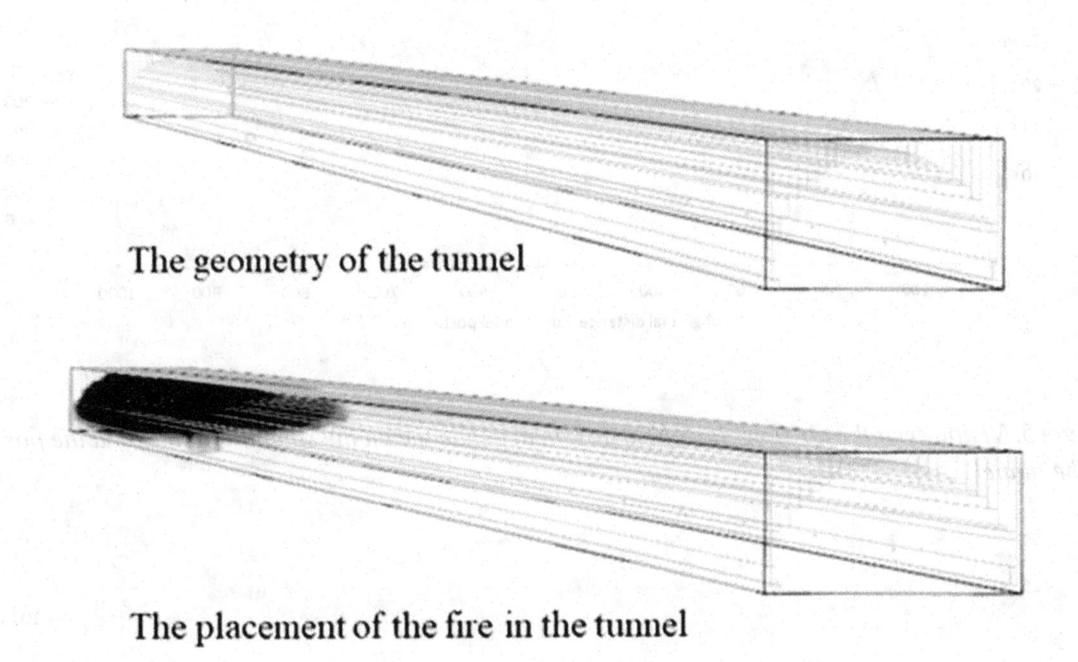

surface (the usual limit value when assessing the safety of persons from the viewpoint of the possible threat by smoke) (Standard CEN/TR 12101-5, 2005).

The moment when the stratification of smoke is disrupted was determined by the decline of visibility to the level of 15 m. This is a usual limit value when it is assumed that people not acquainted with the environment will panic and the process of evacuation will become more difficult (Folwarczny & Pokorný, 2006; Hurley, 2015).

In general, the safe time for evacuation along an unprotected escape route is 2.5 minutes (Hosser, 2013). The movement of persons along an unprotected escape route is usually characterized by the speed of 30 $m.min^{-1}$ (standard ČSN 73 0804, 2010). Based on the aforementioned assumptions, the safe distance for the movement of persons along an unprotected escape route is approximately 75 m. If the visibility drops to the level of 15 m for a distance longer than 75 m, the situation was assessed as unsuitable.

The Results of the Study

The visibility was assessed for the times of free development of the fire 100 s, 200 s, 300 s, 400 s, 500 s and 600 s.

The drop in visibility for the heat output of the fire of 30 MW in relation to the time of simulation and the position in the tunnel is depicted in figures 4 to 7.

Figure 4. Visibility without ventilation in relation to the simulation time and the position in the tunnel

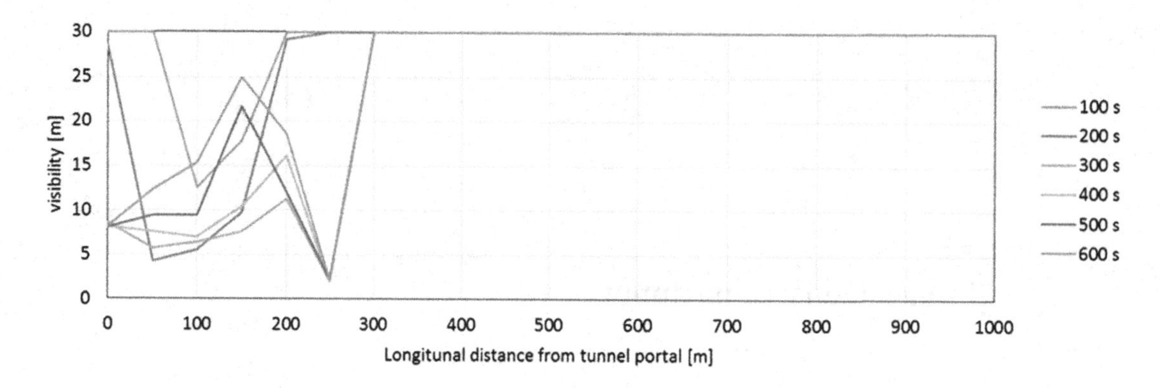

Figure 5. Visibility at the air flow rate of 1.0 m.s^{-1}in dependence on the simulation time and the position in the tunnel

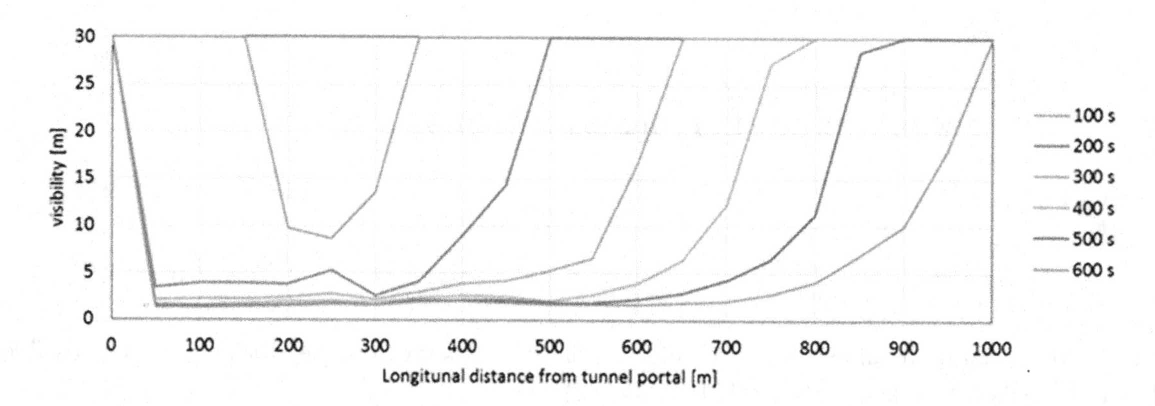

Figure 6. Visibility at the air flow rate of 2.0 m.s^{-1} in relation to the simulation time and the position in the tunnel

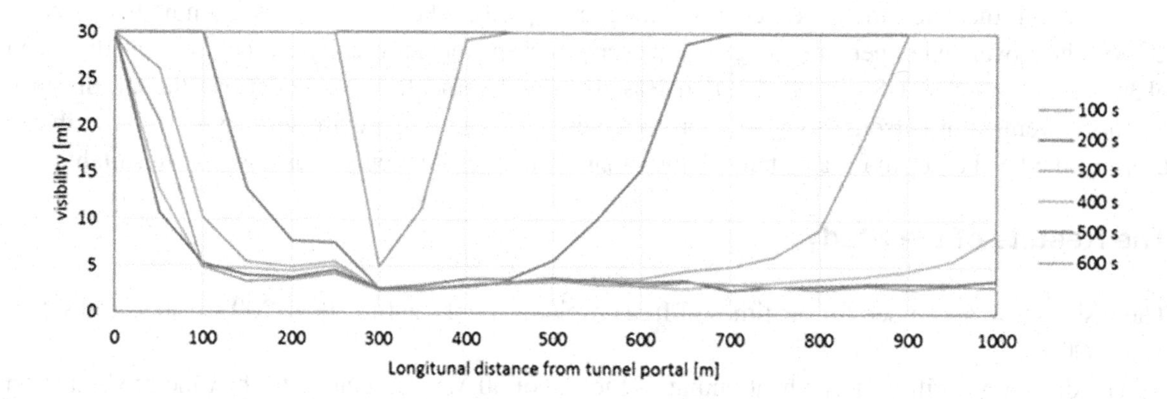

Figure 7. Visibility at the air flow rate of 3.5 $m.s^{-1}$ in dependence on the simulation time and the position in the tunnel

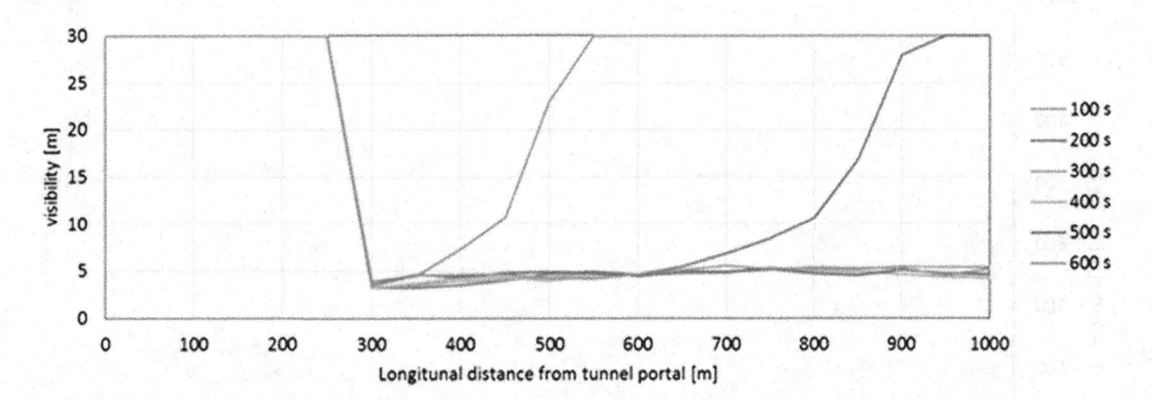

Table 5 depicts distances at which the visibility in the tunnel dropped to the level of 15 m or less (critical level of visibility) in dependence on the heat output of the fire, the air flow rate and the time of the fire development. The distances are determined for the times of 300 s and 600 s. The stated times were selected as times that can clearly represent the results achieved.

Table 5. Comparing the distances of the smoke level in the tunnel pipe

Flow rate [$m.s^{-1}$]	The distance in the tunnel with critical visibility [m]	
	Time of fire development	
	300	600
0.0	250	435
0.5	435	615
1.0	585	980
1.5	700	975
2.0	745	925
2.5	790	870
3.0	725	735
3.5	720	720
5.0	720	720

Distances with critical visibility are depicted in figure 8.

Table 5 and figure 8 clearly show that in the case of heat output of 30 MW, the critical visibility is exceeded for all air flow rates along the distance of more than 75 m. However, the results partially differ for times of 300 s and 600 s. At the time of 300 s the shortest distances of critical visibility were achieved in cases when ventilation was not put into operation and at very low flow rates (up to 1 $m.s^{-1}$). At the time of 600 s, the shortest distances of critical visibility were achieved in cases when ventilation was not put

Figure 8. Distances in the tunnel pipe with critical visibility

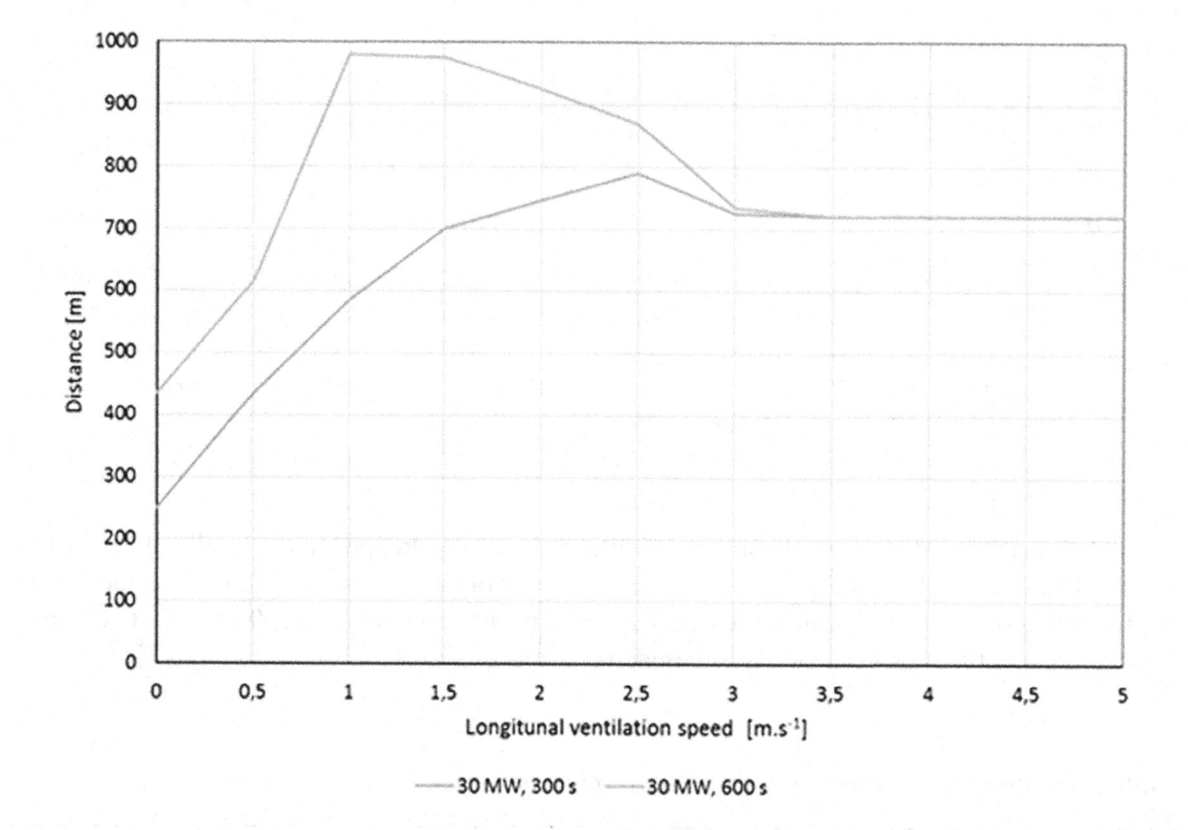

into operation, at very low flow rates (up to 0.5 $m.s^{-1}$) and high flow rates. The results achieved at low air flow rates make an unconvincing impression from the viewpoint of securing smoke stratification.

From this viewpoint it is disputable whether the strategy of longitudinal ventilation that is used in one-way tunnels with a high probability of congestion and in bidirectional tunnels when the design flow rates are below the critical air flow rate are justified.

SOLUTIONS AND RECOMMENDATIONS

The scope of the equipment of tunnels with safety devices is derived from the assessment of risks that occur in the tunnels. The comparison of the requirements for the installation of safety devices in tunnel construction in individual countries clearly shows that the scope of these requirements significantly varies. The specifying conditions for the installation of these devices also differ. These differences may be caused by a number of influences (e.g. by historical reasons, specific local conditions, the relation to other existing regulations of a national nature, the system, facilities and preparedness of rescue services). However, the differences in requirements need not be necessary perceived as negative.

In order to secure the safety of road tunnels it is necessary to verify the functionality of the devices by tests that may be divided into more stages. In the long run, especially the tests at the commissioning of the construction and during its operation have been discussed. In some cases, tests with "a realistic energy source" are preferred, in other cases "modified procedures" are developed that get close to the real situation in the case of fire. Considering the lack of unequivocal evidence as to the suitability of modified methods their use is disputable.

The option to use asphalt surfaces of roads in the construction of road tunnels has been discussed and examined by safety experts for a long time. The problematic aspects are considered to be the possible contribution to the fire intensity and possibly the softening of asphalts. The results of the research show that asphalt surfaces of roads may catch fire and may release a relatively significant amount of heat. It has also been proved that at certain distances asphalt surfaces may soften, which may lead to difficulties during the intervention of rescue units. However, the softening usually occurs only in the near vicinity of the centre of the fire. A positive aspect, to a certain extent, may be considered the fact that the thickness of the bitumen surface usually does not exceed 50 mm and the possible "firemen getting bogged down" in the asphalt mix will probably not be crucial (Pokorny et al., 2018). As a result, the use of asphalt surfaces on roads is not recommended in tunnels longer than 1000 m.

One of the most important safety devices in the equipment of road tunnels is ventilation that provides safety also in the case of fires. One of the types of ventilation is the longitudinal ventilation of tunnels. The strategy of the longitudinal ventilation depends on the character of the tunnels from the viewpoint of traffic direction (one-way or bidirectional) and the probability of the occurrence of congestion. With one-way tunnels with a high probability of congestion and with bidirectional tunnels the air flow rates selected are below the critical flow rate.

The case study shows that even at normally envisaged heat outputs of fires that are normally presumed in the designs of fire ventilation the stratification of smoke is disturbed for a significant length of the tunnel at lower flow rates. On the other hand, at higher flow rates the results may be more favorable. This positive result is related to the fact that the critical air flow rate prevents reverse flow of smoke under the ceiling structure. Thus, the positive effect of reducing the air flow rate below the critical flow rate is disputable as far as the preservation of smoke stratification is concerned.

It is obvious that the extent of smoke in the tunnel gets worse with the increasing time of the fire's development. Therefore, the evacuation and rescue of people and the rescue works shall take place in the shortest time possible.

The case study has been created for a concrete tunnel and a situation that is less favorable for the tunnel has been selected as the subject of examination.

FUTURE RESEARCH DIRECTIONS

The different scope of safety devices in tunnels need not be necessary perceived as a crucial shortcoming that calls for unification. Even though unified procedures would bring certain simplification for designing the construction of road tunnels, the differences may be caused by significant influences that shall not be ignored.

From the viewpoint of the verification of the correct functions of safety devices at the commissioning of the construction and during its operation, it is recommended that the "modified procedures" be researched further to get clear results for the deviations from the tests with "realistic energy sources". The influence of deviations shall be taken into account when assessing the functionality of the devices.

The current knowledge of the behavior of asphalt mixes during fires in tunnels does not provide clear answers to the described problematic areas. In this context it is therefore possible to assume that the requirements from the regulations of some countries for the creation of road surfaces in tunnels from the cement-concrete cover are justified. In order to assess the influence of asphalts during fires in tunnels, further research is necessary. A good way may be the creation of new types of asphalt mixes or the modification of the existing ones with fire retardants.

The strategy of longitudinal ventilation in the structure of tunnels leads in some cases to requirements for lower air flow rate than the critical flow rate. However, the performed simulations led to unconvincing results from this viewpoint and the justification of this requirement is not unequivocal. It is necessary to perform other simulations and experimental measurements in the future.

CONCLUSION

The building of road tunnels makes up an important part of traffic infrastructure. Their operation is historically accompanied by a number of emergency incidents. Even though the frequency of emergency incidents is not high, the consequences of these events may often be very serious.

Immense requirements are imposed on the construction of tunnels from the viewpoint of the prevention and solving of emergency incidents these include the preservation of the supporting capacity of the structure in the case of fire, the limitation of fire spreading outside and inside the structure, providing for the evacuation and rescue of persons and providing for the safety of rescue corps.

Even though the issue of safety in road tunnels has been considered for decades very seriously and a number of supranational and national standards have been created, even nowadays there are many questions closely related to the construction and operation of road tunnels.

The discussed questions include the scope of the equipment with safety devices and the verification of their functionality, the influence of bitumen surfaces on the increase in the amount of heat released and the complications for the intervention of rescue units because of the softening of these surfaces under heat or the strategies for the design of longitudinal ventilation for the creation of suitable conditions for the evacuation of persons and the activity of rescue units.

This chapter of the book presents the approaches of selected countries to the mentioned questions and the results of research works. However, in some cases it is not yet possible to give clear answers to these questions. The authors recommend the performance of further research in these cases.

ACKNOWLEDGMENT

This work was supported by the Ministry of the Interior of the Czech Republic, project no "VH20182020042 The protection of the population within zone planning and in defining technical conditions for the design of constructions".

REFERENCES

Ministry of Transport, the Land roads division. (2013). *A methodological instruction, Ventilation of road tunnels, Selection of the system, designing, operation, the quality assurance of ventilation systems of road tunnels.*

AEPA. (2008). *Asphalt pavements in tunnels*. Retrieved from https://eapa.org/wp-content/uploads/2018/07/asphalt_pavements_tunnelsMay2008-1.pdf

ASTRA. (2008). *Lüftung der Strassentunnel, Systemwahl, Dimensionierung und Ausstattung, ASTRA 13001* (2.03). Retrieved from www.astra.admin.ch

ASTRA. (2010). *Sicherheitsanforderungen an Tunnel im Nationalstrassennetz, ASTRA 74001* (01.08.2010 V1.02). Retrieved from www.astra.admin.ch

Brekelmans, J., & Bosh, R. (2003). Summary of Large Scale Fire Tests in the RUNEHAMAR Tunnel in Norway. Retrieved from http://www.vlada.cz/cz/media-centrum/aktualne/audit-narodni-bezpecnosti-151410/

Carvel, R. O., & Torero, J. L. (2006). *The contribution of asphalt road surfaces to fire risk in tunnel fires: Preliminary findings*. Retrieved from https://www.era.lib.ed.ac.uk/bitstream/handle/1842/895/Carvel%20Torero%20Hong%20Kon;jsessionid=F1F323A8F25A2DE73E34343958E27CC7?sequence=1

CI. (2000). *Annexe n° 2 à la circulaire interministérielle n° 2000- 63 du 25 août 2000 relative à la sécurité dans les tunnels du réseau routier national—Instruction technique relative aux dispositions de sécurité dans les nouveaux tunnels routiers (conception et exploitation).*

Comite Europeen de Normalisation. (2002). *EN 1991-1-2 Eurocode 1: Actions on structures—Part 1-2: General Actions—Actions on structures exposed to fire*.

Comite Europeen de Normalisation. (2005). *Standard CEN/TR 12101-5* Smoke and heat control systems - Part 5: Guidelines on functional recommendations and calculation methods for smoke and heat exhaust ventilation systems.

Comite Europeen de Normalisation. (2009). *EN 12591 Bitumen and bituminous binders – Specifications for paving grade bitumens.*

Czech Office for Standards, Metrology and Testing. (2010). ČSN 73 0804 Fire safety of constructions – Production structures.

Czech Office for Standards, Metrology and Testing. (2013). ČSN 737507 The designing of road tunnels.

DMRB. (1999). *Highway structures: Design (substructures and special structures) material. Special structures. Design of road tunnels (BD 78/99).* London: The Highways Agency.

FDS-SMV. (2019). Fire Dynamics Simulator. Retrieved from https://pages.nist.gov/fds-smv/

Folwarczny, L., & Pokorný, J. (2006). *Evacuation of persons*. Ostrava: The Association of fire and safety engineering.

Government of the Czech Republic. (1997). Act no. 13/1997 Coll., on land roads, as amended. Retrieved from https://www.zakonyprolidi.cz/cs/1997-13

Høj, N. P. (2004). Guidelines for fire safe design compared fire safety features for road tunnels. In *Proceedings of the 1st International Symposium on Safe & Reliable Tunnels* (Vol. 133). Academic Press.

Höj, N. P. (2005). *Fire in tunnels, Thematic network, Technical Report - Part 2 Fire Safety Design - Road Tunnels*. Brussels: European thematic network on fire in tunnels.

Hosser, D. (2013). *Leitfaden Ingenieurmethoden des Brandschutzes*. Retrieved from http://www.kd-brandschutz.de/files/downloads/Leitfaden2013.pdf

Huijben, J. W., Fournier, P., & Rigter, B. P. (2006). *Aanbevelingen ventilatie van verkeerstunnels*. Utrecht: Ministerie van Verkeer en Waterstaat, Rijkswaterstaat, Bouwdienst.

Hurley, M. (2015). SFPE handbook of fire protection engineering. New York, NY: Springer Science+Business Media.

HZS CR. (2019). Hasičský záchranný sbor České republiky. Retrieved from https://www.hzscr.cz

Jofré, C., Romero, J., & Rueda, R. (2010). *Contribution of concrete pavements to the safety of tunnels in case of fire*. Eupave. Retrieved from https://www.eupave.eu/wp-content/uploads/eupave-safety-of-tunnels-in-case-of-fire.pdf

Kasa, J. (2012). The issue of flammable bitumen surfaces (asphalts) on roads in tunnels in the CR. Czech Technical University in Prague. Retrieved from http://people.fsv.cvut.cz/~wald/edu/134SEP_Seminar_IBS/2012/06_SP12_Kasa_Zivice.pdf

Kudrna, J., Bebčák, P., Bebčáková, K., Šperka, P., & Stoklásek, S. (2018). *Asphalt road surfaces in tunnels*. Retrieved from http://www.silnice-zeleznice.cz/clanek/asfaltove-vozovky-v-tunelech/

Li, J., Li, Y. F., Cheng, C. H., & Chow, W. K. (2019). A study on the effects of the slope on the critical velocity for longitudinal ventilation in tilted tunnels. *Tunnelling and Underground Space Technology, 89*, 262–267. doi:10.1016/j.tust.2019.04.015

Li, X., Zhou, Z., Deng, X., & You, Z. (2017). Flame Resistance of Asphalt Mixtures with Flame Retardants through a Comprehensive Testing Program. *Journal of Materials in Civil Engineering, 29*(4), 04016266. doi:10.1061/(ASCE)MT.1943-5533.0001788

Li, Y. Z., & Ingason, H. (2018). Overview of research on fire safety in underground road and railway tunnels. *Tunnelling and Underground Space Technology, 81*, 568–589. doi:10.1016/j.tust.2018.08.013

Ministry of Transport, the Road infrastructure division. (2010). Tests of fire safety devices for road tunnels.

NFPA. (2017). NFPA 502, Standard for road tunnels, bridges, and other limited access highways (National Fire Protection Association, Technical Committee on Road Tunnel and Highway Fire Protection).

Norwegian Public Roads Administration Standard Road Tunnels 03.04. (2004). Norwegian Public Roads Administration Road Tunnels.

NV. (2006). Government decree no. 344/2006 Col. On minimum safety requirements for tunnels in the road network. Retrieved from https://www.slov-lex.sk/pravne-predpisy/SK/ZZ/2006/344/20060601

NV. (2009). Government decree no. 264/2009 Col., on safety requirements for road tunnels longer than 500 meters. Retrieved from https://www.zakonyprolidi.cz/cs/2009-264/zneni-20090901

PIARC Committee on Road Tunnels. (1999). *Fire and smoke control in road tunnels*. La Défense.

PIARC Committee on Road Tunnels. (2019). World Road Association. Retrieved from https://www.piarc.org/en/

Pokorny, J. & Gondek, H. (2016). Comparison of theoretical method of the gas flow in corridors with experimental measurement in real scale. *Acta Montanistica Slovaca*, *21*(2), 146–153.

Pokorny, J., Malerova, L., Wojnarova, J., Tomaskova, M., & Gondek, H. (2018). Perspective of the use of road asphalt surfaces in tunnel construction. In *Proceedings of the 22nd International Scientific on Conference Transport Means 2018* (pp. 290-296). Academic Press. Retrieved from https://www.scopus.com/inward/record.uri?eid=2-s2.0-85055566722&partnerID=40&md5=b5478e2d8440ed6be2fca65db63ed873

PPJK. (2018). The policy of the quality of land roads. Retrieved from http://www.pjpk.cz/

Qiu, J., Yang, T., Wang, X., Wang, L., & Zhang, G. (2019). Review of the flame retardancy on highway tunnel asphalt pavement. *Construction & Building Materials*, *195*, 468–482. doi:10.1016/j.conbuildmat.2018.11.034

RABT. (2006). *Richtlinien für die Ausstattung und den Betrieb von Straßentunneln (RABT 2006)*. Köln: Forschungsgesellschaft für Straßen- und Verkehrswesen e.V.

Rimac, I., Šimun, M., & Dimter, S. (2014). Comparison of pavement structures in tunnel. *Elektronički časopis građevinskog fakulteta Osijek*, 12–18. doi:10.13167/2014.8.2

RVS. (2008). *RVS 09.02.31 Tunnel Ventilation—Basic Principles*. Austrian Research Association for Roads, Rail and Transport.

Slovak Office for Standards, Metrology and Testing. (2008). STN 737507 Designing road tunnels.

The EP and Council. (2004). Directive of the EP and of the Council 2004/54/ES on minimum safety requirements for tunnels of the Trans-European Road Network. Retrieved from https://eur-lex.europa.eu/legal-content/CS/TXT/PDF/?uri=CELEX:32004L0054&from=CS

The Regulation of the EP and of the Council. (2011). *Regulation of the European Parliament and of the Council (EU) no. 305/2011 dated 9th March 2011, which defines harmonized conditions for the introduction of building products to the market and which revokes the Directive of the Council 89/106/EEC. Text significant for the EEC.*

TP 049. (2018). *TP 049 Technical conditions for the ventilation of road tunnels*. Bratislava: Ministry of Transport, construction and regional development of the Slovak Republic road transport and land roads division.

TP 13. (2015). *TP 13 Technical conditions, Fire safety of road tunnels*. Bratislava: Ministry of transport, construction and regional development of the Slovak Republic Road transport and land roads division.

Yang, D., Li, P., Duan, H., Yang, C., Du, T., & Zhang, Z. (2019). Multiple patterns of heat and mass flow induced by the competition of forced longitudinal ventilation and stack effect in sloping tunnels. *International Journal of Thermal Sciences*, *138*, 35–46. doi:10.1016/j.ijthermalsci.2018.12.018

ADDITIONAL READING

Carvel, R., & Beard, A. (Eds.). (2005). *The handbook of tunnel fire safety*. London: Thomas Telford.

Ingason, H., Li, Y. Z., & Lönnermark, A. (2015). Tunnel Fire Dynamics. New York: Springer.

Maevski, I. (2017). *Guidelines for emergency ventilation smoke control in roadway tunnels*. Washington, DC: Transportation Research Board of the National Academies. doi:10.17226/24729

Maevski, I. Y. (2011). *Design fires in road tunnels*. Washington, D.C: Transportation Research Board. doi:10.17226/14562

Ntzeremes, P., & Kirytopoulos, K. (2019). Evaluating the role of risk assessment for road tunnel fire safety: A comparative review within the EU. *Journal of Traffic and Transportation Engineering*, *6*(3), 282–296. doi:10.1016/j.jtte.2018.10.008

Rhodes, N., & Allemann, M., Brandt, R., Carlotti, P., Del Rey, I., Drakuluić, M., … Yumsteg, F. (2011). *Road tunnels: Operational Strategies for Emergency Ventilation*. Retrieved from http://publications.piarc.org/ressources/publications_files/6/7723,WEB-2011R02.pdf

Xu, Z., Zhao, J., Liu, Q., Chen, H., Liu, Y., Geng, Z., & He, L. (2019). Experimental investigation on smoke spread characteristics and smoke layer height in tunnels. *Fire and Materials*, *43*(3), 303–309. doi:10.1002/fam.2701

Yang, X. (2016). Design Environmental Protection, Energy Saving and Safety Ventilation System of Long Highway Immersed Tunnel. *Procedia Engineering*, *166*, 32–36. doi:10.1016/j.proeng.2016.11.533

Zhao, H., Li, H.-P., & Liao, K.-J. (2010). Study on Properties of Flame Retardant Asphalt for Tunnel. *Petroleum Science and Technology*, *28*(11), 1096–1107. doi:10.1080/10916460802611465

KEY TERMS AND DEFINITIONS

Critical Flow Velocity: The velocity of the air flow in the tunnel, which controls the flow of smoke in the selected direction and prevents its spreading against stationary vehicles.

Critical Visibility: The limited value of visibility, which, when is achieved, assumes the possible panic and considerable difficulties in the evacuation process for people who are unfamiliar with the environment.

Emergency Incidents: Harmful effects of forces and phenomena caused by human activities, natural influences or accidents that endanger life, health, property or environment and require the execution of rescue and liquidation work.

Fire: Undesirable combustion during which there was death, injury or endangering people, animals, material values or environmental damage.

Functionality of Safety Devices: The ability of safety devices to perform their functions in the event of an emergency.

Heat Output of Fire: The amount of heat released during a fire per unit of time.

Longitudinal ventilation: One of the forms of ventilation of road tunnels, which is intended to ensure the flow of smoke in the selected direction.

Road Tunnel: A transport structure that runs beneath the surface of the earth, under the sea or river and serves for road transport.

Safety Devices in Tunnels: A set of devices that participates in the creation of safe conditions in tunnels.

Chapter 11
Materials for Safety and Security:
Materials for Shielding, Protective Suits, Electrical Insulation, and Fire Protection

Jozef Martinka
https://orcid.org/0000-0002-0060-5785
Slovak University of Technology in Bratislava, Slovakia

Janka Dibdiakova
Norwegian Institute of Bioeconomy Research, Norway

ABSTRACT

This chapter deals with materials used in safety and security engineering. The most commonly used materials in this field include shielding materials, materials for protective suits, electrically insulating materials and materials for fire protection. The first part of the chapter describes the properties of materials used in the above applications. The second part of the chapter focuses on characteristics of materials that accurately describe their fire risk. The fire risk of a material is quantified by its resistance to ignition (determined generally by critical heat flux and ignition temperature) and by the impact of the fire on the environment. The impact of fire is usually determined by the heat release rate, toxicity of combustion products (primarily determined by carbon monoxide yield and for materials that contain nitrogen, also through the hydrogen cyanide yield) and the decrease of visibility in the area (depending on the geometry of the area and the smoke production rate).

DOI: 10.4018/978-1-7998-3059-7.ch011

INTRODUCTION

Every object or product is made of a particular material. Therefore, material engineering is one of the most widespread and most advanced disciplines. Material engineering currently provides safety and security engineering with a wide range of problem-solving capabilities for issues that had virtually no solution a few decades ago or solutions that were not feasible from a financial, technical, or time point of view.

Given the wide range of issues mentioned above, it is not possible to summarize all the questions and problems concerning materials used in safety and security engineering in a single book and certainly not in a single chapter. Therefore, this chapter will only focus on the most widespread and important materials used in safety and security engineering.

The most important and widespread materials in safety and security engineering include:

- **Shielding Materials**
  - Materials for ionizing radiation shielding
  - Materials for non-ionizing radiation shielding
  - Materials for acoustic shielding
- **Materials for Protective Suits**
  - Materials for bulletproof vests
  - Materials for bomb suits
  - Materials for chemical suits
  - Materials for fire proximity suits
- **Electrical Insulating Materials**
  - Materials for electrical insulators
  - Materials for electrical cable insulation
  - Insulating materials for personal protective equipment in electrical engineering
- **Materials for Fire Protection**
  - Flame retardants
  - Materials that increase the fire resistance of structures

Each material has various characteristics. The relevant characteristics for an assessment of the suitability of a material application in the safety and security field include:

- Mechanical Properties
- Electrical Properties
- Chemical Properties
- Thermal Properties
- Physical Properties
- Magnetic Properties
- Fire Properties

BACKGROUND

Before describing the materials used in safety and security engineering, it is necessary to mention at least the basic divisions of materials. The materials are divided into organic and inorganic. Organic and inorganic materials are further divided into natural and artificial. Artificial materials are further subdivided into materials produced by the physical or chemical processing of natural raw materials without synthesis (e.g. glass is produced by rapidly cooling molten silica with admixtures) and materials produced by synthesis (synthetic polymers). An example of an inorganic natural material is stone, organic natural materials include wood and straw, artificially produced materials through the physical or chemical processing of natural raw materials include glass, concrete and metals, and artificially materials made by synthesis include e.g. polyethylene and polypropylene.

The most important materials used in safety and security engineering included:

- Metals
- Synthetic Polymers

Metals are divided into ferrous, non-ferrous and noble. Ferrous metals include iron, cobalt and nickel. Non-ferrous metals are further divided into light and heavy (they are divided by density or proton number, although there is no agreement in the literature from which density or proton number metal is considered heavy; generally light metals include, e.g. aluminum and magnesium, and heavy metals include, e.g. lead and tungsten). Noble metals, in particular, include gold, silver and the platinum group of metals (ruthenium, rhodium, palladium, osmium, iridium and platinum). Alloys are a separate category of metals (alloys of at least two different metals – a typical example is steel).

Metals are generally characterized by high electrical and thermal conductivity. In safety and security engineering, they are mainly used in shielding materials and as an important component of some protective suits (where it is their high reflectivity at certain wavelengths that is mostly used).

Mleziva and Suparek (2000) divide synthetic polymers into polyolefins (mainly polyethylene and polypropylene) polydienes (mainly caoutchouc or rubber), polystyrene plastics, polyhalogenated olefins (mainly polyvinylchloride and polytetrafluorethylene), polyvinyl esters (mainly polyvinyl acetate), polyvinyl ethers, polymers and copolymers of acrylic and methacrylic acid and their derivatives (e.g. polymethylmethacrylate), polyethers (e.g. polyethylene oxide and polypropylene oxide), polyacetals (mainly polyoxymethylene), polyesters (e.g. polyethylene terephthalate), polyamides, polyimides and polyimidazoles, polysulfides and polysulfones, phenoplast, aminoplastics, furan resins, silicones, epoxy resins, polyurethanes and cellulose and its derivatives.

Polymers (both natural and synthetic) tend to have very low electrical conductivity and low thermal resistance. In safety and security engineering, they are mainly used in shielding materials (especially for acoustic shielding), materials for protective suits and electrical insulating materials.

Composite materials represent a special group. They are made up of at least two phases (components) of matrix and reinforcement. Generally, only artificial materials are considered to be composites. Composites are also considered to be relatively new materials, but it is true that many natural materials (e.g.: bones or wood) have a composite structure. Composite and layered materials are very often used in safety and security engineering (especially as shielding materials or for protective suits).

In addition to metals, synthetic polymers and composite materials, we often encounter natural polymers, especially wood, in safety and security engineering. As such, wood is rarely used in safety and security engineering (in the aforementioned applications – shielding materials, materials for protective suits, electrical insulating materials and materials for fire protection). In most cases, however, it is a material whose properties must be improved (e.g. improved fire performance).

Radiation

Before describing the properties of materials used for shielding, it is necessary to provide some basic information relating to radiation. Radiation is classified according to its wavelength into:

- Electromagnetic Radiation: The wavelength increases from gamma radiation to radio wave.
  - Gama radiation
  - X-ray
  - Ultraviolet radiation
  - Light
  - Infrared radiation
  - Microwaves
  - Radio waves
- Alpha Radiation: High speed helium nucleus
- Beta Radiation: High speed electrons
- Neutron Radiation

In the terms of safety and security engineering, the division of radiation into ionizing and non-ionizing is very important.

Ionizing radiation is further divided into:

- Higher Ultraviolet Radiation
- X-Ray
- Gamma Radiation
- Alpha Radiation
- Beta Radiation
- Neuron Radiation

Non-ionizing radiation is further divided into:

- Radio Waves
- Microwaves
- Infrared Radiation
- Light
- Lower Ultraviolet Radiation

Ionization is considered to be radiation that has sufficient energy to ionize atoms (the loss of one or more electrons from the electron shell of an atom). In terms of its impact on human health and safety, ionizing radiation is more dangerous than non-ionizing radiation. The so-called germicidal radiation (UV radiation with a wavelength of 253.7 nm) is, in practice, used for surface disinfection.

Figure 1. The reflection, transmission, absorption and dissipation of an incident beam of radiation

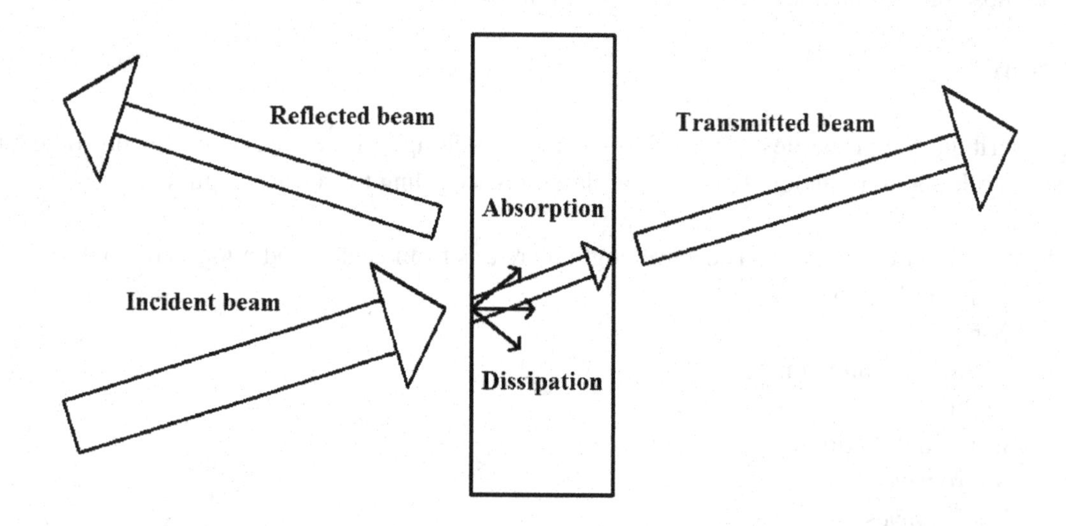

The properties of electromagnetic radiation are described by Planck's law, Stefan-Boltzmann's law and Wien's displacement law. The mathematical expression of Stefan-Boltzmann's law is shown by Equation (1) and the mathematical expression of Wien's displacement law is shown by Equation (2).

$$I = \varepsilon \cdot \sigma \cdot T^4 \quad (1)$$

Where I is radiation intensity (W/m^2), ε is emissivity (-), σ is the Stefan–Boltzmann constant ($5.67\ 10^{-8}\ W/m^2K^4$) and T is thermodynamic temperature (K).

$$\lambda_{max} = \frac{b}{T} \quad (2)$$

Where λ_{max} is the peak at given wavelength (mm) and b is Wien's displacement constant (2.897 mm/K).

Radiation Protective Materials

Parts of the radiation incident on the material are reflected into the environment, dispersed in the material, absorbed and transmitted through the material as shown in Figure 1.

Materials that have a high reflectivity (reflect a large part of the radiation) or absorbance (absorb a large part of the radiation) are used as radiation protective materials. In technical practice the problem is that the reflectivity and absorbance of a material depends on the type (wavelength) of radiation. Therefore, different types of shielding materials are often used for different types of radiation (despite the fact that there are materials capable of shielding more than one type of radiation – e.g. lead, we must also take into account other properties, due to which they are often used only for shielding of radiation where other materials are ineffective). A typical disadvantage of lead is its high density and toxicity; therefore, it is usually used only for shielding ionizing radiation with a high material penetration capability (X-ray, gamma radiation and neuron radiation).

Ultraviolet radiation is shielded by virtually all opaque materials and some transparent ones (e.g. specific types of transparent plastic shield ultraviolet radiation, while pure silica glass transmits ultraviolet radiation well). The ability of transparent materials to shield ultraviolet radiation usually decreases as the wavelength of radiation decreases.

The ability of materials to shield (reflect) infrared radiation is given by the emissivity provided by Equation (1). The lower the emissivity of the material, the greater the proportion of the incident infrared radiation that is reflected. Typical shielding materials for infrared radiation (radiation shields) are high-gloss polished metals (e.g. aluminum).

A Faraday's cage made from metal with high electrical conductivity (low electrical resistance) is used for shielding microwave and radio wave radiation. The ideal material is copper. As the wavelength decreases so the effectiveness decreases. The ideal Faraday's cage is one without any holes (such a cage can completely shield microwave and radio wave radiation of any wavelength – only X-ray radiation and shorter wavelengths can penetrate such a cage). The ability of real Faraday's cages made from mesh or screens to shield microwave and radio wave radiation is reduced as the wavelength becomes shorter, the mesh/screen size increases or the conductivity of the mesh/screen material decreases.

Shielding of alpha radiation is relatively simple in practice, since it does not have a high ability to penetrate materials. The shielding of gamma radiation and neutron radiation is much more complicated. Gamma radiation absorption in materials can be expressed by Equation (3), which is similar to the Beer-Lambert law.

$$I = I_0 \cdot e^{-¼l} \quad (3)$$

Where I is the intensity of the transmitted beam, I_0 is the intensity of the incident beam, e is Euler's number, μ is the absorption coefficient of the material and l is the material thickness.

The most commonly used material, with the highest shielding efficiency, for gamma radiation is lead.

When shielding against neutrons, all kinds of interactions between neutrons and the substance, should be considered. One of them is capture, which also excludes high energy photons that are captured most effectively by heavy elements. However, the neutrons themselves are best shielded using hydrogen nuclei due to elastic scattering. This implies that effective neutron protection must either contain multiple layers (the so-called sandwich), some with a high hydrogen core content (e.g. water), some with heavy elements, or it can be homogenized, with the two groups of protection combined (e.g. concrete with an increased water and iron content – heavy concrete) (Racek, 2009 and Koneck, 2013).

Materials for Protective Suits

The variety of materials suitable for protective suits is determined by the hazard that the suit should protect against.

Materials for Bulletproof Vests and Bomb Suits

Modern bulletproof vests contain several layers. Steel, titanium or metal carbide may be used for the first layer (on the surface). The role of this layer is to flatten the projectile (additionally, it can increase the resistance of the vest to knife puncture, since conventional bulletproof vests without this layer do not exhibit knife puncture resistance – which is by default a considerable weakness of a bulletproof vest). The second layer is usually made of Kevlar fiber. Currently, artificial synthetic silk webs are being developed, whose fibers have a higher strength and lower density than steel and are likely to replace Kevlar fibers in body armor in the future. Another promising material is graphene, which has a higher strength and lower density than steel. Another layer (only used in some vests) is an anti-shock layer, the purpose of which is to dampen the impact of the projectile captured by the vest on surface of the body being protected. The vest may also have other layers (a layer to provide thermal comfort and to meet the microclimate conditions of the user, a hydrophobic coating, etc.). Similar materials and forms of construction are also used in bomb suits, with the difference that bulletproof vests only protect the user's torso, while bomb suits protect the entire body. The parts of the suit that must be transparent (the visor on the helmet) are made of bullet-proof glass (usually laminated glass – several layers of glass, between which there is always a layer of polycarbonate, or sometimes are made using polycarbonate itself). The greatest weakness of all bomb suits are those parts that must allow some flexibility (around the knees, elbows, fingers, etc.).

Materials for Chemical Suits and Fire Proximity Suits

Chemical suits are used to protect the body in environments where respiratory protection using breathing apparatus is not sufficient. Typically, chemical suits are used to protect against toxic substances (capable of absorption through the skin) and corrosive substances. The design and material composition depends on the substances, to which resistance is required and the type of suit. According to durability, chemical suits are divided into (EN 943-1:2015+A1:2019 and EN 943-2:2019):

- Single Use
- Limited Life
- Reusable

According to the protection level, chemical suits are in technical standards EN 943-1:2015+A1:2019, EN 943-2:2019, EN 14605:2005+A1:2009, EN ISO 13982-1:2004/A1:2010 and EN 13034:2005+A1:2009 divided into 6 types (type 1 to 6, with type 1 providing the highest level of protection and type 6 the lowest level of protection). The highest level of protection is provided by type 1 (gas-tight protective chemical suit). European standards EN 943-1:2015+A1:2019 and EN 943-2:2019 sub-divides type 1 gas-tight chemical protective suits into type 1a (gas-tight protective chemical suit with self-contained breathing apparatus worn inside the suit), type 1b (gas-tight protective chemical suit with self-contained breathing apparatus worn outside the suit) and type 1c (gas-tight protective chemical suit used in conjunction with breathing apparatus providing positive pressure). According to EN 943-1:2015+A1:2019, EN 943-2:2019, EN 14605:2005+A1:2009, EN ISO 13982-1:2004/A1:2010 and EN 13034:2005+A1:2009 is other types of chemical suits defined as follow: type 2 (not gas tight), type 3 (protects against fluids), type 4 (protect against spray), type 5 (particle protection) and type 6 (limited protection against spray).

By default, chemical suits (like bulletproof vests) are made up of several layers. The first layer is usually a backing layer (textile), to which the layers of chemical-resistant synthetic polymers are applied. Materials, such as rubber, polyvinylchloride, polytetrafluoroethylene, etc. are used as chemical-resistant synthetic polymers.

Fire proximity suits, like chemical suits, have a layered structure. Depending on the temperature and protection method, they are divided into radiant heat-resistant and flame-resistant suits. Radiant heat-resistant suits usually consist of a Kevlar fiber coated with an aluminum layer that reflects thermal (infrared) radiation. These suits are commonly used up to 300 °C. Flame-resistant suits consist of mostly glass fiber, on which a shiny metal alloy is applied. These suits are used from 800 °C to 1050 °C.

Electrical Insulating Materials

The relationship between electrical current, electrical voltage and electrical resistance is expressed by Ohm's law, whose mathematical notation is expressed by Equation (4).

$$I = \frac{U}{R} \tag{4}$$

Where I is electrical current (A), U is electrical voltage (V) and R is electrical resistance (Ω).

Materials with high electrical resistance are called electrical insulators. In addition to electrical resistance, these materials are subject to other requirements depending on their application. The insulating materials for electrical cables (insulating materials for electrical conductors and the sheaths of electrical cables) must also have sufficient flexibility and resistance to external influences (depending on their installation; for a fixed installation, the requirement for flexibility is reduced, but in corrosive environments the requirements for chemical resistance of the sheath increase). This insulation is usually made of polyvinylchloride or cross-linked polyethylene. On the other hand, isolators must have a very high resistance to external influences as well as high electrical resistance, so these products are often made of ceramic. The insulation for personal protective equipment (e.g. insulated screwdriver handles or insulated pliers) must have suitable mechanical properties in addition to high electrical resistance. These products are made of, for instance, polypropylene, cellulose acetate and other similar materials (often a layered handle structure is used where the base of the handle is made of a material with high electrical resistance and strength and is coated with a material to provide an ergonomic and non-slip finish).

Materials for Fire Protection

Virtually all organic (natural and synthetic) polymers are flammable. The fire behaviour of materials can be evaluated in two basic ways.

The first is the reaction to fire, which evaluates the contribution of a construction product to the development of a fire. In the European Union, the reaction of a product to fire is assessed according to EN 13501-1:2018, which, according to the reaction to fire, divides products into 7 classes A1, A2, B, C, D, E and F (class A1 is the least reactive to fire and class F the most reactive). However, the reaction to fire is a much broader term and, in its nature, (as perceived outside the European Union) includes the resistance of the material to ignition and the impact of the fire of the product on the environment (expressed by the released heat and combustion products). Thus, reaction to fire concerns a product (in terms of the terminology used within the European Union) or material (in terms of the terminology used outside the European Union).

In addition to materials and products, structures are also evaluated during a fire. Structures are evaluated in terms of their fire resistance. The fire resistance of a structure represents the time interval for which the structure resists a fire (defined by the fire curve, or by temperatures, or by heat release rate) without losing its defined function (e.g. load-bearing capacity and stability).

The reaction to fire of materials and products is usually improved by flame retardants. Intumescent coatings and linings are frequently used to increase the fire resistance of structures (but in the case of wooden structures, fire resistance may also be increased by flame retardants).

Flame Retardants

Flame retardants are classified according to several criteria. According to the method of application, flame retardants are according to Masarik (2003) divided into:

- Flame Retardants for Wood and Wood-Based Materials
    - Surface coatings
    - Maceration
    - Impregnation
    - Additive flame retardants (for agglomerated wood-based materials e.g. oriented strand board)
- Flame Retardants for Synthetic Polymers
    - Surface coatings
    - Additive flame retardants (added to granulate)
    - Reactive flame retardants (added to reaction mixture during synthesis - polyreactive)

According to the mechanism of action, Petrova, Soudek, and Vanek (2015), divided flame retardants into:

- Flame Retardants that Affect the Process of Pyrolysis
- Flame Retardants that Affect the Chemical Reactions in Flame
- Flame Retardants that Restrict the Access of Oxygen to Polymer Surface

According to chemical composition, Petrova, Soudek, and Vanek (2015), divided flame retardants into:

- Halogenated Flame Retardants
    - Polybrominated diphenyl ethers
    - Tetrabromobisphenol A
- Inorganic Flame Retardants
    - Alkaline earth metal hydroxides
    - Zinc
    - Antimony
- Phosphorous-Based Retardants
    - Inorganic phosphorus retardants

- Organic phosphorus retardants
- Halogenated phosphorus retardants

- Nitrogen-Based Retardants
- Other Flame Retardants

Materials to Increase the Fire Resistance of Structures

Materials that increase the fire resistance of building structures are among the essential materials of safety and security engineering. Structures whose fire resistance may needs to be increased include:

- Steel Structures
- Wood Structures
- Concrete Structures

There is a common misconception that as steel and concrete structures are non-combustible, they are also fire-resistant. This belief is true in terms of the contribution of these structures to the development of fire (since they do not release heat during a fire). However, it is often necessary to improve their properties in terms of maintaining load-bearing capacity and stability (static function) and thermal insulation during a fire. In the case of an unprotected steel structure, there is a risk of a decline in strength (due to the temperature increase caused by fire) and subsequent collapse. In the case of an unprotected concrete structure, the fire temperature can cause the surface to crack (the occurrence of cracks, their timing and extent are mainly determined by the water content of the concrete and the fire temperature). Cracking can cause the cross-section of the concrete structure to fall below the critical limit. In the case of a steel reinforced concrete structure, the situation is even more serious, as cracking of the concrete surface exposes the steel reinforcement allowing rapid overheating. The protection of these structures is therefore mainly based on insulation from fire. The fire resistance of steel and concrete structures is increased by:

- Fire-Resistant Plasters
- Fire Coatings (Mainly Intumescent Coatings)
- Fireproof Linings or False Ceilings (Mainly Plasterboard)

The situation is a little better for wooden structures, which ignite and burn in fire, but the loss of their bearing capacity and stability (fire resistance) occurs only after the cross-section falls below the critical value (depends on the extent, to which the design was oversized). The fire resistance of wooden structures is increased mainly by:

- Fire Coatings (Mainly Intumescent Coatings)
- Fireproof Lining or False Ceiling (Mainly Plasterboard)

At present, however, there are relatively few scientific works that comprehensively address the fire hazard of materials, which is a key feature in safety and security engineering. Therefore, this chapter aims to develop a methodology to assess the fire hazard of materials and to create a fire risk assessment of selected commonly used synthetic polymers (polyethylene and polypropylene).

Fire Hazard of Materials

The fire hazard of a material is expressed by parameters that assess its resistance to ignition and the impact of combustion on the environment (the impact on the health and safety of those exposed plus the environment and property). The fire hazard of a material is thus expressed by the following fire properties:

- Critical Heat Flux
- Ignition Temperature
- Flashover Category
- Heat Release
- Combustion Products Release
- Smoke Release

All of the following fire properties were determined for polyethylene and polypropylene in a cone calorimeter. Both the test procedure and the test equipment were according to ISO 5660-1:2015. The ambient temperature during the tests was between 23 and 25 °C and the relative humidity between 50 and 60%. The tests were performed at heat fluxes of 30 and 50 kW/m^2. The polyethylene repeat unit is $(-\,CH_2 - CH_2\,-)_n$. Density of investigated polyethylene and polypropylene was 940 kg/m^2 and 910 kg/m^2, respectively. Fire characteristics of polyethylene and polypropylene are important for fire safety engineering because they are most common synthetic polymers.

Critical Heat Flux and Ignition Temperature

Critical heat flux is the minimum heat flux that causes piloted ignition of the material. The method used for the determination of the critical heat flux depends on the thermal thickness of the material. The difference between thermally thin and thermally thick material is illustrated in Figure 2. The thermally thin material is heated to approximately the same temperature by the external heat flux and ignites when the ignition temperature is reached, while the thermally thick material heats to the ignition temperature only on its surface and then ignites. The thermal thickness of the material depends on a large number of parameters (external heat flux, material density, heat and thermal conductivity of the material, thermal capacity of the material, material thickness, etc.). In fire engineering, the thermal thickness of a material is determined in a simplified manner using Equation (5), derived by Babrauskas and Parker (1987).

$$L = 0.6 \cdot \frac{\rho}{q} \tag{5}$$

Where L is the thickness above which the material is considered to be thermally thick (mm), ρ is the material density (kg/m^2), q is the external heat flux (kW/m^2).

The basic procedure for calculating critical heat flux has been described by Mikkola and Wichman (1989). The cited procedure is based on a measurement of the time to material ignition at a minimum of three heat flux densities. The statistical dependence of the ignition time on heat flux is calculated from the measured data. For thermally thick materials, the statistical dependence of the ignition time raised to the power of -1/2 is plotted from the heat flux. For thermally thin materials, the ignition time is increased to the power of -1 (corresponding to the inverse of the initiation time) in the statistical dependence of the ignition time on the heat flux. In this statistical dependence, the zero value (theoretically this represents an infinite ignition time) is set for the ignition time (raised to the power of -1/2 or -1) and then the critical heat flux is calculated. The critical heat flux calculated in this way corresponds to an infinite ignition time (theoretically, it would have to act on the material for an infinitely long time), therefore the cited authors recommend adding 3 kW/m^2 to the calculated value. This value represents the heat loss and by adding it, we obtain a critical heat flux corresponding to an initiation time of 30 minutes. A similar process for the calculation is given by Spearpoint and Quintiere (2001) with the difference that the cited authors consider only thermally thick materials and the value calculated from statistical dependence is divided by 0.76 (instead of adding a constant value of 3 kW/m^2).

In addition to critical heat flux, ignition temperature is the second important parameter that expresses the resistance of the material to ignition. The ignition temperature is calculated according to Equation (6), derived by Spearpoint and Quintiere (2001).

$$q_{cr} = \sigma \cdot \left(T_{ig}^4 - T_a^4\right) + h_c \cdot \left(T_{ig} - T_a\right) \quad (6)$$

Where q_{cr} is the critical heat flux (kW/m^2), σ is the Stefan-Boltzmann constant (5.67 10^{-8} W/m^{2}K^4), T_{ig} is the ignition temperature (K), T_a is the ambient temperature (K) and h_c is the natural convective heat transfer coefficient (5 W/m^{2}K).

The statistical dependences of the ignition times raised to the power of -1/2 from the heat flux for the polyethylene and polypropylene examined are shown in Figure 3.

The critical heat fluxes of polyethylene and polypropylene together with the ignition temperatures are shown in Table 1.

The critical heat fluxes and ignition temperatures of both polyethylene and polypropylene shown in Table 1 approximately correspond to the critical heat fluxes and ignition temperatures of organic polymers as reported in scientific papers (Scudamore et al., 1991 and Tewarson, 2002). Therefore, in terms of resistance to fire initiation, the polyethylene and polypropylene examined do not represent a higher fire hazard than most conventional organic polymers.

Flashover Category

Flashover category is one of the most important fire characteristics, as it quantifies the behavior of a material or product in the most unstable and least explored stage of fire development, the flashover phase. The technical standard, ISO 9705:2016, divides materials according to the time they reach flashover into four flashover categories:

- Flashover Category 1: Flashover does not occur.
- Flashover Category 2: Flashover occurs over a time interval of 600 to 1200 seconds.
- Flashover Category 3: Flashover occurs over a time interval of 120 to 600 seconds.
- Flashover category 4: Flashover occurs within 120 seconds.

The test according to ISO 9705:2016 is the only large-scale test that evaluates the behavior of a material or product in the flashover phase that is widely accepted by the professional community. However,

Figure 2. Difference between a) thermally thin and b) thermally thick material (T_{exp} is the temperature of the exposed surface and T_{unexp} is the temperature of the unexposed surface)

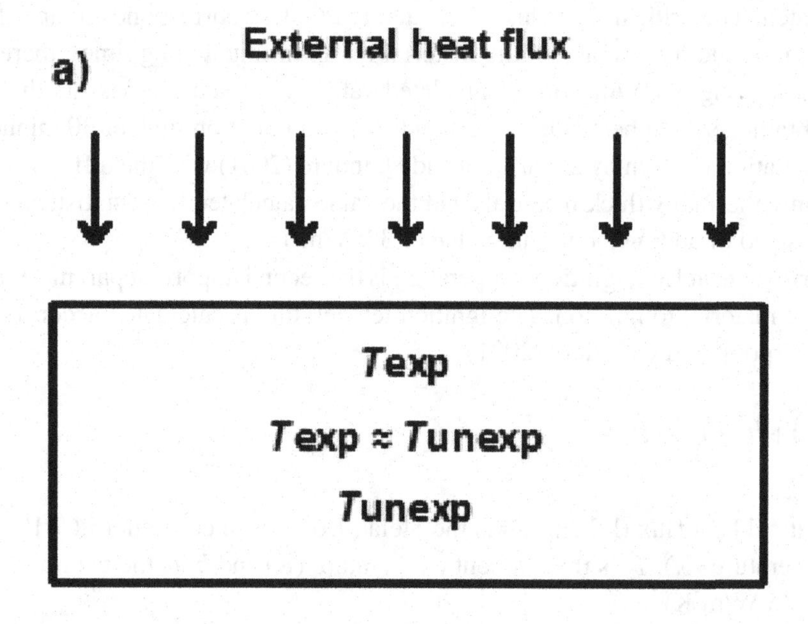

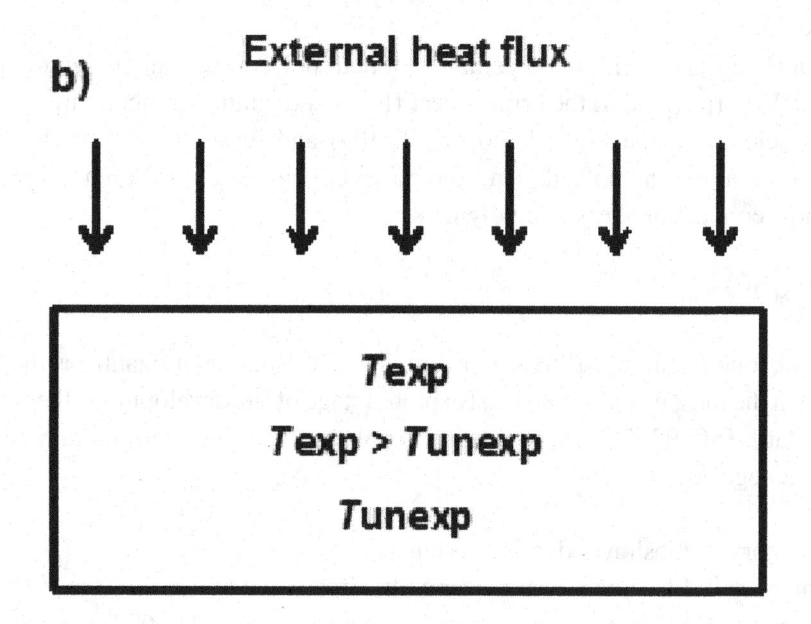

Figure 3. Time to ignition of a) polyethylene and b) polypropylene

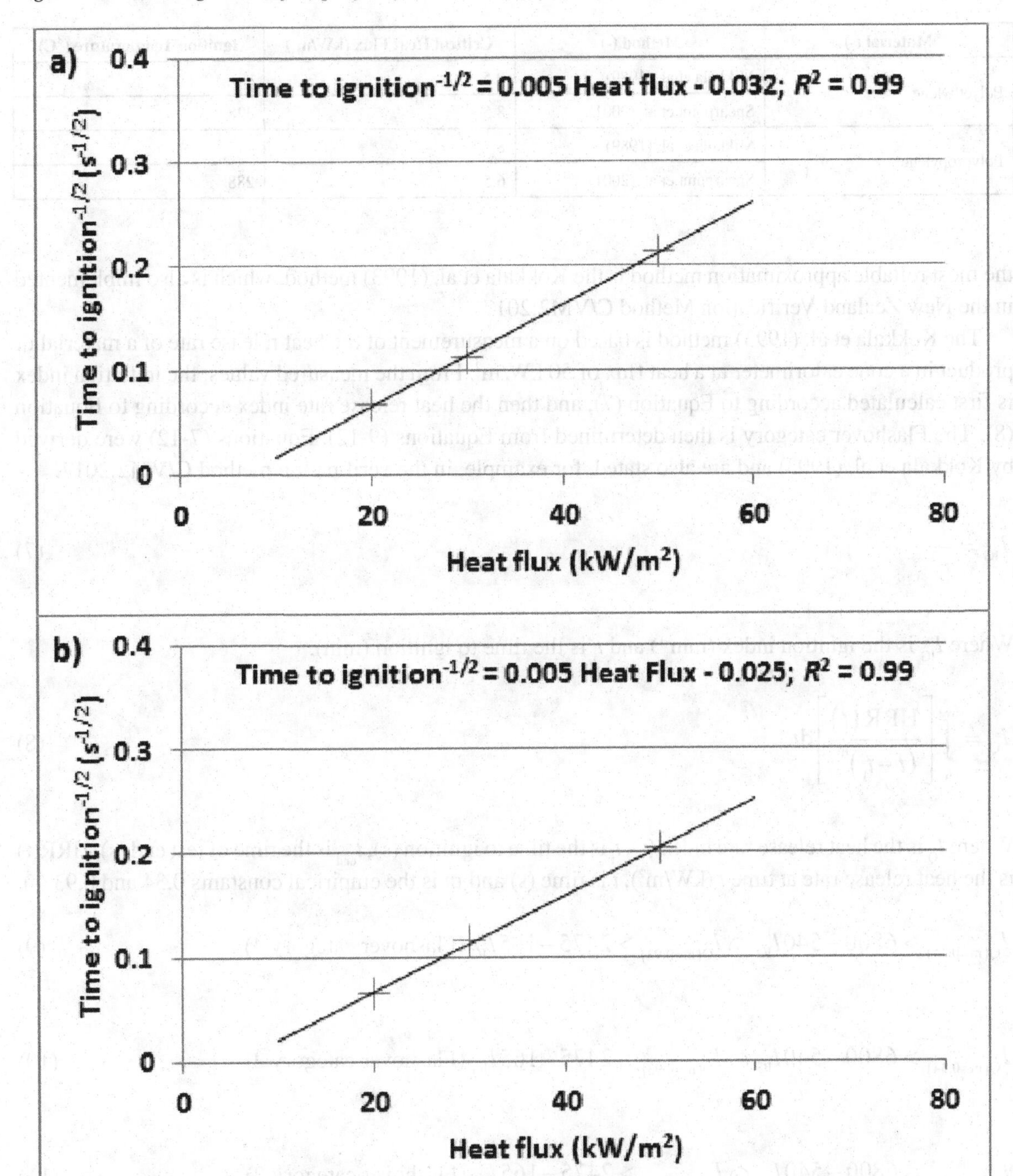

this test is extremely time-consuming and expensive. Therefore, many research institutions and scientists around the world have been looking for a method that can approximate it with a laboratory test. The result has been a relatively large number of more or less reliable test methods and models. At present,

Table 1. Critical heat flux and ignition temperature of polyethylene and polypropylene

Material (-)	Method (-)	Critical Heat Flux (kW/m²)	Ignition Temperature (°C)
Polyethylene	Mikkola et al. (1989)	9.5	-
	Spearpoint et al. (2001)	8.5	328
Polypropylene	Mikkola et al. (1989)	8	-
	Spearpoint et al. (2001)	6.5	288

the most reliable approximation method is the Kokkala et al. (1993) method, which is also implemented in the New Zealand Verification Method C/VM2:2017.

The Kokkala et al. (1993) method is based on a measurement of the heat release rate of a material or product in a cone calorimeter at a heat flux of 50 kW/m^2. From the measured values, the initiation index is first calculated according to Equation (7), and then the heat release rate index according to Equation (8). The Flashover category is then determined from Equations (9-12). Equations (7-12) were derived by Kokkala et al. (1993) and are also stated, for example, in the verification method C/VM2:2017.

$$I_{IG} = \frac{1}{t_i} \quad (7)$$

Where I_{IG} is the ignition index (min^{-1}) and t_i is the time to ignition (min).

$$I_Q = \int_{t_i}^{t_{end}} \left[\frac{HRR(t)}{(t - t_i)^m} \right] dt \quad (8)$$

Where I_Q is the heat release rate index (-), t_i is the time to ignition (s), t_{end} is the time of test end (s), HRR(t) is the heat release rate at time t (kW/m^2), t is time (s) and m is the empirical constants 0.34 and 0.93 (-).

$$I_{Q(m=0.34)} > 6800 - 540 I_{IG} \cap I_{Q(m=0.93)} > 2475 - 165 I_{IG} \text{ (Flashover category 4)} \quad (9)$$

$$I_{Q(m=0.34)} > 6800 - 540 I_{IG} \cap I_{Q(m=0.93)} \leq 2475 - 165 I_{IG} \text{ (Flashover category 3)} \quad (10)$$

$$I_{Q(m=0.34)} \leq 6800 - 540 I_{IG} \cap I_{Q(m=0.93)} > 2475 - 165 I_{IG} \text{ (Flashover category 2)} \quad (11)$$

$$I_{Q(m=0.34)} \leq 6800 - 540 I_{IG} \cap I_{Q(m=0.93)} \leq 2475 - 165 I_{IG} \text{ (Flashover category 1)} \quad (12)$$

The polyethylene and polypropylene examined were classified on the basis of the measured values (in a cone calorimeter) as flashover category 4 materials, consistent with the results of the scientific paper (Xu et al., 2012).

Heat Release Rate

According to Babrauskas and Peacock (1992), the heat release rate is the most important fire characteristic of a material. The rate of heat release can be considered as a two-dimensional parameter as it depends on the effective heat of combustion of the material and its mass loss rate (in addition to the thermal content, it also takes into account the time factor, i.e. the time, during which the heat is released from the material). The unit of measurement of the heat release rate is kW/m^2. In SI units, it can be expressed as kJ/m^2s (i.e. it expresses how many kJ of heat are released from one square meter of material per second).

Other parameters can also be calculated from the heat release rate, in particular:

- Heat Release Rate Peak
- Maximum Average Rate of Heat Emission
- Fire Growth Rate Index
- Total Heat Release

The peak heat release rate represents the maximum HRR achieved during the test. Taking into account the time dependence, the Maximum Average Rate of Heat Emission (MARHE) or the Fire Growth Rate Index (FIGRA) is reported more often than the peak heat release rate. MARHE is defined as the maximum value of the Average Rate of Heat Emission (ARHE), calculated according to Equation (13), from Zhang (2008). FIGRA is calculated according to Equation (14), which is similar to the equation given in EN 50399:2011/A1:2016.

$$\mathrm{ARHE}(t)=\frac{\sum_{t_i}^{t_{end}}\frac{\left(\mathrm{HRR}(t)+\mathrm{HRR}(t-1)\right)}{2}}{t-t_i}\cdot{}_{\Delta}t \tag{13}$$

$$\mathrm{FIGRA}=\max\frac{\frac{\mathrm{HRR}(t-1)+\mathrm{HRR}(t)+\mathrm{HRR}(t+1)}{3}}{t-t_i} \tag{14}$$

Where ARHE is the average rate of heat emission (kW/m^2), FIGRA is the fire growth rate index (kW/m^2s), ${}_{\Delta}t$ is the time interval between two nearest measured values (s), t is time (s), t_i is the ignition time (s) and HRR(t-1), HRR(t) and HRR(t+1) are the heat release rates measured at the time interval from t-1 to t+1 (kW/m^2).

Another important parameter related to the rate of heat release is Total Heat Release (THR). From a mathematical point of view, it is the integral of the heat release rate curve. THR is expressed in MJ/m^2 and expresses the amount of heat released from one square meter of material over a period of time.

From a mathematical point of view, it would be more correct to talk about THR over a time interval (e.g. THR_{fi}^{1200} means total heat release in the interval from ignition to 1200 seconds after ignition).

The HRR together with the THR of the polyethylene and polypropylene examined are shown in Figure 4. The ARHE of the polyethylene and polypropylene examined is shown in Figure 5 and the FIGRA in Figure 6.

The data in Figures 4, 5 and 6 approximately agree with the results in scientific papers by Hirschler (2003) and Babrauskas (2003). The data in the Figures further demonstrate that the polyethylene and polypropylene examined exhibit similar values of heat release rate, maximum average rate of heat emission and fire growth rate index. This is probably due to the fact that polyethylene and polypropylene have a very similar chemical composition (the ratio of carbon to hydrogen is the same in both macromolecules).

Figure 4. Heat release rate and total heat release for a) polyethylene and b) polypropylene

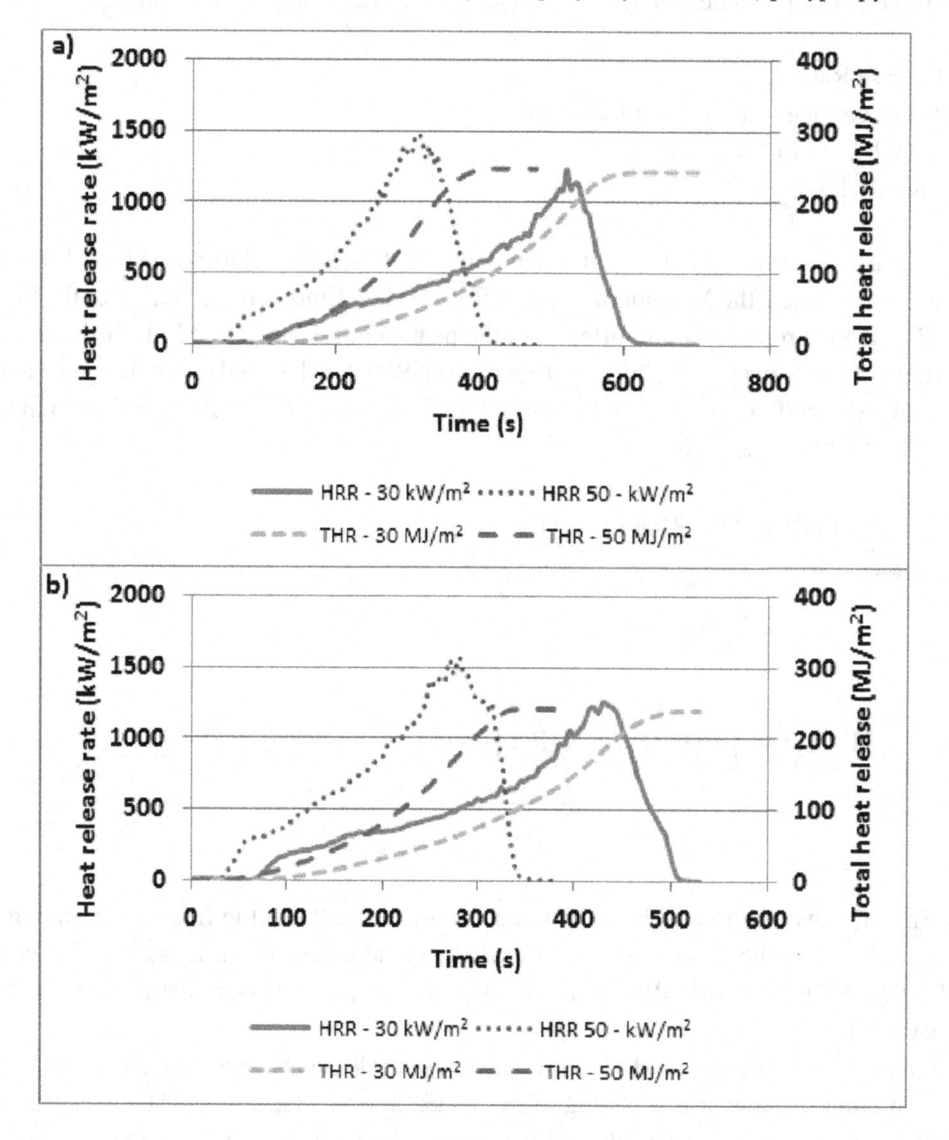

The comparison of Figures 4 to 6 with the results of scientific studies by Babrauskas (2003), Hirschler (2003), Parker (2003) and Tran (2003) show that the polyethylene and polypropylene examined (in terms of heat release rate and derived parameters) exhibit a higher fire hazard than natural polymers and most synthetic polymers. The reason is probably that polyethylene and polypropylene are thermoplastics that melt under thermal stress and behave like flammable liquids during a fire.

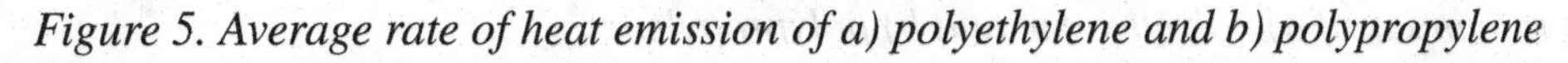

Figure 5. Average rate of heat emission of a) polyethylene and b) polypropylene

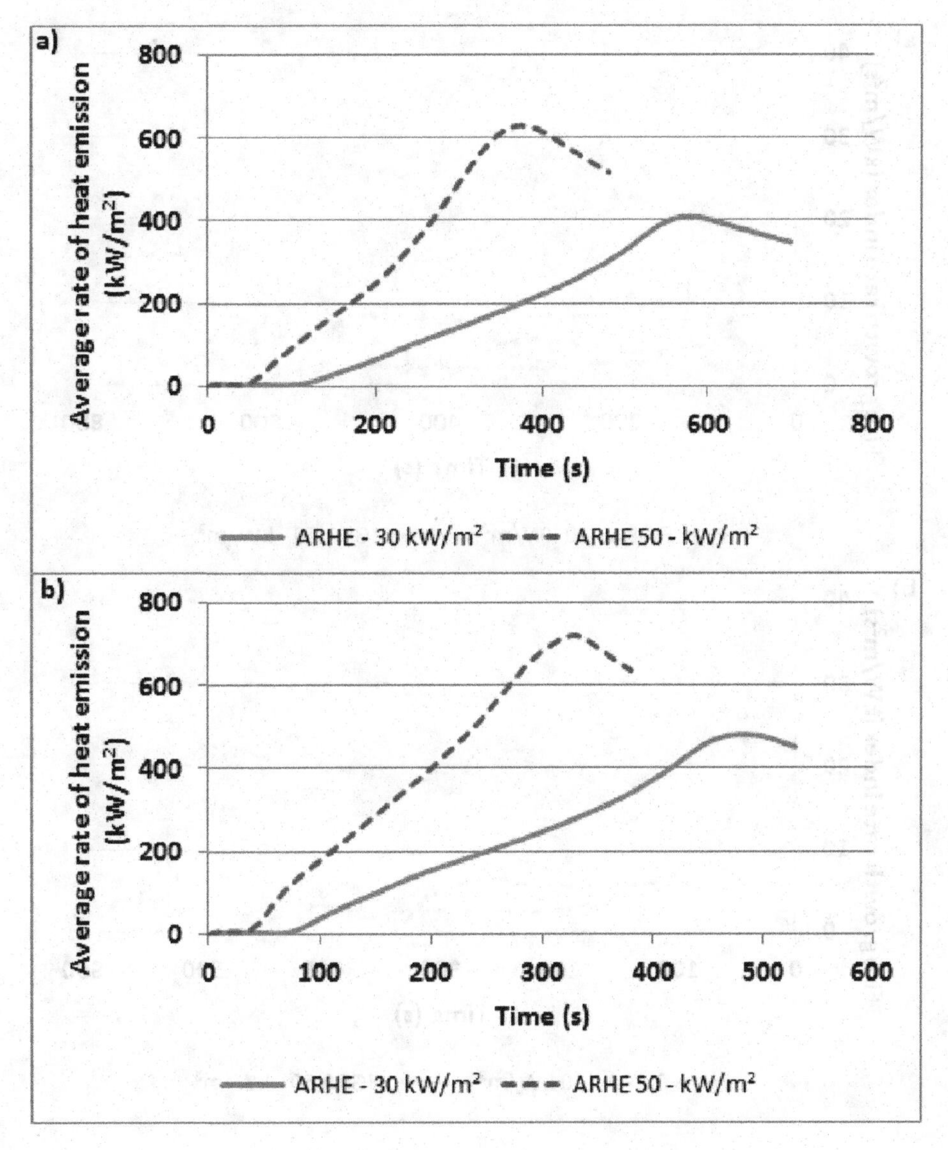

The comparison of the values in Figures 4 to 6 further demonstrates that the fire hazard of polyethylene and polypropylene rises with increasing heat flux (due to faster thermal decomposition of the material due to the higher heat flux). The Figures also show that polypropylene exhibits (in terms of HRR and derived parameters) a slightly higher fire hazard than polyethylene. In practice, however, the difference between these materials is negligible.

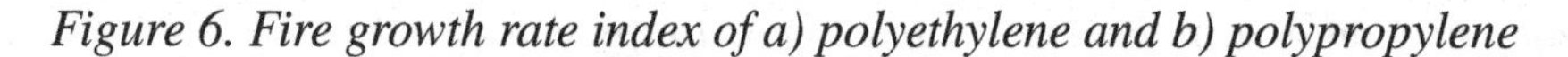

Figure 6. Fire growth rate index of a) polyethylene and b) polypropylene

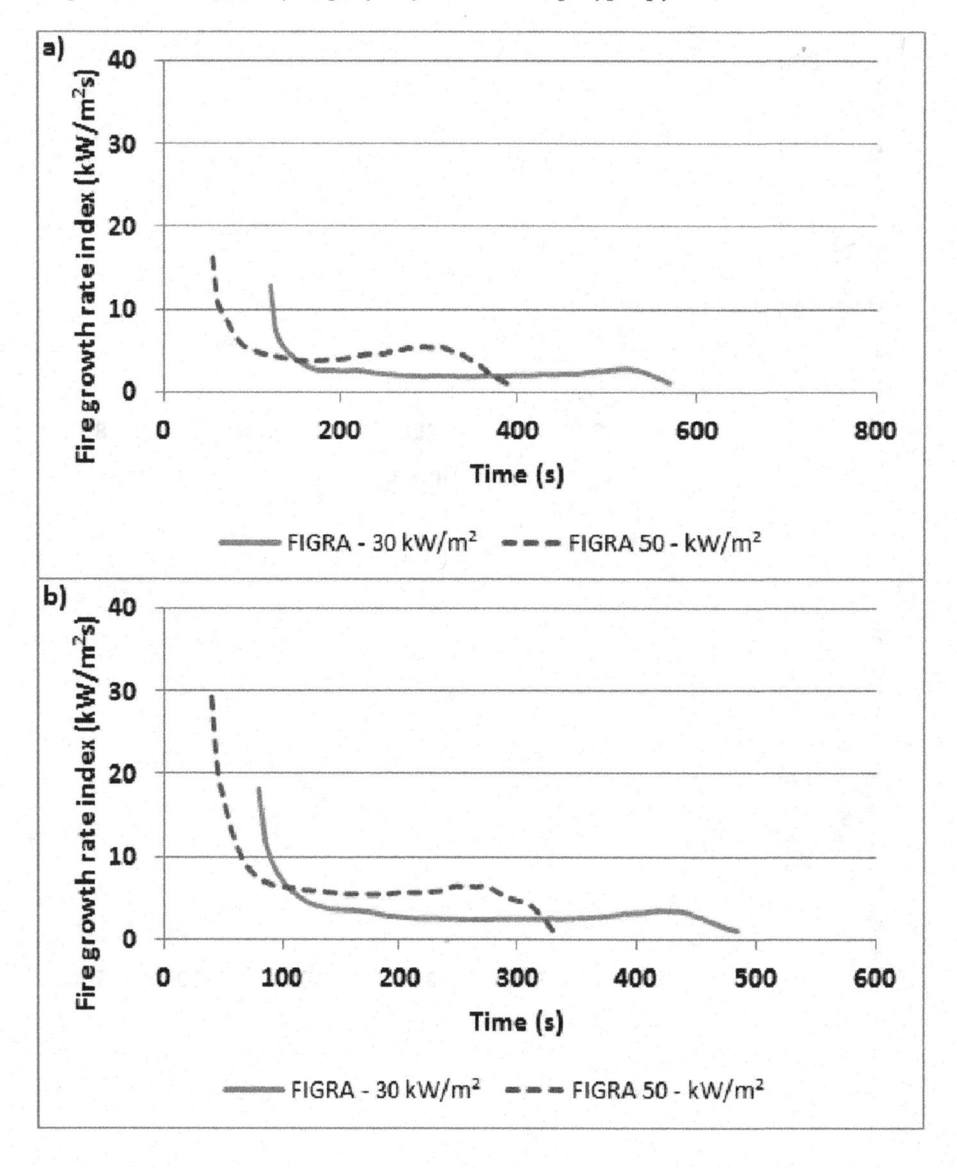

Combustion Products and Smoke Release

Combustion products are the substances that are produced when a material or product burns (or substances produced by fire). The combustion products have the following effects on the environment:

- **Convection Heat Transfer.**
- **Toxicity.**
- **Reduction in Visibility.**

Combustion products are a heat transfer medium and allow heat transfer by convection. Convection heat transfer is especially dominant in the local fire phase, when up to 70% of the energy is transferred by convection. Moreover, according to Masarik, Dvorak, and Charvatova (1999), combustion products cause 70 to 90% of fire casualties. Although, a large amount of gaseous combustion products is generated in a fire, according to C/VM2:2017, carbon monoxide (CO) is the most important from a toxicological point of view. In practice, it is not technically or economically feasible to compare materials in terms of product toxicity by quantifying all the components of the combustion products. Therefore, in terms of toxicity of combustion products, materials are compared by the amount of carbon monoxide released and for materials containing nitrogen in the molecule, also by the amount of hydrogen cyanide released.

The materials examined do not have nitrogen molecules, so their toxicity is compared based on carbon monoxide alone.

Moreover, solid combustion products cause a decrease in visibility in the area affected by fire. In terms of reduced visibility due to smoke release, materials are compared based on the specific smoke production rate and total smoke production.

The amount of carbon monoxide released is expressed as the CO production rate in units of g/m^2s. This parameter expresses the amount of CO released from one square meter of material surface per second. The total amount of CO released is expressed as total CO production (TCO) in g/m^2. This parameter expresses how many grams of carbon monoxide are released from one square meter of material surface over a period of time. The Smoke Production Rate (SPR) m^2/m^2s and Total Smoke Production (TSP) m^2/m^2 have a relatively unconventional unit. The unit (m^2/m^2s) expresses the amount of smoke that is released from the surface of 1 m^2 piece of material per second. Similarly, the unit of m^2/m^2 expresses the total amount of smoke that is released from a surface area of 1 m^2 over a period of time. The particularity of this unit is in expressing the amount of smoke (m^2). The amount of smoke expressed in m^2 can be explained as the area of a beam of light that the released smoke can cover (this approach allows a simple calculation of the decrease in visibility in the smoke-affected area). The units of SPR (m^2/m^2s) and TSP (m^2/m^2) are covered in more detail by Ostman (2003).

The CO production rate and total CO production of the polyethylene and polypropylene examined are shown in Figure 7 and the smoke production rate together with total smoke production are shown in Figure 8.

Figure 7 and 8 show that, similar to the HRR, ARHE and FIGURE, the CO production rate and total CO production increases (Figures from 4 to 7), as the heat flux increases, as does the smoke production rate and total smoke production (Figure 8). The Figures cited above also show that, from the point of view of CO production and smoke production rate, as with the HRR, polypropylene presents a higher fire hazard than polyethylene. In practice, however, the differences between these materials are negligible.

The cause of the higher CO production rate, total CO production (Figure 7) together with the smoke production rate and total smoke production (Figure 8) with increasing heat flux is a higher mass loss rate. A higher mass loss rate causes a greater mass of gaseous fuel in the combustion zone and hence a higher fuel / oxygen ratio, resulting in an increase in imperfect combustion (CO and smoke production).

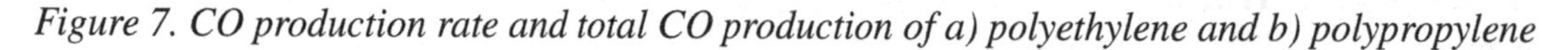

Figure 7. CO production rate and total CO production of a) polyethylene and b) polypropylene

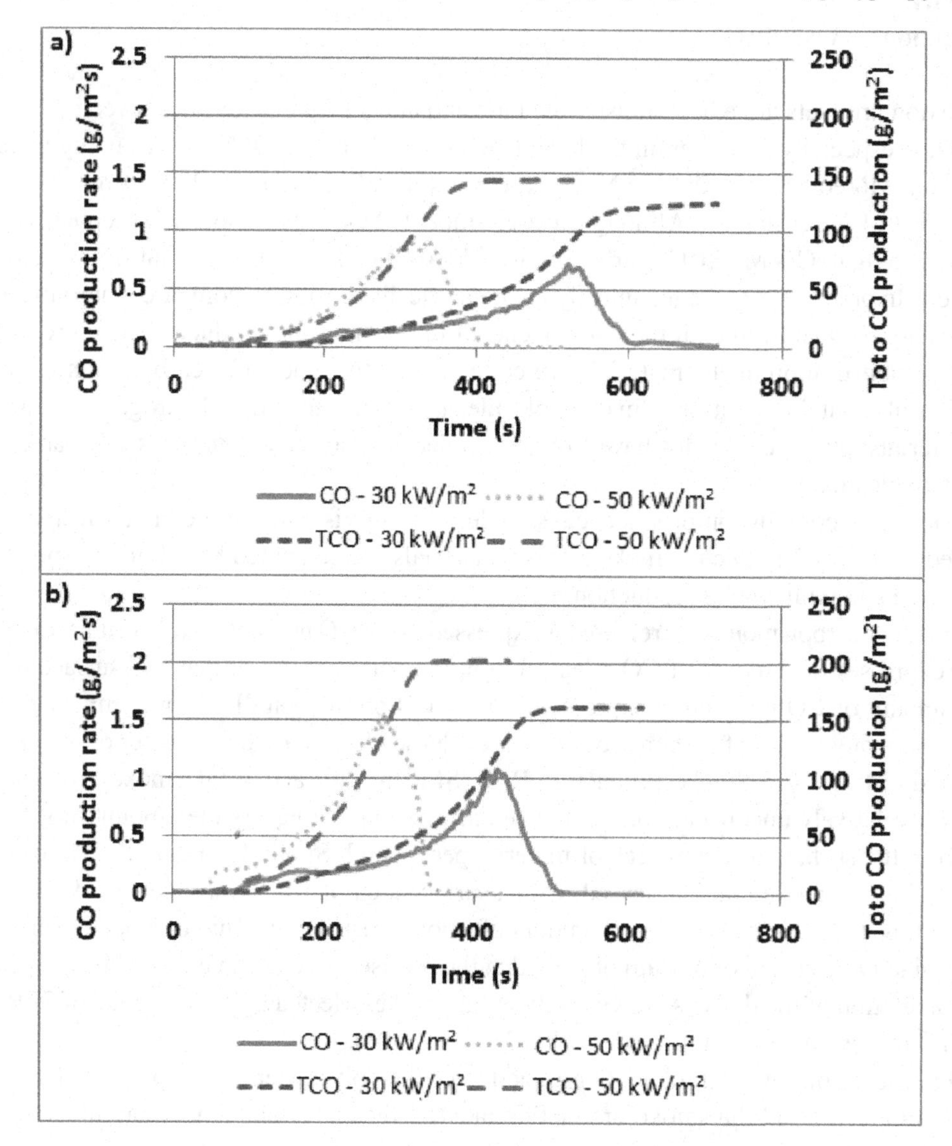

Figure 8. Smoke production rate and total smoke production of a) polyethylene and b) polypropylene

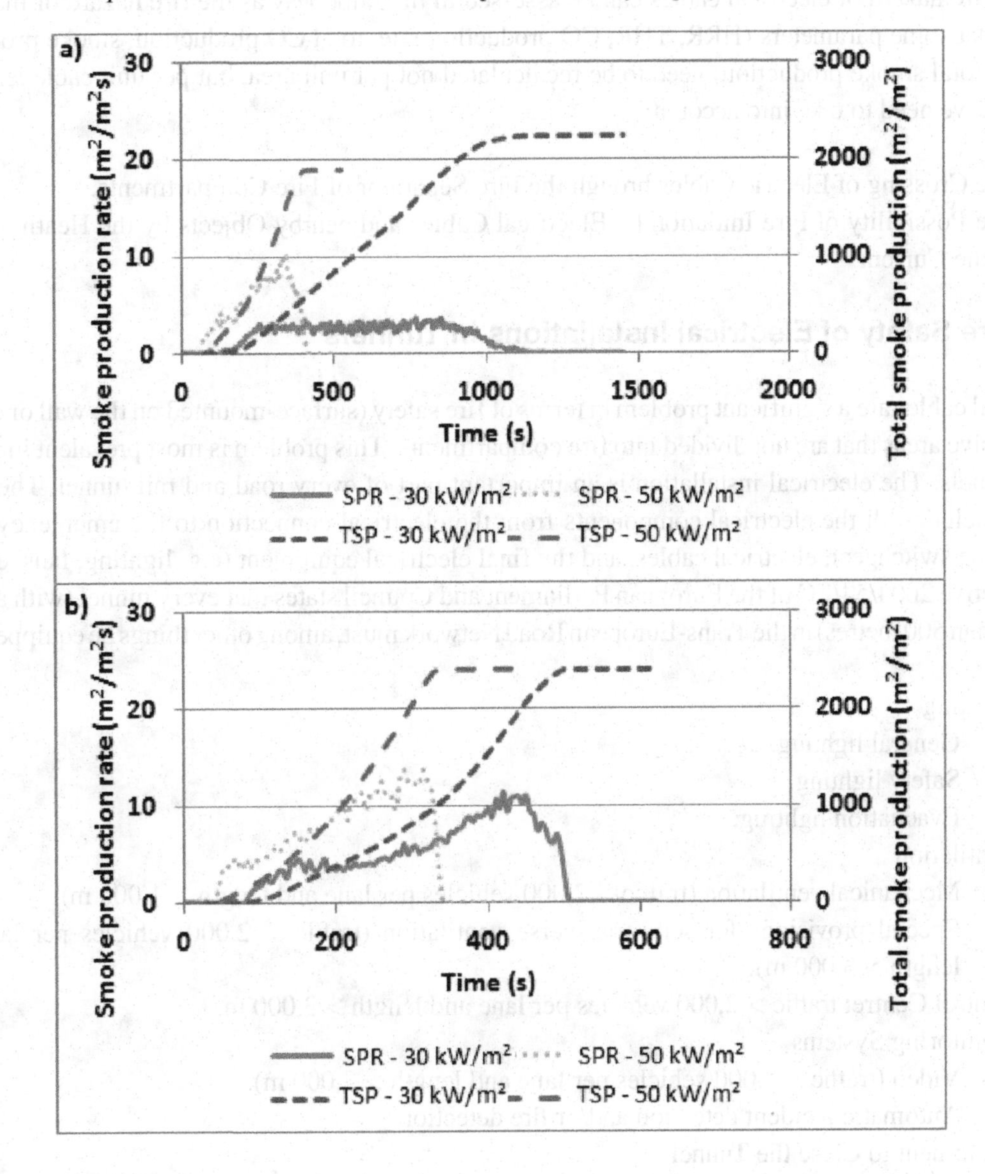

ELECTRICAL CABLES

Electrical cables pose a specific problem in the category of materials for safety and security engineering for several reasons. The main one is that synthetic polymers with high electrical resistance (electrical strength) and sufficient flexibility and resistance to external influences are used for insulation to ensure safe operation without electric shocks. These organic synthetic polymers (the cable sheath and wire insulation) are flammable in almost all cases. Moreover, electrical cables often pass through fire compartments and can become an initiating source as a result of the electrical current they carry. Joule's law clearly states that the heating power of an electrical conductor is proportional to the product of its resistance and the square of the current.

The fire hazard of electrical cables can be assessed in the same way as the fire hazard of materials, except that some parameters (HRR, THR, CO production rate, total CO production, smoke production rate and total smoke production) need to be recalculated not per unit area, but per unit cable length. In addition, we need to take into account:

- The Crossing of Electric Cables hrough the Fire Separator of Fire Compartments.
- The Possibility of Fire Initiation by Electrical Cables and nearby Objects by the Heating. Effect of the Current.

The Fire Safety of Electrical Installations in Tunnels

Electrical cables are a significant problem in terms of fire safety (surface-mounted on the wall or ceiling) in extensive areas that are not divided into fire compartments. This problem is most prevalent in rail and road tunnels. The electrical installation is an important part of every road and rail tunnel. The tunnel wiring includes all the electrical components from the electrical connection to the emergency power supply, the switchgear, electrical cables, and the final electrical equipment (e.g. lighting, fans, etc.).

Directive 2004/54/EC of the European Parliament and Council states that every tunnel (with a length greater than 500 metres) in the Trans-European Road Network must, among other things be equipped with:

- Lighting
  - General lighting.
  - Safety lighting.
  - Evacuation lighting.
- Ventilation
  - Mechanical ventilation (traffic > 2,000 vehicles per lane and length > 1,000 m).
  - Special provisions for semi-transverse ventilation (traffic > 2,000 vehicles per lane and length > 3,000 m).
- Control Centre: traffic > 2,000 vehicles per lane and length > 3,000 m.
- Monitoring Systems
  - Video (traffic > 2,000 vehicles per lane and length > 3,000 m).
  - Automatic incident detection and/or fire detection.
- Equipment to Close the Tunnel
  - Traffic signals before the entrances (length > 1,000 m).
  - Traffic signals inside the tunnel at least every 1,000 m (recommended for traffic > 2,000 vehicles per lane and length > 3,000 m).
- Communication Systems
  - Radio re-broadcasting for emergency services (traffic > 2,000 vehicles per lane and length > 1,000 m).
  - Emergency radio messages for tunnel users.
  - Loudspeakers in shelters and exits.
- Emergency Power Supply
- Fire Resistance of Equipment

All of these devices need electrical energy to operate and this energy is supplied through electrical cables. Thus, electrical cables are a critical part of the electrical installation of the tunnel.

Electrical cables are divided into:

- Power Cables: used to supply energy to electrical devices.
- Communication Cables: used to transmit information as electrical signals.

Electrical cables consist of at least three components: the electrical conductors, the insulation and the sheath. The space between the insulated conductors and the sheath is often filled with a bedding material. Moreover, the construction of the cable can be supplemented with reinforcement (to improve its mechanical properties), shielding (which serves to shield the conductors from electrical disturbances) and other components. Every electrical cable consists of at least two electrical conductors, these are constructed using metals with a high level of electrical conductivity and acceptable mechanical properties (copper or aluminium are most often used). Electrical conductors are constructed as either cables or wires. The insulation of the electrical conductors, the bedding material and the sheath are almost always made from synthetic polymers (especially polyvinyl chloride and/or polyethylene) which have a high electrical resistance and a degree of flexibility. Metal hydroxides (especially aluminium hydroxide and magnesium hydroxide) and/or chalk are often added as a bedding material to the synthetic polymers used in the bedding and sheath. Metal hydroxide-based bedding improves the fire performance of the electrical cables. According to the mutual arrangement of electrical conductors, electrical cables are divided into:

- Axial Cables
- Coaxial Cables

The difference between an axial and coaxial cable is shown in Figure 9.

Figure 9. Cross-section of cable a) axial cable and b) coaxial cable

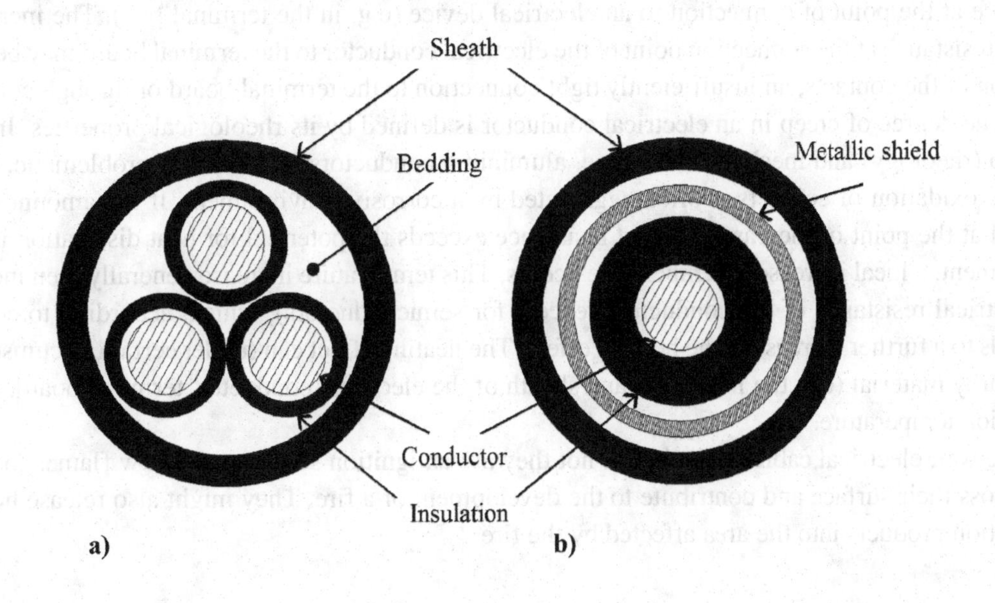

The metallic shield (Figure 9) serves as the second conductor in the coaxial cable.

Optical cables have a slightly different design. The basic difference between electrical and optical cables is in the material composition. Optical cables are made of optical fibres instead of electrical conductors. An optical fibre is normally made of glass or a synthetic polymer (e.g. polymethyl methacrylate). Moreover, optical cables may have a single optical fibre (a significant difference to electrical cables which require at least two electrical wires to function). Another important difference is that optical cables only transmit signals (information) in the form of electromagnetic radiation (they cannot transmit electrical power).

Electrical cables in tunnels are associated with three serious hazards:

- Fire Cause
- Fire Spread and the Impact of Fire on People and Property
- Power Outage

Every electrical conductor is heated by the flow of electrical current. According to the Joule-Lenz law, the thermal power is proportional to the product of the electrical resistance of the conductor and the square of the current, as expressed by equation (15).

$$P = I^2 \cdot R \tag{15}$$

Where P is the heating power (W), I is the electric current (A) and R is the electrical resistance (Ω).

An electrical conductor (cable) can become the cause (ignition source) of a fire if it is loaded with an electrical current that exceeds its rating or through an increase in its electrical resistance. Electrical cables and equipment are protected by protective devices (circuit breakers and/or fuses) to prevent the flow of electrical current above the maximum rating of the cable. Electrical cables are rated according to e.g. CSN 33 2000-5-52:2012. A more serious problem is an increase in the electrical resistance, against which conventional protective devices do not provide protection. The electrical cable resistance can increase through mechanical damage (e.g. through kinking) or more often through an increase in resistance at the point of connection to an electrical device (e.g. in the terminal box). The increase in contact resistance at the connection point of the electrical conductor to the terminal board may be due to oxidation of the contacts, an insufficiently tight connection to the terminal board or through conductor creep. The degree of creep in an electrical conductor is defined by its rheological properties. In terms of creep (rheology) and mechanical damage, aluminium conductors are the most problematic. In tunnels, the oxidation of contacts is often aggravated by a corrosive environment. If the amount of heat released at the point of increased contact resistance exceeds the potential for heat dissipation into the environment, a local increase in temperature occurs. This temperature increase generally then increases the electrical resistance of the conductors (except for semiconductors), which, according to equation (1), leads to a further increase in the heating effect. The heating effect can (under certain circumstances) heat nearby material (e.g. the insulation and sheath of the electrical conductor, terminal board, etc.) to its ignition temperature.

Moreover, electrical cables (whether or not they are an ignition source) can allow flames to propagate across their surface and contribute to the development of a fire. They might also release heat and combustion products into the area affected by the fire.

The last significantly negative phenomenon associated with electrical cables during a fire is their failure and the consequent loss of power to all the electrical equipment. This failure can be caused through the following mechanisms:

- Thermal Degradation of Insulation and the Resultant Short Circuit
- Melting of Conductors
- A Voltage Drop to a Level below the Rated Voltage of Electrical Devices

Optical cables are a different case. The major problem with optical cables is that they cease to function during a fire (due to the increase in temperature). This is mainly due to the Raman Effect. Moreover, optical fibres that are made from synthetic polymers (e.g. polymethyl methacrylate) melt at temperatures that are substantially lower than those found in a fire and the melting temperatures of copper and aluminium.

The first hazard (fire cause) is mainly eliminated by sizing electrical cables according to CSN 33 2000-5-52:2012 and through the periodic inspection of the electrical installation.

The second hazard (flame spread and impact on people and property) is mainly reduced by selecting electrical cables with a suitable fire reaction class.

Similarly, the third hazard is eliminated through the use of cables and cable systems that exhibit properties that maintain the circuit integrity (they are designed to operate in the event of a fire for a specified period of time). At this point, it should be noted that the maintenance of circuit integrity must be exhibited by the entire cable system (it is not sufficient for only the cables to exhibit these properties), as the destruction of the structure, on which the cable is installed, would render the cable itself inoperable.

The optical cables provide the main issue in maintaining circuit integrity, it is extremely difficult to find optical cables that can maintain their functionality at the temperatures found in fires (for the above reasons).

Electrical Cables Reaction to Fire Class

According to the technical standard, EN 13501-6:2018, electrical cables are divided into 7 classes for their reaction to fire: A_{ca}, $B1_{ca}$, $B2_{ca}$, C_{ca}, D_{ca}, E_{ca} and F_{ca}. The contribution to the development of a fire ranges from fire reaction class A_{ca} (no contribution to fire development) to F_{ca} (a very significant contribution to fire development). The assessment criteria and test methods for the classification are specified in EN 13501-6:2018.

In practice, electrical cables classified to A_{ca} class are rarely used. The reason is that this fire reaction class is virtually impossible to achieve with an electrical cable using insulation and a sheath based on organic polymers (ceramic based materials are most often used as the insulation and sheath for this fire reaction class). While these materials exhibit high thermal resistance, they also have many mechanical properties which are unsuitable for the construction and installation of electrical systems. Therefore, in practice, surface-mounted electrical cables with a lower fire reaction class (most commonly $B1_{ca}$, $B2_{ca}$ and/or C_{a}) are used in areas where there is a high level of risk to the life and health of people and property. The classification criteria indicate that in the fire growth stage, electrical cables with fire class $B1_{ca}$ and $B2_{ca}$ virtually prevent the spread of flames (fire) across their surface and only burn at the point of direct contact with flame from another source. The flame only spreads to the entire cable in the fire flashover stage. Electrical cables with fire reaction class C_{ca} also exhibit a relatively high resistance to flame propagation. Moreover, cables within the $B1_{ca}$ and $B2_{ca}$ fire reaction class exhibit a very low maximum heat release rate (as compared to most conventional materials). Similarly, cables within the C_{ca} fire reaction class have a lower maximum heat release rate than most commonly used materials.

From this it may be concluded that the use of surface-mounted electrical cables with fire reaction classes of $B1_{ca}$ to C_{ca} does not constitute an increased contribution to the development of fire in tunnels. However, it should be noted here that the test conditions required by EN 13501-6:2018 differ from the conditions found within a tunnel fire (e.g. when tested according to EN 50399:2011+A1:2016 as required by EN 13501-6:2018, a 30 kW burner is used for the $B1_{ca}$ class and a 20.5 kW burner is used for the $B2_{ca}$ and C_{a} classes). For comparison, research papers by Grant and Drysdale (1997) and Carvel (2016) report significantly higher values of heat release rate (in MWs) during tunnel fires. For completeness, however, it should be added that the heat release rate values stated by the cited authors apply to the overall fire performance in tunnels (the fire is extended to tens or hundreds of square meters), while during the test according to EN 50399:2011+A1:2016, a flame with an output of 20.5 or 30 kW impacts a significantly smaller cable area (depending on the cable diameter, but usually the area directly exposed to the flame is less than one square meter). The heat flux density incident on the electrical cable surface during the pre-flashover fire stage in a tunnel and the heat flux density, to which the electrical cable is exposed during the test according to EN 50399:2011+A1:2016, are usually comparable.

It is essential that if the cable does not allow the propagation of flames across its surface (in the pre-flashover stage) and only burns in the area directly exposed to the flame, the amount of heat, smoke and combustion products released into the fire affected area is usually negligible (compared to other materials and products burning in the tunnel). Therefore, when installing surface-mounted electrical cables in tunnels (which are required to be resistant to surface flame propagation), it is necessary to assess whether they retain this resistance even under real fire conditions (in the pre-flashover stage) within tunnels.

When assessing the safety of electrical cables in tunnels, it is also necessary to take into account the possibility that the fire may be initiated by Joule heat (an overload of the electrical cable), as with this mechanism, the electrical cable may start to burn after a very short time interval over its entire surface. This fire scenario is significantly different to that where an electrical cable is exposed to a local fire. However, the probability of such a fire scenario in tunnels (assuming that the electrical cables are protected by a protective device – fuse or circuit breaker) is very low.

Maintain Circuit Integrity

In addition to the contribution of the electrical cables to the development of the fire, the maintenance of circuit integrity is also a very important parameter, as it ensures that the cable withstands the effects of fire for a specified length of time without failure (especially short circuit) that would interrupt the power supply to electrical equipment. Therefore, cable systems able to maintain their circuit integrity are used to power emergency devices in tunnels. It is not sufficient that the cable maintains its integrity but also the whole cable system.

In this case it should also be noted that the conditions during the tests to determine the ability of the cable to maintain circuit integrity (e.g. according to EN 50200:2015) differ from the real conditions within a fire (especially those within a tunnel). Therefore, when providing power to emergency devices, it is necessary to assess whether the cable system can maintain circuit integrity even under real tunnel fire conditions.

FUTURE RESEARCH DIRECTIONS

The development of the use and application of materials in safety and security engineering primarily reflects advances in material engineering. In safety and security engineering, the use and application of new materials with new or improved properties continues to grow. We may expect a particular increase in the use of modern materials (especially graphene, bio-polymers and bio-semiconductors), or relatively older materials which, due to their high cost of production in the past, were hardly used in practice (e.g. aerogel). Progress in nanotechnologies and nanomaterials, composite and laminated materials will play an important role. Furthermore, it is expected that the application of data on fire hazard materials in fire safety engineering will be further expanded. The application of fire hazard data for materials in fire engineering has been described by e.g. Pokorny et al. (2019). The development of special materials for all areas of safety and security engineering (mainly shielding materials, materials for protective suits, electrical insulating materials and materials for fire protection) will also continue. New synthetic spider silk is likely to be crucial in the further development of bulletproof vests. According to Zemanova, Klouda and Zeman et al. (2011), the potential of fullerene C_{60} and its derivatives is interesting for radio protective applications.

At the end of this subchapter, we should note that safety and security engineering is one of the major areas that drive the development of material engineering (e.g. the space program was and is one of the priority areas that has driven the development of new materials needed to ensure crew safety).

CONCLUSION

It is not possible to list and completely describe all the materials used in safety and security engineering and their applications within the scope of one chapter (or even within one publication). This chapter therefore has focused on a description of the most commonly used materials in safety and security engineering and their applications. The chapter describes the properties and applications of shielding materials, protective suit materials, electrical insulating materials and fire protection materials. The second part of the chapter deals with the fire hazard of materials. In this section, we formulated the research and evaluation methodology of the fire hazard of materials. The methodology described is based on an assessment of the properties of materials that express their resistance to ignition (e.g. critical heat flux and ignition temperature) and properties that express the impact of their combustion on the health and safety of people, the environment and property. The method described was used to assess the fire hazard of the most widely used synthetic polymers (polyethylene and polypropylene). The data obtained demonstrates that polypropylene shows a slightly higher fire hazard in all the assessed criteria. In practice, however, the difference between the materials examined is negligible. At the end of the chapter, we summarized the specificities in the examination and assessment of the fire hazard of electric cables. Electrical cables differ from common materials and products (also used in safety and security engineering), as they can become a fire initiation source due to the heating effect of the current.

Surface-mounted electrical cables are particularly critical for the fire safety of large areas that are not divided into smaller fire compartments (especially road and rail tunnels). An important aspect of the fire safety of surface-mounted electrical cables is their fire reaction class $B1_{ca}$ and $B2_{ca}$ class electrical cables virtually does not spread flames (fire) over their surfaces in the pre-flashover stage. However, since the conditions used to determine the fire reaction class are different from those within a tunnel fire, it is necessary to assess whether an electrical cable with the relevant fire reaction class retains its flame propagation resistance during a real fire (such an assessment is necessary for a specific fire scenario). If the electrical cable (in the pre-flashover phase) does not spread flame over its surface and burns only on the surface directly exposed to a source of fire, its contribution to the development of a fire within the fire compartment is negligible. Moreover, to ensure the protection of life and health of people, it is necessary to supply emergency devices using cable systems that can maintain circuit integrity (even under real fire conditions).

ACKNOWLEDGMENT

This work was supported by the Slovak Research and Development Agency under the contract No. APVV-16-0223.

REFERENCES

Babrauskas, V. (2003). The effects of FR agents on polymer performance. In V. Babrauskas & S. J. Grayson (Eds.), *Heat release in fires* (pp. 423–446). London, UK: Interscience Communications.

Babrauskas, V., & Parker, W. J. (1987). Ignitability measurements with the cone calorimeter. *Fire and Materials, 11*(1), 31–43. doi:10.1002/fam.810110103

Babrauskas, V., & Peacock, D. (1992). Heat release rate: The single most important variable in fire hazard. *Fire Safety Journal, 18*(3), 255–272. doi:10.1016/0379-7112(92)90019-9

Carvel, R. (2019). A review of tunnel fire research from Edinburgh. *Fire Safety Journal, 105*(1), 300–306. doi:10.1016/j.firesaf.2016.02.004

Czech Office for Standards, Metrology and Testing. (2012). *Low-voltage electrical installations. Part 5-52: Selection and erection of electrical equipment – wiring systems* (CSN 33 2000-5-52:2012).

European Committee for Standardization. (2004). *Protective clothing for use against solid particulates - Part 1: Performance requirements for chemical protective clothing providing protection to the full body against airborne solid particulates (type 5 clothing)* (EN ISO 13982-1:2004/A1:2010).

European Committee for Standardization. (2005). *Protective clothing against liquid chemicals - Performance requirements for chemical protective clothing offering limited protective performance against liquid chemicals (Type 6 and Type PB [6] equipment)* (EN 13034:2005+A1:2009).

European Committee for Standardization. (2005). *Protective clothing against liquid chemicals - performance requirements for clothing with liquid-tight (Type 3) or spray-tight (Type 4) connections, including items providing protection to parts of the body only (Types PB [3] and PB [4])* (EN 14605:2005+A1:2009).

European Committee for Standardization. (2011). *Common test methods for cables under fire conditions. Heat release and smoke production measurement on cables during flame spread test. Test apparatus, procedures, results* (EN 50399:2011/A1:2016).

European Committee for Standardization. (2015). *Method of test for resistance to fire of unprotected small cables for use in emergency circuits* (EN 50200:2015).

European Committee for Standardization. (2015). *Protective clothing against dangerous solid, liquid and gaseous chemicals, including liquid and solid aerosols - Part 1: Performance requirements for Type 1 (gas-tight) chemical protective suits* (EN 943-1:2015+A1:2019).

European Committee for Standardization. (2016). *Common test methods for cables under fire conditions. Heat release and smoke production measurement on cables during flame spread test. Test apparatus, procedures, results* (EN 50399:2011+A1:2016).

European Committee for Standardization. (2018). *Fire classification of construction products and building elements - Part 1: Classification using data from reaction to fire tests* (EN 13501-1:2018).

European Committee for Standardization. (2018). *Fire classification of construction products and building elements - Part 6: Classification using data from reaction to fire tests on power, control and communication cables* (EN 13501-6:2018).

European Committee for Standardization. (2019). *Protective clothing against dangerous solid, liquid and gaseous chemicals, including liquid and solid aerosols - Part 2: Performance requirements for Type 1 (gas-tight) chemical protective suits for emergency teams (ET)* (EN 943-2:2019).

European Parliament and the Council. (2004). *Directive 2004/54/EC of the European Parliament and of the Council of 29 April 2004 on minimum safety requirements for tunnels in the Trans-European Road Network* (Directive 2004/54/EC).

Grant, G. B., & Drysdale, D. D. (1997). Estimating heat release rates from large-scale tunnel fires. [). London, UK: International Association for Fire Safety Science.]. *Proceedings of Fire Safety Science*, *5*, 1213–1224. doi:10.3801/IAFSS.FSS.5-1213

Hirschler, M. M. (2003). Heat release from plastic materials. In V. Babrauskas & S. J. Grayson (Eds.), *Heat release in fires* (pp. 375–422). London, UK: Interscience Communications.

International Organization for Standardization. (2015). *Reaction-to-fire tests - Heat release, smoke production and mass loss rate - Part 1: Heat release rate (cone calorimeter method) and smoke production rate (dynamic measurement)* (ISO 5660-1:2015).

International Organization for Standardization. (2016). *Reaction to fire tests - Room corner test for wall and ceiling lining products - Part 1: Test method for a small room configuration* (ISO 9705:2016).

Kokkala, M. A., Thomas, P. H., & Karlsson, B. (1993). Rate of heat release and ignitability indices for surface linings. *Fire and Materials*, *17*(5), 209–216. doi:10.1002/fam.810170503

Koncek, R. (2013). *Characteristics and shielding of ionizing radiation*. Brno, Czech Republic: Brno University of Technology.

Masarik, I. (2003). *Plastics and its fire hazard*. Ostrava, Czech Republic: SPBI.

Masarik, I., Dvorak, I., & Charvatova, V. (1999). *Investigation of combustion products toxicity*. Prague, Czech Republic: Technical Institute of Fire Protection.

Mikkola, E., & Wichman, I. S. (1989). On the thermal ignition of combustible materials. *Fire and Materials, 14*(3), 87–96. doi:10.1002/fam.810140303

Ministry of Business, Innovation & Employment. (2017). *Verification method: Framework for fire safety design* (C/VM2:2017).

Mleziva, J., & Snuparek, J. (2000). *Polymers: production, structure, properties and using*. Praha, Czech Republic: Sobotales.

Ostman, B. L. A. (2003). Smoke and soot. In V. Babrauskas & S. J. Grayson (Eds.), *Heat release in fires* (pp. 233–250). London, UK: Interscience Communications.

Parker, W. J. (2003). Prediction of the heat release rate from basic measurements. In V. Babrauskas & S. J. Grayson (Eds.), *Heat release in fires* (pp. 333–356). London, UK: Interscience Communications.

Petrova, S., Soudek, P., & Vanek, T. (2015). Flame retardants, their use and environmental impact. *Chemical Papers, 109*(9), 679–686.

Pokorny, J., Mozer, V., Malerova, L., Dlouha, D., & Wilkinson, P. (2019). A simplified method for establishing safe available evacuation time based on a descending smoke layer. *Communications - Scientific Letters of the University of Zilina, 20*(2), 28-34.

Racek, J. (2009). *Nuclear installations*. Brno, Czech Republic: Novpress.

Scudamore, M. J., Briggs, P. J., & Prager, F. H. (1991). Cone calorimetry – a review of tests carried out on plastics for the association of plastics manufacturers in Europe. *Fire and Materials, 15*(2), 65–84. doi:10.1002/fam.810150205

Spearpoint, M. J., & Quintiere, J. G. (2001). Predicting the piloted ignition of wood in the cone calorimeter using an integral model - effect of species, grain orientation and heat flux. *Fire Safety Journal, 36*(4), 391–415. doi:10.1016/S0379-7112(00)00055-2

Tewarson, A. (2002). Generation of heat and chemical compounds in fires. In P. J. DiNenno (Ed.), *The SFPE Handbook of Fire Protection Engineering* (pp. 618–697). Quincy, US: National Fire Protection Association.

Tran, H. C. (2003). Experiment data on wood materials. In V. Babrauskas & S. J. Grayson (Eds.), *Heat release in fires* (pp. 357–372). London, UK: Interscience Communications.

Xu, Q., Majlingova, A., Zachar, M., Jin, C., & Jiang, Y. (2012). Correlation analysis of cone calorimetry test data assessment of the procedure with tests of different polymers. *Journal of Thermal Analysis and Calorimetry, 110*(1), 65–70. doi:10.100710973-011-2059-7

Zemanova, E., Klouda, K., & Zeman, K. (2011). C60 fullerene derivative: Influence of nanoparticle size on toxicity and radioprotectivity of water soluble fullerene derivative. In *Proceedings of 3rd International Conference NANOCON 2011* (*Vol. 1*, pp. 1-10). Tanger.

Zhang, J. (2008). *Study of polyamide 6-based nanocomposites*. New York: Polytechnic University.

ADDITIONAL READING

Babrauskas, V. (2003). *Ignition handbook*. Issaquah: Fire Science Publishers.

Babrauskas, V., & Grayson, S. J. (2003). *Heat release in fires*. London, UK: Interscience Communications.

Beard, A., & Carvel, R. (2005). *The handbook of tunnel fire safety*. London, UK: Thomas Telford Publishing. doi:10.1680/hotfs.31685

Cote, A. E. (2008). *Fire protection handbook*. Quincy: National Fire Protection Association.

Hurley, M. J. (2016). *SFPE Handbook of fire protection engineering*. New York: Springer. doi:10.1007/978-1-4939-2565-0

Jakobczak, D. J. (2016). *Analyzing risk through probabilistic modeling in operations research*. Hershey, US: IGI Global. doi:10.4018/978-1-4666-9458-3

Koradecka, D. (2010). *Handbook of occupational safety and health*. Boca Raton: CRC Press. doi:10.1201/EBK1439806845

Ronk, R., White, M. K., & Linn, H. (1984). Personal protective equipment for hazardous materials incidents: a selection guide. Morgantown, US: U.S. Department for health and human services.

Thue, W. (2012). *Electrical power cable engineering*. New York: CRC Press.

Tong, X. C. (2009). *Advanced materials and design for electromagnetic interference shielding*. Boca Raton: CRC Press.

KEY TERMS AND DEFINITIONS

Carbon Monoxide Production Rate: A characteristic of combustion that quantifies the amount of carbon monoxide produced by a surface area of one square meter of material per second.

Electrical Cable: A construction product intended for the transmission of electricity or electrical signals.

Electrical Installation: The interconnected components intended for electricity production, transmission and use.

Electrical Insulator: A material with high electric resistance.

Heat Release Rate: The heat released from a surface area of one square meter of material per second.

Fire Risk of Material: A characteristic of combustion that quantifies the resistance of the material to ignition (e.g. by critical heat flux, ignition temperature and the impact of combustion of the material on the surrounding area (the health and safety of people, the environment and property)).

Material for Security and Safety: Material commonly used in safety (e.g. fire protection) or security (e.g. bulletproof vests) applications.

Protective Suit: A suit that protects professionals against specific hazard (e.g. the protection of fire fighters from the high temperatures produced by fires).

Shielding Material: A material that shields a person from one or more types of radiation (e.g. X-radiation).

Smoke Production Rate: A characteristic of combustion that quantifies the amount of smoke produced by a surface area of one square meter of material per second.

Chapter 12
Nanotechnology Safety and Security:
Nanoparticles and Their Impact on the World

Janette Alba
University of Guanajuato, Mexico

Petra Roupcova
VSB – Technical University of Ostrava, Czech Republic

Karel Klouda
Occupational Safety Research Institute, Czech Republic

ABSTRACT

Because of their unique characteristics, research and development of new nanomaterials is one of the major disciplines of the twenty-first century, and examining their special properties, especially toxicity, is therefore necessary. As well as their benefits (technological improvements, specific material properties, improved resistance to natural effects), new materials also bring new risks requiring assessment in terms of occupational health and safety and their abuse as potential biological carriers or other materials. The study presents general information about nanoparticles and their distribution and properties in relation to entering aquatic, soil, and atmospheric environments. The study describes and cites examples of measurements conducted on the exposure of different nanomaterials to the work environment. Risk assessments of nanomaterials according to the available methodologies, measures to protect against nanoparticles, and importantly, the abuse of nanoparticles as a potential threat to the CZ population are also described.

DOI: 10.4018/978-1-7998-3059-7.ch012

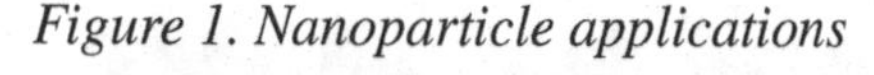
Figure 1. Nanoparticle applications

INTRODUCTION

There's Plenty of Room at the Bottom, was a notion expressed on the subject by American physicist Richard Feynman in 1959 in a lecture given at the annual meeting of the American Society of Physicists, although the future phenomenon of the scientific field had not yet been named (Klouda, Frišhansová, & Senčík, 2016). The European Commission Regulation 2011/696/EU defines nanomaterials (NMs) as follows: "Nanomaterial means a natural material, by-product or material produced containing particles in an unbound state or as an aggregate or agglomerate in which 50% or more of the particle size distribution in one or more external dimensions is in the size range of 1–100 nm" (Larena, 2017). Though primarily concerned with developing nano-sized materials (NMs), nanotechnology is an interdisciplinary field that examines the properties of NMs and involves classic disciplines such as chemistry, physics, biochemistry, electronics and quantum mechanics (Figure 1). NMs behave specifically as a result of the different physicochemical properties in substances that are not the same on the surfaces of the solids as the material inside. After reducing the components of the material to below the magical limit of 100 nm, the physicochemical properties of its surfaces dominate, and the particle begins to behave as if it were only its surface. For example, the material increases in chemical reactivity, which may result in a change in its toxicity (Vampati & Uyar, 2014). Nanoparticles can penetrate directly into a cell nucleus and interact directly with DNA molecules or proteins, and activated reactions leading to apoptosis (programmed cell death) may also occur (Cagno et al., 2018; Huang et. el, 2018).

According to their origin, nanoparticles (NPs) may be categorized as follows (Klouda, Frišhansová, & Senčík, 2016):

- **Natural:** E.g., volcanic fumes, pollen, forest fires.

- **Anthropogenic:** For industrial production (joining materials for metal, wood, stone processing).
- **Synthetically Produced:** Laboratory-prepared materials such as carbon NMs (graphene), metal nanoparticles (TiO_2, ZnO, Al_2O_3), pigments).

According to toxicity, NPs may be categorized as follows (Filipová, Kukutschová, & Mašláň, 2012; Huang et al., 2018):

- **Primary:** Targeted, synthesized engineered nanomaterials (ENMs).
- **Secondary:** Particles produced during material degradation in a mechanical or thermal process.

History of Nanoparticles

Nanoparticles have existed around us for eons, arising from natural phenomena such as volcanic eruptions or forest fires. Industry can produce NMs by burning fossil fuels or through the chemical decomposition of organic substances. In recent years, targeted production in a laboratory environment has produced NM products called ing-nanoparticles. People even used nanoparticles in times when they did not understand the physical world around them (Roupcová, 2018). For example:

- The unique Lycurgus Cup from the fourth century A.D. Glassmakers mixed metal powders into the glass.
- Glossy surfaces (glaze) on ceramics from the thirteenth to sixteenth centuries.
- Metallurgy (copper clusters)
- Carbon black production
- Colloidal solutions
- Chemical catalysis (zeolites)

NP sizes are beyond the limit of visibility to the human eye. The invention of the electron microscope last century was therefore especially important in being able to identify them. People could at long last observe the structure of nanoparticles in detail and identify many unique, unknown materials through them (Roupcová, 2018).

As a phenomenon of the late twentieth and early twenty-first centuries sufficiently supported by research projects in relevant fields, we can therefore assume that nanotechnologies will have a major impact on the world economy. Interested companies are not shy of using risk capital. In Japan and the US, we see a balanced ratio between public and private investment, while public funding in Europe is massive. Today, it is difficult to find a field in which NM research is not conducted (Roupcová, 2018).

Characteristics of Nanoparticles

NMs are miniscule yet complicated molecules. They consist of three layers: a surface layer, shell and core. Nanoparticles also fall under several categories depending on their morphology, size or chemical properties. NMs are characterized by common features that dependent on their size. Their basic structure consists of building units (Skřehot & Roupcová, 2011; Sirovátka, 2018):

- Nanoparticles of defined properties – these properties include shape, chemical composition, atomic structure, etc. These particles range from molecules to solid particles less than 100 nm.
- Diverse clusters in macroscopic materials – particles with chemically identical compositions can be tightly arranged, separate or coalescence, and indicate a process of merging dispersed particles into larger units. They can form nanowires, nanotubes or other thin films or layers.
- Coherent units for use in creating larger structures and materials suitable for technical applications.

The properties of NPs are primarily influenced by their size and ability to form larger units or aggregates. This principle applies to both gaseous (i.e. atmospheric) and liquid environments. To an extent, liquid environments are limited because the particles are coated with liquid molecules and share its surface tension. In atmosphere, dispersed nanoparticles tend to behave similarly to gas molecules. Their very small size and mass results in no sedimentation, and they therefore remain airborne for a long time. This is especially true at low concentrations, with higher concentrations subject to aggregation. The basis of the aggregation process is mutual precipitation of individual particles into larger units. In this situation, the average particle size increases. Such aggregation, in which clusters of loosely bound particles are formed, is called flocculation. Flocculation can be understood as a reversible reaction in which aggregates can be converted back into a colloidal system (the general term colloid is used for nanoparticles in an aqueous system). This process is called peptization. Otherwise, more solid clusters may form. In a process called coagulation, aggregates of tightly bound particles form. However, coagulation is irreversible and the coagulates formed cannot be re-peptized, i.e. converted back into a colloidal system. When the aggregate reaches a certain size, a sedimentation process takes place, thereby scattering the dispersion system. Aggregation itself is very quick and spontaneous, and depending on the aerosol concentration, individual nanoparticles may undergo aggregation for several seconds after their formation. Consequently, many larger aggregates are formed at the same time, and a large number of males disappear from the aerosol. (Skřehot & Roupcová, 2011; Sirovátka, 2018)

Agglomerate is a group of particles weakly bound by Van der Waals or electrostatic forces. Aggregate, however, is strongly bonded particles by covalent or metal bonding [4]. As mentioned above, NMs with the same chemical composition may show different toxic effects. Factors that can significantly affect particle toxicity are particle size and surface modification (surface charge, surface chemistry, surface shape, defects and porosity), particle size decrease is associated with overall surface growth and increased Gibbs energy, resulting in significantly greater chemical reactivity in smaller particles. (Filipová, Kukutschová, & Mašláň, 2012)

RISKS OF NANOPARTICLES

NMs carry four basic potential risks:

- Health hazards (toxicity + cell reactions)
- Environmental hazards
- Physicochemical hazards (fires, explosions, adverse reactions)
- Unethical use of nanotechnologies and nanoparticles by third parties

The first two potential hazard sources are closely related, because the presence of nanoparticles in the environment allows contact with living organisms.

In general, several parameters affect particle toxicity, including particle type, particle size, shape, concentration and distribution in the environment, water solubility, chemical reactivity, frequency and exposure time, interaction with other chemicals in the environment, pulmonary ventilation and immunological status of the individual (Medinsky & Bond, 2001).

In operations where nanoparticles form (especially dusty operations), employees are particularly at risk from their inhalation. Inhaled air containing particles (macro, micro and nano) is filtered in the successive parts of the respiratory tract, these being the nasopharynx, trachea and bronchi and pulmonary system. The pulmonary region is the main gateway through which inhaled nanoparticles enter the extra-pulmonary organs and tissues for subsequent toxic action. The absorption of nanoparticles depends on their physicochemical properties (chemical composition, size, shape, etc.). (Certified methodology, 2016). Compared to other materials, nanomaterials have an exceptional ability to travel through a living organism.

Responses to inhaled materials range from immediate, immediate, to long-term, chronic negative effects, at the level of single tissue action to systemic diseases (Nel et al., 2009). Large-scale epidemiological studies have shown that exposure to air-polluting and ultra-fine particles contributes significantly to the incidence of respiratory and cardiovascular diseases and affects mortality.

Health Hazards (Toxicity + Cell Reactions)

The presence of nanoparticles in a living organism induces a number of interactions between their surface and biological systems. These interactions can lead to protein corona formation, particle coatings, intracellular absorption and biocatalytic processes that may have beneficial or adverse effects in terms of toxicity. The organic world blends with the synthetic world of nanomaterials produced. Nano-bio interfaces are associated with dynamic physicochemical interactions, kinetic and thermodynamic exchanges between surfaces of nanomaterials and surfaces of biological components (proteins, membranes, lipids, DNA, biological fluids). Research on interaction at the nano-bio interface shows (Bakand, Hayes, & Dechsakuthorn, 2012) that we have little information about what happens to nanoparticles inside a cell. Nanoparticles can cause a wide range of intracellular reactions depending on their physicochemical properties, intracellular concentrations and contact duration. (Table 1).

Oxidative stress is seen as a key mechanism of toxicity. It is an ideal balance between reactive oxygen species and antioxidants. ROS is the essence of a nanomaterial and determines the reactivity of transition metals on the surface, which is described in the Fenton reaction.

Environmental Hazards

Analyzing the behavior of nanoparticles towards cells and its outcome in the environment is complicated, as there are many unknown factors and effects. Besides size, which represents the above-mentioned magic boundary, each type of nanoparticle can be described and characterized by certain properties (Kolářová, 2014):

- Chemical composition
- Surface functional groups (hydrophilicity, lipophilicity)

Table 1. Characteristics of nanomaterials and their possible biological effects (Frišhansová & Klouda, 2017)

Characteristics of NMs	Possible biological effects
Miniscule size (less than 100 nm)	Penetration through tissues and cell membranes Cell damage Disruption of phagocytosis, collapse of defense mechanisms Migration to other organs Transportation of other pollutants
Large surface / mass ration	Increased reactivity Increased toxicity
Surface properties	Generation of reactive oxygen species (ROS) Oxidative stress Inflammation Cytokine production Loss of glutathione Mitochondrial exhaustion Cell damage Damage to proteins and DNA
Insolubility or low water solubility	Bioaccumulation within living systems such as human cells, tissues and lungs Potential long-term effects
Aggregation	Disruption of cellular processes Cell damage

- Shape
- Particle size distribution
- Density
- Crystalline structure (topographic defects in the carbon lattice, e.g., in graphene oxide, cycles 5–7, 5–8-5, 6–5)
- Zeta potential, aggregation capabilities, etc.

These are the properties nanoparticles enter the external environment with. In the atmosphere, abiotic factors play a role, including temperature, humidity, ambulance, solar radiation intensity, smog pollutants of inorganic and organic origin, etc. In aquatic environments, the physicochemical characteristics of surface water, river and sea water, temperature, pH, ionic strength, divalent ion concentration, natural organic matter concentration and sediment composition must be taken into account. (Kolářová, 2014)

Aqueous environments, nanoparticles are affected by the following:

- Surface hydrophilicity (solubility)
- Hydrolysis (oxidation-reduction)
- Adsorption
- Aggregation
- Heteroaggregation
- The presence of aquatic organisms (bacteria, algae, protozoa, plankton, larvae, fish, etc.)
- Sedimentation rate
- Sediment composition and its reactivity (adsorption, reduction of phytoextracts and biological material, presence of soil organic matter, presence of pollutants – PAH, heteroaggregation, etc.).

Even a soil environment must consider the physicochemical characteristics of nanoparticles. Soil can be in solid, liquid or gaseous form. The solid phase (component) contains the mineral fraction (primary and secondary minerals) of different grains (separation according to the fraction content 0.01 mm) and the organic fraction – soil organic matter. This substance is based on humic substances, which are mixtures of high molecular weight polydisperse compounds with aromatic and aliphatic moieties with functional groups -COOH, -OH, phenolic -OH, -NH2, N-heterocycles, etc. The basic classification of humic substances is according to solubility in pH. Fulvic acids are soluble in water, humic acids are alkaline and humins are completely insoluble in water. Soil water, also called soil solution, may contain fulvic acids, phytoextracts with polyphenol structure substances, carbohydrates, ions, etc. The effect of nanoparticles in a soil solution is similar to that in an aqueous environment.

In soil environments, nanoparticles are affected by the following (Kolářová, 2014):

- Surface hydrophilicity and lipophilicity
- Hydrolysis
- Reduction (green reduction by phytoextract substances)
- Heteroaggregation (clay soils)
- The presence of soil micro-organisms and animals (bacteria, larvae, earthworms, etc.)
- Adsorption (pi-bonding interactions, hydrogen bonds, electrostatic interaction, acid-base interaction)

C-nanoparticles and their Impact on the Environment

A special group of nanomaterials are derived from graphite (C-nanoparticles) Graphite is a natural layered compound with sp2 hybridization with a layer spacing of 3.38 Å. A flat layer of graphite consists of six-membered C-cycles (honeycomb). From this natural compound, a series of compounds with a carbon skeleton can be prepared, such as graphene, graphene oxide, fullerene, single-wall carbon nanotubes, multiple wall carbon nanotubes, etc.

The movement and ultimate effect of these C-nanoparticles can be simply summarized into three independent actions that may combine or have a synergistic effect (Roupcová, 2018):

- Modification (adsorption plays a major role)
- Change in composition (interaction of functional groups with organic or inorganic substances)
- Degradation (may be physical, chemical or biological)

The question remains how the ultimate effect and movement of nanoparticles changes, especially the toxicity of C-nanoparticles after, for example, degradation or after the change of functional groups on the surface. In our case, we can assume that the least inert to the environment will be graphene oxide (GO). It is characterized by strong adsorption ability, susceptibility to reduction (subject to so-called green reduction) and reactivity of its functional groups on a C-skeleton and susceptibility to photochemical decomposition. The interaction of GO with cell membranes can lead to indirect toxicity, i.e. obstruction of ion and gas exchange. Their incorporation into cells can cause oxidative stress, damage to DNA (deoxyribonucleic acid) and induce mitochondrial dysfunction (Roupcová, 2018).

Physicochemical Hazards (Fires, Explosions, Adverse Reactions)

Hazards in terms of physicochemical properties are predominantly fires, explosions and uncontrolled reactions. It is complicated to predict the behavior of nanoparticles, as they differ in several basic factors (Klouda, Frišhansová, & Senčík, 2016):

- Chemical composition
- Shape
- Density
- Crystalline structure
- Particle distribution

This type of risk is represented mainly by a TOP-DOWN particle production system, in which nanoparticles are produced mechanically by grinding, cutting and super-grinding, for example a certain method of carving a statue out of a large block of marble (Klouda, Frišhansová, & Senčík, 2016).

Many TOP-DOWN methods work on the principle of forming semiconductor structures on silicon and germanium substrates, whose density is increased while continually removing details. These methods are collectively nicknamed lithography, which is a form of image transfer to a pre-prepared substrate. The quality of the structure and speed of its formation is at the heart of this production method. This method consists of three parts (Klouda, Frišhansová, & Senčík, 2016; Nanoparticles and their toxicity, 2015):

- The substrate is covered with a sensitive polymer layer (resistor).
- The resistor is exposed to a laser, electrons, or ion beam.
- Resistor recall – either the exposed or unexposed parts are removed.

Nanoparticles are present in all technologies that create a dusty environment, such as metalworking, woodworking and welding. The amount and reactivity of these nanoparticles is mainly affected by mutual aggregation and agglomeration. These dust mixtures are not homogeneous in time or space. Particle size mainly determines the explosiveness of dust, and if we reduce its size, we can increase the risk of explosion (Klouda, Frišhansová, & Senčík, 2016).

Na, Fe, Ni, Al, Mn and Co nanoparticles are self-igniting because of their large surface area, thereby initiating an uncontrolled exothermic reaction. Nanoparticles are much more reactive than substances of the same composition but larger in size (Klouda, Frišhansová, & Senčík, 2016).

Special Risk – Aerosols

An aerosol is as a collection of solid, liquid or mixed particles with a size of 1 nm to 100 μm dispersed in air. Aerosols are found, for example, in dust, smoke, smoke, fog, clouds and smog. Airborne particles have a diverse range of size and origin and contain organic and inorganic compositions with complex toxic and carcinogenic potential. It is not necessarily the risk associated with the chemical nature of the particles themselves but also their ability to bind dangerous chemicals or microorganisms to their surfaces. (Skřehot & Roupcová, 2011; Sirovátka, 2018)

Aerosols can be categorized as primary or secondary according to their formation in the atmosphere.

- Primary aerosols are those with direct sources. These sources may be, for example, mineral dust particles, which are formed in nature by weathering and soil erosion, particles from combustion processes, ocean salt particles or pollen and spores from biological sources.
- Secondary particles are produced in the atmosphere through the conversion of gases into particles. This includes many particles comprising inorganic substances such as sulfates and nitrates. Sulfates form through the oxidation of sulfur oxides, mainly sulfur dioxide (SO_2), and nitrates through the oxidation of nitrogen oxide (NO_x). (Sirovátka, 2018)

This type of particle poses a specific risk because it is dispersed in the air and enters our bodies through inhalation. Depending on their size, surface and structure, they can then enter the organs. (Klouda, Frišhansová, & Senčík, 2016)

However, most substances are absorbed by the lungs, where the size of the substance is the most important factor. The number of particles absorbed in this manner is referred to as the respirable fraction. Particles are deposited in the airways after contact. For particles larger than 500 nm, the main mechanism is sedimentation when air flow comes into contact with the mucosa. For particles less than 500 nm in size, deposition is controlled by diffusion, and the shape of the molecule and its diffusion coefficient are important aspects. For very fine charged particles, sorption is facilitated by electrostatic interaction, and strong van der Waals forces then allow the particles to form aggregates, which means groups of particles of a given size can behave as particles of an order of magnitude higher (Tenschert, 2016).

The size of nanoparticles allows them to interact only with cell membranes, and the surface of a nanoparticle plays a major role. However, mechanical or chemical damage to these membranes may occur, or their function may change. Proteins are then inhibited (slowing activity), and DNA can have genotoxic or mutagenic effect. In genotoxicity, it is important whether the effect is direct or indirect. Direct damage occurs when a nanoparticle reacts directly with a DNA molecule. Indirect damage occurs through the interaction of DNA with oxygen and nitrogen compounds, which reduces the body's ability to remove these harmful substances. A secondary toxic effect is induced by an inflammatory reaction and leads to DNA damage (Tenschert, 2016).

In carbon nanotubes, which are very similar in property to asbestos, much emphasis is placed on the shape of the particle and therefore the production process. Single-walled tubes often form spherical aggregates, while multi-wall tubes form fibrous aggregates. These characteristics are very important in pulmonary exposure to aerosols since the genotoxicity of the particles is strongly dependent on the length of the fibers, their diameter, rigidity, surface area, density and shape. Carbon fiber rigid tubes with a length / thickness ratio greater than 3:1 can lead to lung cancer. Particle size and toxicity can vary as nanomaterials form or dissociate aggregates (Tenschert, 2016).

Unethical Use Of Nanotechnologies And Nanoparticles By Third Parties

Whenever a fundamental change occurs in history in technology, the response is felt in economic, ethical, moral, human and environmental (or counter) fields. It should be borne in mind that nanotechnology and its wide range of applications across different industries brings with it a number of risks. Currently, major nations are competing to find more modern and efficient facilities. This is also not only the case for military production. For example, Kim Eric Drexler (1955), one of the most prominent promoters of molecular nanotechnology and an employee of Nanorex (engaged in molecular engineering), warned the public: "An army of nanorobots with their own artificial intelligence and reproductive capabilities could very easily get out of hand." Opponents of this assertion argued that this could be prevented by coded self-destruction commands for several generations of self-replicable nanorobots, but the question is, what would prevent hyperintelligent mini-machines from modifying the software? Another problem not mentioned is computer viruses and hackers, who could very easily modify the self-destruct command. Consequently, these, though very intelligent yet especially dangerous technologies, could be exploited by terrorist and other groups. In general, the use of stronger and lighter nanotechnology-based materials is contributing to very rapid developments in the military industry. Faster projectiles, lighter and stronger armor and more durable and lighter equipment for soldiers can be made with nanomaterials. For soldiers, movement is lighter and less strenuous (polymers, metals, ceramic nanocomposite substances, nanofoams). However, some conventional weapons may potentially become weapons of mass destruction. The misuse of these types of weapons could affect international arms control agreements, international environmental protection laws (Kyoto Protocol), the political and economic stability of nations, society and the environment. Furthermore, to limit the production and use of conventional weapons using new technologies, international humanitarian law will have to consider the introduction of autonomous combat systems, as they will not be able to reliably distinguish non-military persons or wounded soldiers (Examples of nanotechnology misuse; 2015; Lauterwasser, 2016; Frišhansová & Klouda, 2016).

From a global perspective, there may be a risk of losing control of nanotechnology applications or acquiring technological capabilities, numerous disciplines changing or disappearing, prolonged ageing accompanied by a loss of social balance, potential increase in social, cultural and economic diversification that may cause unequal access to nanotechnology applications and the beneficial and adverse effects of nanotechnologies. (Examples of nanotechnology misuse, 2015)

Communication technologies, MEM and NEM systems or bio and electronic systems represent another group of military development in nano applications. These include various micro and nano systems for assembling nanostructured materials, systems for producing and maintaining electricity, sensory networks for monitoring the presence of different substances or the status and position of soldiers, various imaging and communication systems, including bioimplants that allow soldiers and their surroundings to be monitored, information to be shared and their activities and even biological processes to be managed. Examples of these technologies can be found in autonomous combat systems that replace human soldiers with machines of different size and effect, from combat robots to miniature insect-sized weapon systems known as "biological hornets", which are capable of monitoring and destroying selected targets, and "micro-robot clay", which is a monitoring network of freely scattered autonomous microsystems that provide information about events in their surroundings (Frišhansová & Klouda, 2016; He, Liu, Wamer, & Yin, 2014).

Finally, nanotechnologies can enhance the range of functions and effect of a soldier's weapon systems. This is primarily represented by the development of high-power lasers, for example, quantum dot disk lasers, which can be used either as a weapon or as a detonator for new types of nuclear devices. A strong explosive known as nano-thermite has also been developed. This comprises metastable intermolecular composites of oxidizing and reducing agents, such as $Al\text{-}KMnO_4$, $Al\text{-}Fe_2O_3$, $Al\text{-}MoO_3$ and AlCuO, and exhibit a highly exothermic reaction upon ignition. These materials find use, for example, as pyrotechnic cartridges, fuel to increase the performance of jet and internal combustion engines and the basis for new types of weapons such as thermobaric bombs. They and also used in the construction of nuclear charges.

All these innovations carry significant risks that must not be ignored, as they could cause catastrophic consequences in the wrong hands (He, Liu, Wamer, & Yin, 2014).

In addition to the risks posed by continual innovation of military technology, it is important to identify the environmental and health risks of nanoparticles. Since the toxic consequences of nanoparticle use are not yet well known or sufficiently researched, the precautionary principle must be followed. It is vital to monitor, for example, the effect of contaminating nanostructured substances in air, watercourses or soil (Lauterwasser, 2016; He, Liu, Wamer, & Yin, 2014).

Mechanisms of Nanoparticle Toxicity

Three basic mechanisms of toxicity potentially inducible by nanoparticles are defined and demonstrable (Filipová, Kukutschová, & Mašláň, 2012):

- Induced oxidative stress
- Inflammatory processes
- Genotoxicity

Oxidative Stress

This is considered as a key mechanism of cell toxicity. The ideal balance between reactive oxygen species (not having paired electrons) and antioxidants is caused.

Reactive Oxygen Species (ROS)

Many biological applications and effects of nanomaterials are attributed to their ability to facilitate the formation of reactive oxygen species (ROS). Electron spin resonance (ESR) spectroscopy is a direct and reliable method for identifying and quantifying free radicals in chemical and biological environments. In this review, we discuss the use of ESR spectroscopy to study the generation of ROS-mediated nanomaterials that have different applications in biological, chemical and material science. The rapid development of nano sciences and nano technologies has introduced a range of nanomaterials that offer revolutionary benefits in electronic, energy, medical and healthcare applications. However, the widespread use of nanomaterials has raised concerns about their potentially dangerous effects on biological systems and their associated short and long-term risks, which are not well known. Many nanomaterials can produce reactive oxygen species under certain experimental conditions. Among the various toxic effects, the oxidation process mediated by ROS has been extensively studied. (Cagno et al., 2018; Das et al., 2018; Tang et al., 2018).

ROS, such as superoxide, hydroxyl radical, singlet oxygen, and hydrogen peroxide, are potent oxidants that can selectively damage cellular targets. Free radicals, including ROS, are short-lived and represent a broad range of chemically distinct bases. As a result, it is difficult to detect these species in dynamic environments such as biological systems. (Huang et al., 2018)

The use of fluorescent probes (e.g. dichlorodihydrofluorescein, hydroethidine and dihydrorhodamine) and chemiluminescence assays is a simple and easy way to detect free radicals and ROS in cellular systems, but there are some inherent limitations. Electron spin resonance (ESR) spectroscopy has become a powerful and direct method for detecting free radicals generated chemically or generated in biological systems. There is a long-standing interest in using ESR techniques to identify and quantify free radicals in biological systems and to study the mechanisms of interaction between biologically relevant systems and nanomaterials, metal ions, and organic molecules. Some sources demonstrate that ESR spectroscopy is a powerful tool for investigating the ability of nanoparticles to generate ROS. (Koshed, Li, Liu, & Mironov, 2018)

By conducting electron spin resonance (ESR) tests, most research has found that nanomaterials, especially metal nanoparticles, metal oxide nanoparticles, and carbon-based nanomaterials, have the ability to generate ROS in cells and induce oxidative stress in cultured cells. Due to their small size, nanomaterials easily penetrate into cell membranes and directly interact with organelles and proteins or lead to intracellular release of metal ions. Together with high nanosurface reactivity, intracellular nanomaterials can interfere with other biochemical reactions in cells or cause protein modification, such as induced ROS formation, glutathione depletion and damage to oxidative stress; all of these examples were discovered experimentally. Although the formation of intracellular ROS through interactions with nanomaterials is a common phenomenon, the chemical mechanisms of nanomaterials induced by intracellular ROS formation are less studied and still remain a major challenge for chemists. The priority is to understand the mechanisms of nanotoxicity from a chemical point of view. It is likely that chemical research will cover the biological aspects of nanotoxicity in order to provide an explanation of biological mechanisms. The chemical mechanisms of nanotoxicity are particularly useful for creating safe functional nanomaterials through chemical design in terms of material sciences, chemistry, chemical engineering, biomedical sciences and engineering. (Huang et al., 2018)

Various metal nanoparticles have been reported exhibiting intrinsic activity in generating or capturing ROS. The following metals, for example, were investigated: Ag, Au, Pt, Cu, Fe, Co, Ni, Fe, and Co.

Ag, Au, Cu, Fe, Ni, and Co nanoparticles have been observed for their ability to induce ROS formation under certain experimental conditions. Most of the individual studies used fluorescent probes instead of ESR to measure ROS and reported only total oxidative stress associated with ROS production. Oxidative stress is reported as an imbalance between the formation of reactive oxygen (otherwise also free radicals), which is a by-product of oxygenation and metabolism, and the body's ability to rapidly break down and detoxify reactive intermediates (Yan, Gu, & Zhao, 2013; Schüllerová, Adamec, & Benco, 2019).

These studies suggest that physiochemical factors, including the size, shape, composition and surface coating of metallic nanoparticles, significantly affect ROS levels. ROS are produced mainly through two mechanisms (Filipová, Kukutschová, & Mašláň, 2012):

- Fenton reaction 1) $H_2O_2 + Fe^{2+} \rightarrow HO\bullet + OH^- + Fe^{3+}$

  $HO\bullet$ (highly reactive radical).

- Increase in surface plasmon resonance.

Inflammatory Processes

Inflammation is the sum of physiological responses to the integrity of the body. Various nanomaterials are capable of eliciting an inflammatory response. ROS mediate an inflammatory response (Korshed, Li, Liu & Mironov, 2018). Inflammation and oxidative stress are closely linked. ROS is considered a secondary vector whose main task is to induce inflammation (Filipová, Kukutschová, & Mašláň, 2012; Huang et al., 2018, Tang et al., 2018).

Genotoxicity

This is the ability to alter cell genetic material and cause oxidative damage to DNA and proteins through the interaction with ROS (e.g., generation of electrophilic unsaturated alpha, beta aldehydes). Nanomaterials can damage DNA by either direct interaction with the DNA or through ROS (Das et al, 2018; Tang et al., 2018). Generally, everything that applies to chemical substances applies to nanoparticles, but their specific properties must not be ignored, which is cell penetration, etc. The real torsional effect is rather chronic or delayed toxicity than acute toxicity (Filipová, Kukutschová, & Mašláň, 2012; Korshed, Li, Liu, & Mironov, 2018).

Methods of NP risk assessment

As long as the production of nanomaterials is not included, no restrictions or prohibitions apply. The current risk assessment methodology for nanomaterials is based on generally accepted principles, research, hazard and risk identification, risk characterization and regulatory mechanisms for reducing hazard. Nanotechnology safety policy works with a risk analysis covering three fundamentally related elements (Filipová, Kukutschová, & Mašláň, 2012):

- Risk assessment
- Risk management
- Risk communication

Nanomaterial risk assessment comprises four main steps:

- Hazard identification
- A description of the hazard
- Exposure assessment
- Risk estimation

It can be stated that the current methods and methodologies are based on the general principle of risk assessment. Currently, the most common is control banding with various modifications. Preventive measures in the form of 'safe by design' have also been recently developed and applied. Its principle is based on three pillars: safe by-products, safe-use products and safe industrial production. The aim of the third pillar is to acquire the knowledge and risk assessment tools in order to ensure safe industrial production and, most importantly, workplace safety (Methods, 2018).

In the Czech Republic, a methodology for assessing personal protective equipment in an environment with nanoparticles has been adopted. This methodology is intended to help employers decide whether their employees need the personal protective equipment or not. However, it deals only with the protection of respiratory organs from the effects of nanoparticles in the workplaces where the presence of nanoparticles is proved and in those workplaces, where it is reasonable to assume the presence of nanoparticles. This applies for insoluble nanoparticles only and it is more of an informative than precise method. There are not enough toxicology records, knowledge of nanoparticles, possible ways of using nanoparticles and many more missing parts for the methodology to give exact numbers.

The application of this methodology has five steps and includes assessing exposure time and classifying hazard groups (nanoparticle concentration measurement). First of all, there must be a specific work process defined, list of workplaces and identification of risks made. Then the definition of the exposure time must be done. Step 3 is an assignment of nanoparticles to the specific group based on their concentration in the air. Next is the assessment of risks based on the effects of nanoparticles and the last step is eliminating or decreasing the risks and providing the personal protective equipment (Certified methodology for providing personal protective equipment in workplaces with a presence of nanoparticles: (focused on employers in need of protective equipment with respiratory protection, 2016). These form inputs for the nanoparticle risk assessment matrix (Nel et al., 2009).

Control Banding

With the advent of modern technology, there are also new risks that need to be addressed in order to declare a workplace safe. Workplace risk assessment methods have been developed specifically for these purposes and facilitate the risk assessment process. These workplace risk assessment methods comprise several stages, such as risk analysis, risk identification and risk evaluation. Control banding is one possible method for workplace risk assessment (Zalk & Nelson, 2003).

Control banding is a qualitative risk assessment method that aims to minimize human exposure to hazardous chemicals and other risk factors occurring in the workplace by providing easy-to-understand and practical ways of managing these risks reliably. The very first model of this method was developed for the needs of the pharmaceutical industry, which manages hazardous substances, mixtures and vapors of many hazardous chemicals. Its principle is to precisely identify the measures required in order to prevent people from harming their health at work. The greater the potential danger, the more measures that must be taken to identify the risk as acceptable (Current control, 2008).

This method can also be defined as a strategy where a selected control, such as ventilation or preventing the spread of hazardous substances, is applied to a selected group of substances. Chemicals that possess certain toxic or irritating properties through inhalation are examples. The method primarily concerns simplifying the process of selecting the most appropriate measure. Its use facilitates greatly the process of deciding which measures to apply, as companies themselves do not need to measure the occupational exposure to the substances in question, compare them to the permissible limits, or determine the volatility of a substance to an exposed part of the body in order to propose the best and most effective measures (Control banding, 2015).

When should control banding be applied? Although the aforementioned method has recently been very popular in managing nanomaterials and generally hazardous substances, it is always advisable to assess whether this is the most appropriate method for conducting a quality risk assessment. The Canadian Center for Safety and Health at Work recommends evaluating five important factors (Control banding, 2017):

- **Toxicity:** On a scale of 1 to 4, indicating the toxicity of the material or chemical managed at a plant.
- **Method of Exposure:** Are employees easily exposed to toxic materials, gases, dust or fumes? Which chemicals or materials increase exposure during the production process?
- **Exposure Time:** How long are employees exposed to toxic materials or chemicals? How long does work proceed during which employees must be directly exposed to harmful substances? If

it is a highly toxic process, how could exposure times be reduced so that the company remains equally productive?

- **Quantity of the Substance:** This factor relates to the elimination and substitution of dangerous substances. Can the same type of product and same amount be produced if a less toxic substance is used? Can the amount of this hazardous substance be reduced in order to reduce exposure and the resulting risk? Or can the time it is handled be reduced and still achieve the results the company wants?
- **Work Specification:** Are there steps in the workflow that put employees at high risk, such as preparing chemical mixtures or transporting volatile chemicals? Has an appropriate risk analysis been conducted to identify activities with risk? Is the activity performed only by employees with the necessary knowledge and training?

Because employee perceptions of what constitutes danger can vary significantly according to personal experience and job position, communication between employees and management is important. In any case, these five factors should be evaluated, as the selected method may not always be the most appropriate (Control banding, 2017).

Advantages and Disadvantages of Control Banding

Although this method is relatively new and many measurements and calculations are still needed, there are clear advantages and disadvantages (Control banding, 2015).

Advantages:

- Clear overview of risk reduction measures for research laboratory with repeating production processes.
- Good for risk communication and training; provides a logical way to assess risks and systematically apply controls.
- Useful tool for understanding how protection strategies are linked to chemical risk management.
- Straightforward tool for laboratory and maintenance staff in order to better understand essential requirements.
- Provides a way to assess risks and select the appropriate control measures in order to reduce workplace exposure. It also allows control recommendations to be drawn up for products that do not have occupational exposure limits.
- Principle of this method can also be used in facilities other than laboratories where hazardous chemicals are managed (art studios, theaters, research stations) (Control banding, 2015).

Disadvantages:

- Not recommended for isolated or high-risk activities that could present unforeseeable hazards.
- Lack of harmonized terminology between enterprises, because, for example, Category 4 toxicity labeling in some cases may be the lowest category, whereas at another enterprise it may be the highest.
- The method is still under development, and there is no generally accepted control banding method. In this way, the testing and development of an ideal method must continue.

- Control banding recommendations may require expert review to ensure that measures are appropriate, adequately designed, properly installed and maintained and that employee exposure is within acceptable limits.
- Not all hazard types are covered. Ensuring that the method used covers those risks faced by the undertaking is necessary. For example, this may involve the safe handling of certain chemicals with a specific toxic effect, but the hazards of flammability and reactivity may not be addressed. Again, seeking expert advice for these types of hazards is recommended.

NM Legislation

Although the year is 2019 and the topic of nanomaterials and nanoparticles is a global issue, there is currently not enough legislation in the Czech Republic that would directly address it. For this reason, Czech legislation especially relies on foreign legislation. The basic legislation is REACH and CLP, under which the relevant provisions apply to nanoparticles. Act No. 350/2011 Coll., on chemical substances and mixtures, as amended, implements the requirements of these regulations into Czech legislation (Bařinová, 2018).

Many projects for nanomaterial research, security research and operational programs are currently underway to explore and develop new nanotechnologies and to measure harmful concentrations. It is worth mentioning projects under the auspices of the Ministry of Labor and Social Affairs, for example “Assessment of Nanoparticle Risk in the Work Environment and Potential Prevention”, which is the subject of the Research Institute of Occupational Safety (VÚBP).

In general, a common position in EU member states and at an international level is that further research is necessary. If the level of risk is not entirely known yet the concerns are so great that it is considered absolutely necessary to adopt risk management measures, which is currently the case for nanomaterials, the measures must be based on the precautionary principle (Bařinová, 2018; Lustyk, 2018).

The precautionary principle is mentioned in Article 191 of the Treaty on the Functioning of the European Union. The role of the principle is to guarantee a high level of environmental protection through preventive decisions in the case of hazards. In reality, however, the scope of the principle is much broader and also extends to consumer policy, EU food law, and human, animal and plant health. The principle must be defined in a way so that ensuring adequate protection levels for the environment and health have an equal benefit at an international level and are promoted at international negotiations. Many international conventions are already in principle (The precautionary principle, 2000).

According to the Statutes of the European Commission, the precautionary principle may be applied where a particular phenomenon, product or process can potentially exhibit dangerous effects as ascertained by expert and objective assessment. It is assumed that such assessment does not allow the risk to be determined with adequate certainty. The application of the principle therefore falls under a general risk analysis procedure, namely risk management. However, the Commission notes that this principle can be applied only in the case of potential risk but may not be applied as a valid reason for arbitration. Generally, it may be applied provided that three preconditions are met (The precautionary principle, 2000):

- Determination of potentially adverse effects.
- Evaluation of available scientific data.
- The extent of scientific uncertainty.

Competent authorities with risk management expertise may decide whether to implement certain procedures in view of the level of risk. Where the risk is high, several categories of measures may be taken. These may include relevant legal acts, funding of research programs, public information measures, etc. (The precautionary principle, 2000)

Although the EC legal framework is generally concerned with nanomaterials, the implementation of legislation needs to be extended in greater detail. Significant constituents are test methods and risk assessment methods, which provide a basis for implementing legislation and administrative decisions and defining the obligations of manufacturers and employers. There is currently no comprehensive scientific base for fully understanding all the properties and risks of nanomaterials. Indeed, many testimonials have been published revealing "knowledge gaps" (Nel et al., 2009; Danihelka, 2016).

EXAMPLES OF NANOPARTICLE MEASUREMENTS

As mentioned above, several research studies correlate the occurrence of nanoparticles with an increasing risk of respiratory and cardiovascular diseases. It is therefore important not to underestimate the presence of nanoparticles in the workplace and to continuously monitor all nanoparticles, whether they are produced specifically or simply arise from an anthropogenic activity. Not only identifying their sources at individual workplaces but also finding and subsequently adopting appropriate measures to reduce exposure to humans and animals is therefore necessary. These measures may be technological in nature and include the use of appropriate personal protective equipment. In order to achieve these goals, obtaining sufficient data from different types of work environments, operations (e.g., manufacturing, food or agricultural production) and other human activities (e.g., smoking, shooting, lawn maintenance, etc.) is necessary.

The research task v.v.i., No. VUS4_02_VÚBP of the Occupational Safety Research Institute was therefore delegated for activities relating to human pursuits. The research identified how activities are distributed in technological workspaces and determined the factors that play a role in changing the concentration or properties of nanoparticles and their distribution in these spaces. After the results were processed and in consultation with operators, technological and operational measures were proposed for reducing contact with nanoparticles and thereby minimizing the effects of toxicity.

The following is a list of examples of the selected activities where results were obtained from measurement and subsequently evaluated:

- Engineering procedures
- Mined stone processing (sandstone, granite)
- Construction materials production
- Manufacturing of electrical equipment for electricity plants
- Food production
- Vehicle repairs
- Brewing
- Industrial processing of straw
- Food mills
- Powder coating
- Logging

- Agricultural production
- Small buildings
- Human activities (smoking, dental hygiene, sports shooting)
- Scented candles
- Hairdressing and nail treatment

Measuring Equipment Used

For basic measurement, the Testo DiSCmini instrument was used. The device provides simple and fast measurement of the quantity and size of nanoparticles per volume unit. Only nanoparticles

10–750 nm in size were counted, and the mean particle size was also measured. Generally, nanoparticles can be further defined e.g. by: Scanning Electron Microscopy (SEM) – to see NMs morphologies, Brunauer, Emment and Teller's (BET) isotherm- to determine NMs surfaces, Dynamic Light Scattering (DLS) for measuring size of NMs etc.

Measurement Limitations

In examining the properties that may affect the quality of methods for assessing nanomaterial risk, the following were considered important (Bařinová et al., 2018):

- High number of toxic properties
- Different toxicokinetics (laboratory animal, human)
- Heterogeneity of nanomaterials (mostly not homogeneous)
- Unclear dosimetry (which dose unit to bind / surface, number of particles)
- Dynamics of behavior depending on environment
- Batch differentiation (influence of production processes, storage)
- Self-organization (formation of secondary aggregation structures, agglomeration)
- Lack of reliable background differentiation methods
- Specific behavior in the environment
- Current exposure and number of exposed persons in the workplace and environment

Measurement Results and Initial Conclusions

The highest levels of nanoparticle concentration in the working environment were found in a modernized semi-automated bakery and the air-conditioned cabins of a new tractor and combine harvester.

A low and stable concentration of nanoparticles was measured in a dairy cowshed (30,000 # $\cdot$ cm^{-3}, diameter 35–40 nm), and, by contrast, increased concentration was measured in a horse stable, especially during horse grooming.

At a machinery plant for processing steel components for wood shredders, the largest source of nanoparticle concentration was in the welding shop. In the CNC machine operations for cutting and burning materials (lasers, water jets, flames, band and circular saws), the highest concentration was measured during laser cutting (Frišhansová et al., 2018).

Very low concentration of nanoparticles was measured in a mill with modern, closed technology. The average value in the milling environment was 2,500–5,000 # · cm^{-3}. The riskiest activity was the process of bagging flour in the mixing room, where the concentration was 5,000–30,000 # · cm^{-3}.

A risk of increased concentration was demonstrated at a plant for preparing mortar mixtures during bagging and dosing into a closed process line.

High concentrations of nanoparticles of approximately 1,000,000 # · cm^{-3} were measured in a production hall for aluminum smelting and casting (Larena, 2017). Small diameter nanoparticles (25–35 nm) increase the risk of cell penetration. Concentrations of nanoparticles up to 500,000 # · cm^{-3} were measured in a hall where copper bundles are soldered for electric motors (Figures 2 and 3). In both cases, a greater effect of nanoparticle toxicity was assumed due to their chemical composition of aluminum oxides and oxyhydroxides and copper oxides.

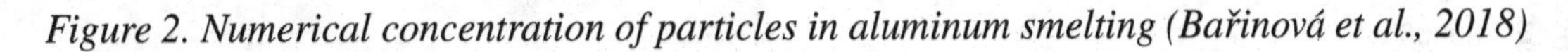

Figure 2. Numerical concentration of particles in aluminum smelting (Bařinová et al., 2018)

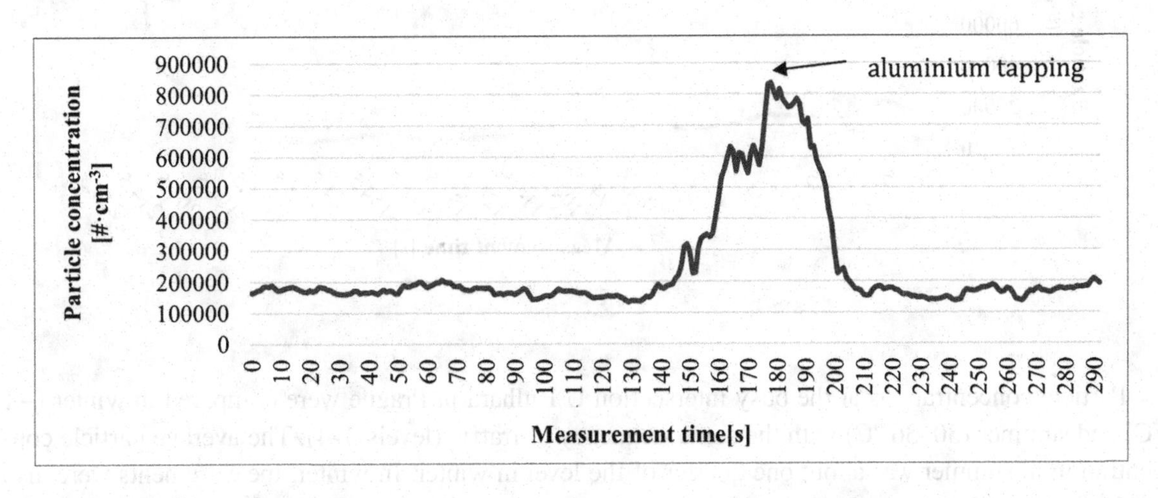

Figure 3. Mean particle size in aluminum smelting (Bařinová et al., 2018)

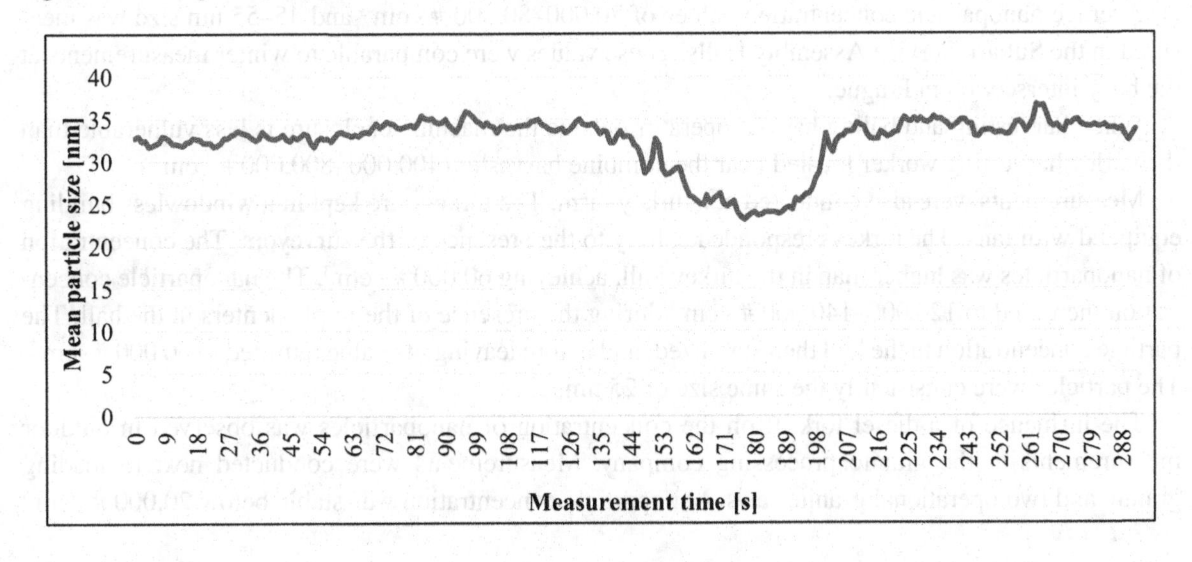

The measurements from the stone processing operation show that sprayed water had a beneficial effect on the concentration and size of nanoparticles released into the work area as the stone was cut with different technologies (band saws, cutters, catheters) compared to dry processing (milling). Concentrations of 83 000–142 000 # · cm^{-3} for sandstone and 15 000–25 000 # · cm^{-3} for granite were measured during water spraying (Figure 4). (Frišhansová et al., 2018)

Figure 4. Particle concentrations measured outdoors during the processing of Lipno granite (Frišhanšová et al., 2018)

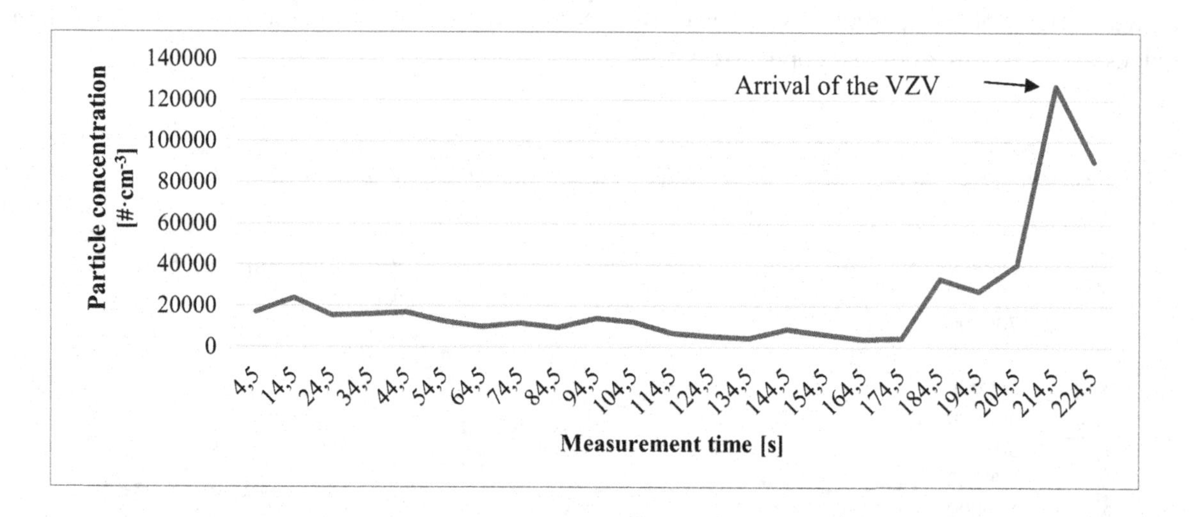

Particle concentrations at the busy intersection U Bulhara in Prague were compared in winter (−1 °C) and summer (30–36 °C) with the same frequency of traffic (levels 3–4). The average particle concentration in summer was about one quarter of the level in winter. In winter, measurements were also made at an intersection in Ostrava under the same traffic conditions as Prague, but the measurements showed twice the concentrations (Sirovátka, 2018).

Average nanoparticle concentration values of 70,000–80,000 # · cm^{-3} and 45–55 nm size was measured in the Subaru Service Assembly Halls. These values were comparable to winter measurements at the busy intersection in Prague.

When harvesting and baling hay, an operator outside the machine enclosure is less vulnerable than the barley harvesting worker located near the combine harvester (400,000–800,000 # · cm^{-3}).

Measurements were also conducted at a turkey farm. The turkeys are kept in a windowless building equipped with fans. The turkeys responded calmly to the presence of the surveyors. The concentration of nanoparticles was higher than in the turkey hall, achieving 60,000 # · cm^{-3}. The nanoparticle concentration increased to 120,000–140,000 # · cm^{-3} during the presence of the implementers in the hall. The particle concentration in the hall then stabilized, and before leaving, its value returned to 60 000 # · cm^{-3}. The particles were consistently the same size of 25 nm.

The influence of a diesel forklift on the concentration of nanoparticles was observed in outdoor measurements at the granite processing company. Measurements were conducted next to loading granite and two operational granite saws. Nanoparticle concentration was stable below 20,000 # · cm^{-3}

(Figure 4), disturbed by the arrival of a forklift truck (a "VZV") with granite material. The mean size of nanoparticles was also disturbed with the arrival of materials (Figure 5). Throughout the production hall, nanoparticle concentration was relatively low at 15,000–25,000 # · cm^{-3}, with a small dispersion of nanoparticle diameter of 50–65 nm.

Figure 5. Mean particle size measured at the outdoor area for processing Lipno granite (Frišhansová et al., 2018)

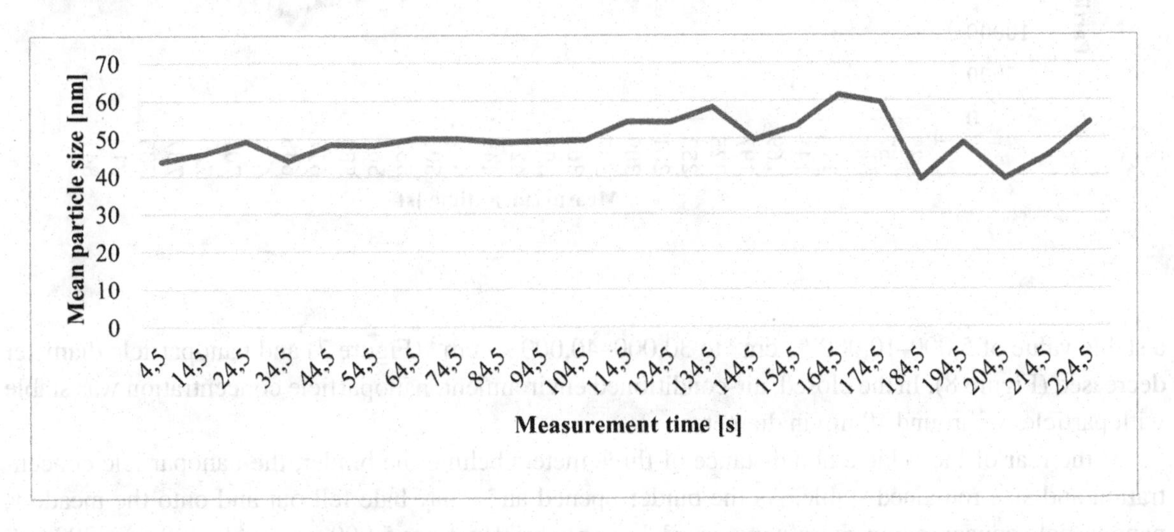

The increase in nanoparticle concentration caused by the entry of the diesel VZV was also monitored when the alternative use of straw as a building material (Ekopanel) was measured. Consistent values of 13,000–15,000 # · cm^{-3} nanoparticle concentration with a mean size of about 45 nm were measured. At the end of the hall where the panels were stored, a diesel VZV with panels entered and increased the nanoparticles concentration to 400,000 # · cm^{-3} with a mean size of 36–37 nm in the vicinity of the measuring apparatus.

Manual powder coating in an open box and closed box were also measured and compared. The values in the paint box were in the range 15,000–45,000 # · cm^{-3} and the nanoparticle concentration during varnishing of the panel (Figure 6) was in the range of 10,000 - 22,000 # · cm^{-3}.

Hay Harvesting Measurements

Nanoparticle concentration as hay from a meadow was processed was also measured. The dried hay in rows was collected by a tractor (John Deere baler), then towed and subsequently bundled into cylindrical bales. During the first measurement, nanoparticle concentration for both the closed, air-conditioned cab for the driver and the machine operator's area was measured. Additional measurements were conducted three meters behind the baler. Finally, measurements were taken in the middle of the cut meadow with a distance of 300 m between the baling operation and the measurement device.

During the measurement procedure inside the cab, rubber tension belts burned out and it was necessary to vacate the tractor cab. As the cabin opened, nanoparticle concentration increased sharply from

Figure 6. Particle concentration during painting in an open box

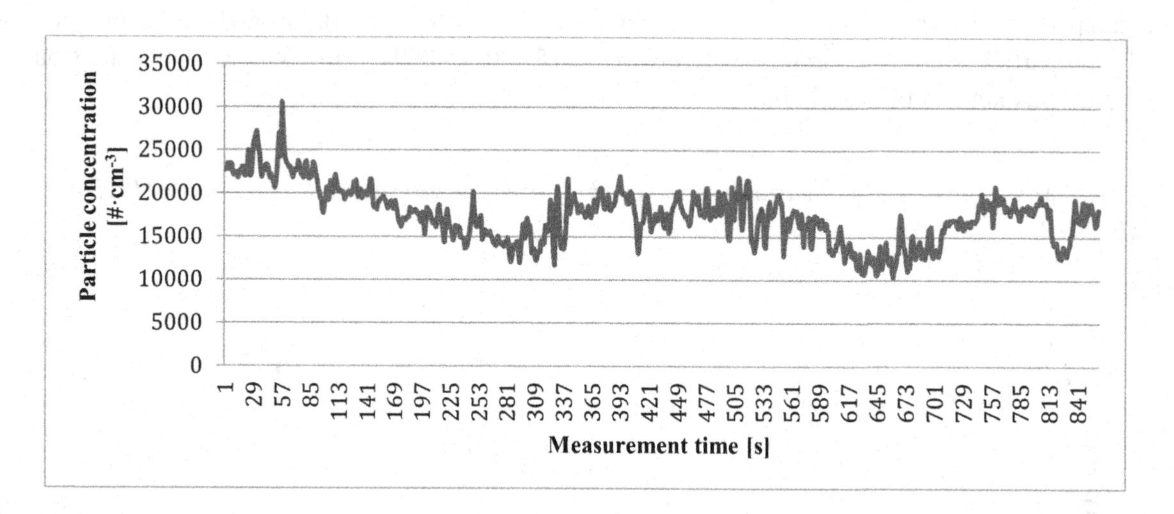

a stable value of 5,000–10,000 # · cm^{-3} to 30,000–40,000 # · cm^{-3} (Figure 7) and nanoparticle diameter decreased (Figure 8). In the closed, air-conditioned environment, nanoparticle concentration was stable with particles of around 40 nm in diameter.

At the rear of the vehicle at a distance of three meters behind the binder, the nanoparticle concentration and size remained stable. As the binder opened and a hay bale fell out and onto the meadow, nanoparticle concentration sharply increased, but only briefly, from 5,000 # · cm^{-3} to 200,000–300,000 # · cm^{-3}. The change in particle size was negligible.

The open meadow environment with hay rows during moderate winds showed relatively higher concentrations (30,000–40,000 # · cm^{-3}) but particle size remained effectively constant (23–24 nm). The distance from the technician and worker to the measuring instrument was 300 m.

Figure 7. Particle concentration during baling (Bátrlová et al., 2019)

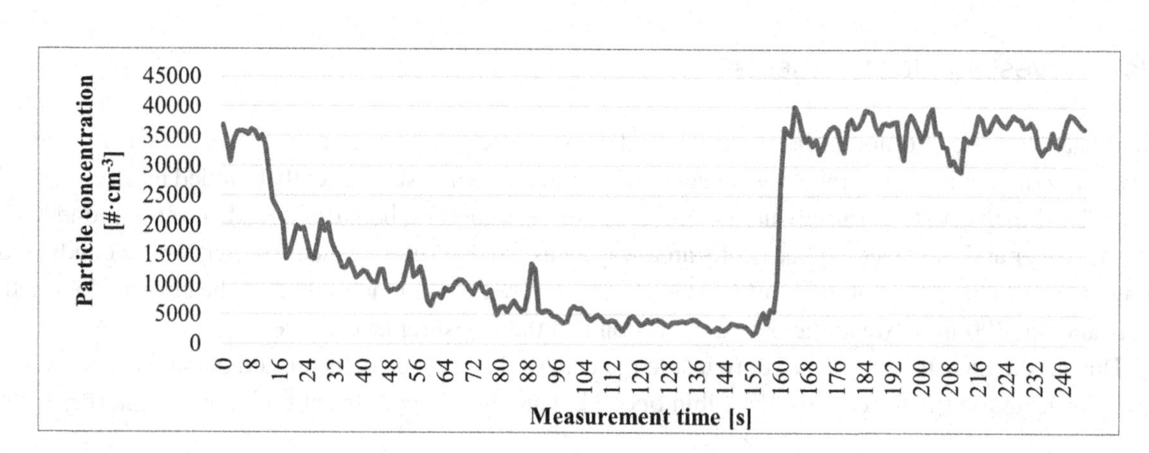

Figure 8. Mean particle size during baling (Bátrlová et al., 2019)

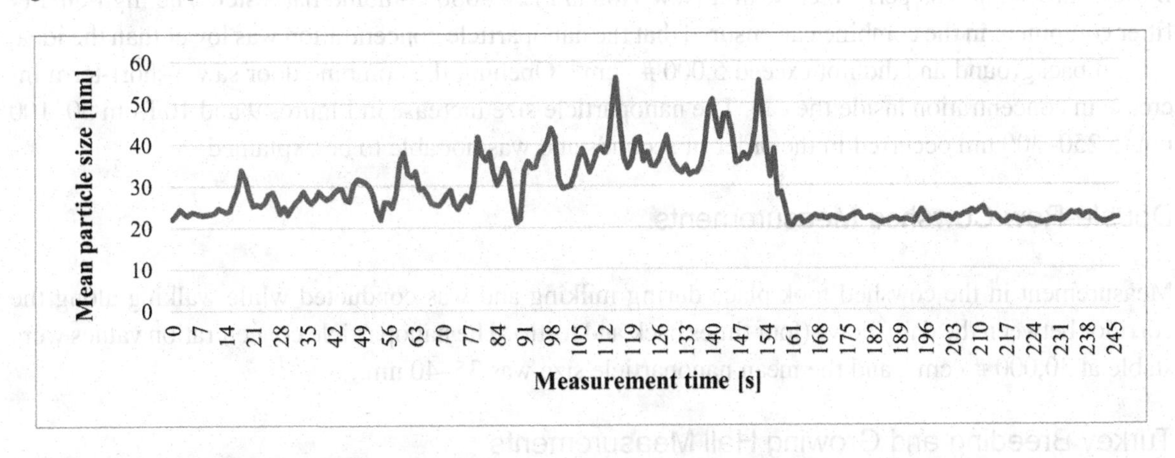

Figure 9. Nanoparticle concentration inside the combine cab (Bátrlová et al., 2019)

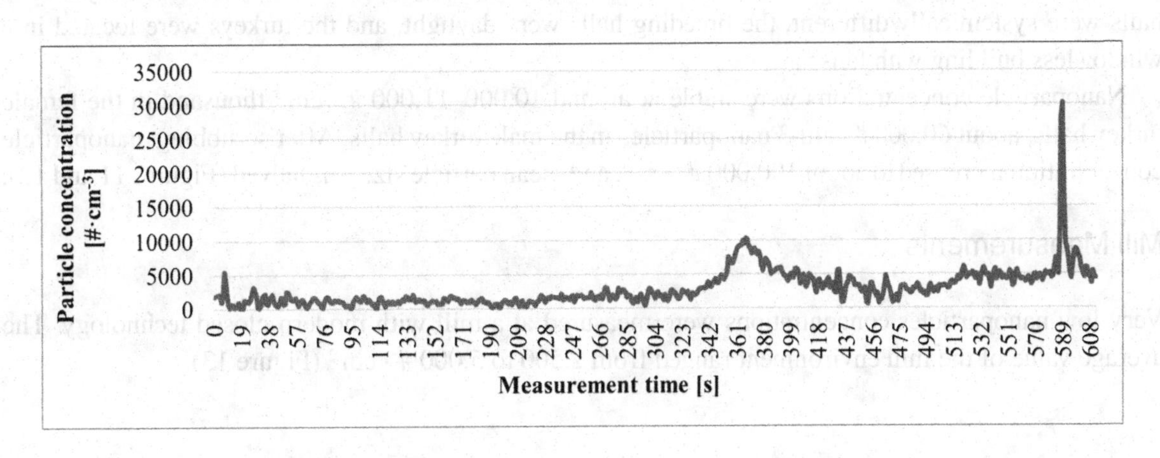

Barley Harvesting Measurements

Figure 10. Mean particle size as measured in the combine cab (Bátrlová et al., 2019)

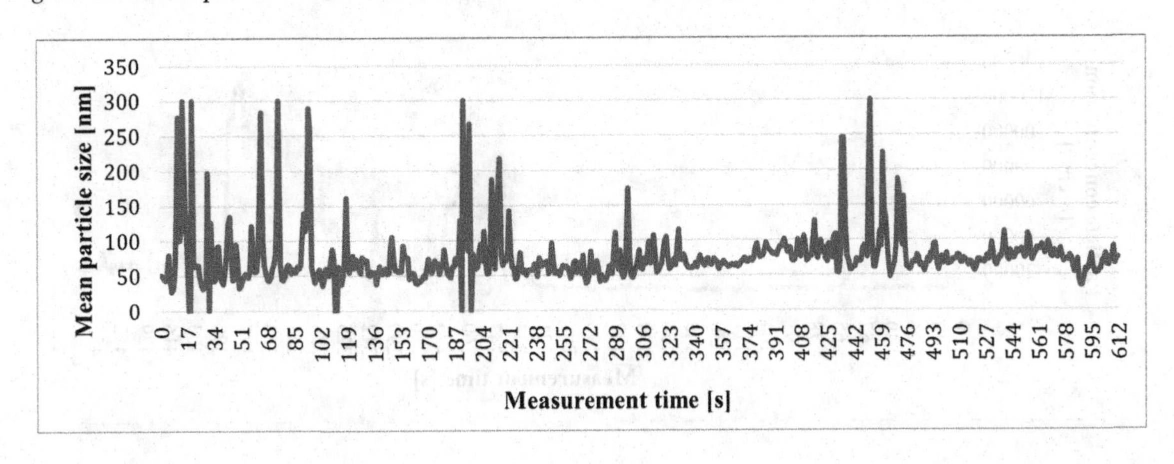

Barley harvesting was performed with a New Holland LX 8080 combine harvester. The high-quality filter equipment in the combine cab ensured that the nanoparticle concentration was lower than the ideal natural background and did not exceed 5,000 # · cm^{-3}. Opening the combine door saw a short-term increase in concentration inside the cab. The nanoparticle size increase in Figures 9 and 10 from 50–100 nm to 250–300 nm occurred in the order of seconds and was not able to be explained.

Double-Row Cowshed Measurements

Measurement in the cowshed took place during milking and was conducted while walking along the corridor between the dairy cows (four times back and forth). The nanoparticle concentration values were stable at 30,000 # · cm^{-3}, and the mean nanoparticle size was 35–40 nm.

Turkey Breeding and Growing Hall Measurements

Measurements were conducted in the breeding halls populated by approximately 3,500 turkeys. Both halls were systemically different, the breeding halls were daylight, and the turkeys were located in a windowless building with fans.

Nanoparticle concentrations were stable at around 10,000–11,000 # · cm^{-3} thousand in the female turkey halls, about 60,000 # · cm-3 nanoparticles in the male turkey halls. After scrubbing, nanoparticle concentration increased to about 100,000 # · cm^{-3}, and mean particle size was halved (Figures 11 and 12).

Mill Measurements

Very low nanoparticles concentrations were measured at a mill with modern closed technology. The average value of the mill environment ranged from 2,500 to 5,000 # · cm^{-3} (Figure 13).

Figure 11. Nanoparticle concentrations in the turkey breeding shed (Bátrlová et al., 2019)

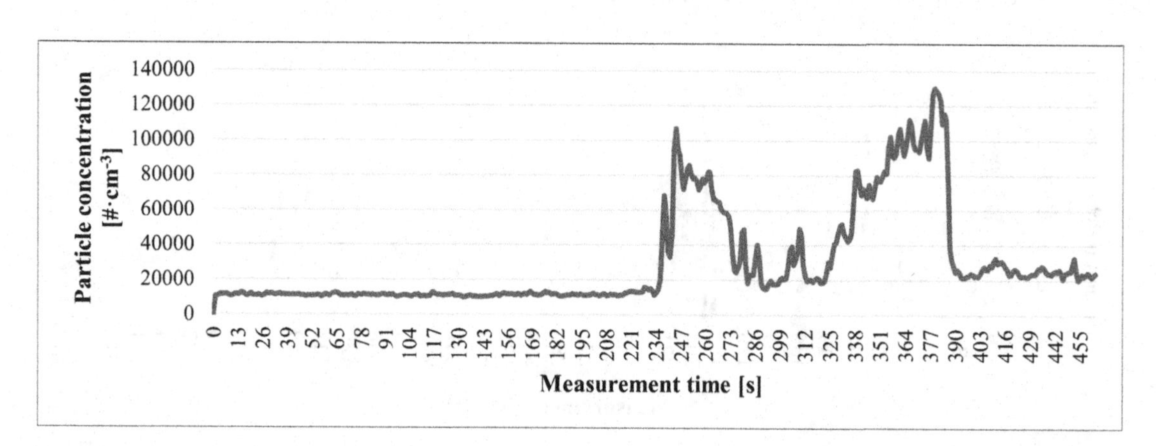

Ekopanel Measurements

The research task also involved visiting the production site for straw construction panels known as Ekopanels. The panels are formed by pressing wheat straw on a production line. The measured nanoparticle concentrations and mean size are shown in Figure 18 and 19. Values of 11,000–17,000

$\# \cdot cm^{-3}$ for nanoparticle concentration and around 55 nm for mean size were measured near the straw bales. The particles were stable in diameter. In the section of line inside the hall, nanoparticle concentration was 16,000–20,000 $\# \cdot cm^{-3}$ and size was 41–48 nm. Constant nanoparticle concentrations of 13,000–15,000 $\# \cdot cm^{-3}$ and mean size of around 45 nm were measured in the other section of the line. At the end of the hall where the panels were stored, a diesel forklift with panels was operated. The environment around the measuring apparatus in this area had a nanoparticle concentration of 400,000 $\# \cdot cm^{-3}$ with a mean size of 36–37 nm (Figures 14 and 15).

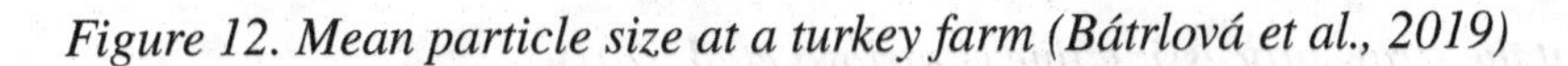

Figure 12. Mean particle size at a turkey farm (Bátrlová et al., 2019)

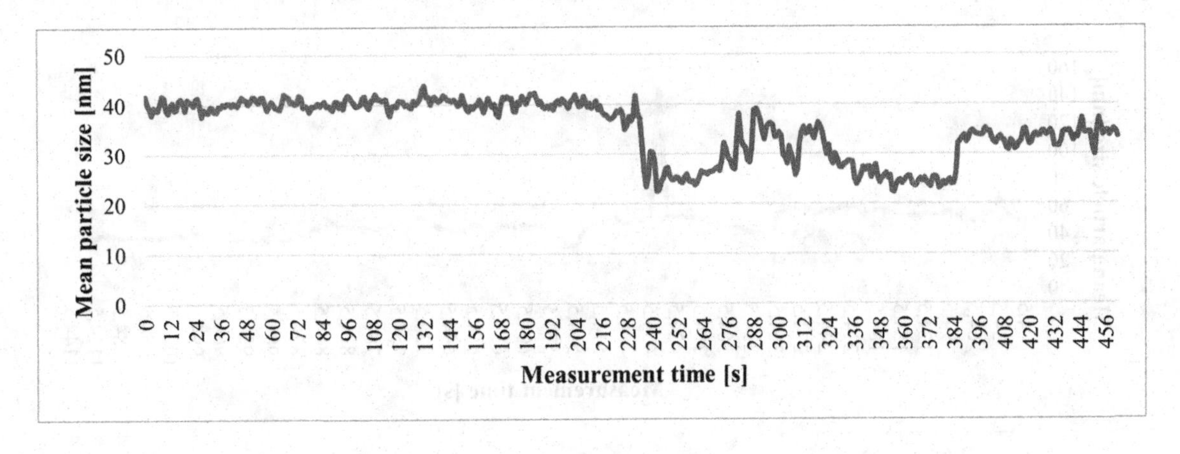

Figure 13. Measurement scheme at a mill and measured values (Chromečka, 2019)

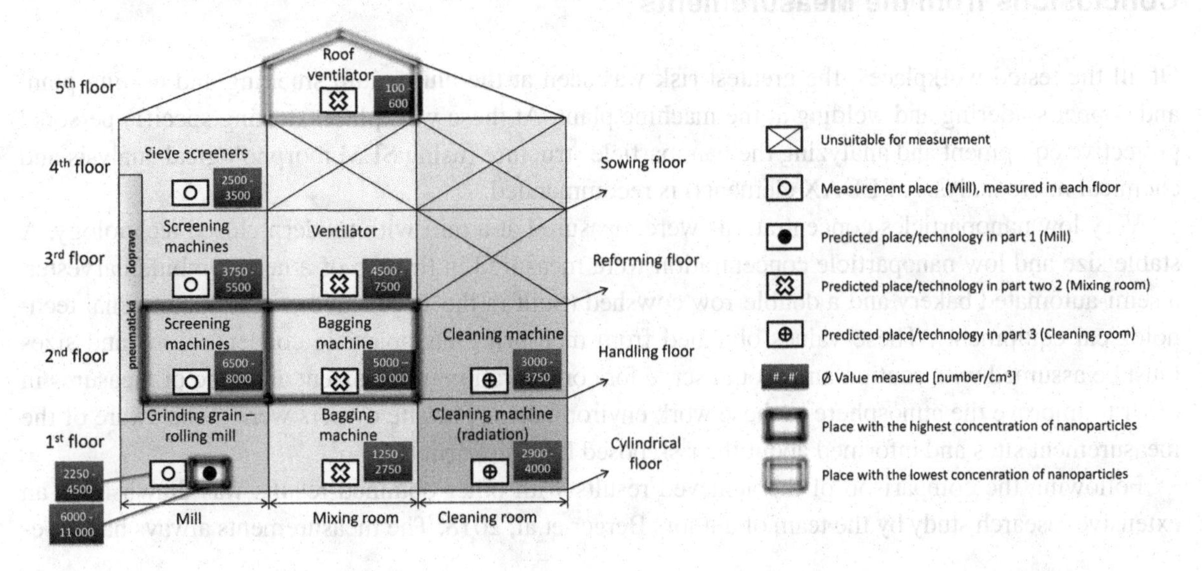

Figure 14. Nanoparticle concentrations at the straw processing line (Bátrlová et al., 2019)

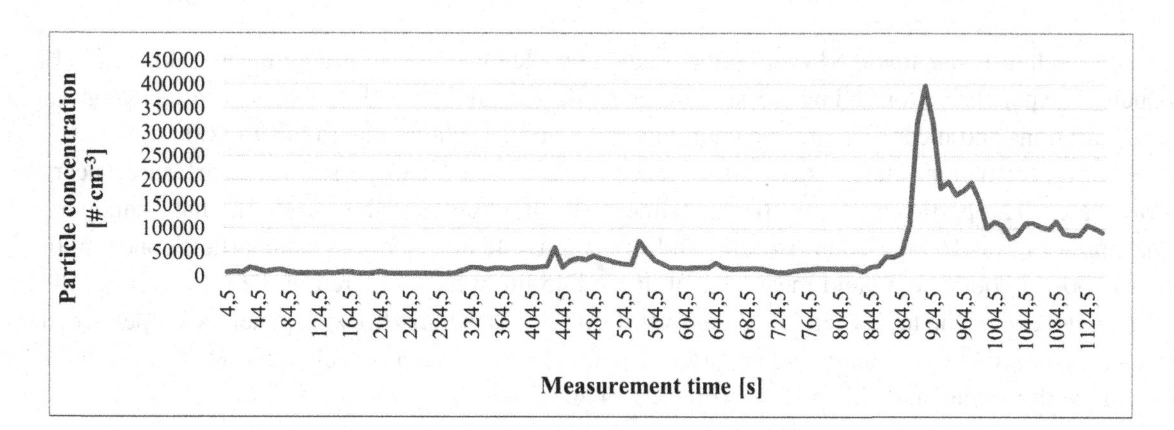

Figure 15. Mean particle size at the straw processing line (Bátrlová et al., 2019)

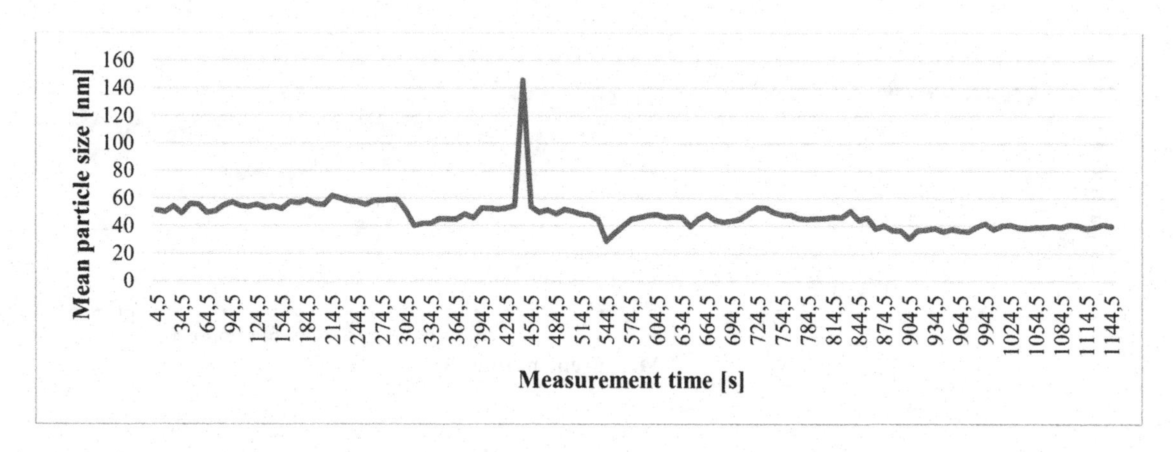

Conclusions from the Measurements

Of all the tested workplaces, the greatest risk was seen at the aluminium smelting and casting plant and copper soldering and welding at the machine plant. At these workplaces, testing specific personal protective equipment and analyzing the nanoparticle structure (using SEM morphological analysis and chemical microanalysis of EDAX elements) is recommended.

Very low nanoparticles concentrations were measured at a mill with modern closed technology. A stable size and low nanoparticle concentration were measured in the cab of a new combine harvester, a semi-automated bakery and a double-row cowshed (built in the 1970's and containing original technological equipment). These values obtained from measuring nanoparticle concentrations and sizes must be assumed with caution and should serve for consideration in the extent and type of measures in order to improve the atmosphere in these work environments. The site owners were made aware of the measurement sites and informed about the risk posed by nanoparticles.

Following the comparison of the achieved results with other obtained results was drawn from an extensive research study by the team of authors Berger et.al, 2018. The measurements always have specific conditions, so it is very difficult to compare achieved results with different results with respect to the behavior of nanoparticles in environments. The measurement places were chosen for comparison: steelworks, crematorium and exhaust nanoparticles. The particle size measurement results measured in steelworks were presented in the form of average particle sizes divided into 14 fractions ranging from 10 nm to 8.126 μm. The highest numerical concentrations measured for the 10 nm fraction, about 130,000 particles per cm^3 for both the ladle furnace and continuous casting. (Berger et al., 2019). Second study assessed the exposure of workers of nanoparticles released in the crematorium during combustion processes due to high temperatures. The concentration at the time of opening the crematorium door was 500,000 particles / cm^3. (Berger et al., 2019). The most common activities producing exhaust particles in the depot include brake control, steering, transmission system, operating fluid levels and also car washing. Most of the particles were measured during the shift at approximately 130,000 particles / cm^3 in the 10-1000 nm range, with 2/3 of these particles measuring between 10 and 100 nm. (Berger et al., 2019).

CONCLUSION

This study work examined the topic of nanoparticles. It explained their origins and history and the distribution and general properties that distinguish nanomaterials from other materials. A specific group of nanoparticles comprises manufactured carbonaceous materials, and combined production of large quantities, burden the planet with waste products. The study inspected in detail four potential nanomaterial risks: health hazards (NM and nanoparticles toxicity towards cells), environmental hazards (NM behavior in aquatic, soil and atmospheric environments), physicochemical hazards (fires, explosions, special risk of aerosols), and importantly, the potential misuse of nanomaterials for criminal purposes. The fundamental mechanisms that can elicit NOS (ROS, genotoxicity and inflammatory response) were presented.

The second part of the thesis analyzed NM risks through available methodologies (control banding) and summarized the legislative requirements relevant to the issue. In general, the precautionary principle applies to NMs in order to ensure a minimum standard for protecting health and the environment through preventive decisions in the case of risk. The final part of the work presented concrete measurements conducted at selected operations under the research project. Measurements concentrated on assessing the exposure of nanoparticles employees come into contact with during work. Given how effectively these particles can penetrate all organs of the human respiratory system, addressing this issue specifically is very important, as employees come into contact with these particles on a daily basis. Pilot conclusions and suggested precautions for eliminating the risks caused by NM were made.

ACKNOWLEDGMENT

This research was supported by the Occupational Safety Research Institute, [grant number VUS4_02_VUBP].

REFERENCES

Bakand, S., Hayes, A., & Dechsakuthorn, F. (2012). Nanoparticles. A review of particle toxikology following inhalation exposure. *Inhalation Toxicology*, *24*(2), 125–135. doi:10.3109/08958378.2010.642021 PMID:22260506

Bařinová, E. (2018). *Application of Certified Methodology, for Providing Personal Protective Equipment, in a Nanoparticle Environment at Selected Operatoíons* [Diploma thesis]. VSB-Technical university of Ostrava, Ostrava, Czech Republic.

Bařinová, E., Klouda, K., Lustyk, K., Nechvátal, M., & Senčík, J. (2018). Presentation and analysis of results from field measurement of nanoparticles in a machine plant. *Paper presented at XVIII. International conference: Occupational health and safety*. Academic Press.

Bátrlová, K., Klouda, K., Kubátová, H., Roupcová, P., & Nechvátal, M. (2019). Monitoring of the occurrence of nanoparticles in the working environment focusing on agricultural activities. *Josra*. Retrieved from https://www.bozpinfo.cz/josra/monitorovani-vyskytu-nanocastic-v-prostredi-prvotni-zavery-mereni

Berger, F., Bernatíková, Š., Přichystalová, R., Schreiberová, L., Šigutová, L., Vavrečková, K. & Tymráková, L. (2019). *Professional exposure and health risks of nanostructured materials and possibilities of their prevention at workplaces across sectors*. Academic Press.

Boček, J. (2012). *Uplatnění nanotechnologií v zemědělství*. Jihočeská univerzita in České Budějovice.

Brzicová, T. (2019). *Nanoparticle safety management to occupational safety* [Doctoral dissertation]. VSB-Technical university of Ostrava, Ostrava, Czech Republic.

Cagno, V., Andreozzi, P., D'Alicarnasso, M., Silva, P. J., Mueller, M., Galloux, M., ... & Weber, J. (2018). Broad-spectrum non-toxic antiviral nanoparticles with a virucidal inhibition mechanism. *Naturematerial*, *17*, 195-203. Retrieved from: https://www.nature.com/articles/nmat5053?fbclid=IwAR0O0PjOFI5sWG8NpeML3JfsFSfYSZ9QTPsTsyg_xebnTQXR6RolzsTXKfE

Certified methodology for providing personal protective equipment in workplaces with a presence of nanoparticles: (focused on employers in need of protective equipment with respiratory protection). (2016). Prague, Czech Republic: Occupational Safety Research Institute.

Chromečka, O. (2019) *Measurement of dust particles less than 10 μm in working atmosphere*. (Bachelor thesis), VSB-Technical university of Ostrava, Ostrava, Czech Republic.

Control banding. (2015) ACS. Retrieved from https://www.acs.org/content/acs/en/chemical-safety/hazard-assessment/ways-to-conduct/control-banding.html

Control banding. Canadian centre for occupational health and safety. (2017). Retrieved from: https://www.ccohs.ca/oshanswers/chemicals/control_banding.html

Current control banding approaches for nanotechnologies. (2008). Safenano. Retrieved from https://www.safenano.org/knowledgebase/guidance/banding/

Danihelka, P. (2016). *Legislation of nanomaterials within the EU, environmental protection.* Retrieved from https://www.enviprofi.cz/33/legislativa-nanomaterialu-v-ramci-eu-ochrana-zivotniho-prostredi-uniqueidmRRWSbk196FNf8-jVUh4EpGY5V1LxZIbDfbtq18xpwk/

Das, P., Barua, S., Sarkar, S., Chatterjee, S. K., Mukherjee, S., Goswami, L., ... Bhattacharya, S. S. (2018). Mechanisms of toxicity and transformation of silver nanoparticles: Inclusive assessment in earthworm-microbe-soil-plant system. *Geoderma, 314*, 73–84. doi:10.1016/j.geoderma.2017.11.008

Examples of nanotechnology misuse. (2015). Retrieved from http://www.seminarky.cz/Priklady-moznosti-zneuziti-nanotechnologie-30621

Filipová, Z., Kukutschová, J., & Mašláň, M. (2012). *Risk of nanomaterials.* Olomouc, Czech Republic: University of Palacký.

Frišhansová, L., & Klouda, K. (2016). *Ambivalence of nanoparticles.* Retrieved from http://www.odpadoveforum.cz/TVIP2017/prispevky/220.pdf

Frišhansová, L., & Klouda, K. (2017). Ambivalence of nanoparticles. In Sborník příspěvků ze symposia Týden vědy a inovací pro praxi a životní prostředí. Academic Press.

Frišhansová, L., Klouda, K., Nechvátal, M., Roupcová, P., & Barták, P. (2018). Analysis of unique operation for processing of sandstone from the view of released nanoparticles. *Paper presented at Sborník přednášek XVIII. ročníku mezinárodní konference. Ostravice VŠB-TU 2018.* Academic Press.

He, W., Liu, Y., Wamer, W. G., & Yin, J. (2014). Electron spin resonance spectroscopy for the study of nanomaterial-mediated generation of reactive oxygen species. *Journal of Food and Drug Analysis. 22*(1), 49-63. Retrieved from: https://linkinghub.elsevier.com/retrieve/pii/S1021949814000052

Huang, Z., He, K., Song, Z., Zeng, G., Chen, A., Yuan, L., ... Chen, G. (2018). Antioxidative response of Phanerochaete chrysosporium against silver nanoparticle-induced toxicity and its potential mechanism. *Chemosphere, 211*, 573–583. doi:10.1016/j.chemosphere.2018.07.192 PMID:30092538

Katsumiti, A., Thornley, A. J., Arostegui, I., Reip, P., Valsami-Jones, E., Tetley, T. D., & Cajaraville, M. P. (2018). Cytoxicity and cellular mechanisms of toxicity of CuO NPs in mussel cells in vitro and comparative sensitivity with human cells. *Toxicology In Vitro, 48*, 146–158. doi:10.1016/j.tiv.2018.01.013 PMID:29408664

Klouda, K., Frišhansová, L., & Senčík, J. (2016). Nanoparticles, nanotechnologies and nanoproducts and their connection with occupational health and safety. In *Occupational health and safety & life quality 2016*. Academic Press.

Kolářová, L. (2014). *Introduction to nanosciences and nanotechnologies.* Retrieved from: http://mofychem.upol.cz/KA4/Nanotechnologie.pdf

Korshed, P., Li, L., Liu, Z., Mironov, A., & Wang, T. (2018). Antibacterial mechanisms of a novel type picosecond laser-generated silver-titanium nanoparticles and their toxicity to human cells. *International Journal of Nanomedicine, 13*, 89–101. doi:10.2147/IJN.S140222 PMID:29317818

Larena, G. I. (2017). *Antimicrobial activity of graphene and its viability.* Retrieved from http://diposit.ub.edu/dspace/handle/2445/112186

Lauterwasser, C. (2016) *Opportunities and risks of Nanotechnologies* OECD. Retrieved from https://www.oecd.org/science/nanosafety/44108334.pdf

Lustyk, K. (2018). *Exposure to Fine and Ultrafine Particles during Smoking and Comparison of Used Measurement Techniques* [Diploma thesis]. Vysoká škola báňská – Technická univerzita Ostrava.

Medinsky, M. A., & Bond, J. A. (2001). Sites and mechanisms for uptake of gases and vapors in the respirátory tract. *Toxicology, 160*(1-3), 165–172. doi:10.1016/S0300-483X(00)00448-0 PMID:11246136

Methods and way of risk assessment in the workplace. Documentation OSH. (2018) Retreived from https://www.dokumentacebozp.cz/aktuality/metody-hodnoceni-rizik-bozp/

Nanoparticles and their toxicity. Chemical papers. (2015). Retrieved from: http://ww.chemicke-listy.cz/docs/full/2015_06_444-450.pdf

Nel, A. E., Mädler, L., Velegol, D., Xia, T., Hoek, E. M. V., Somasundaran, P., ... Thompson, M. (2009). Understanding biophysicochemical interactions at the nano-bio interface. *Nature Materials, 8*(7), 543–557. doi:10.1038/nmat2442 PMID:19525947

Roupcová, P. (2018). *Monitoring of ecotoxicity of carbon based nanoparticles* [Doctoral dissertation]. VSB-Technical university of Ostrava, Ostrava, Czech Republic.

Schüllerová, B., Adamec, V., & Benco, V. (2019). Current approaches of risk assessment of nanomaterials. *Paper presented at XIX. International conference: Occupational safety and health,* Ostravice, *2019*. Academic Press.

Sirovátka, J. (2018). *Measurement of dust Particles of less than 10 μm in Ambient Air in Transport with regard to Particle Phytotoxicity* [Diploma thesis]. VSB-TUO, Ostrava, Czech Republic.

Skřehot, P., & Rupová, M. (2011). Nanosafety. Praha, Czech: Occupational Safety Research Institute.

Tang, J., Wu, Y., Esquivel-Elizondo, S., Sørensen, S. J., & Rittmann, B. E. (2018). How Microbial Aggregates Protect against Nanoparticle Toxicity. *Trends in biology, 36*(11), 1171-1182. Retrieved from: https://www.sciencedirect.com/science/article/abs/pii/S0167779918301732?fbclid=IwAR3eWRkxiAW0sQSzXJ7gwYKBOVaJv_ukliP85j-WVUzi4XeFfae8pXFt9gI

Tenschert, J. (2016). NANOaers - Fate of aerosolized nanoparticles: the influence of surface active substances on lung deposition and respiratory effects. *Nanopartikel.* Retrieved from https://nanopartikel.info/en/projects/era-net-siinn/nanoaers-en

The precautionary principle. (2000) EUR-Lex: Acess to European Union law. Retreived from: https://eur-lex.europa.eu/legal-content/CS/TXT/?uri=LEGISSUM:l32042

Tůma, M. (2004). Are nanotechnologies the salvation of humanity with safety risks?

Vempati, S., & Uyar, T. (2014). Fluorescence from graphene oxide and the influence of ionic, π–π interactions and heterointerfaces: Electron or energy transfer dynamics. *Journal of Physical Chemistry Chemical Physics, 16*(39), 21183–21203. doi:10.1039/C4CP03317E PMID:25197977

Yan, L., Gu, Z., & Zhao, Y. (2013). Chemical Mechanisms of the Toxicological Properties of Nanomaterials: Generation of Intracellular Reactive Oxygen Species. *Chemistry - An Asian Journal*, 8(10), 2342-2353. Retreived from: http://doi.wiley.com/10.1002/asia.201300542

Zalk, D. M., & Nelson, D. I. (2003). *History and evolution of control banding*. Retrieved from https://www.researchgate.net/publication/5502902_History_and_Evolution_of_Control_Banding_A_Review

ADDITIONAL READING

Klouda, K., & Kubátová, H. (2009). Engineered nanomaterials: risk analysis of their production and their impact on human health and environment. *Josra.* Retrieved from https://www.bozpinfo.cz/josra/vyrabene-nanomaterialy-analyza-rizik-jejich-pripravy-dopadu-na-zdravi-zivotni-prostredi

Klouda, K., Kubátová, H., & Večerková, J. (2010). Deliberately produced nanomaterials. Proposal of risk management methodology in production and handling them. *Spektrum, 1*, 41–45.

Kubátová, H. (2018). *Industrial toxicology and the environment*. Ostrava: SPBI.

Roupcová, P., & Klouda, K. (2017). Hybrid compounds based on graphene oxide and their compounds with metals. *Paper presented at IX. International conference: Nanomaterials- Research & Application 2017.* Academic Press.

Roupcová, P., Kubátová, H., Klouda, K., & Lepík, P. (2016). Phytotoxicological tests – applications of foils based on graphene (graphene oxide). *Transactions of the VSB- Technical University of Ostrava, 12*(2), 6-14. 10.1515/tvsbses-2016-0011

Roupcová, P., Lichorobiec, S., Klouda, K., Bednářová, L., & Gembalová, L. (2018). Analysis of the way to increase safety in terms of possible inhalation of nano and microparticles during gun shooting. *Paper presented at the 10th International conference: Nanomaterials- Research & Application 2018.* Academic Press.

KEY TERMS AND DEFINITIONS

Nanomaterials: A natural material, by-product or material produced containing particles in an unbound state or as an aggregate or agglomerate in which 50% or more of the particle size distribution in one or more external dimensions is in the size range of 1–100 nm.

Nanotechnology: An interdisciplinary field of science that studies, creates and uses technologies in the nanometer scale.

Risk: can be defined as a function of probability and magnitude of losses.

Risk Assessment: is a term used to describe the overall process or method where you: Identify hazards and risk factors that have the potential to cause harm (hazard identification). Analyze and evaluate the risk associated with that hazard (risk analysis, and risk evaluation).

Safety: A condition in which the hazard is at an acceptable level.

Security: A situation where threats to an object and its interests are effectively reduced, and the object is effectively equipped and willing to cooperate to reduce existing and potential threats.

The Precautionary Principle: The role of the principle is to guarantee a high level of environmental protection through preventive decisions in the case of hazards.

Toxicity: is a property of chemical substances or compounds that causes poisoning of persons or animals that have ingested, inhaled or absorbed through the skin.

Chapter 13
Security of Infrastructure Systems:
Infrastructure Security Assessment

Ladislav Mariš
https://orcid.org/0000-0003-4952-9055
University of Žilina, Slovakia

Tomáš Loveček
University of Žilina, Slovakia

Mike Zeegers
Security Risk Watch, The Netherlands

ABSTRACT

The security of infrastructure systems is increasingly associated and ties to ensuring a company's basic functional continuity. Increasing security and ultimately the resilience of infrastructure systems is significantly linked to the process of infrastructure security assessment. It is obvious that the basic pillar of ensuring the required level of security and resilience of infrastructure systems is the level of physical security. Therefore, the chapter will discuss the methods for physical security assessment with a link to the different nature of selected infrastructure systems. The basic logic will be the exploitation of qualitative-quantitative methods, assessing an existing or proposed security system, based on certain measurable values such as probability of detection, response force time, delay time and probability of correct and timely guard communication, where based on this data, the probability of interruption is estimated.

INTRODUCTION

Today's world is a complex system, more complex than ever. Its functionality is based on interdependent, technologically and energy interconnected systems. One system is dependent on the functionality of another system, the functionality of which is ensured by another system, and so everything is directly or

DOI: 10.4018/978-1-7998-3059-7.ch013

indirectly connected to everything. The interconnectedness and interdependence of the complex structure of contemporary society makes it at the same time vulnerable to the impact of negative events of different nature. There is an absolute functional interdependence of all society subsystems. The functionality of one subsystem is dependent on the functionality of all others, failures in one subsystem, in one part of society, have a dramatic effect on the whole.

There are concerns that the ever-increasing complexity and interconnectedness of today's world will sooner or later result in tragic consequences. Political scientist Thomas Homer - Dixon (2013) writes in The Upside of Down on the current world margo that increasing the number of linkages of Its subsystems creates such a close interconnection that a failure in one part will shake the whole system. This is because the complex networks that connect the world and which are used to ensure the movement of people, material, information, money and energy transmit and multiply every upheaval that arises within them. Therefore, even a seemingly insignificant change or failure in one part of the system can cascade through the entire network and cause, at another point in time, a change in another part of the system.

The security, economic and social stability of the state, its functionality, but also the protection of citizens' lives and property is dependent on the proper functioning of many state infrastructure systems. The infrastructure elements needed to ensure living conditions - energy, water, food, communications, transport, health, finance, defence - are so closely interconnected that events that may arise in one of the infrastructure sectors have the potential to shake the entire social system. (Soltes, Kubas, & Stofkova, 2018)

If the disruption, lack or destruction of physical and virtual systems, institutions, facilities and other services could cause a disruption of the social stability and security of the state, trigger a crisis or seriously affect the functionality of state administration and self-government in crises - we call this critical infrastructure. Such a perception of critical infrastructure expresses its content and its importance not only for the functionality of the state in normal conditions, but also in crisis, which may be caused intentionally or unintentionally, whether by external or internal, resp. natural, technological or social agents.

It is in the interest of the state that critical infrastructure be effectively protected. Investigating the protection of CI objects does not only mean a description of the protection methods and the structure of the systems for their protection, but is also conditional on seeking, finding and detecting the causality of phenomena and events giving rise to the need for protection.

SECURITY THREATS RELATING TO THE PHYSICAL SECURITY OF THE CRITICAL INFRASTRUCTURE SYSTEM

A system that incorporates individual subsystems of tangible, resp. intangible assets protection under the administration or ownership of the entity, and which are created by the expedient arrangement and application of security measures, are in practice referred to various attributes. The most commonly used terms are a security system, a property / object protection system, a physical protection system, a security system or an integrated security system.

Risk is a potential opportunity for something to happen that has a negative impact on the organisation's objectives, measured by its consequences or probability estimation. While the term threat means the designation of a specific, physically existing entity, phenomenon or event (security incident) capable of causing damage or harm. Thus, the process of identifying security risks consists in detecting possible negative events and phenomena in various forms and forms in the security environment that may cause a breach of the security status of the protected interest. The basic breakdown of possible security risks is shown in Figure 1.

Figure 1. Basic threat distribution scheme (Hofreiter et al., 2005)

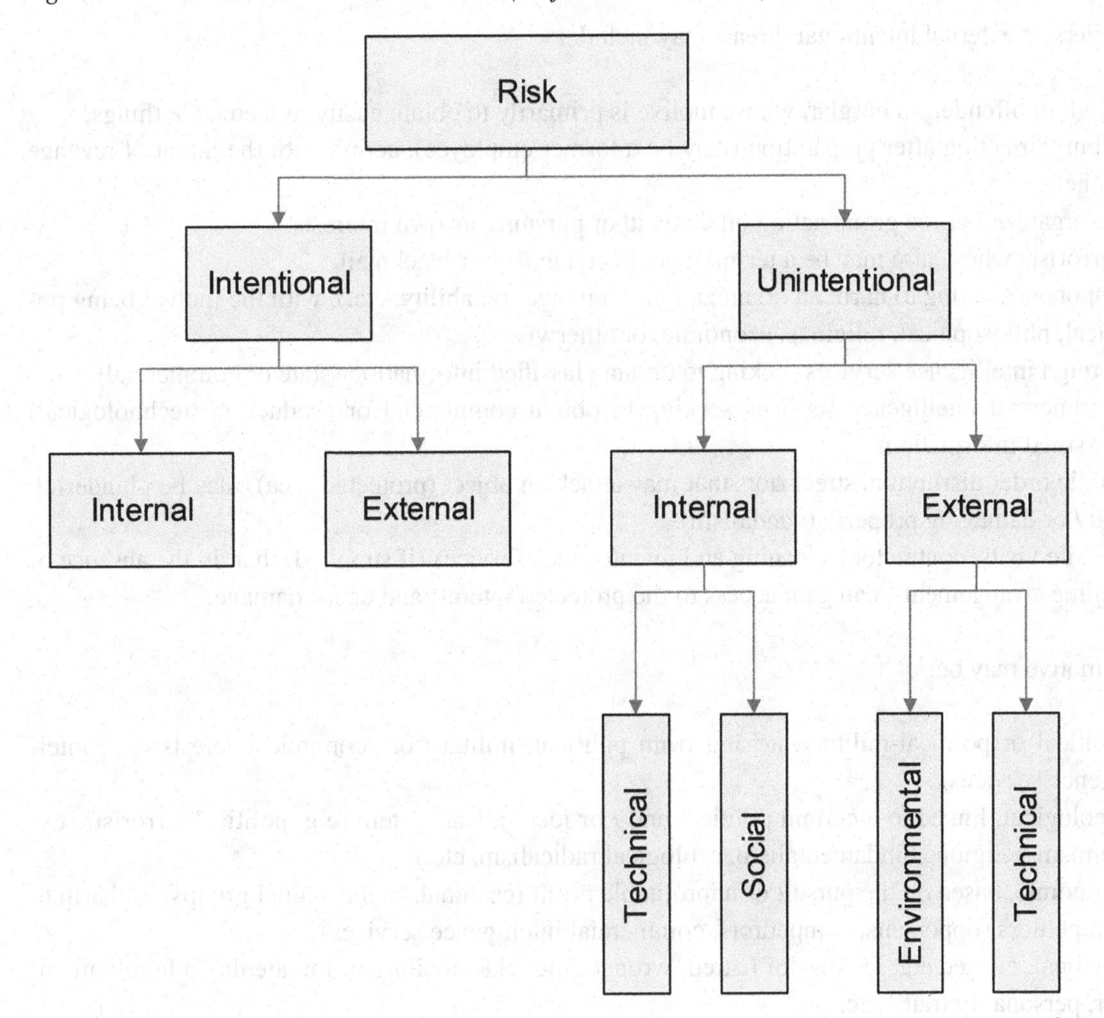

Intentional Security Risks

Intentional security risks are those that are the source of an adversary - intruder, intentionally acting to tamper with an entity's assets (e.g. damage, alter, steal, misuse, destroy).

The most common intentional external security risks include burglary, vandalism, sabotage, terrorist acts, industrial espionage, or any other offense affecting the entity's assets. Depending on the nature of the crime, it is possible to discuss of the source of the security risk. A person (e.g. the entity's own employee) who intends to enter one of the protected area, where he or she is not allowed access is also considered to be a source of intentional external risk. If the source of the security risk is a deliberately acting natural person who is characterized by his or her experience, knowledge, physical propositions, financial and technical capabilities, these persons can be defined according to the design basis threat typology.

Carriers of External Intentional Threats

The carriers of external intentional threats may include:

- Random offender - a burglar, whose motive is primarily to obtain easily monetizable things.
- A burglar acting after preparation (may be a former employee), acting with the intent of revenge or theft.
- An organized crime group acting on demand or pursuing its own interests.
- Terrorists whose aim may be a terrorist act to create fear or blackmail.
- Opponents, acting to harm an organization (its image, reliability, etc.), with the motive being political, philosophical, religious, economic, or otherwise.
- Foreign intelligence services seeking to obtain classified information (state or commercial).
- Commercial intelligence services seeking to obtain commercial or production (technological) classified information.
- Public order disruption, street riots that may attack an object (protected area), may be plundering and / or damaging property (vandalism).
- On-site visits, contractors, cleaning and maintenance services (if supplied) that, in the absence of regime arrangements, can gain access to the protected interest and cause damage.

The motive may be:

- Political or political-military, arising from political, military or economic interests (e.g. intelligence services).
- Ideological, linked to a certain political and / or ideological system (e.g. political terrorism, extremism, religious fundamentalism, ecological radicalism, etc.).
- Economic, based on the pursuit of unprofitable profit (criminals and criminal groups), or harm to competitors (opponents, competitors, commercial intelligence services).
- Personal, evoked e.g. feelings of hatred, wrongdoing, blackmailing or threatening a family member, personality traits, etc.

Carriers of Internal Intentional Threats

Among the carriers of internal intentional threats may include:

- Own employees.
- Service staff and service personnel.
- Physical protection staff (guard) who can:
    - Alienate or harm the protected interest.
    - To assist the external offender to penetrate the protected area and gain access to the protected interest.
    - Deliberately revenge cause an accident in the facility and thereby cause a hazard.

The motive for such action may be:

- Feeling wrong (real or imagined).
- Character traits (profanity, covetousness, dependence, gambling tendency, difficult way of life, etc.).
- Revenge.
- Blackmail by another person, etc.

Unintentional Security Risks

Unintentional risks are those that can arise independently of the will of man. Their sources can be either outside or directly inside the protected area. External unintentional risks are those whose sources are outside the protected area and operate from outside. They are divided into environmental and technical security risks. Environmental external risks are in particular natural disasters (e.g. floods, fires, storms, floods, landslides) in which undesirable releases of accumulated energies or masses occur as a result of adverse effects of natural forces or destructive factors affecting assets in the protected area.

Unintentional technical external risks are in particular accidents (e.g. explosion, fire, leakage of chemical and radioactive substances) that result in the release of hazardous substances or the effects of destructive agents with a negative impact on assets. Unintentional security risks also include disasters (e.g. nuclear facility accidents, disruption of water works, major air / rail / road accidents involving the release of dangerous substances or fire) that result in an increase in destructive factors and their subsequent accumulation as a result of a natural disaster or accidents with a negative impact on assets (Simak, 2015).

Internal unintentional security risks are those whose sources are inside or are part of a protected area. We can divide them into social and technical. Internal unintentional social risks are mainly due to beneficiaries who, by virtue of lack of knowledge, forgetfulness or negligence, may themselves endanger the value of the assets. Internal unintentional technical risks are mainly intentional unintended accidents of technical, technological and energy equipment, that are part of the protected area, and which may cause damage to assets (e.g. explosion, fire, flooding of premises).

REQUIREMENTS FOR THE PROTECTION OF CRITICAL INFRASTRUCTURE OBJECTS

The subject of protection may take various forms depending on the issue. In the broadest sense, we can talk about the protection of tangible or intangible assets, which may be a thing or information that is owned or managed by a physical or non - legal person.

In most cases, the subject of protection (e.g. information on a carrier, information system, service, technology, object, structure, substance, etc.) is bound to a space having a clearly defined perimeter. Such an area in which the object of protection is located may be, for example, a territory, a site, a land, a zone, an object or a room in an object. (Zvakova, 2018; Stoller, 2017)

Requirements for the protection of objects from intentionally acting unauthorized persons with the aim of damaging, destroying or theft of protected tangible / intangible assets located in a given facility, which is owned or managed by a physical entity, legal entities are primarily determined by generally binding legal regulations, technical national / international standards / standards, requirements of insurance companies or other third parties such as parent companies or strategic customers.

Generally, binding regulations can be divided into two groups:

- The first set of regulations is of a proclamatory nature and stipulates the obligation of the employee, owner, resp. the lessee to protect his property which is in his possession or administration.
- The second set of regulations specifies specific requirements for the protection of buildings that, by their activity, have a significant impact on the functionality of the state or under certain conditions endanger the lives and health of the population (e.g. critical infrastructure elements, objects of particular importance for the storage and handling of classified information, SEVESO enterprises, financial institutions, etc.).

From the point of view of designing the object protection system, the planning and design phases are the most important. The fact that the design of protection systems must be primarily based on third party security requirements is also confirmed by the technical standard (EN 16763, 2017), where in the initial planning phase it places the identification of relevant generally binding legislation, regulations and standards. The European Standard (EN 16763, 2017) defines the basic life cycle framework of the object protection system in Figure 2. The basic framework in Figure 2 of the life cycle of the object protection system specifies the objective / purpose and the framework of the protection system based on risk identification and knowledge of border conditions. While these risks need to be managed.

RISK MANAGEMENT IN THE PROCESS OF PLANNING AND DESIGNING OF CRITICAL INFRASTRUCTURE PROTECTION SYSTEMS

In many cases, the establishment of a minimum level of protection is associated with a risk management process where the requirements for security guards are increasing in scope and / or are tightened with increasing risk (e.g. the security class of alarm systems is increasing).

If the risk management process also does not affect the resulting minimum level of protection (i.e. it is set in a directive), it has a significant impact in determining the dislocation of security measures elements (e.g. cameras, mechanical barriers, etc.).

Requirements for the process of assessing the risks related to the protection of objects against anthropogenic intentional threats are given by international and national generally binding regulations, standards and standards aimed at a certain field of application.

Risk Management According to ISO 31000

General principles and guidance on how to approach the risk management process are defined in ISO 31000 from 2018 Risk management - Guidelines (ISO 31000, 2018). According to this standard, risk management is an organized and coordinated set of activities and methods to guide and manage an organization in relation to risks that may affect its ability to achieve its objectives. The term risk management refers to an effective risk management architecture that includes:

- Risk Management Principles.

Figure 2. Life cycle of the object protection system (EN 16763, 2017)

Planning	Specification of the objective / purpose and framework of the protection system based on the identification of risks and knowledge of boundary (minimum) conditions.
Proposal	Selection and dislocation of elements of the protection system so that the resulting system meets the stated objectives and purpose.
Installing	Implementation of the proposal, namely the assembly, installation and connection of the relevant elements of system protection.
Commissioning and verification	Activating and testing of the protection system according to the prepared proposal.
Third Party Approval	The process of confirming that the ordered system meets the requirements of planning, design, installation and commissioning.
Operation	The process of transferring responsibility to the owner / operator of the protection system.
Maintenance	Combination of preventive and corrective activities throughout the system life cycle.

- Risk management structure (framework).
- Risk management process.

The standard can be used by any public, private or social organization, association, group or individual. Therefore, this standard is not specific to any species or industry, for convenience different users of the standard are referred to by the generic term organization.

The standard can be used during the existence of any public or private organization, in associations and individuals for a wide range of activities and processes related to strategy and decision making, operation (production, services), project preparation and, last but not least, property protection. It can be applied to any kind of risk and of any nature, regardless of whether it has positive or negative consequences. An example of how to deal with unacceptable risk at different levels of decision is shown in Table 1.

From the point of view of object protection systems designing, the risk assessment process can be used at various levels, namely:

- Setting a minimum level of protection for the entire building protection system.

Table 1. An example of how to deal with unacceptable risk at different levels of decision (Lovecek et al., 2011)

Risk	Level	Process owner	Purpose	How to deal with unacceptable risk
Security risks to the basic activities of the state.	Macro	Security analyst	Decision making process on the significance of potential risks, based on an evaluation of the potential likelihood of the risk and its impact on the main objectives of the state.	• Avoiding risk. • Accepting or increasing risk to seize the opportunity. • Removing the source of risk. • Change the risk level. • Share risk with another party or parties. • Maintaining risk.
Interruption of the organization's activities due to various security risks.	Micro	Security manager	Decision making process on the significance of a particular risk, based on an evaluation of the potential likelihood of the risk and its impact on the organization's processes (eg production).	• Changing the likelihood and consequences of risk by applying protective measures. • Risk-sharing with an insurance company.
Interruption of the supply of building materials to production due to theft from the warehouse C2 of object H8.	Micro	Designer	Decision making process on the specific location of protection measures, based on an assessment of the likelihood of possible attack scenarios.	• Location of CCTV cameras in the area of the organization.

- Determination of the minimum level of protection of the selected protection measure (e.g. in the case of determining the security degree of the Video Surveillance System (VSS), based on the requirements for its functionality and the parameters of individual components).
- Dislocation of protective measures systems and their elements (e.g. in the case of placing security cameras in the building).
- Determination of the risk of failure in the preparation and implementation of the protection system project in the given building (e.g. in case of risks assessment related to project management).

In the case of a requirement to consider a risk when setting the minimum level of protection, either in whole or in part, a methodological guideline is elaborated in most cases by the competent authority (e.g. National Security Authority or CEN / CENELEC). In the event of a decision to deploy individual elements of systems, in addition to the manufacturers' instructions and technical standards defining general instructions for the use of these systems in practice. The risk assessment process mentioned above, including the risk identification, analysis and evaluation, plays an important role. (Bologna, 2016)

In the case of securing an area with an alarm system, the technical standard (EN 50131-1, 2006) defines four levels of security only related to the anticipated capabilities and skills of the potential intruder. In this case, although a range of possible risk sizes is declared, the standard no longer specifies how to set a specific level and leaves it to the ad hoc assessment of the designer. The standard also specifies that a security assessment of the secured area should be carried out so that the required level can be determined, but a closer procedure is not specified (EN 50131-7, 2010). The Annexes to Standards B to D define which factors are to be taken into account when determining the level of the risk (EN 50131-7, 2010):

- The value of the property (e.g. type of property, value of property, quantity and size, history of theft, etc.).
- Construction areas (construction, openings, building operation modes, key mode, location, existing security, generally binding legal regulations, environmental safety).
- Effects on alarm systems and originating in the protected area (water pipes, ventilation, heating, air-conditioning systems, lifts, electromagnetic disturbance, external sounds, storage arrangements, wild or domestic animals, etc.).
- Impacts on alarm systems and originating outside the protected area (long-term factors - construction of road, energy infrastructure, short-term factors - temporary construction, weather effects, radio frequency interference, adjacent areas, etc.).

In the case of a VSS system, the standard for this type of alarm system (EN 62676-1, 2014) also defines four levels of security, but already in relation to the type of object, resp. the risk level. In this case, the determination of the level of collateral should be based on the level of the risk given by the possible consequences and the likelihood of a negative event defined in Figure 3.

Figure 3. Risks and security levels of CCTV (VSS) systems (EN 62676-1-1, 2014)

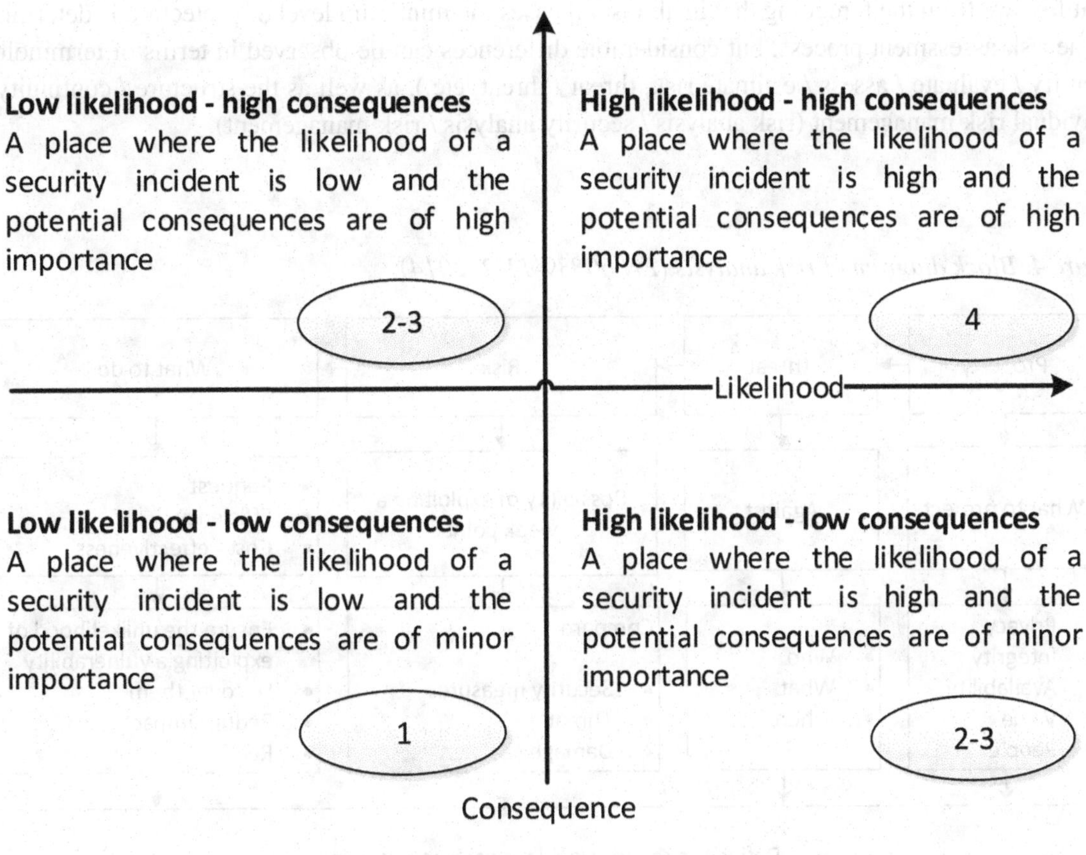

Low likelihood - high consequences
A place where the likelihood of a security incident is low and the potential consequences are of high importance

2-3

High likelihood - high consequences
A place where the likelihood of a security incident is high and the potential consequences are of high importance

4

Likelihood

Low likelihood - low consequences
A place where the likelihood of a security incident is low and the potential consequences are of minor importance

1

High likelihood - low consequences
A place where the likelihood of a security incident is high and the potential consequences are of minor importance

2-3

Consequence

According to the standard (EN 62676, 2014), a threat assessment and risk analysis should be carried out before the design of the VSS. Threats to the premises should be identified and their likelihood and impact assessed. A risk assessment should be carried out and the VSS should be designed to mitigate these risks. All risk-related processes should be implemented in accordance with ISO 31000 (2018), which defines the basic general principles for risk management, taking into account the following factors:

- The cost of possible losses, such as: What is the value of things in the locality? What is the impact of disrupting site processes?
- Location, e.g.: Is the site situated in a high-risk environment? Are there unfavourable climatic conditions?
- Settlement, e.g.: Is the site unpopulated for a long time? Are there security services in this area?.
- History of theft, robbery and threats.

In the case of securing an area with an access control system, the technical standard for these alarm systems (EN 60839-11-1, 2013) defines four levels of security in relation to the abilities and skills assumed by the intruder type. In this case, although the range of potential risks is again determined, the standard no longer specifies how to select a specific level of risk and leaves it to the ad hoc assessment of the designer. Although the standard specifying the guidelines for application to practice (EN 60839-11-2, 2014), a block diagram is given describing which factors should be taken into account (Figure 4).

It follows from the foregoing that in almost all cases the minimum level of protection is determined by the risk assessment process, but considerable differences can be observed in terms of terminology (identify / evaluate / assess / estimate risk, threat / threat, etc.), as well as the structure / continuity of individual risk management (risk analysis / security analysis / risk management).

Figure 4. Block diagram of risk analysis (EN 60839-11-2, 2014)

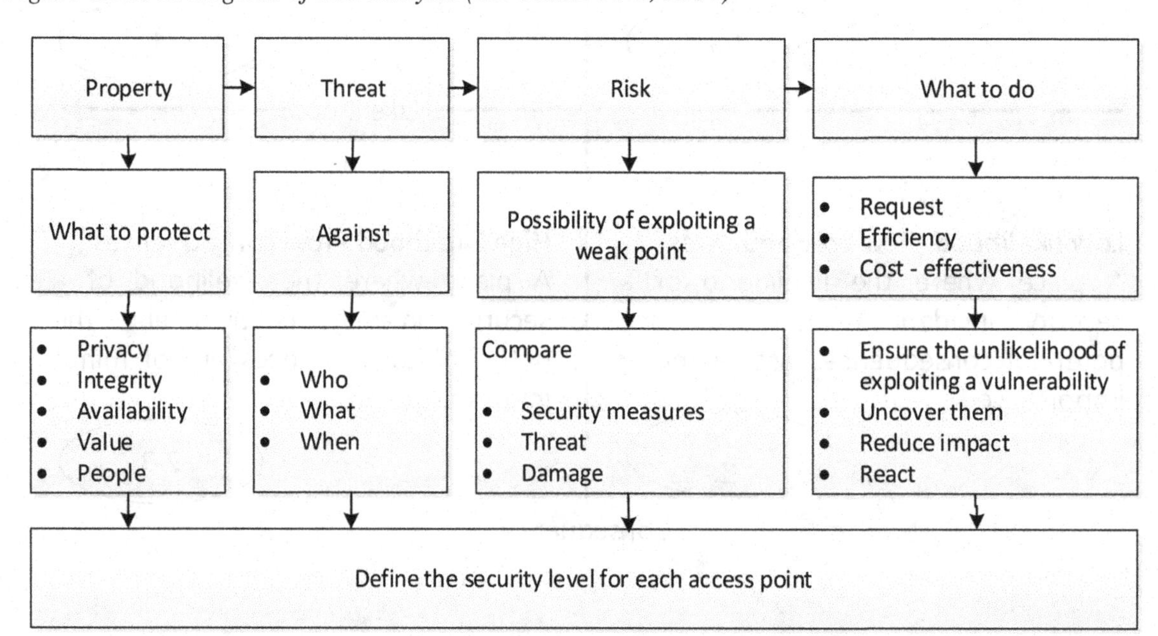

CRITICAL INFRASTRUCTURE PROTECTION SYSTEM

From the requirements of generally binding legal regulations, technical standards, insurance conditions, respectively the requirements of in-house facility protection regulations imply the need to take certain security or protective measures, which should be arranged to ensure the protection of the property of their owner or operator.

If property protection is a process of inducing a state of security using protective measures to prevent or stop any unwanted activities or events (e.g. electric short-circuiting and subsequent fire) that is contrary to the interests of the owner or manager of the property, then the protection system is a tool used to achieve this.

Security System

In the literature, the most commonly used terms such as security system, integrated security system or object protection system (Lovecek & Reitspis, 2011). According to Hofreiter (2003), the security system represents an integrated set of real elements that create a tool to ensure security in a given time and space. In terms of systemic approach, it can be considered as a synergistic system with target behaviour. According to (Tallo, Rak, & Turecek 2006), the basic components of an integrated security system are technical protection in the form of mechanical, electrical and electronic or other equipment, such as a security system capable of creating a solid obstacle to an attacker and an alarm system for reporting integrity breaches. (Mach & Boros, 2017)

In Anglo-Saxon countries, two terms are most commonly used to describe a property protection system. These are the Physical Protection System (Garcia, 2001) and the Security System. Already in the 1970s, the world's first to use the term Physical Protection System, in connection with the comprehensive protection of atomic and military equipment against deliberate anthropogenic threats (EN 50131-1, 2006).

A common requirement of the above definitions is to establish a system of protection, which as an effective way of organizing security measures, allows the intentionally acting unauthorized person to be prevented from achieving its aim, which may be theft, damage or destruction of the protected interest. (Collins, Petit, Buehring, Fisher, & Whitfield, 2011)

Such a protection system can be understood as a system implemented by technical and regime protection measures or elements that can be divided into

- Alarm systems:
  - Intrusion and hold-up systems (EN 50131-1, 2006).
  - Video surveillance systems (EN 62676-1-1, 2014).
  - Electronic access control systems (EN 60839-11-1, 2013).
  - Social alarm systems (EN 50134-1, 2002).
  - Alarm transmission systems and equipment (EN 50136-1, 2012).
- Mechanical barriers:
  - Perimeter protection, e.g. fencing, walls, barriers, burrow barriers, entrance and entry, gates, tourniquets, etc.
  - Sheath protection, e.g. various building structures and their openings. Building structures are classified according to: production technology (masonry, monolithic, prefabricated, etc.), material (concrete, reinforced concrete, steel, etc.), static and functional arrangement, etc.

Openings are most often divided into windows, doors, grilles, shutters, blinds, fitting of such panels (frames) locks, sills, visors, ventilation elements, foils, lintels, etc.
    - Subject protection, e.g. storage facilities - stable and mobile safes, cabinets, safety briefcases, bags, ATM, containers, cash registers etc.
- Security services.
- Regime measures.

Mechanical Barriers

Mechanical barriers are a set of physical barriers, the primary purpose of which is to act on the intruder in order to abandon his intention. If they do not abandon their intention, then their task is to create a sufficient time delay between the moment of attacking the guarded area and the achievement of its objective, make it difficult or virtually impossible for an intruder to enter a protected interest.

Alarm Systems

While alarm systems are used for subsequent detection and alarm condition. In some cases, alarm systems may also substitute for the role of mechanical barrier means. For example, a VSS system can act as a deterrent, resp. security fog devices (EN 50131-8, 2019) delay function.

Physical Protection And Security Services

An integral part of the protection system are elements of physical protection that ensure timely intervention and detention of the intruder. The regime protection ensures the correct functioning of the mentioned protective measures (Lovecek & Reitspis, 2011).

Depending on the direction and type of activity, security services can be divided according to whether they perform any of the organizational-tactical forms, such as operational-search activity or prevention. In addition, we know two spheres of security services, namely the fight against crime and the protection of public order (Berzi, 1996). All these security forces, armed forces and guards carry out some physical protection. However, in real life, anyone can exercise the physical protection of their property, regardless of their physical predispositions, unless it protects it commercially on the basis of an employment relationship. Physical protection as such can also be carried out by family members who, for example, guard old people or neighborhood patrols, respectively patrolling parents.

The basic terminology of security service providers is governed by a technical standard (EN 15602, 2008) where such services are understood to mean services provided by a security firm intended to protect persons, property and other assets. Within these services, the standard describes various positions such as Chief Security Officer, Authorized Security Officer, Officer Security Officer, Handler, Pedestrian Patrol, Head of Entry Control, Monitoring and Alarm Receiving Center Operator, Mobile Object / Area Patrol, Business Center Detective, static watch, armed or security officer. The field of airport and aviation security services is governed by a technical standard (EN 16082, 2011), which, in addition to basic terminology, describes the scope of training and staff selection. The area of maritime and port security services is covered by a technical standard (EN 16747, 2015).

Regime Measures

According to Belan (2015), regime protection constitutes a summary of administrative-organizational measures to security guard protected interests and values. It consists of a system of order and regime, its security and regular monitoring.

The object protection regime measures are procedures for the effective application of the object protection system, which determine:

- Conditions for the entry of persons and the entry of vehicles into the building and conditions for the exit of persons and the exit of vehicles.
- Conditions of movement of persons, means of transport in the building during working hours and outside working hours.
- Determining the conditions of use of mobile phones, camcorders, cameras, etc.
- Conditions for the protection of premises where significant assets are stored.
- Measures to protect meeting rooms.
- Terms of use, assignment, marking, storage and registration of originals and copies.
- Security keys.
- Terms of use, assignment, marking, custody and registration of code settings and passwords.
- Conditions of handling of mechanical barriers and technical security devices and conditions.
- Procedure in the event of an emergency, including a plan for protection and evacuation, together with the identification of responsible persons. If an emergency is imminent, or if an emergency has already occurred, the manager is authorized to allow persons providing or performing rescue operations to enter the facility; in such cases, measures must be taken to prevent unauthorized activities before, during and immediately after the rescue operation.

The design of object protection systems should be based on already existing regime measures (e.g. operational changes), and at the same time newly established / proposed regime measures must ensure the proper functioning of the proposed protection system. In many cases, security documentation (e.g. security policy, security directive, operating rules, rules for the performance of physical protection of the building, emergency plan, etc.) is used to implement the regime measures in practice.

ASSESSMENT OF THE CRITICAL INFRASTRUCTURE PROTECTION SYSTEM

Protection for the state of strategic facilities / elements, respectively objects are individually solved in different legislation but with different approach to their protection. These include, for example, objects of particular importance and other important objects, nuclear facilities, objects and premises for the storage and handling of classified information or objects of financial institutions. However, critical infrastructure may also include other elements. It means objects whose specific protection has remained unnoticed by law (e.g. line or nodal objects and elements of road, air, water and rail transport, chemical plants, energy suppliers, water structures, food businesses, industrial businesses, mobile operators, healthcare facilities, etc.). The responsibility for their protection to be with the public authorities, together with the owners and operators of the various elements of critical infrastructure.

Existing legislation in the EU treats the physical and object-based protection of critical infrastructure elements proclamatory and does not specify concrete solutions to it. While the Green Paper on a European Program for Critical Infrastructure Protection (Commission of The European Communities, 2005) lists possible ways (tools) to improve prevention, protection, preparedness and response in critical infrastructure protection under EU conditions, it does not further specify them. The question therefore remains how to ensure the protection of critical infrastructure elements in particular if the economic and financial coverage by insurance is secondary and essentially inapplicable as the threat to vital functions for the State is crucial and no legislation specifies in detail the manner and scope of implementing element protection system. (Rehak, Markuci, Hromada, & Barcova, 2016)

There are three basic approaches used in common to design and assess the degree of securing the strategic state objects protection system:

- A directive approach where the entity must adopt a precisely specified protection system, regardless of the specificity of its operation and the environment in which it is located.
- Alternative approach, where the entity can choose from a finite number of alternative solutions, combining various technical security means and organizational, respectively regime measures.
- Variable approach where the entity must take such measures within the protection system that take into account the breakthrough resilience of the mechanical barriers, the response times of the intervention unit and the probability of detection of the alarm systems.

The most effective approach is currently considered to be a variable approach, based on the assumption that so much technical security is needed to ensure that the intruder is detected and detained by the intervention unit before reaching its target. However, a methodology, standard, or software tool has not yet been developed in the EU to define in detail and allow a variable quantitative approach to protection to be applied. Existing software tools that were designed primarily for the protection of nuclear material and equipment in the US, Russia, or South Korea offer a certain starting point. (Siser, Maris, Rehak, & Pellowski, 2018)

In general, we can discuss of a qualitative and quantitative approach. Qualitative approaches to design, respectively to assess the object protection system are based on expert estimation, where it is not possible to prove the efficiency, reliability, respectively efficiency of these systems. It is necessary to rely on the professional competence of the authors of technical standards, generally binding legislation, resp. software applications (e.g., RISKWATCH - Campus Security, RISKWATCH - Nuclear Power, RISKWATCH - Phys. & Homeland Security, RISKWATCH - NERC, from Risk Watch International develops, USA).

Quantitative approaches are based on mathematical and statistical methods, which enable to accurately prove the efficiency, reliability, resp. efficiency of the security system. Software tools using these methods include SAVI / ASSESS (Sandia National Laboratories, USA), Sprut (ISTA, Russia), Vega-2 (Eleron, Russia), SFZ Analyzer (FRTK MFTI, Russia), SAPE (Korea Institute) of Nuclear Non-Proliferation and Control, South Korea) or SatANO (University of Žilina, Slovakia).

According to technical definitions, efficiency is a dimensionless number that expresses how close to an ideal process the process is in the system or equipment under assessment. The ideal process is 100% efficient. (Lovecek & Reitspis, 2011) The effectiveness of the object protection system can then analogously express how close the real processes in the object protection system are to the ideal processes, where ideal process shall, to an acceptable extent, eliminate the risks identified and for which the protection system has been designed. However, to express the effectiveness of the protection system, it is necessary to use specific output quantities that are ad hoc for this purpose and their definition is unambiguous. In our case, for example, it is a coefficient of protective measures, which gives the ratio of breakthrough resistance of mechanical barriers and protection times of the intervention unit. (Lovecek & Reitspis, 2011)

From an economic point of view, the effectiveness of the system can be defined as the effectiveness of the funds invested in the system and assessed in terms of its results. The economic effectiveness of the object protection system can be defined as a relationship, which using economic indicators, expresses the dependence between the economic benefits of the system of reducing economic losses due to crime and the economic costs of its design. (Kampova & Makka, 2018)

The technical reliability of a system is characterized by its complex property, expressing the general ability to maintain functional properties at a given time and under specified conditions.

Reliability is an indicator often expressed as the probability that a system (e.g. electrical alarm system, VSS) or an element (e.g. detector, control panel, communicator) will perform the required function for a specified period of time and under predetermined conditions. In practice, reliability is reported as the number of failures per unit of time during the reporting period.

The reliability of the alarm system, in particular the VSS, can be determined using the Mean Time Between Failures (MTBF) parameter, given in hours. This parameter indicates, respectively should be reported by the camera manufacturer on the basis of accurately recorded statistics of the faults. (Lovecek, Velas, Kampova, Maris, & Mozer, 2013)

In many cases, the reliability of alarm systems also depends on a human factor (e.g. an operator of an urban surveillance system). The most important types of human errors and their causes are:

- Errors caused by lack of concentration or distraction by other stimuli.
- Errors due to lack of training, errors due to lack of physical or mental abilities.
- Errors due to lack of motivation.
- Errors due to non-compliance with work procedures.
- Errors caused by poor management, preparation, training, etc. preparation and use of plans.
- Mistakes and shortcomings in training, lack of experience. (Lovecek & Reitspis, 2011)

The quality of a protection system is a collection of features of the whole system that make it capable of meeting the legitimate and anticipated needs of a particular entity (e.g. owner, operator, administrator), thereby ensuring security in a given environment, time and for a specified purpose.

Finding an optimal security solution means finding a security solution that is technically efficient, reliable, economically efficient and at the same time meets the essential requirements of a functional building protection system.

In order to quantify the level of protection of any protected area / object, resp. of the protected interest / CI element, it is necessary to design its formalized description using a mathematical model. Intrusion of protected area takes place as a process in which its individual objective and subjective elements (protected area, protected interest, mechanical barriers, alarm systems, intruder, security guards, as well as accidental influences such as weather, random participants, witnesses) whose relationships and properties can be characterized by measurable physical quantities and mathematical relationships. Each object protection system model must have its input and output variables / parameters. Using the output parameters of the models it is possible to quantify the level of the object protection from the technical and economic point of view, depending on the input parameters, which take into account:

- Intention of the intruder (attack to damage or destroy an object of protection, a CI element, or an attack to steal it).
- Nature (e.g. attractiveness, structure, mobility) and value of the protected interest.
- Deciding an intruder under conditions of certainty, uncertainty or uncertainty.
- Breakthrough resistances of mechanical barrier systems, expressing the degree of ability to resist over time, varying intensity (depending on the type of tool used) and how the attack was performed (destructive / non-destructive).
- Method of detection and assessment of intrusion of protected area.
- Possibility to divide protected space into zones.
- Dislocation of protected area.
- Investment and operating costs in relation to expected losses, resp. yields.

Among the input variables we can include breakthrough resistance of individual mechanical barriers (T_R), tool, respectively equipment parameters intended to overcome mechanical barrier systems (equipment performance and consumption, necessary instrument control skills, power supply options, noise and mobility, availability / registration and anonymity of the owner, unambiguous identification of the instrument when used). Also total response time of the intervention unit (T_{IU}), Intruder Transfer Times (T_{TRAN}), Intervention Unit Transfer Times (T_{tran}), Intruder Attack Time (T_{AT}), Intruder Break Time (T_{ESC}), Intervention Time (T_{INT}), Alarm Time (T_{ALM}) Attack Verification Time (T_{VER}). Probability of correct detection by an alarm system (e.g. PIR detector) in the i-th detection zone during intruder path (P_{Di}), probability of failure free of the alarm system (P_{FFA}), probability of failure free of transmission of the alarm signal through the alarm transmission path to the remote alarm receiving center (P_{FFT}), reliable human factor (P_{HF}), probability of timely and correct evaluation of the alarm state (P_{EVA}), detection characteristics of alarm systems (A_{DET}), investment, respectively the operating costs of the protection system (C) and, last but not least, the value of the protected interest (V).

Due to the use of the model, among its output variables (parameters) we include the coefficient of effectiveness of protective measures (Q_{PRO}), the total duration of attack by the intruder, from the moment of detection at t_{DET} by alarm elements to its exit from the protected area (T_{EXT}). Total mechanical barrier system breakthrough resistance time (T_{RES}), minimum mechanical barrier system breakthrough resistance time (T_{RESmin}). Intruder elimination probability (P_{IE}), cumulative intruder detection (P_{CDET}) probability, intervention unit (P_{INT}) hit probability, critical the value of the maximum interval between two guards (ΔT_{BTW}), average and discounted annual costs (AAC / DAC), total annual operating costs (T_{AC}), expected losses due to the negative consequences of the occurrence of a risk phenomenon (LOSS) or the payback period financial system.

An effective system of protection of an object is considered to be a system which fulfils the basic condition that the attack time of the T_{EXT} or the total breakthrough time of the mechanical barrier means T_{RES} is greater than the reaction time of the T_{IU} intervention unit, ie. $T_{EXT} > T_{IU}$ or $T_{RES} > T_{IU}$ ($T_{EXT} / T_{IU} > 1$ and $T_{RES} / T_{IU} > 1$ or, the system is effective if the ratio of times is greater than one). Meeting this condition may not always be sufficient. However, to assess the effectiveness of the protection system is a prerequisite.

In the first case of this condition, it is sufficient to detain the intruder during the time of the attack T_{EXT}, which also considers the time of the attack T_{AT} and the time of his escape T_{ESC}. This increases the disposition time of the intervention unit T_{IU} and thus increases the likelihood of detention. In the latter case, the intruder must be detained before the breakthrough time of the mechanical barrier systems T_{RES}. This is the case when an attack with the aim of industrial espionage, sabotage, or a terrorist attack (eg an attack on a CI element) is envisaged.

Explanation of some variables:

- T_{EXT} is the total invasion time from the moment of detection at the time t_{DET} by the alarm system until it leaves the protected area.
- T_{RES} is the total time required to break mechanical restraint means, i.e. the time required to approach an intruder to the CI element at a distance that could immediately endanger it. This T_{RES} time consists of the mechanical barrier systems breakthrough time of the intruders T_R and the total time it takes to move the T_{TRAN} to the CI element from the moment it was detected by the alarm system at t_{DET}.
- T_{IU} is the total response time of the intervention unit, consisting of:
  - T_{ALM} alarm time, i.e. the time elapsed from the moment of detection at t_{DET} to the alarm condition. The T_{ALM} alarm time includes:
    - The time that is preset for the eventual release of the system from the armed state to the non-armed state.
    - The time it takes for the alarm system to be armed once the power supply has reached its nominal value after a power failure.
    - Time required to change the alarm system from arming to alarm.
    - The transmission time required to transmit a signal / message to the guarded station in the protected area, the transmission time is the time from when the alarm system status change occurs on its communicator interface and when the status change occurs on the alarm interface status and indicating-display device at the Monitoring Alarm Receiving Center (ARC).

In the event that an intruder attempts to trigger an alarm condition of the alarm system (applies to the Intruder Alarm System), the message reporting time shall be added to the time T_{ALM}. The time interval from the moment the fault occurred in the alarm transmission system and the transmission of this status information to the ARC.

- T_{VER} verification time, i.e. the time required to assess whether it is indeed an intruder attack or a false alarm caused by an unintentional fault or failure of the alarm system, i.e. the time needed to decide to take action.
- Time to move to T_{TRAN} hit point, i.e. the time needed to move the intervention unit to the place of intervention against the intruder.
- Time of intervention against the intruder T_{INT}, i.e. the time required for effective intervention against the intruder, deterrence, detention or destruction of the intruder, or the security of the CI element, in general, those activities of the intervention unit which prevent the intruder from achieving its objective.

In the case of intruder detection by the guards, it is necessary to determine the time interval between two searches performed by the guard ΔT_{BTW} to calculate the T_{IU} response time. This is the time interval in which the guards are carried out and the integrity of the mechanical barriers is checked.

In order to fulfil the basic condition $T_{EXT} > T_{IU}$, resp. $T_{RES} > T_{FO}$, so the maximum interval between two inspections ΔT_{BTW} must not exceed the critical value $T_{max.}$

T_{AT} is the time of attack, i.e. the time it takes the intruder to reach his goal with which the attack has taken place (e.g. stealing, damaging, or destroying a protected interest).

T_{ESC} is the time of escape, i.e. the time the intruder needs to leave the protected area.

It is important to note that individual times are counted from the time of detection of the t_{DET} intruder, which has several reasons. The first reason is that at some point in time, the intervention unit is also activated. The second plea is based on the very essence of the system. Where there is no detection, there can be no intervention and where there is no intervention, it is unnecessary to take into account, respectively invest in the mechanical barriers protection. Each such device has its limited breakthrough resistance, the overcoming of which is only a matter of time. If the intruder is not, or has not yet been detected by the alarm system, it is irrelevant to account for its resistance to the total time of T_{EXT} or T_{RES}. For example, a number of objects have boundaries of land bounded by a fence, which serve as a definition of private property and not as a mechanical barrier to protect the property in the building. The time t_{DET} also plays an important role in the calculation of the T_{RES} time, in particular when counting the T_{R1} time, i.e. the breakthrough time of the first barrier on the path to the protected interest. Intruder detection can occur just at or in front of the detection zone boundary, during transfer between zones, or up to intrusion entering the detection zone. In the first case, the T_{R1} time value does not change within the total T_{RES} time. In the second case, it is necessary to multiply that time by coefficient (in the case of assessing the variance of the random variable T_{R1}, multiply it by 1/4) and in the third case the T_{R1} time is 0 (the breakthrough resistance of the first mechanical barrier is not included in the total T_{RES}).

The relationship between the time of the T_{EXT}, or T_{RES} and T_{IU} time can be therefore considered as a basic assessment criterion of the effectiveness of the object protection system. In order to establish the correlation between them, it is necessary to introduce parameters / coefficients that better reflect the efficiency of the system. An important parameter that describes the effectiveness of a given system is the coefficient of effectiveness of Q_{PRO} protective measures. In the case of intruder detection by alarm systems, the coefficient of effectiveness of Q_{PRO} protective measures can be defined by the following formulas (1), (2)

$$Q_{PRO} = \frac{T_{EXT}}{T_{IU}} = \frac{T_R + T_{TRAN} + T_{AT} + T_{ESC}}{T_{ALM} + T_{VER} + T_{tran} + T_{INT}} \; for \; T_{EXT} > T_{IU} \quad (1)$$

$$Q_{PRO} = \frac{T_{RES}}{T_{IU}} = \frac{T_R + T_{TRAN}}{T_{ALM} + T_{VER} + T_{tran} + T_{INT}} \; for \; T_{RES} > T_{IU} \quad (2)$$

In the case of intruder detection by the guard, the coefficient of effectiveness of protective measures Q_{PRO} can be defined by the following formulas (3), (4).

$$Q_{PRO} = \frac{T_{EXT}}{T_{IU}} = \frac{T_R + T_{TRAN} + T_{AT} + T_{BR}}{T_{BTW} + T_{tran} + T_{INT}} \; for \; T_{EXT} > T_{IU} \tag{3}$$

$$Q_{PRO} = \frac{T_{RES}}{T_{IU}} = \frac{T_R + T_{TRAN}}{T_{IU} + T_{tran} + T_{INT}} \; for \; T_{RES} > T_{IU} \tag{4}$$

It follows from the above that if Q_{PRO} is less than 1, the protective measures are inadequate, making the whole system ineffective. Conversely, the Q_{PRO} is greater than 1, the effectiveness of the protective measures, and hence the overall system is greater. It is stated in the scientific literature that this coefficient of protective measures effectiveness should be between 6 and 12.

Another important parameter that describes the technical effectiveness of the object protection system is the probability of eliminating the intruder P_I. This parameter defines how likely the intruder will be detained, or otherwise, eliminated during his route to the protected interest. This parameter, unlike the previous parameters, also takes into account the probability of intruder detection, the probability of the intervention unit's successful response, and the stochastic effects.

The probability of P_I is based on the basic assessment criterion of the effectiveness of the object protection system. And that the T_{EXT} attack time, or the total breakthrough time of the T_{PRL} mechanical barrier means, must be greater than the response time of the T_{FO} intervention unit, ie. $T_{EXT} > T_{FO}$ or $T_{RES} > T_{FO}$, respectively. This criterion implies the conditions set out in the following relationship. Since both will be analogous, we will only consider the $T_{RES} > T_{IU}$ criterion.

$$T_{EXT} - T_{IU} > 0 \; or \; T_{RES} - T_{IU} > 0$$

The T_{IU} and T_{RES} times are independent random variables having a normal distribution with the parameters μ, σ^2. This creates a new random variable X, which will have the same distribution as the times indicated. The following relations apply to the new random variable X (5)

$$X = T_{RES} - T_{IU}$$

$$\mu = E\left(T_{RES} - T_{IU}\right) = E\left(T_{RES}\right) - E\left(T_{IU}\right)$$

$$\sigma^2{}_X = Var\left(T_{RES} - T_{FO}\right) = Var\left(T_{RES}\right) + Var\left(T_{IU}\right) \tag{5}$$

where: μ – mean value of random variable X,

σ – dispersion (variation) of a random variable X,

T_{RES} – the total breakthrough time of the mechanical barrier systems [s],

T_{IU} – total response time of the intervention unit [s].

As already indicated, the probability of eliminating an intruder P_I is based on the probability of detecting an intruder and the probability of a successful response by the intervention unit (6).

$$P_{IE} = P_D * P_{FFA} * P_{FFT} * P_{R|D} \tag{6}$$

where: P_{IE} – probability of eliminating the intruder,

P_D – probability of detection by the alarm system,

P_{FFA} – the probability of failure free of the alarm system,

P_{FFT} – the probability of transmitting the alarm signal via the alarm transmission path to the remote alarm receiving center,

$P_{R|D}$ – the probability of the intervention unit's successful response.

For a positive assessment of the intervention unit's success, it is necessary that the random variable X be greater than 0 (X> 0). Based on this statement, we can calculate the probability of a successful response of the intervention unit (7).

$$P_{R|A} = P(X > 0) = \int_0^{\infty} \frac{1}{\sqrt{2\pi\sigma^2_X}} * e^{\frac{-(X-\mu_X)^2}{2\sigma^2_X}} dx \tag{7}$$

where: $P_{R|D}$ – the probability of the intervention unit's successful response,

μ – mean value of random variable X,

σ – dispersion (variation) of a random variable X.

The relationship $P_1 = P_D * P_{FFA} * P_{FFT} * P_{R|D}$ describes the probability of elimination an intruder for only one detection by the alarm system. However, as mentioned above, an intruder may cross several detection zones during his route to the protected interest. If an intruder crosses two detection zones during his / her route, the probability of eliminating the intruder can be calculated by relation (8).

$$P_1 = P_{D1} * P_{FFA1} * P_{FFT1} * P_{R|D1} + (1 - P_{D1}) * P_{D2} * P_{FFA2} * P_{FFT2} * P_{R|D2} \tag{8}$$

Detection can occur with some probability on the first detector and with some probability on the second, third to n-th detector. If the first detector is detected and the response unit responds adequately, it will no longer be necessary to respond to the alarm / message from the second detector. Although the probability of detection by the alarm system is relatively high ($P_D \cong 1$), there is still the possibility that the first detector will not respond. In this case, additional detectors are available in a sequence that will depend on the intruder's specific path (cumulative probability arises). For this reason, the variable $(1\text{-}P_{D1})$ is introduced in the second part of the previous relationship. In general, the probability of eliminating an intruder can be calculated by relation (9).

$$P_{IE} = P_{D1} * P_{FFA1} * P_{FFT1} * P_{R|D1} + \sum_{i=2}^{n} \left(P_{Di} * P_{FFAi} * P_{FFTi} * P_{R|Di} * \prod_{j=1}^{i-1} (1 - P_{Di}) \right) \tag{9}$$

Alarm systems, regardless of the technology used, also serve to detect unauthorized intrusion of protected area. The intruder, from overcoming the perimeter of the protected area, passes through individual zones (e.g. rooms, premises, objects) to the protected interest. In the case where a given zone is both a detection zone and an active zone, there is a probability P_i that the intruder will be detected and an alarm condition will be triggered at the same time. This is the state of the alarm system or its components, which is the result of the system's response to the presence of a risk.

A detection zone is one where an intruder is detected with a certain probability (P_{Di}) when an unauthorized entry or passage occurs. The extreme detection zone is the zone in which intruder detection occurs first at the t_{DET} time. The active zone can be considered as a zone where all the implemented technical security means fully perform their function at a given time. Otherwise, the zone is temporarily inactive. The temporarily inactive zone status may occur, for example, because of the particular mode of operation being performed. It is therefore necessary to distinguish whether the protected area has different operating modes (e.g. outside working hours, working hours, extra operating modes). If so, the effectiveness of the protection system should be assessed from the point of view of each such scheme.

From overcoming the perimeter of the protected area to reaching the protected interest, the intruder often crosses several detection zones, with different probabilities of P_{Di} detection. In this case of multiple detection, we are talking about the cumulative probability of intruder detection P_{CDET}, which represents the overall probability of intruder detection before reaching its target. This quantity can be calculated by relation (10).

$$P_{\text{CDET}} = \left[1-\prod_{i=1}^{n}\left(1-P_{Di}\right)\right] * P_{FFT} * P_{FFA} * P_{HF} \tag{10}$$

where: P_{CDET} – cumulative probability of intruder detection,

n – number of detection zones during the intruder's route,

P_{Di} – probability of correct detection by an alarm element (e.g. PIR detector) in the i-detection zone during the intruder's route,

P_{FFA} – the probability of the alarm system or alarm transmission system failure,

P_{FFT} – the probability of transmitting the alarm signal via the alarm transmission path to the remote monitoring alarm receiving center,

P_{HF} – probability of human factor reliability.

Using the probability of P_{CDET} it is then possible to deduce the probability of the intervention by the intervention unit P_{INT} (11) (This probability does not take into account the success of the intervention).

$$P_{INT} = P_{CDET} * P_{EVA} \tag{11}$$

where: P_{INT} – the probability of intervention of the intervention unit,

P_{CDET} – cumulative probability of intruder detection,

P_{EVA} – the probability of timely and correct evaluation of the alarm condition.

The last parameter we can use to assess the effectiveness of an object protection system is the Critical Detection Point (CDP). The CDP (Figure 5) reflects the feature of the system that for the system to be effective, it must be detected at the latest at this point / element / moment, or on a previous alarm system.

The position of the point / alarm element k-1 is determined by the fact that the condition that the intruder detention time of moving and overcoming the mechanical barrier systems (T_{RESmin}) from this point to the protected interest (point n) has just exceeded the T_{IU} (14).

Figure 5. Graphical representation of the probability of elimination of the intruder P_1 (Lovecek at al., 2011)

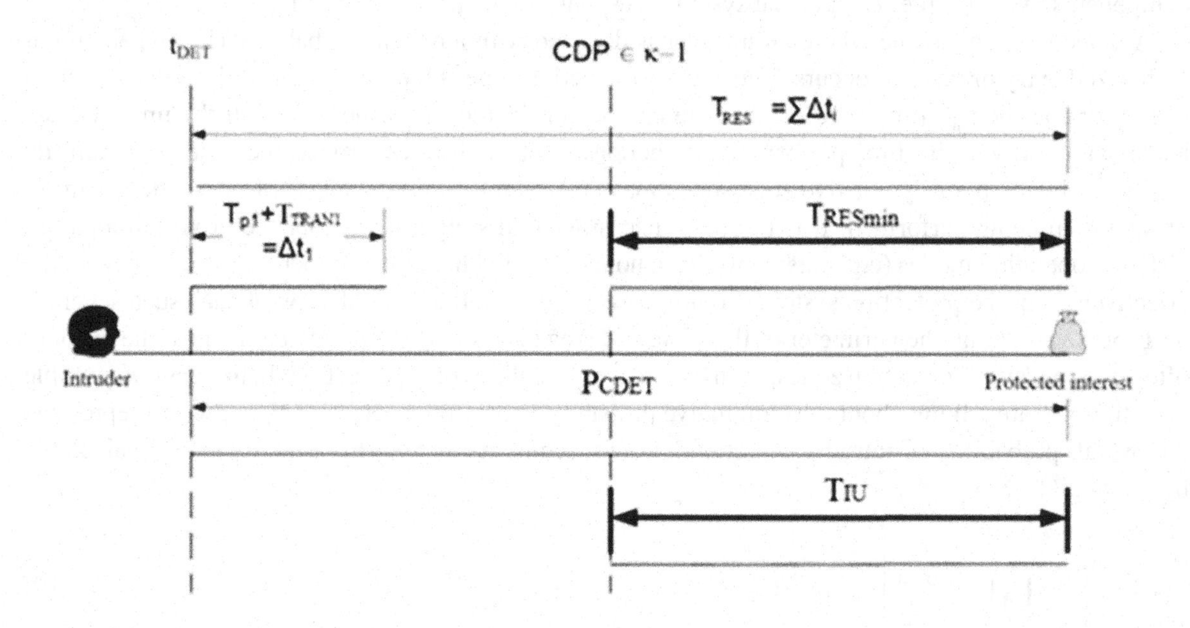

$$T_{RES_{min}} = \sum_{j=k}^{n} \Delta t_i > T_{IU} \tag{14}$$

where: T_{RESmin} – minimum total breakthrough time of mechanical barrier systems restraints [s],

Δt_i – the sum of all breakthrough times of the mechanical barrier systems and the times of transfer of the intruder on the route to the protected interest, which satisfy the condition that their sum is just greater than the response time of the intervention unit [s],

T_{IU} – total response time of the intervention unit [s],

n – total number of mechanical barrier systems on the route [number],

k – mechanical barrier system, before the intruder must be detected for the system to be effective (in terms of breakthrough system resistance and response time).

Parameters have been defined above to assess the technical effectiveness of the object protection system. In order to determine the level of protection of any protected area, it is necessary to create a formalized description, for example using a mathematical model.

Mathematical models can be deterministic or stochastic. Deterministic models have individual input parameters determined unambiguously, while stochastic models have input parameters determined using probabilistic value distribution. Normally the normal probability distribution is used. The deterministic and stochastic approach to modelling is also evident in the compilation of intruder routes. Between the unprotected surroundings of the facility and the intruder's target, the quantitative assessment of the efficiencies shall identify the routes that the intruder may use to overcome the object's protection system. Since this route determination is performed in an algorithmized form, the term route generation is also used. The deterministic approach is manifested by the fact that all possible routes determined by the surroundings and the destination are generated. The set of generated routes is thus clearly determined. The stochastic approach uses a random number generator to generate specific routes, each element being characterized by a certain probability of the intruder selecting that element during the process. The deterministic and stochastic approach can thus be applied to input parameters, but also to the way of route generation (Lovecek, Velas, Kampova, Maris, & Mozer, 2013).

Thus, the mathematical model represents a system of mathematical relations that clearly describe the phenomenon or process under investigation. It is a wide area of mathematical expression, which includes various algebraic equations, derivatives and their systems, as well as various relationships from set theory, algebra, probability theory, mathematical logic and other areas of mathematics. However, these equations and relationships become mathematical models only when they are clearly associated with a physical process or phenomenon. The process of disruption of protected space takes place between real objects (guard service, protected interest, protected space, mechanical barrier systems, alarm systems, intruder) whose relationships and properties can be characterized by measurable physical quantities (e.g. time, speed, pressure, weight, torque, sound, light). To describe the object protection system using a mathematical model, we need to define its input variables / parameters (T_R, T_{TRAN}, T_{tran}, T_{AT}, T_{ESC}, T_{INT}, T_{ALM}, T_{VER}, P_{Di}, P_{FFA}, etc.), output variables / parameters (e.g., T_{RES}, T_{EXT}, T_{IU}, Q_{PRO}, P_{IE}, P_{CDET}) and their functional dependencies.

In the future, it can be expected that there will be a requirement to establish a standard that not only obliges operators to take certain protective measures, but also sets out a specific procedure to achieve a certain minimum level of CI element protection. Approaches in other areas of social life (e.g. protection of nuclear facilities, classified information, banking entities), where mentioned, two basic approaches to assessing the level of object protection system (quantitative and qualitative approach) may be inspiration.

In practice, more use is made of quality-based methods using mainly expert estimates, although quantitative-based methods are more objective, accurate, but mainly verifiable. There are several reasons for this:

- There is a lack of scientific and technical debate on the subject in European fora.
- Tool catalogues / databases are missing that are useful for overcoming mechanical barrier systems.
- The catalogues / databases of mechanical barrier systems with their breakthrough resistance characteristics are missing.
- Missing alarm catalogues / databases that contain data on the probability of intrusion detection.
- There are no experimental measurement laboratories, techniques and methodologies in Europe

SECURITY MEASURES TO ENSURE TECHNICAL SECURITY OF CRITICAL INFRASTRUCTURE ELEMENTS

Technical Standard (EN 50136-1, 2012) defines an Alarm System as an electrical installation that responds to the manual or automatic detection of the risk presence. A similar definition of an alarm system can also be found in the Alarm Systems Technical Standard (CLC/TS 50398, 2009), by the European Committee for Electrotechnical Standardization - CENELEC, where the alarm system is an electrical device responding to manual initiation or automatic risk detection. However, this standard introduces the term Alarm Application, which refers to an application intended to protect life, property or the environment. Such an application could be, for example:

- Hold up and alarm system.
- The system of summoning assistance.
- Lift alarm system.
- Environmental alarm system.
- Closed circuit or camera security system.
- Access control system.
- Fire Detection and Fire Alarm System.

In the case of the definition of an alarm application, the purpose of the alarm device is understood in a broader context than the mere detection of the risk presence. An example is CCTV or VSS, the purpose of which (EN 62676-1-1, 2014), in addition to detection, may also be monitoring, observation, recognition, identification, resp. investigation.

If the standards (EN 50136-1, 2012) and (CLC/TS 50398, 2009) under the alarm system understand a device responding to manual input or automatic detection, the standard (EN 50130-4, 2011) distinguishes up to three types of alarm systems:

- Intruder Alarm System, an alarm system designed to detect and signal the presence, intrusion or attempt of intruder intrusion into protected areas.
- Hold-up Alarm System, a system that provides the user with the option of deliberately triggering an alarm condition.
- The Social Alarm System is a system that provides a means to call for assistance and is intended for persons who may be considered living in danger.

Electrical Alarm and Emergency Alarm Systems

Intruder Alarm System (IAS) is an alarm system for detecting and indicating the presence, entry or attempt of an intruder to enter a protected area. The Hold-up Alarm System (HAS) is an alarm system that provides the user with the means to deliberately generate an emergency alarm. The Intrusion and Hold-up Alarm System (IHAS) is a combined electrical alarm and emergency alarm system. The Intrusion and Hold-up Alarm System can consist of multiple subsystems. A subsystem is a part of an IHAS located in a clearly defined part of a protected area capable of operating independently (EN 50131-1, 2006).

Alarm Transmission System

Alarm Transmission System (ATS) is a system used to transmit information relating to the status of one or more IAS or HAS to one or more alarm receiving centres. The alarm transmission system shall comprise:

- Alarm transmission path.
- Alarm transmission equipment located in the monitoring and alarm receiving centre.
- Alarm transmission equipment located in a protected object (EN 50136-1, 2012).

Video Surveillance Systems

VSS or Video Surveillance System is a system consisting of camera equipment, monitoring and associated equipment for transmission and control purposes that may be necessary for the surveillance of the protected area (EN 62676-1-1, 2014).

CCTV (VSS) systems and their elements are graded according to required security level. The levels of protection shall take into account the level of risk, which depends on the likelihood of the incident and the potential harm caused by it. There are four levels of security known as low risk, low to medium risk, medium to high risk, and high risk (as with IHAS).

Fire Detection and Fire Alarm System

The Fire Detection and Fire Alarm System (FDFAS) is a family of components including a control panel that is capable of detecting and indicating fire and signalling for the relevant operations (EN 54-1, 2011; EN 54-2, 1997).

The Control and Indicating Equipment is a component of FDFAS through which other components can be powered and used for (EN 54-1, 2011):

- For receiving signals from connected fire alarms or pushbuttons.
- To identify the status of received signals (e.g. alarm condition, fault condition).
- For acoustic and optical indication of each alarm condition, fault condition.
- To identify the hazard location.
- Transmission of system status messages to the receiving centre.

The control panel is able to transmit a fire alarm signal (EN 54-1, 2011):

- For acoustic or optical fire alarms or voice alarms.
- For a fire alarm or fire protection unit.
- For automatic fire extinguishers.

Access Control Systems

Access Control Systems (ACS) is a system containing all design and organizational measures related to the equipment required for access control (EN 60839-11-1, 2013). ACS is used to:

- Decision:
  - Who has access.
  - Where access can be obtained.
  - When access is obtained.
- Minimizing the risk of unauthorized entry.

An Access Point is a place where access can be controlled using the Access Point Actuator and Sensor (APAS) controls and sensors. The controls include: electronic locks, tourniquets, barriers. Access point sensors include: switches, pressure signalling devices and door switches.

Security Locks are specially designed security door cabinet. An electric lock and an additional mechanical lock on the operator's side may supplement the security lock.

An Access Point Reader is a device for retrieving recognition data / information from an identifying element, biometry, or memory.

The identification element / token is a card, key, label that contains identification data (EN 60839-11-1, 2013), while biometrics is information relating to the unique physiological characteristics of the user.

An Access Control Unit is a device that decides to release one or more access points and controls the sequence of related devices. (Velas, Zvakova, & Svetlik, 2016)

In systems with a distributed database, a copy of the user identity and access privileges to each part of the object is stored in all access control units that control the APAS through the access point interface.

CONCLUSION

Object protection can be perceived as part of a direct and situational strategy of prevention against threats of various nature. The essence of the protection of objects and premises is the planning and implementation of measures that reduce the likelihood of security threats by changing the conditions of those assumptions that enable their activation.

This is the implementation of measures that prevent the occurrence of security threats (security incidents), they affect the amount of costs and profits of the potential offender / attacker and increase the risk of detecting and detaining the offender / attacker.

If we consider protection to be an activity aimed at creating a secured environment for objects and premises, then such activity will be aimed in particular at collection and evaluation of information on the security situation in the security environment of protected objects. In addition, it will be aimed at taking measures to prevent threats, creation of a flexible and effective system of objects and premises (protected interests) and at protection and preparation of forces and means for resolving security incidents and crises arising in the objects protection.

Such an understanding of security also requires a change in approaches to ensure the protection of objects and premises. The point is that it is no longer sufficient to respond to security risks and threats and eliminate their negative consequences.

Nowadays, when many security risks and threats do not seem so obvious, the probability of early detection of potential sources of threats is given by the protection entity's ability to apply a systemic approach, to understand the relationship between part and total, cause and effect. Responding effectively to security challenges implies the ability to perceive them, correctly assess and respond adequately. Deciding on effective protection of an object (space) means to recognize the structure of the protected object (space) as well as the structure and factors of the relevant external and internal security environment, to assess the state and level of safety and security situation in the external and internal environment of the building, to identify and evaluate those factors that have an impact (positive or negative) on the security of the facility and assess the possibility of their consequences for the protection subject and to evaluate and correctly interpret available facts that may indicate the evolutionary tendencies of the object's security in the given environment or the evolutionary tendencies of the security environment itself.

Accordingly, the more secure the object will be, the higher the capabilities and capabilities of the subject will be to ensure that the object is protected from threats of all kinds. However, this implies an active approach of the subject of protection especially in the direction of identification of potential threats, their sources and carriers, discovering the proximate (immediate) causes of security threats, thus revealing what can happen and discovering the ultimate (ultimate) causes of security threats, thus finding out why this may happen.

Object protection is a creative activity. There is no universal all approach or method that applies everywhere.

ACKNOWLEDGMENT

This research was supported by the Scientific Grant Agency of the Ministry of Education, Science, Research and Sports of the Slovak Republic [VEGA 1/0628/18 Minimizing the level of experts' estimations subjectivity in safety practice using quantitative and qualitative methods].

REFERENCES

Belan, L. (2015). *Security management. Security and risk management. (Bezpečnostný manažment. Bezpečnosť a manažérstvo rizika)*. Žilina, Slovakia: EDIS University of Žilina.

Berzi, L. (1996). *Teória policajno-bezpečnostných služieb [Theory of police security services]*. Bratislava, Slovakia: Academy of the Police Force in Bratislava.

Bologna, S. (2016). *Guidelines for Critical Infrastructure Resilience Evaluation*. Roma, Italy: Italian Association of Critical Infrastructures' Experts.

CLC/TS 50398 Alarm systems. Combined and integrated alarm systems. General requirements. (2009).

Collins, M., Petit, F., Buehring, W., Fisher, R., & Whitfield, R. (2011). Protective measures and vulnerability indices for the Enhanced Critical Infrastructure Protection Programme. In International Journal of Critical Infrastructures (vol. 7, pp. 200-219). Retrieved from doi:10.1504/IJCIS.2011.042976

Commission of The European Communities. (2005). *Green Paper on a European programme for critical infrastructure protection*. Retrieved from https://eur-lex.europa.eu/legal-content/EN/TXT/?uri=CELEX:52005DC0576

EN 15602 (2008). *Security service providers. Terminology*.

EN 16082 (2011). *Airport and aviation security services*.

EN 16747 (2015). *Maritime and port security services*.

EN 16763 (2017). *Services for fire safety systems and security systems*.

EN 50130-4 (2011). *Alarm systems. Electromagnetic compatibility. Product family standard: Immunity requirements for components of fire, intruder, hold up, CCTV, access control and social alarm systems*.

EN 50131–1 (2006). *Alarm systems. Intrusion and hold-up systems. Part 1: System requirements*.

EN 50131-7 (2010). *Alarm systems. Intrusion and hold-up systems. Part 7: Application guidelines*.

EN 50131-8 (2019). *Alarm systems. Intrusion and hold-up systems. Part 8: Security fog devices.*

EN 50134-1 (2002). *Alarm systems. Social alarm systems. Part 1: System requirements.*

EN 50136-1 (2012). *Alarm systems. Alarm transmission systems and equipment. Part 1: General requirements for alarm transmission systems.*

EN 54-1 (2011). *Fire detection and fire alarm systems. Introduction.*

EN 54-2 (1997). *Fire detection and fire alarm systems. Control and indicating equipment.*

EN 60839-11-1 (2013). *Alarm and electronic security systems - Part 11-1: Electronic access control systems - System and components requirements.*

EN 60839-11-2 (2014). *Alarm and electronic security systems - Part 11-2: Electronic access control systems - Application guidelines.*

EN 62676-1-1 (2014). *Video surveillance systems for use in security applications. System requirements. General.*

Garcia, M. L. (2001). *The Design and Evaluation of Physical Protection Systems*. USA: Elsevier.

Hofreiter, L. (2003). Nové determinanty ochrany objektov [New determinants of object protection]. In *Proceedings of the Solving crisis situations in a specific environment.* (vol. 8, pp. 155-160). Žilina, Slovakia: EDIS University of Žilina.

Hofreiter, L., Lovecek, T., & Velas, A. (2005). Zásady a princípy analýzy rizík v oblasti fyzickej a objektovej bezpečnosti [Principles of risk analysis in the area of physical and object security]. National Security Agency of Slovak Republic.

Homer-Dixon, T. (2013). *The upside of down: Catastrophe, creativity and the renewal of civilization.* Toronto, Canada: Knopf.

ISO 31000 (2018). *Risk management. Guidelines.*

Kampova, K., & Makka, K. (2018). Economic aspects of the risk impact on the fuel distribution enterprises. In *Proceedings of the International Conference Transport Means* (pp. 231-235). Academic Press.

Lovecek, T., & Reitspis, J. (2011). *Projektovanie a hodnotenie systémov ochrany objektov [Design and evaluation of physical protection systems]*. Slovakia: University of Žilina.

Lovecek, T., Velas, A., Kampova, K., Maris, L., & Mozer, V. (2013). Cumulative Probability of Detecting an Intruder by Alarm Systems. In *Proceedings of the IEEE International Carnahan Conference on Security Technology*. IEEE Press. 10.1109/CCST.2013.6922037

Mach, V., & Boros, M. (2017). Perimeter protection elements testing for burglar resistance. In Key Engineering Materials (vol. 755, pp. 292-299). Retrieved from doi:10.4028/www.scientific.net/KEM.755.292

Rehak, D., Markuci, J., Hromada, M., & Barcova, K. (2016). Quantitative evaluation of the synergistic effects of failures in a critical infrastructure system. International Journal of Critical Infrastructure Protection, 14, 3-17. doi:10.1016/j.ijcip.2016.06.002

Simak, L. (2015). *Crisis management in public administration (Krízový manažment vo verejnej správe)*. Žilina, Slovakia: University of Žilina.

Siser, A., Maris, L., Rehak, D., & Pellowski, W. (2018). The use of expert judgement as the method. to obtain delay time values of passive barriers in the context of the physical protection system. In *Proceedings of the IEEE International Carnahan Conference on Security Technology* (pp. 126-130). IEEE. 10.1109/CCST.2018.8585718

Soltes, V., Kubas, J., & Stofkova, Z. (2018). The safety of citizens in road transport as a factor of quality of life. In *Proceedings of the International Conference Transport Means* (pp. 370-374). Academic Press.

Stoller, J., & Dubec, B. (2017). Designing protective structure using FEM simulations. In *Proceedings of the International Conference on Military Technologies (ICMT)* (pp. 236-241). Academic Press. Retrieved from 10.1109/MILTECHS.2017.7988762

Tallo, A., Rak, R., & Turecek, J. (2006). *Moderné technológie ochrany osôb a majetku [Modern technologies of protection of persons and property]*. Bratislava, Slovakia: Academy of the Police Force in Bratislava.

Velas, A., Zvakova, Z., & Svetlik, J. (2016). Education and lifelong learning opportunities in the private security services in Slovak republic. In *EDULEARN16: 8th international conference on Education and new learning technologies* (pp. 6517-6523). Barcelona, Spain: IATED Academy.

Zvakova, Z., Velas, A., & Mach, V. (2018). Security in the transport of valuables and cash. In *Proceedings of the International Conference Transport Means* (pp. 1209-1214). Academic Press.

ADDITIONAL READING

Bologna, S. (2016). *Guidelines for Critical Infrastructure Resilience Evaluation*. Roma, Italy: Italian Association of Critical Infrastructures' Experts.

Broder, J. F. (2006). *Risk Analysis and the Security Survey*. Amsterdam, Netherland: Elsevier.

Fay, J. J. (1993). *Encyclopedia of Security Management. Techniques and Technology*. Newton, MA: Butterworth-Heinemann.

Garcia, M. L. (2001). *The Design and Evaluation of Physical Protection Systems*. Elsevier.

Lovecek, T., Ristvej, J., Sventekova, E., Siser, A., & Velas, A. (2016). Currently Required Competencies of Crisis and Security Managers and New Tool for Their Acquirement. In *Proceedings of the 3rd International Conference On Management Innovation And Business Innovation (ICMIBI 2016)*. Academic Press.

Lovecek, T., Velas, A., & Durovec, M. (2016). Level of protection of critical infrastructure in the Slovak Republic. In Production Management And Engineering Sciences (pp. 163–168). Academic Press.

Svetlik, J., & Velas, A. (2016). The safety training in the municipality. In *Proceedings of the EDULEARN16: 8th International Conference On Education And New Learning Technologies* (pp. 1350–1355). Academic Press. 10.21125/edulearn.2016.1271

Vidrikova, D., Boc, K., Dvorak, Z., & Rehak, D. (2017). *Critical Infrastructure and Integrated Protection*. Ostrava: Association Of Fire And Safety Engineering.

KEY TERMS AND DEFINITIONS

Critical Infrastructure: The assets that are essential for the functioning of a society and economy especial in some country.

Physical Security: Security measures that are designed to deny unauthorized access to facilities, equipment and resources and to protect personnel and property from damage or harm.

Probability of Detection: The level of possibility of detection in a protected area with security measures. **Risk:** The effect of uncertainty on objectives.

Risk Assessment: The act of deciding the level of risk.

Risk Management: The techniques of minimizing and preventing loss.

Security Measures: Precautions taken to protected infrastructure against crime.

Chapter 14
Uncertainties in Safety and Security:
Uncertainties in Critical Infrastructure Protection and Human Factors

Tünde Anna Kovács
https://orcid.org/0000-0002-5867-5882
Donát Bánki Faculty of Mechanical and Safety Engineering, Óbuda University, Hungary

Zoltán Nyikes
Donát Bánki Faculty of Mechanical and Safety Engineering, Óbuda University, Hungary

ABSTRACT

In today's world, critical infrastructure encompasses facilities vital to the economy, politics, and population. Their maintenance and safe operation can ensure the supply for the population. These facilities are at risk due to climate change, natural disasters, terror attacks, or wars which are increasingly affecting countries around the world. In addition, the human factor can also cause uncertainty and damages. The function of the world depends on human activities. In this chapter the uncertainties in safety and security are introduced. Security is the most important part as it is the critical infrastructure protection and human safety . The important pillars of safety and security and these uncertainties are introduced in this chapter.

INTRODUCTION

In the world, the governs need to assure all goods what peoples needs for the citizens. The needs can interpret on the base of the pyramid of Maslow's. Maslow's pyramid structure contains human needs at different levels (Szakali & Szűcs, 2017). Security and safety are the basic needs of this structure. The citizen's life quality and simultaneously the economy increasing of the countries depends strongly on security and safety. Even that the human activity is not directly part of the critical infrastructures, the

DOI: 10.4018/978-1-7998-3059-7.ch014

infrastructures are not independent of the personal activities (Szakali & Szűcs, 2018). Critical Infrastructure Protection (CIP) is a strategic task of the governs. The strategic task also to assure the security and safety of human operators of the infrastructures. The goal of this chapter to introduce the human factors and it's a risk in Critical Infrastructure operations. Critical Infrastructure us the most important and most sensible for the citizens of the world because the Critical Infrastructure segments include all necessary parts of life. This is the reason why it needs to understand human factors and risk analyses methods. The critical infrastructure protection uncertainties are the most important in the life of the world peoples. In this chapter, authors want to introduce some most important parts of the uncertainties and the importance of human factors because the human errors cause and caused high-level damages. In this chapter, it can not answer and introduce treatment for all events but discover some possibilities of the risk identification, analysis and the risk level decreasing the view of the human factors.

Critical Infrastructures Protection

Critical infrastructure (CI) is assigned in the literature by different definitions. It can define as those assets and systems, that are essential for the maintenance of the social functions, health, safety, security, and economics or social well-being of people (Moteff & Parfomak, 2004). This definition covers the key government services including energy, utilities, emergency services, banking and finance, transport, health, food supply and communication systems (Brunner & Suter, 2008). They are vulnerable from the number of factors including from the physical attacks to natural disasters, and human error such as poor design or operator error (Moteff, 2014). During recent decades, the urbanization and also climate change greatly increased natural hazards damages. (Freeman, 2003; Leaning, 2013).

The CI sectors defined separately but it's not independent, typically are there a special relationship between the CI elements. This chapter tries to introduce a general overview of understanding potential uncertainties in the selection of CIP policy and some shows on useful uncertainty analysis approaches during the decision-making process. Figure 1 shows a basic critical infrastructure protection process.

Identification of Critical Infrastructure

Elements of critical infrastructure are identified by the government with the owners of the infrastructure in the criteria of the nation. Elements of critical infrastructure are defined according to national criteria, and the qualitative and quantitative effects of the destruction of each element are taken into account by governments. (Commission of the EU Communities, 2006; Commonwealth of Australia, 2011).

These criteria of CI has two categories:

- **Scope:** Determined by the unavailability and loss of the geographic area of the particular critical infrastructure.
- **Severity:** Determined by the economic loss, degradation of the products and services, the political and environmental effects for the population number.

Figure 1. Process of critical infrastructure protection (Moshadeghi, 2017)

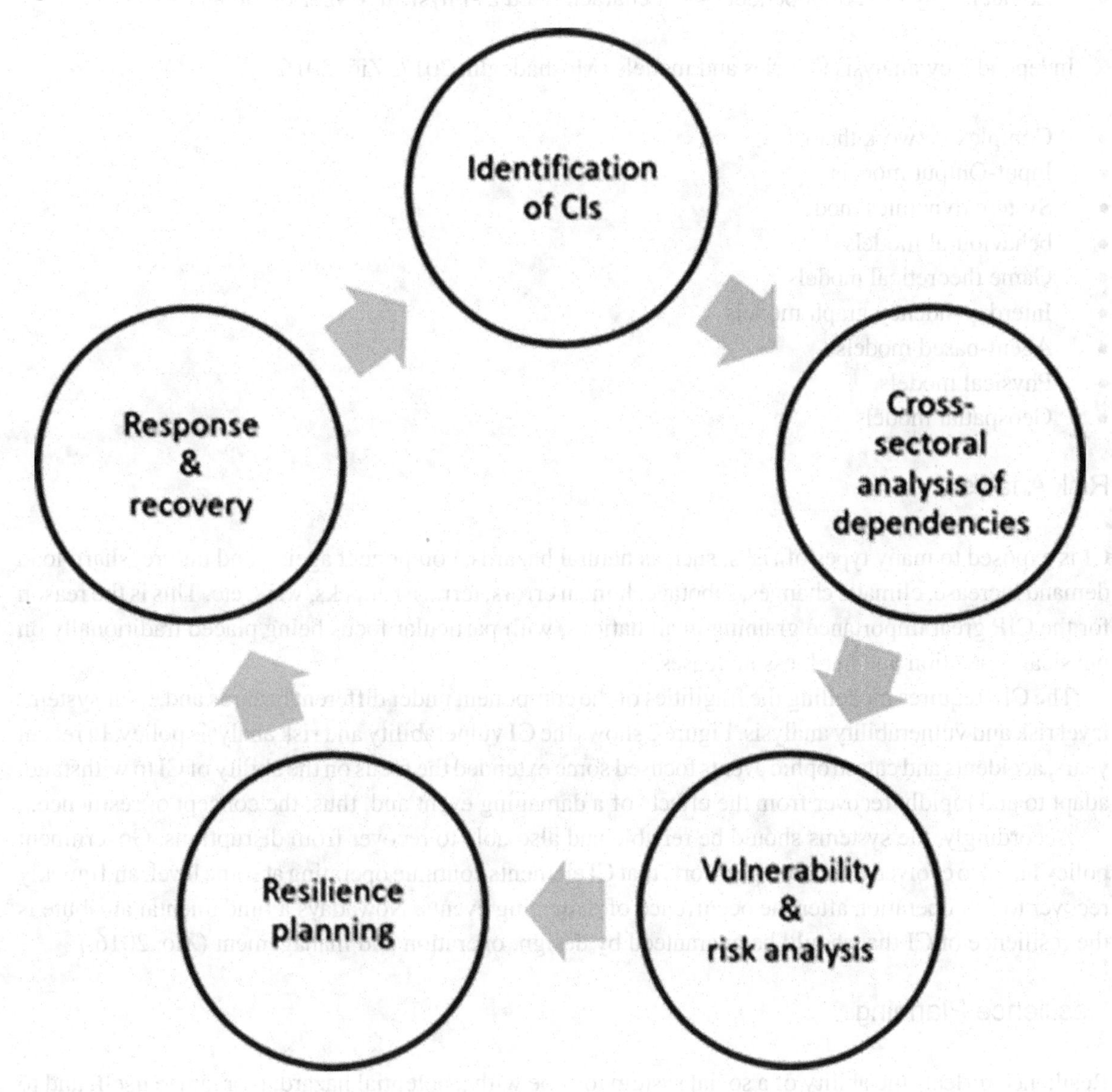

Cross-Sectoral Analysis of Dependencies

Critical infrastructures are not single sectors or location, they can't exist independently, so they depend on each other. For instance, electricity is the basic service for many different infrastructures like the operation of water infrastructure, and the communications industry both rely on electricity.

CI interdependencies have four principle classes

- Physical: The infrastructure dependence from the output material of the other.
- Cyber: Information infrastructure dependency from transmitted information.
- Geographic: infrastructure Dependency on local environmental effects.

- Logical: Any kind of dependency not characterized as Physical, Cyber or Geographic.

Independency analysis theories and models (Moshadeghi, 2017; Zio, 2016):

- Complex network theory
- Input-Output models
- System dynamics models
- behavioural models
- Game theoretical models
- Interdependency graph models
- Agent-based models
- Physical models
- Geospatial models

Risk Analysis

CI is exposed to many types of risks, such as natural hazards, component ageing and failure, sharp load demand increase, climatic changes, sabotage, human errors, terrorist attacks, wars, etc. This is the reason for the CIP great importance graining in all nations, with particular focus being placed traditionally on physical protection and hardness increases.

The CIP requires modelling the fragilities of the component under different hazards and, their system-level risk and vulnerability analysis. Figure 2 shows the CI vulnerability and risk analysis policy. In recent years, accidents and catastrophic events focused some extended the focus on the ability of CI to withstand, adapt to and rapidly recover from the effects of a damaging event and, thus, the concept of resilience.

Accordingly, the systems should be reliable and also able to recover from disruptions. Government policy has also evolved to encourage efforts that CI elements continue operating at some level, and quickly recover to full operation after the occurrence of damaging events. Nowadays a fundamental attribute is the resilience of CI that should be guaranteed by design, operation and management (Zio, 2016.)

Resilience Planning

Resilience reflects the ability of a social system to cope with a potential hazard, reorganize itself, and to improve through adaptive processes and through learning from experience (Adger et al., 2005; Folke, 2006; Cutter et al., 2008). Adaptive capacity, which is defined as the capacity of a system to adjust to change, moderate the effects, and cope with a disturbance (Burton et al., 2002; Brooks et al., 2005; Cutter et al., 2008) is therefore intrinsically linked with resilience. Hazard mitigation is another concept that is nested within resilience planning. Hazard mitigation is any action taken to reduce or avoid risk or damage from hazard events (Mileti, 1999; Godschalk, 2003).

In the context of CI, resilience can be defined as the ability of CI and the associated social system to survive and cope with a disaster with minimum disruption and damage (Berke & Campanella, 2006; National Research Council, 2006).

Figure 3 shows the resilience mechanism, include the prevention and the preparedness parts. The prevention in this task is very important to be ready in the case of any hazardous event. The preparedness is the ability in the case of unexpected hazardous events.

Figure 2. Critical infrastructure risk analysis (Zio, 2016)

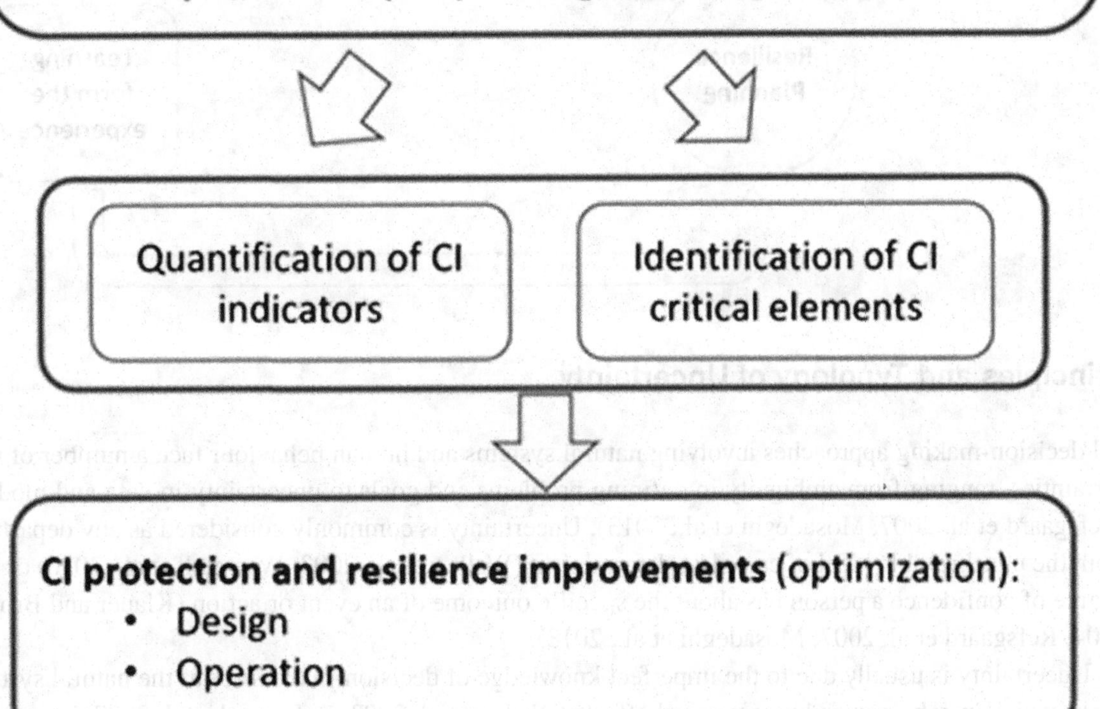

Response and Recovery

Response and recovery in the CIP means to actions taken during, and immediately after, a disaster. Response actions are effective measures to ensure the CI sustained the least damage and are able to be restored rapidly after the hazardous event has occurred. That is, these actions are designed to prevent or minimize disruption to CI and to ensure that the affected community are given immediate support. For the recovery activities is necessary the collaboration of governments, the community, along with the owners and operators of the CI.

Figure 3. Resilience mechanism (Kovács, 2015)

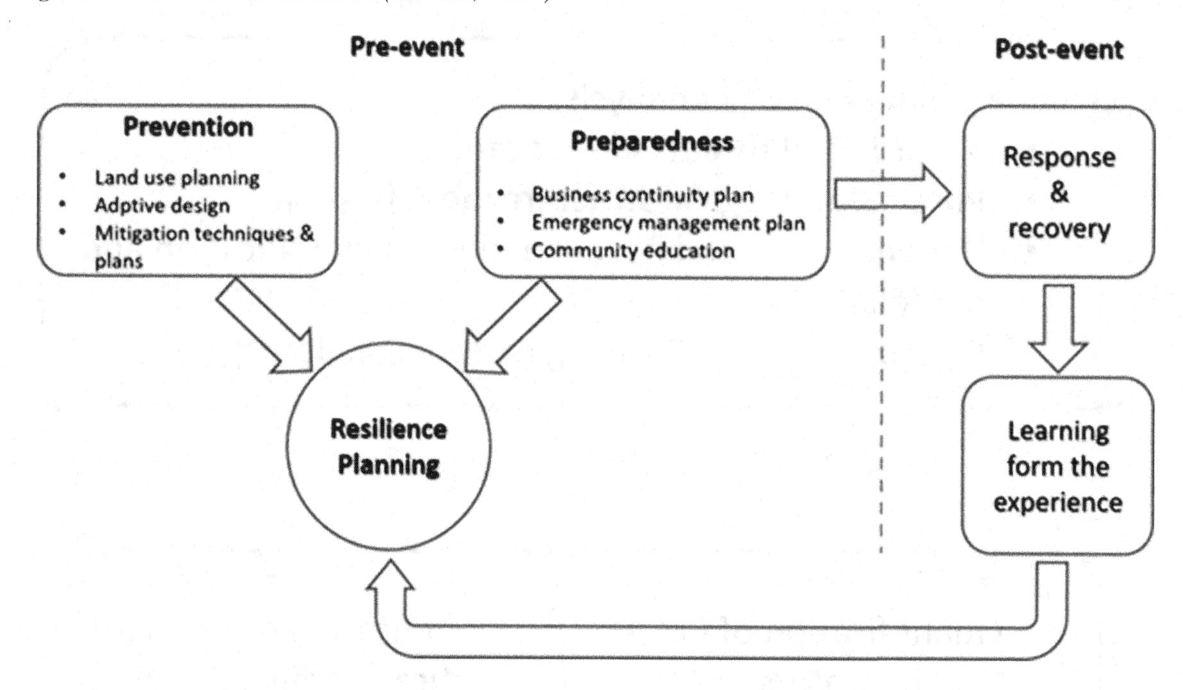

Principles and Typology of Uncertainty

All decision-making approaches involving natural systems and human behaviour face a number of uncertainties, ranging from ambiguity in defining problems and goals to uncertainty in data and models (Refsgaard et al. 2007, Mosadeghi et al., 2013). Uncertainty is commonly considered as any departure from the unachievable ideal of complete determinism (Walker et al., 2003; Warmink et al., 2010) or the degree of confidence a person has about the specific outcome of an event or action (Klauer and Brown 2004, Refsgaard et al. 2007; Mosadeghi et al., 2013).

Uncertainty is usually due to the imperfect knowledge of decision-makers about the natural system itself, which may be reduced by empirical efforts (Walker et al. 2003, Refsgaard et al. 2007).

The most commonly used typology and terminology distinguish three dimensions of uncertainty (Walker et al., 2003; Refsgaard et al., 2007; Warmink et al., 2010; Mosadeghi et al., 2013):

1. The location of uncertainty – where the uncertainty manifests itself within the model complex;
2. The level of uncertainty – where the uncertainty manifests itself along the spectrum between deterministic knowledge and total ignorance;
3. The nature of uncertainty – whether the uncertainty is due to the imperfection of our knowledge or is due to the inherent variability of the phenomena being described.

The four main locations of uncertainty are identified as:

- **Context Uncertainty:** Includes uncertainty about the external economic, environmental, political, social, and technological situation that forms the context for the problem being examined (Walker et al., 2003; Refsgaard et al., 2007; Warmink et al., 2010; Mosadeghi et al., 2013). Considering this uncertainty can help to avoid problems arising from incorrect problem framing (Dunn, 2001);

- **Model Structure Uncertainty:** Arises from a lack of sufficient understanding of the system (past, present, or future) that is the subject of the policy analysis, including the behaviour of the system and the interrelationships among its elements;
- **Model Technical Uncertainty:** Is the uncertainty generated by software, errors in algorithms or hardware errors (Walker et al., 2003);
- **Input Uncertainty:** Is related to data that describe the reference system (i.e. land use maps, data on infrastructure and climate data, and the external driving forces that have an influence on the system and its performance).

The various levels of uncertainty are distinguished as statistical, scenario, qualitative and recognized uncertainties.

Uncertainty in Critical Infrastructure Protection (CIP)

Input Uncertainty

Similar to any other strategic decisions, the process of CIP is subject to uncertainty. Public and private officials responsible for CI rely on expert elicitation to estimate the likelihood of certain events occurring, their consequences, possible damages, and appropriate protective measures (Grossi & Kunreuther, 2001; Barker & Haimes, 2009; Giannopoulos et al., 2012; Lickley et al., 2013).

Considering the input uncertainty is particularly important in CIP strategies as many existing CI is located in vulnerable locations, including coastal zones and river flood plains. In recent years, lack of local information on how CI may cope and adapt to climate change has also increased the level of uncertainty in appropriately planning response and recovery actions (Voice et al., 2006). This means small increases in extreme weather events can potentially increase damages to the CI and, more resources will be required to restore the infrastructures and recover from the hazardous event (Auld & MacIver, 2007).

Analyzing these uncertainties can assist stakeholders to understand the role of uncertainty in the estimates of losses from natural disasters and in evaluating alternative disaster management (Grossi & Kunreuther, 2001). The next section reviews uncertainty analysis approaches that can be used in the process of CIP.

Uncertainty Analysis Approaches

Careful consideration of uncertainty in the CIP process provides an estimation of the robustness of disaster management decisions as well as analysis of the gaps in data collection and actions (Hall & Solomatine, 2008). The problem of uncertainty in the CIP process can be addressed in three principally different ways:

- Improving the accuracy of data;
- Quantitative uncertainty analysis; and
- Improving planning policies

In constructing new infrastructure, it is important that the climatic values used for CI design be regularly updated and approaches such as the use of a Climate Change Adaptation Factor should also be considered on a regional basis (Auld and MacIver, 2007).

To incorporate uncertainties inherent in the decision models, uncertainty and sensitivity analysis techniques have also continued to expand. To incorporate probability distribution information into decision-making, numerical and analytical methods can be used (Tung, 2009; Chen et al., 2011). For example, Monte Carlo simulation allows exploration of the full range of variation in the input factors and does not require an assumption about the model structure (Madani and Lund, 2011).

Although quantitative uncertainty analysis techniques have been widely used and the accuracy of the input data has been improved, methods for addressing the problem of uncertainty are not only limited to the technical evaluations. Even if data and technical information were perfect, there would still be uncertainty in the protection and management of CI due to governance and planning issues (Voice et al., 2006). Consequently, more resources are required for local governments to plan for long term horizons and address the uncertainties within their decision model. This can be achieved through planning scheme amendments and measures such as increasing safety factors, forensic analyses of extreme events and incorporating climate change projections into engineering codes, standards and practices (Voice et al., 2006; Auld and MacIver, 2007).

Human Factors, Risks And Uncertainties

This chapter introduces some part of human factors. The measuring and knowledge of human body size science of ergonomics is the anthropometry. The word 'anthropometry' means a measurement of the human body. It is derived from the Greek words 'Anthropos' (man) and 'metron' (measure). Anthropometric data are used in ergonomics to specify the physical dimension of workspaces, equipment, furniture and clothing to ensure that physical mismatches between the dimensions of equipment and products and the corresponding user dimensions are avoided.

The assumptions are beliefs derived from concepts of logic, science, and theory. Assumptions are more abstract, more general than principles; principles reflect more directly customary practices. Principles reflect more directly customary practices.

Human Error Caused Damages in CI

It can find in the world many accidents caused by human errors. Some well-known example can highlight the risk of human operations. The Lenin nuclear plant accident in 1986 happened near two Ukrainian cities, Chernobyl and Pripjaty. This event was the most dangerous nuclear catastrophe near Europe. In this catastrophe can find human errors, the decision-makers had a low knowledge level of nuclear study and the situation was strange. After the human error induced accident Unfortunately the decision-makers made more and more fault.

A guided cyber attack in Iran (2010). An innocent operator used flash drive caused event, when the Stuxnet virus destroyed 20% of the centrifuge of the uranium dressing process. The accident reason, Iran stopped the nuclear program. The reason for the human errors come from the low competency and low safety awareness of the task (Nyikes 2017). The low safety awareness and task knowledge competency are increasingly by training.

Physical Factors

Physical factors are the human body capabilities and limits during operations in static posture or motion. Defined two major physical activity factors, manual handling and cumulative trauma disorders. The human body is limited in the amount of force it can apply and continually exceeding that force can be injurious. Manual handling is defined in terms of the types of tasks performed in operations and maintenance, from the basics of lifting through to the more complex tasks involving unusual or dynamic body positioning.

To identify the critical risk factors, it has conceptually categorized the physical work activities into two types, those that consist mainly of single exertions involving the entire body and those that are more repetitive and most likely involve more intensive use of the upper body or arms and hands.

Environmental Features

It has to talk about the conditions of the work environment and human behaviours. Conditions of the work environment that could contribute to a musculoskeletal disorders injury, such as a back injury resulting from a slip, trip, or fall, include good housekeeping and adequate illumination of work areas. (Kovács 2015)

Physical factors of the human are the capabilities and limits of the human body. An individual has a body aptitude and also knowledge of the work motion and posture.

The physical work activities that involve whole-body exertions typically involve carrying or moving an object, so they are called manual handling tasks. Manual handling tasks are activities during which workers move objects from place to place by lifting, lowering, carrying, pushing, or pulling. When we plan or study a physical workplace it needs to take account of the risk factors, capability, limits of the human body and human behaviour.

Human Errors

Human error is a term describing a planned sequence of mental or physical actions that fail to achieve the intended objective, and that this failure cannot be attributed to chance. Norman (1983) made a distinction among the variety of possible failures by suggesting that failure can be related to a problem either with the intention and/or planning of the actions or to the actual execution of the actions.

Knowledge-based behaviour is typical of new, non-familiar situations that require basic understanding and knowledge of the system and the situation. Handling such situations requires conscious cognitive processes such as retrieval of information from long-term memory, search for new information, analytical problem solving, and decision-making. Planning and intention errors are typical of knowledge-based behaviour and are associated with the inappropriate or incomplete search of information, and making the wrong diagnosis or decision based on incomplete knowledge. When attempting to understand and analyse the possible causes of planning and execution errors, external and internal Performance Influencing Factors should be considered. Performance Influencing Factors are usually associated with the behavioural aspects of human error and the underlying psychological mechanism. External factors are various circumstances of any mission ranging from organizational and task characteristics to human-system interface aspects such as controls, displays, and ergonomics.

This framework is used to

1. Classify the errors in the empirical study of related human errors
2. Theoretically account for and understand the possible causes of those errors
3. Derive practical implications (Kovács, 2015)

Human Error Analysis

Predicting Errors

- Task Analysis and Error Identification

Preventing Errors

- Specifying Training Requirements
- Equipment Design (E.G. Pressure Gauge)
- Detailed Procedures (Administrative Control)

Ultimately:

- Reduce Risk
- Save Money
- Justify Design Decisions (Kovács, 2015)

Human Reliability Assessment (HRA)

Human reliability assessment (HRA) involves the use of qualitative and quantitative methods to assess the human contribution to risk. There are many and varied methods available for HRA, with some high hazard industries developing 'bespoke', industry-focused methods. (Kovács, 2015)

HRA Process

Task analysis is used to describe and understand human interactions with the system.

The results of the task analysis are used with an error taxonomy (classification scheme) to allow error identification.

The identified errors are analysed either qualitatively or quantitatively.

The process is repeated each time a design iteration occurs (Figure 4).

Task Analysis

The task analysis method was developed essentially focusing on the assessment and reduction of human error, but currently, they are used in several other ways which are explained later. There are a wide variety of different methods and also these methods have other techniques and variations.

Task analysis is the analysis of how a task is accomplished, including a detailed description of both manual and mental activities, task and element durations, task frequency, task allocation, task complex-

Figure 4. Human Reliability Assessment Process (Kovács, 2015)

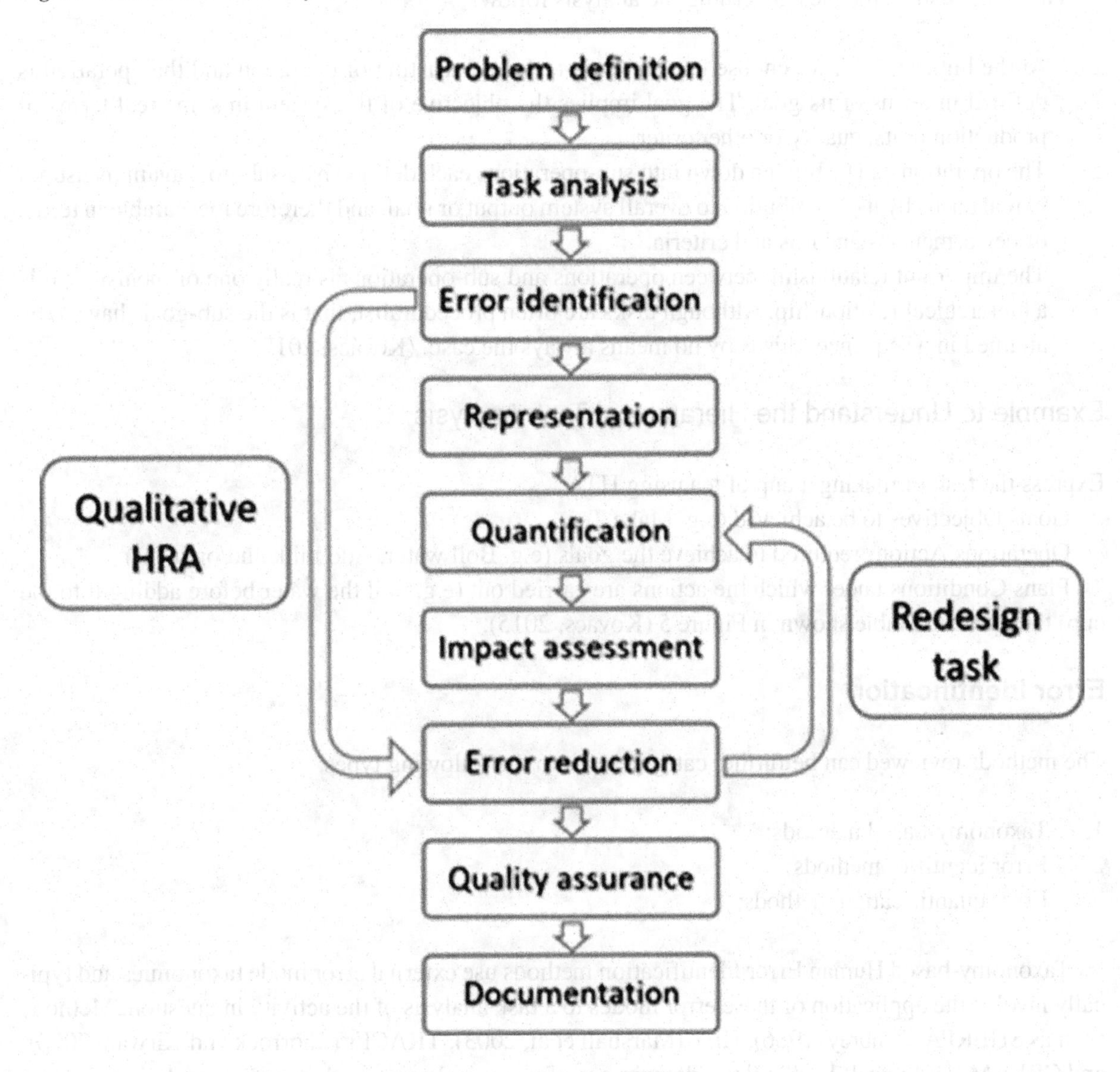

ity, environmental conditions, necessary clothing and equipment, and any other unique factors involved in or required for one or more people to perform a given task.

The term Task Analysis can be applied very broadly to encompass a wide variety of human factors techniques.

Hierarchical Task Analysis (HTA)

It is the most popular method. Originally developed in response to the need greater understanding of cognitive tasks. The mainline is based on a hierarchical structure of the actions. It is an organization of the operations schematically drawn taking a tree shape. An operation may be broken down into suboperations, these suboperations in other and so on.

The three main principles governing the analysis follow:

1. At the highest level, we choose to consider a task as consisting of operation and the operation is defined in terms of its goal. The goal implies the objective of the system in some real terms of production units, quality or other criteria.
2. The operation can be broken down into sub-operations each defined by a sub-goal again measured in real terms by its contribution to overall system output or goal, and therefore measurable in terms of performance standards and criteria.
3. The important relationship between operations and sub-operations is really one of inclusion; it is a hierarchical relationship. Although tasks are often proceduralist, that is the sub-goals have to be attained in a sequence, this is by no means always the case. (Kovács, 2015)

Example to Understand the Hierarchical Task Analysis

Express the task of making a cup of tea using HTA

Goals Objectives to be achieved (e.g. Make Tea)

Operations Actions required to achieve the goals (e.g. Boil water, Add milk and/or sugar)

Plans Conditions under which the actions are carried out (e.g. boil the water before adding it to the cup) the operation table shown in Figure 5 (Kovács, 2015).

Error Identification

The methods reviewed can be further categorized into the following types:

1. Taxonomy-based methods;
2. Error identifier methods;
3. Error quantification methods;

Taxonomy-based Human Error Identification methods use external error mode taxonomies and typically involve the application of these error modes to a task analysis of the activity in question. Methods such as SHERPA (Embrey, 1986), HET (Marshall et al, 2003), TRACEr (Shorrock and Kirwan, 2000), and CREAM (Hollnagel, 1998) all use domain-specific external error mode taxonomies designed to aid the analyst in identifying potential errors. Taxonomic approaches to HEI are typically the most successful in terms of sensitivity and are the quickest and simplest to apply and with only limited resource usage.

Error identifier HEI methods, such as Human Error Identification in Systems Tool (HEIST) and THEA use a series of error identifier prompts or questions linked to external error modes to aid the analyst in identifying the potential human error.

The Human Error Identification in Systems Tool (HEIST; Kirwan 1994) uses a set of error identifier prompts designed to aid the analyst in the identification of potential errors.

Error quantification methods are used to determine the numerical probability of error occurrence. Identified errors are assigned a numerical probability value that represents their associated probability of occurrence.

Figure 5. Operation table (Kovács, 2015)

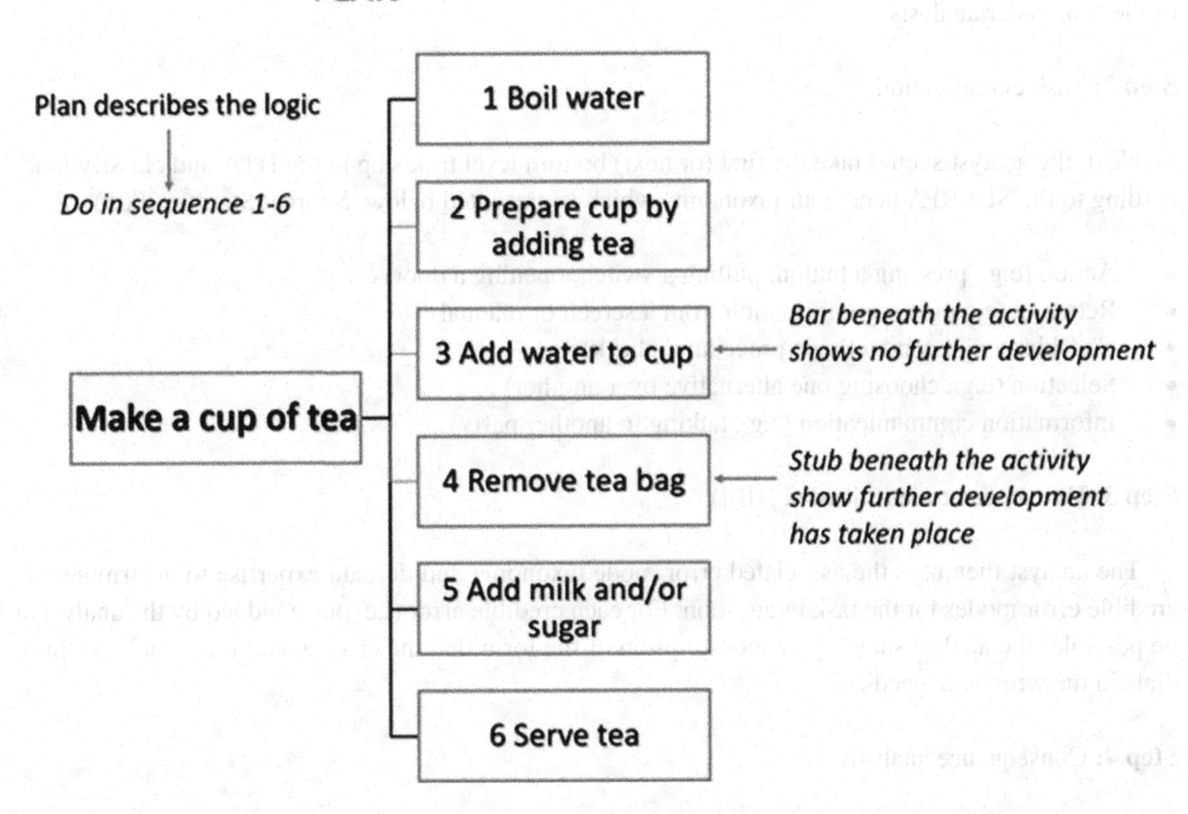

Systematic Human Error Reduction and Prediction Approach (SHERPA)

Background and Applications

SHERPA comprises of an error mode taxonomy linked to a behavioural taxonomy and is applied to an HTA of the task or scenario under analysis in order to predict potential human or design induced error. As well as being the most commonly used of the various HEI methods available, according to the literature, it is also the most successful in terms of accuracy of error predictions.

Despite being developed originally for use in the process industries, the SHERPA behaviour and error taxonomy is generic and can be applied in any domain involving human activity.

Procedure and Advice

Step 1: Hierarchical task analysis (HTA)

The first step in a SHERPA analysis involves describing the task or scenario under analysis. For this purpose, an HTA of the task or scenario under analysis is normally conducted. The SHERPA method works by indicating which of the errors from the SHERPA error taxonomy are credible at each bottom level task step in an HTA of the task under analysis. A number of data collection techniques may be used in order to gather the information required for the HTA, such as interviews with SMEs and observations of the task under analysis.

Step 2: Task classification

Next, the analyst should take the first (or next) bottom level task step in the HTA and classify it according to the SHERPA behaviour taxonomy, which is presented below (Source: Stanton 2005).

- Action (e.g., pressing a button, pulling a switch, opening a door)
- Retrieval (e.g., getting information from a screen or manual)
- Checking (e.g., conducting a procedural check)
- Selection (e.g., choosing one alternative over another)
- Information communication (e.g., talking to another party)

Step 3: Human error identification (HEI)

The analyst then uses the associated error mode taxonomy and domain expertise to determine any credible error modes for the task in question. For each credible error (i.e. those judged by the analyst to be possible) the analyst should give a description of the form that the error would take, such as, 'pilot dials in the wrong airspeed'.

Step 4: Consequence analysis

The next step involves determining and describing the consequences associated with the errors identified in step 3. The analyst should consider the consequences associated with each credible error and provide clear descriptions of the consequences in relation to the task under analysis.

Step 5: Recovery analysis

Next, the analyst should determine the recovery potential of the identified error. If there is a later task step in the HTA at which the error could be recovered, it is entered here. If there is no recovery step then 'None' is entered.

Step 6: Ordinal probability analysis

Once the consequence and recovery potential of the error has been identified, the analyst should rate the probability of the error occurring. An ordinal probability scale of low, medium or high is typically used. If the error has not occurred previously then a low (L) probability is assigned. If the error has occurred on previous occasions, then a medium (M) probability is assigned. Finally, if the error has occurred on frequent occasions, a high (H) probability is assigned.

Step 7: Criticality analysis

Next, the analyst rates the criticality of the error in question. A scale of low, medium and high is also used to rate error criticality. Normally, if the error would lead to a critical incident (in relation to the task in question) then it is rated as a highly critical error.

Action Errors

A1 – Operation too long/short
A2 – Operation mistimed
A3 – Operation in the wrong direction
A4 – Operation too little/much
A5 – Misalign
A6 – Right operation on the wrong object
A7 – Wrong operation on the right object
A8 – Operation omitted
A9 – Operation incomplete
A10 – Wrong operation on the wrong object

Checking Errors

C1 – Check omitted
C2 – Check incomplete
C3 – Right check on the wrong object
C4 – Wrong check on the right object
C5 – Check mistimed
C6 – Wrong check on the wrong object

Retrieval Errors

R1 – Information not obtained
R2 – Wrong information obtained
R3 – Information retrieval incomplete

Communication Errors

I1 – Information not communicated
I2 – Wrong information communicated
I3 – Information communication

Selection Errors

S1 – Selection omitted
S2 – Wrong selection made

Step 8: Remedy analysis

The final stage in the process is to propose error reduction strategies. Normally, remedial measures comprise suggested changes to the design of the processor system. According to Stanton (2005), remedial measures are normally proposed under the following four categories: Equipment (e.g. redesign or modification of existing equipment); Training (e.g. changes in training provided); Procedures (e.g. provision of new, or redesign of old, procedures); and Organizational (e.g. changes in organizational policy or culture).

Advantages

The SHERPA method offers a structured and comprehensive approach to the prediction of human error. The method is exhaustive, offering error reduction strategies in addition to predicted errors, associated consequences, probability of occurrence, criticality and potential recovery steps. The SHERPA error taxonomy is generic, allowing the method to be used in a number of different domains.

Disadvantages

Can be tedious and time consuming for large, complex tasks. The initial HTA adds additional time to the analysis. SHERPA only considers errors at the 'sharp end' of system operation. The method does not consider a system or organizational errors (Kovács, 2015).

Human Error Template (HET)

The human error template (HET) (Marshall et al., 2003) method was developed by the ErrorPred consortium specifically for use in the certification of civil flight deck technology. Along with a distinct shortage of HEI methods developed specifically for the civil aviation domain, the impetus for HET came from a US Federal Aviation Administration (FAA) report entitled 'The Interfaces between Flight crews and Modern Flight Deck Systems' (Federal Aviation Administration, 1996), which identified many major design deficiencies and shortcomings in the design process of modern commercial airliner flight decks.

For example, the report identified a lack of human factors expertise on design teams, which also had a lack of authority over the design decisions made. There was too much emphasis on the physical ergonomics of the flight deck and not enough on the cognitive ergonomics.

Human error is a term describing a planned sequence of mental or physical actions that fail to achieve the intended objective, and that this failure cannot be attributed to chance. The systematic human error reduction and prediction approach (SHERPA) was originally developed for use in the nuclear reprocessing industry and are probably the most commonly used Human Error Identification approach, with further applications in a number of domains, including aviation.

Environmental Factors

Physical environmental factors such as noise, vibration, lighting, the climate can affect people's safety, health and comfort. Other examples of environmental features that could be encountered at work and may affect behaviour are dangerous chemicals, radioactive materials, atmospheric pressure, and the like. The four environmental features that can affect people in three ways: (1) health, (2) performance, and (3) comfort. The effects of these three aspects are usually combined. For example, poor health can lead to both poor performance and reduced comfort and, thus, reduced work satisfaction. Stressors, arising from illumination, temperature, noise, vibration, or any other aspect of the environment can adversely affect people when they reach a certain level, although the effect may not be apparent either to the person being affected or an observer. The ideal range for performance and comfort is narrow.

Illumination Effect

The light stimulus is characterized by its intensity and wavelength. In our daily environment, the amount of illumination (also called illuminance) that falls on a surface (e.g., desk) depends on the energy of the light source, its distance from the surface, and the angle of the surface to the light source. Illumination is described in terms of the rate of flux (lumens) produced by the source and the surface area over which it spreads. Light may fall on a surface from multiple sources or, for example, as reflections from walls, ceilings, and other objects in the immediate environment. These reflections can be considered individual light sources, and the efficiency of each depends on the angle at which it reflects light on the surface. In summary, the amount of light falling on a surface depends on the:

1. The light source and its luminous intensity.
2. Distance between the light source and the surface.
3. The angle of the light source to the surface.
4. The number of light sources and reflecting sources in the immediate environment.

We do not see the light falling on a surface but we see the light reflected from a surface. Luminance is a measure of the light reflected from the surface and is associated with the subjective sensation of brightness. Keep in mind that, depending on their surface characteristics, different bodies absorb and reflect different amounts and qualities of light. Luminance is expressed in terms of candela per meter squared (cd/m2). (Kovács, 2015)

Lighting Quantity

Studies in the literature demonstrate that increasing illumination can certainly lead to an increase in visual performance in terms of productivity and efficiency. However, the increase depends highly on the type of task performed (Barnaby, 1980; Jaschinski, 1982; Knave, 1984) and age of the person (Bennett, Chitlangia, and Pangrekar, 1977; Ross, 1978). One must be careful in prescribing higher and higher levels of illumination: a continuous increase in the level of illumination results in smaller and smaller improvements in visual performance, until that performance levels off.

Task Factors

The level of illumination needed generally depends on the task performed. The more visually intense the task (i.e., smaller details or lower contrast), the higher is the illumination level required, up to a maximum, where performance levels off no matter how much more the illumination is increased. The point at which this levelling off occurs is different for different tasks. Another approach to enhance visual performance is through changing the features of the task; for example, increasing the size of the object seen or the contrast between the object and its background. The visual tasks that people perform can vary in difficulty from simple ones, such as finding a car in a parking lot at night, to the highly demanding, such as calibrating an instrument.

Age Factors

The ageing process has its effects on the visual system. As we grow older, in general, over 40 years old, we experience the following changes (Hughes and Neer, 1981; Wright and Rea, 1984):

- The lens of the eye atrophies and hardens, decreasing its ability to conform to the shape necessary for quick and efficient focus on objects at different distances from the eyes. These changes are also the contributors to the improvement of the far vision (farsighted) and the deterioration of the near vision (near-sighted).
- The muscles controlling the diameter of the pupil also atrophy, reducing the size of the pupil and decreasing the range and speed of adjusting and adapting quickly and efficiently to different levels of illumination.
- A lower level of illumination reaching the retina leads to a reduction in the level and rate of dark adaptation.
- As the lens hardens, it yellows, leading older people to make errors in sorting or matching colours in the blue-green and red regions.
- The changes in both the lens and pupil flexibility and efficiency lead to a decrease in visual acuity; the ability to see fine details.

These changes should encourage us to account for the elderly population in the design of lighting to ensure visual comfort and efficient visual performance.

For example, we should provide:

- More light for elderly people. This can be achieved by providing an adjustable personal light.
- Higher contrast. For example, for older people to see as well as people 20–35 years old (baseline), the following multiplication factors are needed to ensure higher contrast and maximize visual performance.

Lighting Quality and Colour

A number of factors have a direct effect on task visual performance and subjective impressions of people. These are the lighting colour, glare, and illumination ratio. Each factor is discussed briefly next.

The objects in the visual field and the various types of lamps we interact with daily differ in colour. In terms of lighting, colour is also referred to as characteristic spectral distribution. Lighting explained in Figure 6.

The colour of an object is determined by the:

- Wavelengths of visible light that it absorbs and reflects. For example, a red object absorbs most colours in the incident light except red. Incident light is the light falling or striking on something.
- Colour of the incident light. Two properties of the incident light determine the resulting colour of the object: the colour rendering index (CRI) and the colour "temperature" of the light.

Figure 6. Office lighting, direct and indirect lighting, glare (Kovács, 2015)

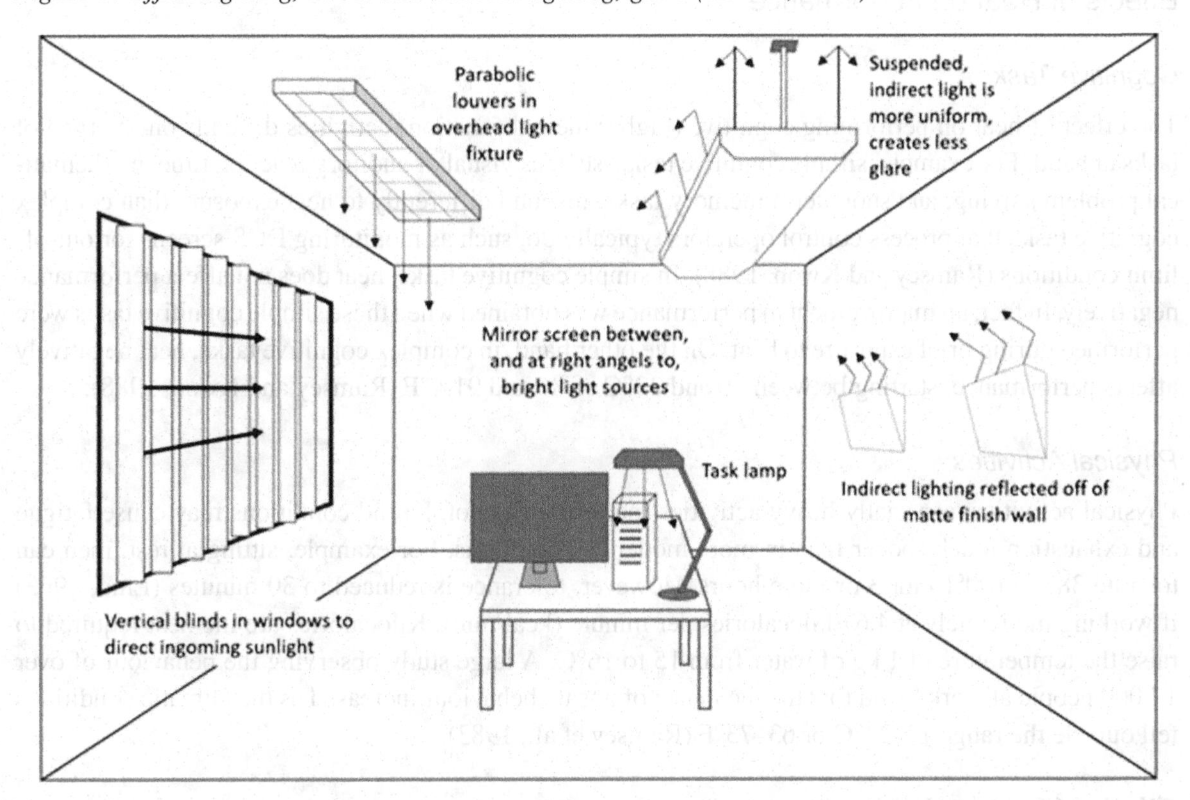

Glare

Glare occurs whenever one part of the field of vision is brighter than the level to which the eye has become adapted, causing a decrease in visual performance, annoyance, and discomfort. Glare can be characterized as direct or indirect.

Disability glare occurs when there is direct interference with visibility and visual performance and often is accompanied by discomfort. Discomfort glare produces discomfort, annoyance, irritation, or distraction but does not necessarily affect visibility or visual performance.

Climate and Temperature Effect

The human body's thermal regulation system tries to maintain a relatively stable internal (core) temperature of between 36.1 and 37.2°C (97 and 99°F). If we add relatively high humidity to the ambient temperature, the resulting condition can lead to fatigue and potential health risks. The human body maintains heat balance by increasing blood circulation to the skin; therefore, we sweat on hot days. When it is cold, the body reduces the blood circulation to the skin and we shiver to keep the extremities warm. The body generates heat through metabolism and physical work.

Effects of Heat on Performance

Cognitive Tasks

The effect of heat on performing cognitive (higher mental function) activities depends on the type of tasks at hand. For example, simple cognitive tasks, such as visual or auditory reaction time, mathematical problem solving, and short-term memory tasks, respond differently to heat exposure than complex cognitive tasks that process control operators typically do, such as monitoring DCS screens for out-of-limit conditions (Ramsey and Kwon, 1988). In simple cognitive tasks, heat does not affect performance negatively. In fact, an improvement in performance was obtained when these simple cognitive tasks were performed during brief exposure to heat. On the other hand, in complex cognitive tasks, heat negatively affects performance, starting between 30 and 33°C or 86 and 91.4°F (Ramsey and Kwon, 1988).

Physical Activities

Physical activities, especially heavy activities, performed in hot, humid conditions may cause fatigue and exhaustion much sooner than in more moderate conditions. For example, sitting at rest, men can tolerate 38°C (100°F) for 3 or more hours. However, tolerance is reduced to 30 minutes (Lind, 1963) if working moderately at 4.67 kilocalories per minute (kcal/min). Kilocalories are the heat required to raise the temperature of 1 kg of water from 15 to 16°C. A large study observing the behaviour of over 17,000 people at work found that the incidence of unsafe behaviour increased as the climatic conditions fell outside the range 17–23°C or 63–73°F (Ramsey et al., 1983).

Effects of Cold on Performance

However, the human body is better able to tolerate increased heat than compensate for heat loss. Examples of jobs where people work in a cold environment are oil and gas extraction, electric line repair, gas line repair, sanitation, trucking, warehousing (especially cold storage), and military (Sinks et al., 1987). Similar to the effects of heat on performance, the studies of the effects of cold on performing different functions are not well understood and therefore lead to no firm conclusions.

The reasons may be related to the following interacting factors:

- Differences in airflow, temperature, and humidity.
- Differences in the duration of exposure.
- The difference in exposure level to the different parts of the body.
- Individual differences (biologically and physiologically).

Cognitive Tasks

The effect of cold on cognitive tasks has been detected in studies on reaction time and complex mental activities. However, the effect depends on the severity of the temperature, a task performed, and skill and experience of the subject with the task and with performing in the cold.

The following is a summary of the results:

- Mental performance decreases when performing complex, demanding tasks requiring high levels of concentration and significant use of short-term memory (Bowen, 1968; Enander, 1987, 1989).

- Cold does not affect the cognitive efficiency of well-motivated subjects (Baddeley et al., 1975).
- Cold acts as a distractor that interferes with some types of mental performance tasks (Enander, 1989).

Physical Activities

Cold significantly affects physical work. The reduction in the limb or whole-body temperature reduces physical capacity. For example, reduced limb temperature (i.e., arm and hand) affects motor ability, causes a loss of cutaneous (touch) sensitivity, affects the limb's muscular control, and reduces dexterity, muscular strength, and endurance abilities (Lockhart, 1968; Morton and Provins, 1960; Enander, 1989). The reduction in whole-body temperature reduces the rate of metabolism within the muscles of the extremities and, in turn, the performance due to shivering.

Effects of Heat on Health

Hot Environment

Exposure to high levels of heat can affect a person's health in two ways. First, increased temperature on the skin can lead to tissue damage from burning (e.g., over 45°C or 113°F). This situation is usually manageable because it is observable and the people react to it by distancing themselves from the heat. The second way heat can affect health is more serious and dangerous. It deals with increased core body temperature to about 42°C (108°F), where heatstroke (hyperthermia) occurs. This rise in core body temperature results in an increase in metabolism, which in turn produce heat that needs to be dissipated. If the cycle is allowed to continue, death can ensue. There are two possible reasons why the heat generated does not leave the body: The body is exposed to heat and humidity that reduces the sweat evaporation, and some protective clothing acts as insulation.

Cold Environment

Exposure to cold can lead to a drop in core or deep body temperature, likely to produce a risk to health. When the core body temperature falls below 35°C (95°F), the state is referred to as hypothermia or cold stress. With core body temperature below 35°C (95°F), the risk of disorientation, hallucination, and unconsciousness increases, leading to cardiac arrhythmia as the core temperature drops even further and subsequent death from cardiac arrest.

Climate Comfort and Discomfort Zones

Comfort in the comfort zone is perceived when the body's physiological responses to the environmental temperature and humidity are within the normal regulatory responses. However, when those physiological responses exceed the normal regulatory responses, individuals perceive discomfort. People's perception of comfort, in the comfort zone, is influenced by the following environmental factors: Air temperature—comfort zone, 20–25°C (68–78°F). Relative humidity—comfort zone, 30–70%.

Air velocity—comfort zone, 0.1–0.3 m/sec (20–60 ft/sec).

Noise

Noise is an aspect of our work and daily living environment produced by equipment, machines, and tools commonly referred to as unwanted sound. Burrows (1960) clearly defines noise as "an auditory stimulus bearing no informational relationship to the presence or completion of the immediate task." However, it depends on the individual's subjective reaction whether to consider a sound source as noise or not.

In addition, sound labelled noise on one occasion may not be noise on a different occasion or environment. Sound is defined in terms of two parameters, frequency and intensity because a sound is a vibration stimulus experienced through the air.

Figure 7. The average daily level should not exceed 80dBA (Kovács, 2015)

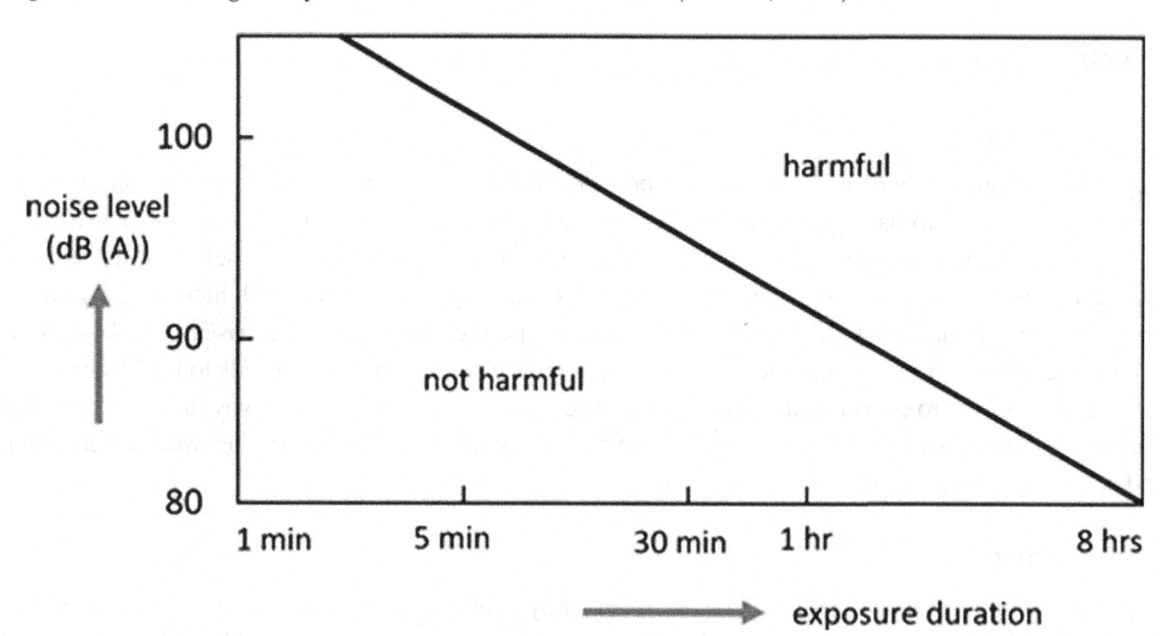

Frequency is acoustic energy reflecting the number of vibrations per second, expressed in Hertz (Hz). One Hertz is equal to 1 cycle per second (shown in Figure 7.).

The range of frequency between 2 and 20,000 Hz defines the frequency limits of the ear. People subjectively perceive frequency as pitch. Intensity is related to the sound pressure variation caused by the sound source.

A noise level which, over an 8-hour working day, exceeds 80 dBA on average, can damage hearing. Assuming constant noise levels, this daily level will be reached, for example, with an 8-hour exposure to 80 dBA, or with a 1-hour exposure to 89 dBA. (Kovács, 2015)

Effects of Noise

Annoyance during thinking and communication tasks can already arise at levels well below 80 dBA. An excess of noise will prove annoying even though the limit for damage to the hearing has not yet been reached. It is mainly noise produced by others, unexpected noise and high-frequency noise that cause annoyance.

Speech and Communication

Speech and communication depend highly on the listener's ability to receive and decode the sounds. This means that the listener must hear the signal (speech or auditory display) and understand the message. Therefore, the speech and communication comprehension depends on the loudness of the voice as well as the level of background noise.

Hearing and understanding the signal may be masked (blocked) by a noisy environment. If the signal intensity remains constant and the masking intensity increases, then more masking occurs. The reverse is also true. Similarly, as the frequency of the masker approaches that of the signal, more masking occurs. From a practical side, it is important to know the intensity and frequency of both the signal and masker.

Cognitive Performance

The relationship between noise and cognitive performance is similar to that between heat or cold and cognitive performance. The results are again mixed. Even so, some reasonable conclusions can be made about the effects of noise on cognitive performance for specific types of tasks:

- Generally speaking, noise clearly has an effect on overall performance, especially when it reaches a level over 80dBA.
- Noise levels over 100dBA can affect monitoring performance over a relatively long duration (Jerison, 1959).
- Tasks with large short-term memory component (i.e., mathematical calculations) are affected by high noise levels (Poulton, 1976, 1977).
- Noise produces more errors in complex rather than simple tasks (Boggs and Simon, 1968).
- Tasks performed continuously without rest pauses between responses are affected by noise (Davies and Jones, 1982).
- Tasks requiring high perceptual or information processing capacity (i.e., high mental demands, complexity, and considerable detail) are affected by noise (Eschenbrenner, 1971).
- Noise does not reduce performance on simple routine tasks or motor performance (i.e., manual work). Positive noise, in the eyes of the beholder, such as music, may facilitate an improvement in performance and satisfaction. The reasoning here is that music can aid in reducing boredom and fatigue associated with these tasks (Fox, 1971).

Nuisance and Distraction

Noise is considered a nuisance (annoying or unpleasant) when it interferes with our ability to carry out an activity. Thus, the response is subjective. This subjective nature indicates that what is annoying to one individual may not be to another. Furthermore, what is a nuisance on one occasion may not be at another time? For example, people are not disturbed by the noise generated by their own activities. A few observations can be made regarding this point:

- Loud, high-frequency noises are considered a nuisance or annoyance.

They contribute to performance decrements by creating a distraction and affecting concentration and effective communication.

- More familiar noises are perceived as less annoying (Grandjean, 1988).
- Sound features that contribute to the different degrees of "noise perceived annoyance" are intensity, frequency, duration, and the meaning to the listener. For example, the annoyance level increases as the intensity and frequency of sound increases, the longer it remains, the less meaningful it is to the listener (Kryter and Pearsons, 1966).
- Overhearing conversations is also considered, by most people, annoyance (Waller, 1969).
- Intermittent noise can reduce performance on mental tasks more than continuous noise.

Effects of Noise on Health

The most important and obvious effect of noise on health is hearing loss and deafness. Hearing loss and deafness can have two causes: conduction deafness and nerve deafness. Conduction deafness is caused by the inability of the sound waves to reach the inner ear and vibrate the eardrum adequately. Conditions that may lead to conduction deafness are wax in the ear canal and infection or scars in the middle ear or eardrum. Conduction deafness is not the product of a noisy environment and does not result in complete hearing loss. Nerve deafness is caused by reduced sensitivity in the inner ear due to damage or degeneration of nerve cells.

The hearing loss here is in the higher frequencies, whereas conduction deafness is more even across frequencies. Establishing a strong relationship between hearing loss and exposure to a specific noise is not easy.

Ageing Hearing Loss

As a result of the normal ageing process, hearing loss can occur due to changes in the structure of the auditory system. The change is more noticeable in men than in women. This hearing loss is also more noticeable at the higher frequencies. The decline in hearing loss is not consistent in all individuals simply because of individual biological and physiological differences and susceptibility.

Noise-Induced Hearing Loss

Noise-induced hearing loss or deafness can be related to continuous or non-contiguous exposure to noise. This noise-induced deafness can be temporary (temporary threshold shift) or permanent (permanent threshold shift). The temporary threshold shift and permanent threshold shift affect the same areas of the ear.

Temporary hearing loss is characterized by the ability to recover normal hearing in a few hours or days after exposure to continuous high-intensity noise (hence, the term temporary threshold shifts).

Vibration

Vibration is any regular or irregular movement of a body about its fixed position. Regular vibration (such as experienced in machinery like pumps, compressors, or power saws) is known to have a predictable waveform that repeats itself at regular intervals. Random vibration is irregular, unpredictable, and the most common type encountered in the real world; for example, vibration experienced in trucks, tractors, trains, and aeroplanes. Vibration is characterized by its direction and amount (frequency and acceleration). The body can vibrate and move in one or more directions.

Effects of Vibration on Performance

The principal effects of vibration on performance are reduced motor (i.e., hand steadiness) and visual (i.e., eye fixation) control. No strong evidence exists to suggest vibration affects information processing, reaction time, pattern recognition, or any intellectual (cognitive) tasks.

Body vibration results in discomfort at certain combinations of exposure duration and average vibration level. Figure 8 shows the limit for various combinations of exposure duration and average vibration level for whole-body vibrations in standing and seated work.

Figure 8. Vibration effect (Kovács, 2015)

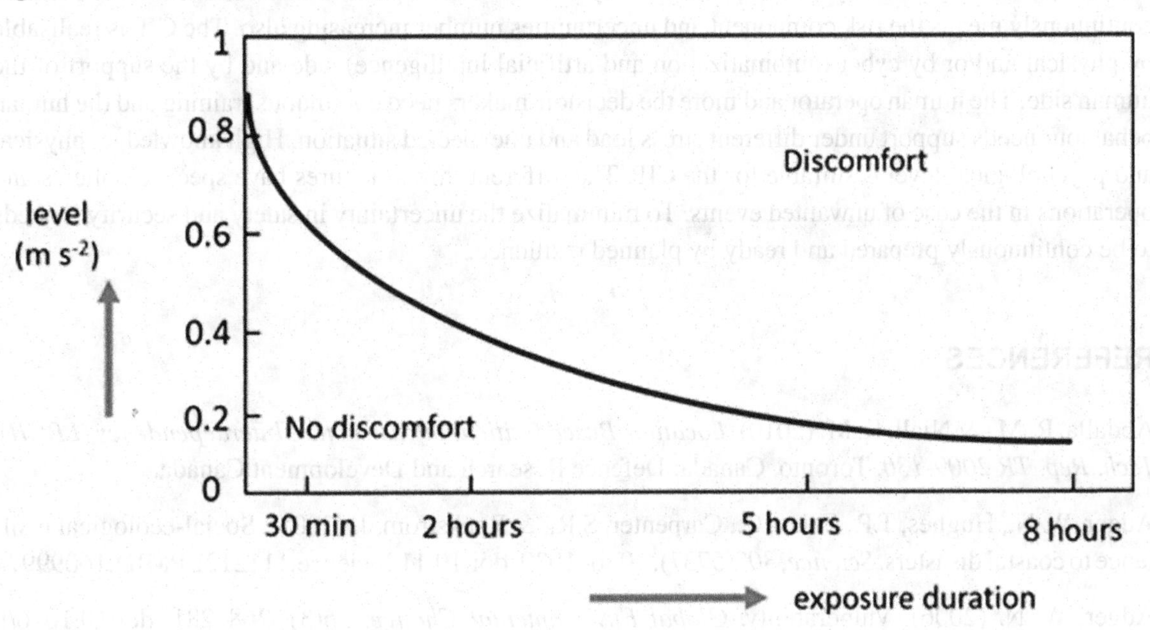

SOLUTIONS AND RECOMMENDATIONS

The CIP task assured by government what it a guarantee for citizens a secure life. For assure safety life and work environment the government by medical infrastructure is responsible. The infrastructures are served and used by citizens the human aspects, needs and abilities must strongly take notice. Unfortunately, the human errors are the cause in several times of the accidents and damages. The human error origin can be the low knowledge level, the environment factors effects for the human mind and physical abilities. On the base of the knowledge of the uncertainties in safety and security, it can conclude that the "human factors" consideration is indispensable to decrease the unexpected human error caused damages.

FUTURE RESEARCH DIRECTIONS

The CIP is a really basic topic of the safe and secure life. Since the operation of the infrastructure is a human task, important to earn higher-level knowledge of human behaviour. In the risk analysis needs to take account of the results of the investigation of the reasons for the human error. On the base of the experimental results needs to devise a modern and investigated education system to train the employees physical, psychological and knowledge level to be able for a suitable behaviour under unexpected or overload conditions.

CONCLUSION

The future challenges in CIP can come by the new technical instruments like drones and individual made devices and hacking. Also, the cyberattacks risk increasing because nowadays everything available via the internet. By this way can stop and destroy a lot of CI component. The technical level increase continuously means the risk component and uncertainties number increasing also. The CIP is realisable by physical and/or by cyber (automatization and artificial intelligence) side and by the support of the human side. The human operator and more the decision-makers need continuous training and the human behaviour needs support under different stress load and unexpected situation. High knowledge, physical and psychological level is suitable for the CIP. The different infrastructures have specific policies and operations in the case of unwanted events. To minimalize the uncertainty in safety and security it needs to be continuously prepared and ready by planned resilience.

REFERENCES

Abdalla, R. M., & Niall, K. M. (2010). *Location-Based Critical Infrastructure Interdependency (LBCII). Tech. Rep. TR 2009-130*. Toronto, Canada: Defence Research and Development Canada.

Adger, W. N., Hughes, T.P., Folke, C., Carpenter, S.R., & Rockstrom, J. (2005). Social-ecological resilience to coastal disasters. *Science, 309*(5737), 1036–1039. doi:10.1126cience.1112122 PMID:16099974

Adger, W. N. (2006). Vulnerability. *Global Environmental Change, 16*(3), 268–281. doi:10.1016/j.gloenvcha.2006.02.006

Auld, H., & MacIver, D. (2007). Changing weather patterns, uncertainty and infrastructure risks: emerging adaptation requirements. Environment Canada.

Baker, G. H. (2005). *A vulnerability assessment methodology for critical infrastructure sites*. Retrieved from http://works.bepress.com/george_h_baker/2/

Barker, K., & Haimes, Y. Y. (2009). Assessing uncertainty in extreme events: Applications to risk-based decision making in interdependent infrastructure sectors. *Reliability Engineering & System Safety, 94*(4), 819–829. doi:10.1016/j.ress.2008.09.008

Brooks, N., Adger, N. W., & Kelly, M. P. (2005). The determinants of vulnerability and adaptive capacity at the national level and the implications for adaptation. *Global Enviro. Change. Part A, 15*(2), 151–163. doi:10.1016/j.gloenvcha.2004.12.006

Bruneau, M., Chang, S. E., Eguchi, R. T., Lee, G. C., O'Rourke, T. D., Reinhorn, A. M., ... von Winterfeldt, D. (2003). A framework to quantitatively assess and enhance the seismic resilience of communities. *Earthquake Spectra, 19*(4), 733–752. doi:10.1193/1.1623497

Brunner, E.M. & Suter, M. (2008). *International CIIP Handbook*. Academic Press.

Bryan, B. A., & Crossman, N. D. (2008). Systematic regional planning for multiple objective natural resource management. *Journal of Environmental Management, 88*(4), 1175–1189. doi:10.1016/j.jenvman.2007.06.003 PMID:17643737

Burby, R. J., Deyle, R. E., Godschalk, D. R., & Olshansky, R. B. (2000). Creating hazard resilient communities through land-use planning. *Natural Hazards Review, 2*(1), 99–106. doi:10.1061/(ASCE)1527-6988(2000)1:2(99)

Burke, D. A. (1999). Towards a Game Theory Model of Information Warfare [PhD thesis]. Air Force Institute of Technology.

Burton, I., Huq, S., Lim, B., Pilifosova, O., & Schipper, E. L. (2002). From impacts assessment to adaptation priorities: The shaping of adaptation policy. *Climate Policy, 2*(2-3), 145–159. doi:10.3763/cpol.2002.0217

Bussey, M., Carter, R. W. B., Keys, N., Carter, J., Mangoyana, R., Matthews, J., ... Smith, T. F. (2011). Framing Adaptive Capacity through a History-Futures Lens: Lessons from the South East Queensland Climate Adaptation Research Initiative. *Futures, 44*(4), 385–397. doi:10.1016/j.futures.2011.12.002

Cain, J. (2001). *Planning improvements in natural resources management. Guidelines for using Bayesian networks to support the planning and management of development programmes in the water sector and beyond*. UK: Centre for Ecology and Hydrology Wallingford.

Charniak, E. (1991). Bayesian networks without tears. *AI Magazine, 12*, 50–63.

Chen, H., Wood, M. D., Linstead, C., & Maltby, E. (2011). Uncertainty analysis in a GIS-based multi-criteria analysis tool for river catchment management. *Environmental Modelling & Software, 26*(4), 395–405. doi:10.1016/j.envsoft.2010.09.005

Commission of the European Communities. (2006). *Communication from the Commission on a European Programme for Critical Infrastructure Protection*. Retrieved from http://europa.eu/legislation_summaries/justice_freedom_security/fight_against_terrorism/l33260_en.htm

Commonwealth of Australia. (2010). *Critical infrastructure resilience strategy*. Retrieved from http://www.tisn.gov.au/

Commonwealth of Australia, (2011). *National guidelines for protecting critical infrastructure from terrorism*.

Cooke, R. M., & Goossens, L. H. J. (2004). Expert judgment elicitation for risk assessments of critical infrastructures. *Journal of Risk Research*, *7*(6), 643–656. doi:10.1080/1366987042000192237

Cutter, S. L., Barnes, L., Berry, M., Burton, C., Evans, E., Tate, E., & Webb, J. (2008). A place-based model for understanding community resilience to natural disasters. *Global Environmental Change*, *18*(4), 598–606. doi:10.1016/j.gloenvcha.2008.07.013

De Porcellinis, S., Oliva, G., Panzieri, S., & Setola, R. (2009, March). A holistic-reductionistic approach for modeling interdependencies. In *Proceedings of the International Conference on Critical Infrastructure Protection* (pp. 215-227). Springer. 10.1007/978-3-642-04798-5_15

Di Giorgio, A., & Liberati, F. (2012). A Bayesian Network-Based Approach to the Critical Infrastructure Interdependencies Analysis. *IEEE Systems Journal*, *6*(3), 510–519. doi:10.1109/JSYST.2012.2190695

Dudenhoeffer, D. D., Permann, M. R., & Manic, M. (2006). CIMS: A Framework for Infrastructure Interdependency Modelling and Analysis. In *Proc. of the 38th Winter Simulation Conference* (pp. 478-485). Academic Press. 10.1109/WSC.2006.323119

Dunn, W. N. (2001). Using the method of context validation to mitigate type III Errors in environmental policy analysis. In M. Hisschemoller, R. Hoppe, W. N. Dunn, & J. Ravetz (Eds.), *Knowledge, power and participation in environmental policy* (pp. 417–436). New Brunswick, London: Transaction Publishers.

Egan, M. J. (2007). Anticipating future vulnerability: Defining characteristics of increasingly critical infrastructure-like systems. *J. Counting. Crisis Manage.*, *15*(1), 4–17. doi:10.1111/j.1468-5973.2007.00500.x

Ezell, B. C. (2007). Infrastructure Vulnerability Assessment Model (I-VAM). *Risk Analysis*, *27*(3), 571–583. doi:10.1111/j.1539-6924.2007.00907.x PMID:17640208

Fenton, N., & Neil, M. (2013). *Risk Assessment and Decision Analysis with Bayesian Networks*. London: CRC Press.

Folke, C. (2006). Resilience: The emergence of a perspective for social-ecological systems analyses. *Global Environmental Change*, *16*(3), 253–267. doi:10.1016/j.gloenvcha.2006.04.002

Ford, J. D., Smit, B., & Wandel, J. (2006). Vulnerability to climate change in the Arctic: A case study from Arctic Bay, Canada. *Global Environmental Change*, *16*(2), 145–160. doi:10.1016/j.gloenvcha.2005.11.007

Freeman, P. K. (2003). Natural Hazard Risk and Privatization. In A. Kreimer, M. Arnold, and A. Carlin (Eds.), Building Safer Cities: The Future of Disaster Risk (pp. 33-44). Academic Press.

Friedlingstein, P., & Solomon, S. (2005). Contributions of past and present human generations to committed warming caused by carbon dioxide. *Proceedings of the National Academy of Sciences of the United States of America*, *102*(31), 10832–10836. doi:10.1073/pnas.0504755102 PMID:16037209

Giannopoulos, G., Filippini, R., & Schimmer, M. (2012). *Risk assessment methodologies for Critical Infrastructure Protection. Part I: State of the art*. Luxembourg: Publications Office of the European Union.

Godschalk, D. R. (2003). Urban hazard mitigation: Creating resilient cities. *Natural Hazards Review*, *4*(3), 136–143. doi:10.1061/(ASCE)1527-6988(2003)4:3(136)

Gonzalez, J. J., Sarriegi, J. M., & Gurrutxaga, A. (2006). A Framework for Conceptualizing Social Engineering Attacks. In J. L'opez (Ed.), *CRITIS 2006* (pp. 79–90). Springer. doi:10.1007/11962977_7

Government of Canada, (2009). National strategy for critical infrastructure.

Kovács, T. (2015). *Ergonomics*. Budapest: Óbuda University.

Nyikes, Z. (2017). *Digital Competence and the Safety Awareness Base in the Assessments Results of the Missle-East-European Generation*. Tirgu-Mures, Romania: INTER-ENG.

Szakali, M., & Szűcs, E. (2017). The defence planning model formation and development, Védelmi tervezésimodellek kialakulása és fejlődése. *Hadmérnök*, *12*(1), 24–40.

Szakali, M., & Szűcs, E. (2018). The price of the security, probably effects of the defence budget expansion, A biztonságára, avagy a védelmiköltségvetésemelésénakhatásai. *Hadmérnök*, *13*(1), 314–325.

Zio, E. (2016). Challenges in the vulnerability and risk analysis of critical infrastructures, *Reliability Engineering and System Safety, 152*, 137-150

ADDITIONAL READING

Beke, É., Kovács, T., & Rajnai, Z. (2020). (article in press). Critical Infrastructure Protection Framework and Trends in Attacks. *Critical Infrastructure Protection Review.*

Dely, P. (2017). The Legislative Environment of the Critical Infrastructure Protection in Hungary. *Hadmérnök*, *12*(3), 188–197.

Kovács, T., & Zhao, F. (2015). *Ergonomics*. Hungary: Óbuda University Press.

Kovács, T. A., Nyikes, Z., & Daruka, N. (2019). Critical Infrastructure Protection in the Historical Urban Region of Eastern European Countries. In M. Banasik (Ed.), Security and Russian Threats (pp. 151-158). The Jan Kochanowski University in Kielce.

Nyikes, Z. (2018). Digital competence and the safety awareness base on the assessments results of the Middle East-European generations. *Procedia Manufacturing*, *22*, 916–922. doi:10.1016/j.promfg.2018.03.130

Nyikes, Z. (2019). *Information Security Development with the User Support Opportunities*. Unpublished doctoral dissertation, Óbuda University, Hungary.

Nyikes, Z., & Rajnai, Z. (2015). Big Data, As Part of the Critical Infrastructure. In *Proceedings of SISY 2015: IEEE 13th International Symposium on Intelligent Systems and Informatics* (pp. 217-222). IEEE Press.

Vukmirovic, A., & Rajnai, Z., Radojičić, M., Vukmirović, J., & Milenković, M.J. (2018). Infrastructure Model for the Healthcare System based in Emerging Technologies. *Acta Polytechnica Hungarica*, *15*(2), 33–48.

KEY TERMS AND DEFINITIONS

Critical Infrastructure: (CI) is assigned in the literature by different definitions. It can define as those assets and systems, that are essential for the maintenance of the social functions, health, safety, security, and economics or social well-being of people.

Environmental Factors: Physical environmental factors such as noise, vibration, lighting, the climate can affect people's safety, health and comfort.

Human Error: Human error is a term describing a planned sequence of mental or physical actions that fail to achieve the intended objective, and that this failure cannot be attributed to chance.

Human Factor: The human factors or ergonomics mean all of the human capabilities, limits and human performance even that these parameters are psychological or physical of human what is need respect during the design engineering.

Physical Factors: Are the human body capabilities and limits during operations in static posture or motion.

Resilience: Resilience is a force against the negative risk impact the critical infrastructure.

Risk Analysis: Risk analysis is the process is identifying and analysing the potential issues that could negatively impact critical infrastructure.

Uncertainty: is commonly considered as any departure from the unachievable ideal of complete determinism.

Chapter 15
Business Continuity of Critical Infrastructures for Safety and Security Incidents

Konstantinos Apostolou
https://orcid.org/0000-0002-7291-8713
Center for Security Studies (KEMEA), Greece

Danai Kazantzidou-Firtinidou
https://orcid.org/0000-0003-3481-6860
Center for Security Studies (KEMEA), Greece

Ilias Gkotsis
https://orcid.org/0000-0003-2228-1387
Center for Security Studies (KEMEA), Greece

George Eftychidis
Center for Security Studies (KEMEA), Greece

ABSTRACT

The chapter is an overview of important timely concepts with a focus on the safety and security of critical infrastructures (CIs). The content is a result of triangulation of sources from the fields of academia, best practices, legislation, and scientific research. The protection of CIs has been a popular topic of discussion through recent years but also a topic for initiative towards the undisrupted function, prosperity and well-being of nations in a world of interconnections and dependencies. In respect to that, the following content offers input which will assist in the understanding of the concepts surrounding the safety and security of CIs while combining theoretical approaches with practical guidelines for the composition of a business continuity plan. The chapter also discusses the factors contributing to the criticality of technical infrastructures as part of a nation or a cross-border network, the threats to which a CI can be exposed to whether these are natural or man-made.

DOI: 10.4018/978-1-7998-3059-7.ch015

INTRODUCTION

Critical infrastructures (CIs) are components of wide networks of technical infrastructures which support the daily activities of every nation and provide vital products and services. These services include distribution of natural gas, electricity, fuels, transportation services, water distribution and others.

CIs can be subjected to damage caused by natural hazards or man-made threats which can be amplified due to the networks that have been created between them and might result into disastrous cascading effects. Such incidents can potentially render entire cities unable to provide vital services and functions, while also threating the societal security and safety. Natural disasters such as the tsunami and earthquake which triggered the nuclear disaster in Fukushima Daiichi, the more recent wildfires in Greece 2018 and the Typhoon Jebi which hit Japan in 2018 were events of very high magnitude. Scientific research has also stressed the importance of the effects of natural hazards due to climate change with potentially devastating effects for the European CIs especially in sectors such as energy, transportation, water management (EU-CIRCLE, 2016). At the same time, the high value of CIs can attract man-made threats such as terrorism, aimed towards attracting the media, causing mass casualties and spreading fear, or antisocial behavior meant to apply political pressure.

Unfortunately, such events are difficult to predict and pinpoint and there is no solution which would create an absolute defense system against such hazards. As such, it is necessary for CI operators to establish and effective Business Continuity Plan (BCP) which will focus not on the absolute protection of an infrastructure, but on the timely response to adverse incidents or emergencies, in order to safeguard the infrastructure's vital operations and assets from future disruptive events.

This chapter will discuss further the significance of the protection European CIs, the possible threats that they can be exposed to, and present an elaborate framework towards Business Continuity Planning based on literature, research and best practices for general application, independently of the infrastructures' field of operations or the nature of hazards.

The objectives of this chapter are:

1. To increase the understanding of risk management concepts around the safety and security of Critical infrastructures.
2. To provide a comprehensive, elaborate and practical guide for Business Continuity Planning which will contribute to an effective emergency response
3. To raise the awareness of Critical Infrastructure stakeholders in regard to natural and man-made risks to which the European Critical Infrastructures may be exposed.

BACKGROUND

The modern approach towards the protection of critical infrastructure against disasters (natural or man-made), is focused on a proactive strategic planning of an infrastructure in order to build resilience against hazards which may occur. Resilience enables an infrastructure (or organization) to respond to disruptions through its ability to absorb damage and adapt to adverse incidents, ensuring the continuation of its critical activities and the safety of its stakeholders. Resilience can be built through an elaborate Business Continuity Management approach (British Standards Institution [BSI], 2018b).

Business continuity (BC) refers to the ability of an organization to deliver goods and services at acceptable predefined levels after a disruptive incident (BSI, 2018b, p. 41). In general, disruptive threats against technical infrastructures, may include natural hazards, man-made disasters, terrorism, IT/IS incidents, disruption to supply chain, disruption to internal support services and others (Smith, 2012). Considering the above categories, a disruptive incident could be a fire outbreak which can cause physical damage to an infrastructure and threaten the safety of the personnel also affecting infrastructure's supply chain, or it could involve a cyberattack targeting the IT systems of an organization which will cripple its operations and prevent delivery of services.

As risks cannot be nullified during an organization's activities, organizations must ensure their viability and continuation of operations when faced with sudden events. This can be possible through a predetermined Business Continuity Plan (BCP) which is designed before and is activated during emergencies (BSI, 2018b). This will allow an organization to practice an effective response to emergencies and avoid the escalation of an incident into a disaster resulting crisis. Business Continuity Management's application is common for both the private and public sector with the only difference being the focus on the continuity of products regarding the private sector, in contrast to the focus on the delivery of services which is crucial in the public sector (Smith, 2012)

In recent years much emphasis has been given to the ability of an infrastructure to continue its operations in the face of all hazards. An all-hazard approach towards safety and security, means that a framework followed for the protection of a Critical Infrastructure, is not threat-specific, but it's designed to account for all types of threats (man-made or natural) that can be imposed onto an infrastructure. Today, the U.S., Australia, and the EU follow risk based all-hazard approaches relying on documented analysis for the protection of their Critical Infrastructure (Commonwealth of Australia, 2015; European Council, 2008; Homeland Security, 2013). That does not necessarily mean that an all-hazard approach is the superior solution, but in the case that an organization follows the guidelines of International Standards (ISOs) on Risk Management, Business Continuity and Emergency Response, duplication of effort can be avoided, since they share a lot of procedures and types of analysis (United Nations [UN], 2018).

CRITICAL INFRASTRUCTURE PROTECTION

Investing in the protection of critical infrastructure based on a proactive risk management approach can be perceived as a waste of resources and time on both national and local level. Especially on a European level, the effort of adopting an integrated approach, can be hindered by national political agendas, budget and resources differences and gaps in the level of expertise. Moreover, it is difficult to convince a management to invest in safeguarding its operations from threats that it is not aware of, or not directly threated by, or that have never been realized in the past in close proximity to an infrastructure's territory. However, proactive planning and informed decision making can safeguard an infrastructure from contingencies and disaster avoidance which could otherwise terminate its operations permanently.

Critical Infrastructure Identification

The following section provides some background on the effort of the EU to underline and initiate a modern integrated approach for the protection of its CI's while providing academic content to highlight the significance of the European Critical Infrastructures as a network, and presents an overview of previous past events to raise the awareness of the reader on the importance of CI protection.

Towards the Protection and Identification of Critical Infrastructures

The first acknowledgement for an official and organized effort for the protection of CIs on a European level took place in 2004, as the European Council requested the designation of an EU integrated strategy for the protection of critical infrastructure (Council of the European Union, 2008).

Following that, in 2006 the EU set the parameters for the implementation of the European Program for the Protection of Critical Infrastructure (EPCIP) which adopted an "all-hazards approach" towards threats, including the protection against natural disasters, technological and man- made threats, while prioritizing and focusing heavily on the prevention preparedness and response towards terrorist attacks. The focus of the program is understandable due to the then recent tragedy caused by the 9/11 attacks and the worldwide shock following the realization of the fragility of a super power's security. Along with EPCIP, the European Commission established a Critical Warning Information Network (CIWIN), a platform for the exchange of best practices and security issues related to Critical Infrastructures (European Commission, 2006). Both EPCIP and CIWIN are still active today.

In 2008, the European Council Directive 2008/113/EC defined CIs in general as an asset, system or part of system which is located in a Member State and is essential for the vital functions of the society, and the health, safety, security, economic or social well-being of people, which (infrastructure) if disrupted or destroyed will have a major impact on the society and functions of a Member State. This includes networks, systems spaces, facilities and operations of an infrastructure. A complementary definition was provided for the designation of a European Critical Infrastructure (ECI) which links the impacts of the disruption and destruction of a CI to at least two Member States while also highlighting the spread of the resulting effects onto different types of infrastructures as a result of cross-sector dependencies (European Council, 2008).

Additionally, the directive identifies as ECI sectors and focuses only on the sector of Energy followed by the sub-sectors of Electricity, Oil and Gas, and the sector of Transportation, which consists of the sub-sectors of Road transport, Rail transport, Air transport, Inland waterways transport, Ocean and short-sea shipping and ports.

On the other hand, the U.S. through the Presidential Policy Directive/PPD-21 (2013), identifies 16 critical infrastructure sectors. The list consists of the Chemical, Commercial Facilities, Communications, Critical Manufacturing, Dams, Defense Industrial Bases, Emergency Services, Energy, Financial Services, Food and Agriculture, Government Facilities, Healthcare and Public Health, Information Technology, Nuclear Reactors, Materials, and Waste, Transportation Systems, Water and Wastewater Systems. The extensive list represents a strong resolution of the U.S. towards its national security and its approach to CIs as a network of networks.

Similarly, the need for a holistic approach towards CI is also underlined by the more recent 2016 EU Directive 2016/1148 (NIS Directive) which focuses on measures for the security of network and information systems in the EU. The directive expands upon the sectors initially introduced by 2008/113/EC Directive by including the sectors of Banking, the Financial market infrastructures, the Health sector, the Drinking water supply and distribution, and the Digital Infrastructure (L 194/27). This expansion is only natural since the functional state and a society's daily activities are heavily dependent on all these services.

EU Legislation for the Protection of Critical Infrastructure

Table 1 presents important initiatives of the European Union for the protection of Critical Infrastructures in recent years.

Table 1. EU initiatives for the protection of European Critical Infrastructure (Center for Security Studies, 2018b)

EU Directives	Main Points
Directive 2008/111/EC	• Identification and Designation of ECIs European Program on CI Protection (EPCIP) • Critical Infrastructure Warning and Information Network (CIWIN) • Operator Security Plan (OSP)
Directive 2016/1148//EC	• Security of Network and Information Systems • EU network Computer Security Incident Response Teams (CSIRT)
EU initiatives and Institution	
European Program on CI Protection (EPCIP)	• Creation of procedures for the identification and designation of ECIs • Creation of Critical Infrastructure Warning and Information Network (CIWIN) • Funding of EU Project for the protection of ECIs
Critical Infrastructure Warning and Information Network (CIWIN)	• E-platform for information exchange and communication for the protection of CIs
Thematic Network on Critical Energy Infrastructure Protection (TNCEIP)	• Information exchange between Energy sector's CI operators (managers) regarding risk assessment, risk management cybersecurity and others.
European Reference Network for Critical Infrastructure Protection (ERNCIP)	• Exchange of knowledge and expertise between experimental facilities and laboratories for harmonization of test protocols in EU
Incident and Threat Information Sharing EU Centre for the Energy Sector (ITIS-EUC)	• CIs of Energy situational awareness- collection and analysis of information about incidents and vulnerabilities
European Union Agency for Cyber Security (ENISA)	• Security Consultancy for EU Member States • Improve security in the EU • Publishing of best practices, recommendations and guidelines
Joint Research Center (JRC) (EU Commission Science and Knowledge Service)	• Independent scientific and technical support for EU policies • Tool development for policy makers

Interdependencies Between Critical Infrastructures

Technical infrastructures can be seen as complicated systems of interconnected components whose connections/links expand outside a single infrastructure (Rinaldi, 2001). Consequently, an infrastructure is part of an interconnected web and constitutes a network within a network (or system within a system).

The interconnection between CIs can occur between CIs of the same sector or it can be cross-sectoral between different types of infrastructure and even cross-border in the case of the ECIs. This interconnection between CIs is characterized by a state of dependency, or interdependency (Rinaldi 2001). This interconnection is beneficiary for the daily operations and productivity of the CIs, as well as the financial, social prosperity and security within a nation. On the other hand, it may also lead to a network of targets with increased vulnerabilities to physical or cyber threats, also susceptible to cascading effects and magnified impact upon realization of threats. Cascading effects can have particularly high impact, when two or more CIs or ECIs are interdependent. In reference to the above interconnections, Hurricane Sandy hit New York in 2005, causing major power outage in the area leading to the disruption of cellular towers and poor cellular service, thus damaging both the energy infrastructures and the telecommunications of the state (Gibbs & Holloway, 2013).

A popular categorization of CI interdependency types is proposed by Rinaldi (2001), which expands beyond the exchange of products and services between infrastructures:

- **Physical Interdependency**: Two infrastructures are physically interdependent if they both rely on the physical connection their inputs and outputs for their operations
- **Geographic Interdependency**: Elements of multiple infrastructure are in close spatial proximity and can all be affected by a local environmental effect such an explosion. However, the state of one infrastructure does not necessarily affect the activities of the other.
- **Cyber Interdependency**: An infrastructure's operability depends on information transmitted through an information infrastructure.
- **Logical Interdependency**: Two infrastructure base their operations on each other but not via a cyber, physical or geographic function.

Based on the above, for the construction of a Business Continuity Plan an analysis of a single infrastructure or its technical components and processes should not be implemented in isolation, but as part of the wider system which affects it and is affected by it. The systemic approach for examining interdependencies, is explained by the theory of "systems thinking" which is a well-established approach for analyzing complex systems and consists of the elements of a system (characteristics), the interconnections between the elements and their function (purpose) (Arnold & Wade, 2015).

Figure 1 provides a general presentation of dependencies and interdependencies between different infrastructures through the exchange of services/output. The categorization starts from the wide sectors of infrastructures (systems), is then narrowed down to the infrastructures (sub-systems), and these systems are furtherly broken down into the service they provide (e.g. SCADA systems support) and the service they receive to support the function of their assets.

Table 6. (see Appendix 1) is indicative of the sectors, sub-sectors and critical services of CIs. In the case of an individual CI's critical services, an analysis should expand even further by prioritizing between services and processes, identifying and prioritizing the critical assets which support these processes, as well as identifying which other infrastructures these services affect by taking into consideration the aforementioned interdependencies resulting in a top-down analysis.

Safety and Security of Critical Infrastructures

Although the terms of safety and security can be approached as one, and use similar analysis framework, their distinction can assist in a more elaborate analysis. In literature, many times, the terms security and safety are used interchangeably, or without clarification about their differences, similarities and their scope. Within this chapter, two different definitions are adopted that reflect the different perspectives of CI protection.

Security in daily operations, describes a framework of proactive activities focusing on the protection of Critical Infrastructure assets, the protection of its operations and the protection of its human resources, against malicious manmade threats (Center for Security Studies, 2018a). Consequently, the effort for

Figure 1. Interconnections between different Critical Infrastructures

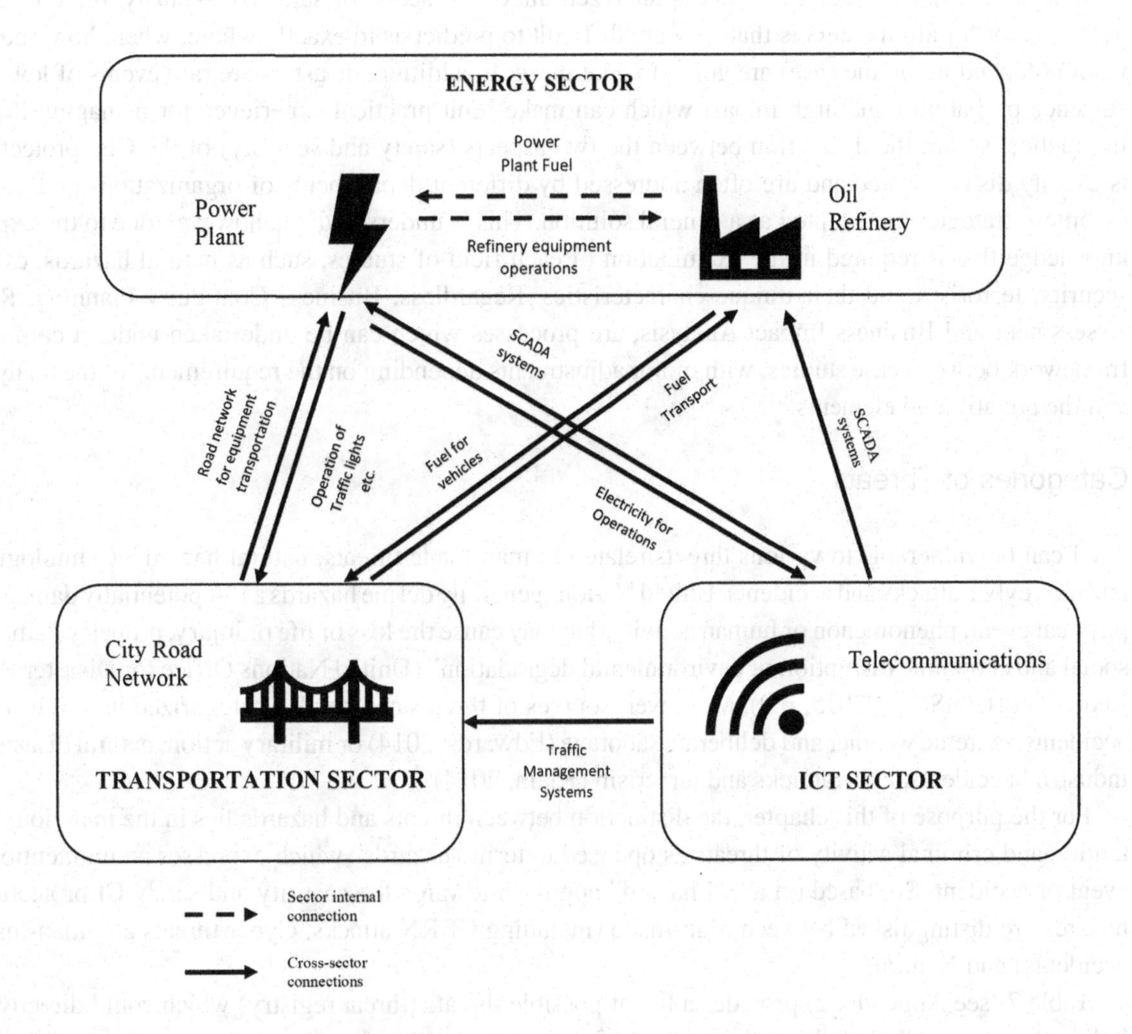

the protection of infrastructure and the mitigation of risk aims at threats such as terrorist attacks, cyber-attacks, acts of sabotage, criminal activities and other malicious acts. Security measures may include alarm systems, surveillance and operations monitoring systems, physical barriers for physical protection or denial of entry to areas, present of security personnel and others. Additionally, cyber security measures may include physical security perimeter of areas with sensitive information, information access restriction, passwords management and others (International Standards Organization [ISO], 2011)

On the other hand, safety describes established measures, policies and procedures in a CI, intended to ensure the protection of a CI and its assets from non-intentional hazards, such as natural hazards, and other threats (accidents), including natural hazards which can be combined with technological accidents and result into what is characterized as Natech hazards (Center for Security Studies, 2018a). Safety polices may traditionally include evacuation plan, fire and earthquake resistant buildings, protective barriers in hazardous areas, controlled access to certain areas etc.

Independently of whether a threat is analyzed under the scope of safety or security, the common denominator for all disasters is that they are difficult to predict as to exactly where, when, how and by whom (depending on the case) are going to take place. In addition, disasters are rare events of low occurrence probability and high impact which can make limit practical experience for managing them. In practice, so far, the distinction between the two aspects (safety and security) of the CIs' protection is usually disaggregated and are often addressed by different departments of organizations and rarely common strategies are adopted as a general solution. This is understandable however, due to the expert knowledge that is required in the examination of each field of studies, such as natural hazards, cyber security, terrorism and their unique characteristics. Regardless, Business Continuity Planning, Risk Assessment and Business Impact Analysis, are processes which can be undertaken under a common framework between case studies, with minor adjustments depending on the requirements of the analysis and the quantifiable elements.

Categories of Threats

A CI can be vulnerable to various threats related to man-made threats, natural hazards, technological hazards, cyber-attacks and accidents. United Nations generally define hazards as "A potentially damaging physical event, phenomenon or human activity that may cause the loss of life or injury, property damage, social and economic disruption or environmental degradation" (United Nations Office for Disaster Risk Reduction [UNISDR], 2015, p.9). Moreover, sources of threats can be also categorized into industrial accidents, extreme weather and deliberate sabotage (Edwards, 2014) or military action, natural disasters, industrial accidents, cyberattacks and terrorism (Klain, 2014).

For the purpose of this chapter, the distinction between threats and hazards lies in the malicious intention and criminal activity of threats as opposed to term "hazards" which expresses an unintentional event or accident. So, based on a "all hazard" approach towards the security and safety CI protection, hazards are distinguished between Man-made (including CBRN attacks, Cyber threats and man-made accidents) and Natural.

Table 7 (see Appendix 2) provides a list of possible threats (threat registry) which could directly or indirectly affect a CI, with the to the purpose of raising awareness regarding the various types of threats that could occur for an infrastructure.

OVERVIEW OF PAST EVENTS

Man-Made Threats

Terrorist Attacks

In 2016, two simultaneous explosions took place at the main terminal of the Zaventem airport in Brussels, involving two suicide bombers, followed by another bomb explosion in the Maelbeek metro station ("Brussels explosions", 2016).

In Madrid 2004, 10 bombs using TNT charges were set off by terrorists in four commuter trains during rush hour, which exploded and killed almost 180 people and injured more than 2000 (Ceballos et al., 2005).

In both cases, the attacks posed a threat to society's health and well-being while disrupting the operations of the transportation sector, also causing financial and physical damage.

CBRN Attack

In 2016 a former Russian Intelligence officer was attacked via a CBRN nerve agent (Novichok). The agent was applied to the front door of the victim's house via a perfume bottle ("Russian Spy Poisoning", 2018; Morris, 2018). The incident exhibits the variety of threats that can potentially occur against the EU and the understandable inability to protect everything at any time.

CBRN threats, refer to attacks which make use of Chemical, Biological, Radiological and Nuclear (CBRN) weapons (European Parliament, 2018). In many cases, the term can be also found as CBRNE with the inclusion of explosives (E).

Cyber Attack

In 2007, Estonia was the victim of a series of high scale cyber-attacks, which disabled major national online services and target Estonia's banks major websites, governing bodies, the police and media outlets. The attacks took place in a period of approximately twenty days. Some the results of the attack included denial of access to information by citizens regarding, news and government updates (denial of communication) and denial of access to online banking services (economic impact) (North Atlantic Treaty Organization Strategic Communications Centre of Excellence, n.d.)

According to the Organization for Security and Co-operation in Europe (2013), among the most critical threats against Industrial Control Systems (ICS) by attackers are:

- Online attacks via office or enterprise networks, because of the office IT connection to the network or network connection from offices to the ICS.
- DoS (Denial of Service) attacks which disrupt the ICS operations and cause a system to fail by impairing network connections and essential resources
- Human error and sabotage (intentional)

(United Nations Counter-Terrorism Committee [UNOCT], Counter Terrorism Committee Executive Directorate [CTED] & Interpol, 2018)

Drone Incident

Drones are a newly introduced threat to the security of critical infrastructures since they have become commercially available and very affordable. The two drones that were spotted flying around the runway of Gatwick airport in December 2018, resulted in over 800 cancelled flights due to security reasons, severely disrupting the airports operations and air traffic around Europe, while also demonstrating how unprepared the EU critical infrastructures' security can be when dealing with newly introduced threats. (Mueller & Tsang, 2018). Additionally, the incident revealed the ease of access drones have into restricted areas.

Antisocial Behavior

In 2012, the center of Athens and more precisely the wider area around the Syntagma square was the stage of violent activities during a clash between the police force and protesters, including vandalism of properties and arson. Improvised Incendiary Devices (IIDs) were used among others, against bank branches (Financial CIs). Furthermore, many traffic lights were destroyed, and the fights isolated the center of Athens denying access to pedestrians and vehicles (Assets of Transportation infrastructure) (Pouliopoulos, 2012).

Man- made threats to critical infrastructure also include potential sabotage from the inside, hijacking of vehicles (as part of the supply chain), hacking into an organizations database, disruption of operations, vandalism or anti-social behavior such as occupation of offices and others (see Appendix 2).

Natural Hazards- Incidents

Typhoon

Japan has suffered many times in the past from Natural Hazards causing many disasters with cascading effects. In 2018 the area of Osaka was hit by Typhoon Jebi which is rated as a category 3 typhoon on the Saffir- Simpson scale. Osaka's Kansai airport is located on an island by Osaka's mainland, and it is connected to the mainland via bridge. The typhoon's strong winds caused a ship (asset) to drift away falling onto the bridge (1st road transportation infrastructure) and damaging it, isolating the island/ airport (2nd air transportation infrastructure) and 3000 people (stakeholders) on it (Takabatake et al., 2018).

Natech (Earthquake, Tsunami, Nuclear explosion)

In 2011 Japan underwent an even more significant disaster which was a combination of natural hazards, combined with technological failure, leading to a Natech disaster. In March of the respective year, Japan suffered the now well-known Great East Japan Earthquake (1st natural hazard) of magnitude 9.0, causing a subsequent massive tsunami (2nd natural hazards), which flooded the Fukushima Daiichi Nuclear plant. The tsunami caused disruption (loss of electric power) in the ability of the nuclear plant to keep its nuclear reactors' cooling system operating as designed (technological failure), also resulting in several explosions (technological disaster) (Labid, and Harris, 2014). Due to safety features in place (back-up power generators to keep cooling system on, the Tokyo Electric Power Company (TEPCO) estimated that absolute power shut down was not possible (Edwards, 2014).

Natech accidents are expected to be more frequent in the future due to the increase in the combination of urbanization, industrialization and climate change (UN, 2018)

Wildfire

In 2018 in Athens- Greece, a fire broke out at Penteli mountain, which is located 5km away from the east coast of Attica. The fire was guided by strong wings with speed of 90 km and spread to the east until it stopped at the coastline, only after having passed through the nearby residential areas, affecting an area of approximately 12.8 km^2. In the area of Mati (around the coast), the area morphology, the transport network design (randomized pattern of buildings and low capacity streets), the weather conditions and the limited reaction time led lower level of service or even to blockage of several transport network links and nodes which entrapped residents and people passing by, with an aftermath of 305 burned vehicles (Lekkas, et al., 2018) and 102 fatalities ("Ενας χρόνος", 2019). Disruption of the water and power network of the wider area took also place, with 50 power substations (out of the 300) being destroyed by the wildfire ("Φωτιά Αττική," 2018).

Extreme Weather Effects and Climate Change

Based on recent research, CIs from various European countries operating in the transportation, energy, water and public sectors have highlighted extreme weather effects stemming from climate change, such as heavy storm, floods, snow and ice etc., leading for example to the following operational impacts (EU-CIRCLE consortium, 2017b):

- Blocked roads
- Tram and track damage
- Harbor closing
- Flood in Transformer/Distribution facilities
- Power lines and pylons out of order

Although the details of the incidents which can cause an impact may vary per infrastructure the following general categories of strategic impacts can be common for all infrastructures independently of their sector of operations and independently of the type of threats which will cause them:

- Contractual compliance exposure, due to inability to fulfill agreed operations
- Regulatory compliance
- Financial viability (on a company level)
- Damage on property
- Consumer confidence
- Effect on reputation

  Life safety and public health

Business Continuity of Critical Infrastructures

The purpose of this section is to provide a series of practical guidelines which can be utilized by an organization in the design of a Business Continuity Plan (BCP) independently of a specific hazard or threat. The information provided below is based on guidelines for good practice such as ISO 31000 (BSI) for risk management, ISO 22301 for Business Continuity (BSI), ISO/TS 22317 (BSI) for Business Impact Analysis, combined with scientific and academic research. Crisis Management is briefly described with the aim of providing background information for better understanding of the content of Business Continuity.

Crisis Management and Business Continuity

Crisis management is a cyclic and proactive process which provides an organization with the capability to respond to irregular and unstable events, which threaten its strategy, reputation and viability. A crisis may be generated from the mismanagement of an incident or emergency (BSI, 2014c). ISO 22300:2018 (BSI) distinguishes between an incident and an emergency by describing an incident as a situation which is not but can lead to an emergency or crisis. Emergency on the other hand is a sudden, urgent and unexpected event in need of immediate effective response, in order to avoid an escalation into a crisis. Consequently, mishandled emergencies or crises can result into irreversible disasters (Borodzicz, 2010). In light of this, the possible escalation of incidents into a disaster can be seen in Figure 2 [the characteristics of the stages are based on ISO 11200: 2014 (BSI) and Borodzicz (2010)]. In some cases, in literature, emergencies and incidents are used as interchangeable terms.

Figure 2. Escalation of incidents to Disasters

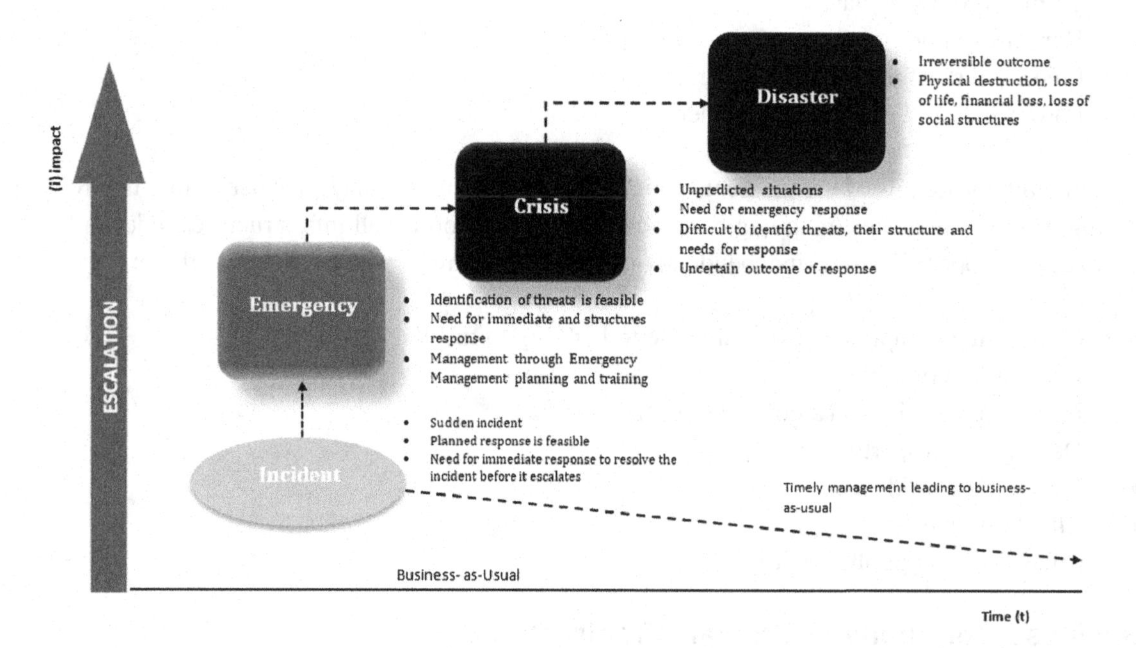

As a proactive process, crisis management is based mainly on the preparation of an organization before a crisis, its quick response capability, adaptability, decision making during a crisis and the recovery of vital functions and assets following the response to a crisis. Following the aftermath of a crisis, it is important to analyze the adverse events and reflect upon implemented actions to review the existing plans and improve further, for future incidents (BSI,2014c). Like so, the following steps are proposed, which form the "crisis management cycle" (Pursianen, 2017):

- Risk Assessment
- Prevention
- Preparedness
- Response
- Recovery
- Learning

The ability of an organization to successfully respond to crises enhances resilience against disruptive incidents, which will allow an organization to be able to endure, adapt and recover from such events. The development of crisis management is built on the same context as risk management, security management and business continuity (BSI, 2014c). A business continuity plan can be considered to take place under a wider corporate strategy of crisis management. The two disciplines share common development processes with risk management such as risk assessment, business impact analysis, vulnerability Thus, if for instance a vulnerability assessment of an infrastructure is performed, it can be then applied in the case of general Risk management or Crisis Management and avoid duplication of efforts. Risk management, Crisis management and Business Continuity Management ultimately complement each other.

Business Continuity Management

Business continuity (BC) is defined as the ability of an organization to deliver goods and services at acceptable predefined levels after a disruptive incident (BSI, 2018b).

The importance and necessity of Business Continuity for the security of CIs is also underlined by Council Directive 2008/114/EC for the identification, designation and protection improvement of the European critical infrastructures.

Business Continuity Management (BCM) is an ongoing cyclic process which starts from risk understanding and impact estimation and comprises the design of the strategy, the development of a business continuity plan, the implementation of the planned actions, preparedness measures, and finally the review and evaluation of the Plan and the action taken during past incidents (BSI, 2014b). It enables an organization to develop its resilience and respond to emergencies through a structured and coordinated manner to achieve a timely restoration of its business-as-usual status (Center for Security Studies, 2018a), focusing especially on the protection of its vital assets, operations and stakeholders. The cyclic process for the design of a Business Continuity Plan is depicted in Figure 3

The result of BCM is the development of a Business Continuity Plan (BCP). From an operational perspective, a BCP consists of a guided procedure which is activated during disruptive events and describes (similarly to the crisis management cycle) the necessary actions for the prevention of disruptive incidents, the response protocol upon an incident, the recovery of assets and operations, and the restoration of critical operations back to their normal function (see "Operational Requirements). It is a "holistic" process which follows a the top-down approach regarding the involvement of an infrastructure's management (initiator) and personnel in the design and execution of a BCP, while at the same time it emphasizes the systemic examination of the infrastructure's components (operations and assets) during the design of the BCP Disruptive events which may cause the activation of a BCP may result into loss of critical facility, operations or equipment, loss of critical material or supplies, loss of communications, loss of power and shortage of personnel (in numbers or value) (EU CIRCLE-consortium, 2017b)

Figure 3. Business Continuity Planning process based on ISO 22313 (BSI, 2015)

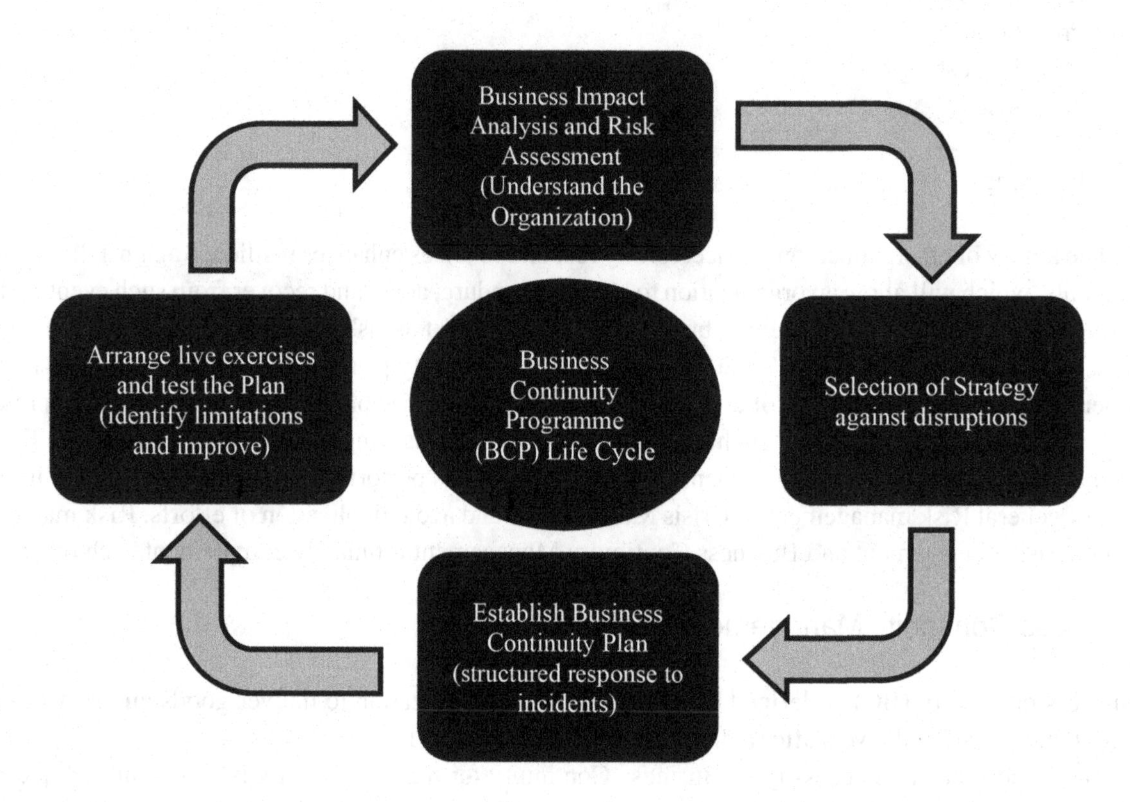

Business Impact Analysis

The initiation of an informed and thorough BCP relies heavily upon the processes of Business Impact Analysis (BIA) and Risk Assessment (RA). BIA focuses on the impact of a disruption to critical operations regardless its source or probability of occurrence. One advantage that BIA offers, is the lack of need for lengthy and time-consuming statistical analysis. Through qualitative analysis, BIA leads to prioritization of actions and allocation of resources without necessarily the need for further complex information, such as statistics (BSI, 2014a). The analysis should be documented with the aim to identify through a systemic approach, the purpose of an infrastructure, its critical products and services, followed by the activities and components (technical and other resources) that support them. It will then conclude to a prioritization of critical operations and assets necessary for planning a focused emergency response and timely recovery of operations after a disruptive incident (BSI, 2012). Responsible for the identification of an infrastructure's critical operations, is its operator (European Union Agency for Cybersecurity [ENISA], 2014).

Regarding the gathering of information related to the functions of an infrastructure, BIA would benefit from the participation of both the management and operator of an infrastructure, the security personnel and departmental managers, engineers, risk, crisis and business continuity managers and/or other relevant experts.

Responsible for the process of BIA is a specific designated Business Continuity Management Team (or other certified/competent experts) who should gather all the necessary information from the stakeholders of the infrastructure (security, engineer, IT and other personnel and the management) via personal or group interviews, questionnaires or workshops (ISO, 2012). The information gathered will later allow the BC managers to assess identified risks based on their impact (during risk assessment) on quantitative and/or qualitative terms for different categories of impacts such as physical, (against humans or buildings), operational, financial, reputational impact etc. and rank them according to criticality levels. The selective or general focus on all or some types of impacts varies depending on the analysis and the infrastructures scope. Most importantly, the BIA impact analysis should include an estimation of the outcome of the impact that the critical processes would have on the infrastructure if they were to be terminated. Moreover, a prioritization between timeframes for resuming each of the activities at a minimum acceptable level is necessary. This should also include the specific impacts that would be fatal for the infrastructures activities if not resumed in time (see Recovery Time Objectives-RTO).

An important part of the BIA is the identification of critical processes, resources and stakeholders which result from the interconnections and interdependencies between CIs as described previously in the chapter (BSI, 2012). As such, a critical operations' registry would benefit from the distinction between internal operations, which include the core operations vital for the production and function of the infrastructure, and external processes, which include critical operations/services in support of other infrastructures, whose failure, could pave the way for serious cascading effects in a network of infrastructures.

Table 6 (see Appendix 1) provides a registry of the sectors, sub-sectors and critical services of CIs. The table can be expanded even further to include the critical assets which support these processes.

Adaptation measures and recovery objectives will be decided based on the nature of the impact and its level, the impact over time and the recovery time needed, as well as the critical dependencies within the system and the effects on interested parties (stakeholders) (BSI, 2014b). A BIA may conclude in a documented database containing all the aforementioned information of critical processes and their impact on the infrastructure if disrupted, including critical assets such as technical components, buildings vehicles etc. and should be reviewed periodically (annually, quarterly etc.) or whenever significant changes occur. Figure 4 presents a diagram of the necessary steps for undertaking a BIA.

In conclusion, BIA is a process which is undertaken as part of the Business Continuity Plan, with the purpose of designating where the emergency response should be targeted, in what order and within which timeframe each operation/asset should be recovered.

Risk Assessment

The Institute of Risk Management (IRM), defines risk, which does not equal to threat, as the combination of the probability and consequences of an event which can vary from positive to negative (Hopkin, 2010, p.12). This definition besides realistic, is also practical since [risk = probability * consequences] is the typical formula used to create a risk matrix for identification of events' risk levels. However, when risk management is examined for security and safety purposes, the analysis of events and design of measures and action, perceives risk mostly as a negative effect. The definition provided by The Institute of Internal Auditors which defines risk as the uncertainty of an event occurring which can impact the achievement of our objectives and follows the same formula as IRM (Hopkin, 2010, p. 12) seems of better application for the protection of CIs, given that our objective is to safeguard an infrastructure from threats. The purpose of Risk Assessment is to examine an infrastructure in a systemic manner and consists of

Figure 4. Steps of BIA process based on ISO 22301 (BSI, 2014b)

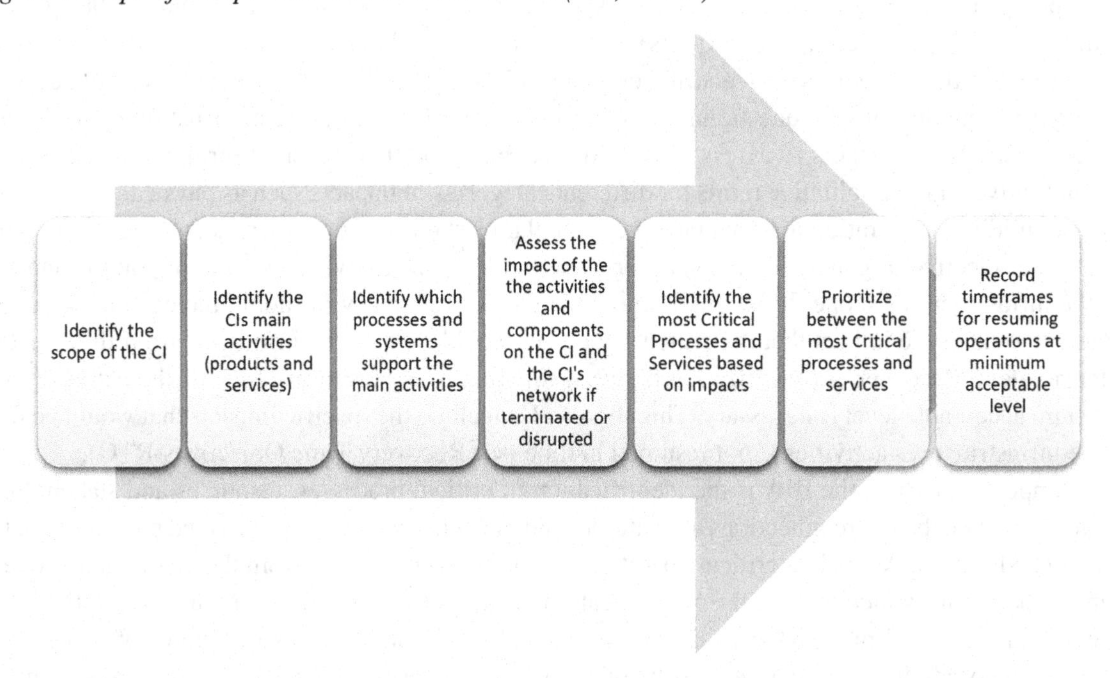

the processes of risk identification, risk analysis, and risk evaluation of disruptive events (BSI, 2018a). Figure 5 displays the necessary steps to be followed during a risk assessment, as has been implemented by the Center for Security Studies for the protection of public spaces. The same steps can be followed as a guideline for assessing the risks against a CI.

Based on ISO 31000 and 31010 A Risk Assessment typically consists of the following steps:

Risk Identification

It is used to identify potential risks based on possible sources and causes of hazards/threats which could inflict damage on a CI and affect its objectives (BSI, 2018a). Risk identification utilizes information gathered from past events/ historical data, or/and teams of experts through methods such as brainstorming workshops or interviews. It is very important to start the process by first examining the infrastructure to understand and document its main functions, operations, system components, existing security and safety measures and external environment. Following the collection of data, hypothetical scenarios of hazards or man-made threats should be designed and discussed as part of a workshop with relevant stakeholders from various expertise, to determine which threats are most likely to occur based on past events (international, national, local), feasibility and vulnerabilities of the infrastructure generated by current security and safety setup. Other risk identification methods may include questionnaires, SWOT analysis, flowcharts (e.g. fishbone diagram) Table 2, Table 3, and Table 4 represent indicative templates for recording hypothetical threats including their possible impacts against an infrastructure or its components. The information which is suggested for the different levels of impact may be adjusted depending on the infrastructure, its size, budget etc. The threat identification can either focus on the examination of one area of interest and the assets it consists of or focus on individual assets per function. The tables provide a sample, but the approach and implementation of analysis depends on the users.

Risk Analysis

Risk analysis (RA) utilizes the feedback from all the risks identified in the previous step to examine specific adverse events against their probability of being realized and their consequences, in order to

Figure 5. Risk Assessment approach for the security of public spaces

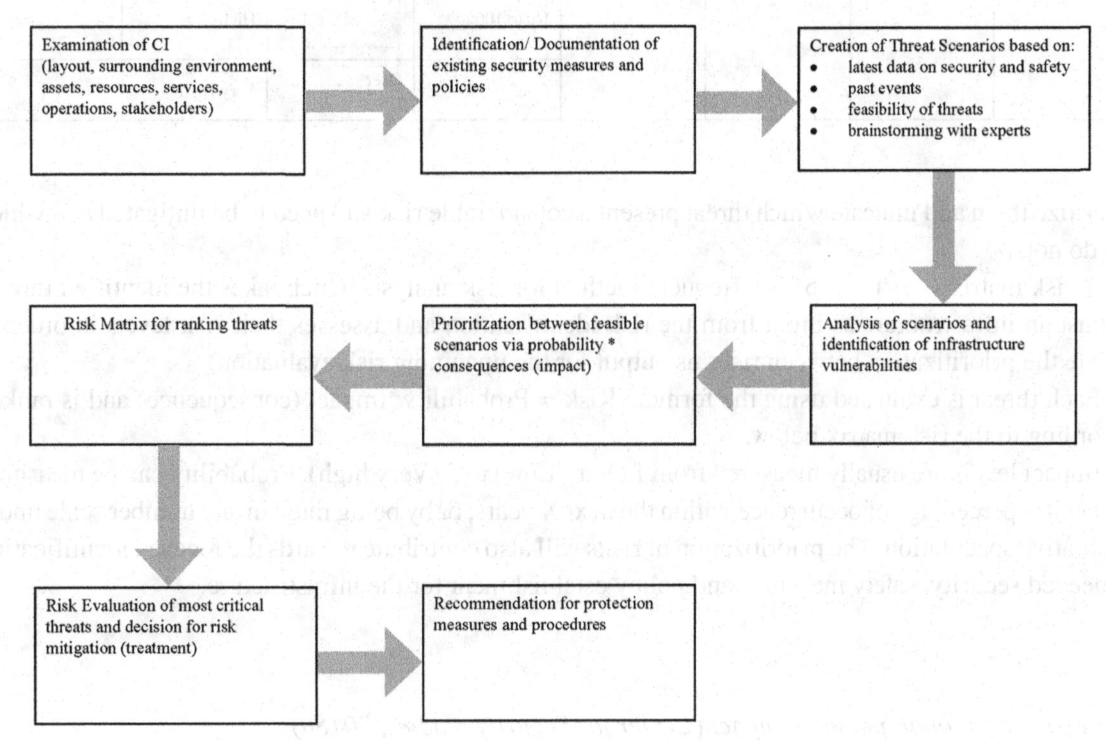

Table 2. Impact in human loss (Center for Security Studies, 2018a)

Type of treat	Sector/ Asset	Affected Function	Time	Deaths	Injuries	Impact Scale	Result/ Impact
Attack with firearms	Bank- Branch A	Occupation of branch X. Robbery attempt with Hostages	Morning (10:30-12:00)	0	Light/Do not need treatment	Very Low	
				0	> 10 manageable on spot	Low	
				0	> 25 manageable on spot and in hospitals	Medium	
				>0	> 50 treated in hospitals	High	
				>5	> 100 (lethal injuries in high numbers)	Very High	

Table 3. Financial loss (Center for Security Studies, 2018a)

Type of treat	Sector/Asset	Affected Function	Time	Gross Domestic Income (GDI) of the infrastructure in (%)	Amount	Impact Scale	Result
Cyber attack	IT network	Denial of access to company funds	23:00	Up to 5%	Limited	Very Low	
				Up to 10%	To …………… €*	Low	
				Up to 30%	From …………€ To …………….€*	Medium	
				Up to 100%	From …………€ To …………….€*	High	
				Over 100%	Over ………… €*	Very High	

prioritize them and indicate which threat present a considerable risk and need to be mitigated and which are do not.

A risk matrix (see Table 5) is a frequent method for risk analysis which takes the identified threats against an infrastructure as input from the risk identification and assesses their risk levels in order to enable the prioritization between risks as output for the upcoming risk evaluation.

Each threat is evaluated using the formula Risk = Probability*Impact (consequence) and is ranked according to the risk matrix below.

Impact levels are usually measured from 1 (Very Low) to 5 (Very high). Probability can be measured in specific percentage of occurrence within the next X years, or by being rated in any number scale under qualitative speculation. The prioritization of risks will also contribute towards the focused identification of needed security, safety measures and policy establishment for the infrastructure.

Table 4. Operational/ physical impact (Center for Security Studies, 2018a)

Type of treat	Sector/ Asset	Affected Function	Time	Damage	Out of order	Impact on personnel	Replacement of service	Impact Level	Result
Sabotage	Power Generator- c	Loss of power in the Control Room and Scada Systems	12:00- 2:00	Negligible	Less than 1 hour	Negligible	None	Very Low	
				Very Low damage	1-6 hours	20-50% until/ end 15 days	Within 1 hour	Low	
				Low damage	6 hours-1 day	20-50% until/ end 1 month	Within 6 hours	Medium	
				Medium damage/ serious disruption	Until 1 month	20-50% until/ end 2 months	Within a few days	High	
				Destruction/ Permanent outage	1 year	More than 50% for 2 months	Special arrangements	Very High	

Table 5. Risk matrix (Dumbrava & Iacob, 2013)

Probability	Will most likely occur	Low	Low	Medium- High	Very High	Very High
	Likely to occur	Very low	Low	Medium	Medium- High	Very High
	May occur at some point	Very low	Low	Medium	Medium- High	Medium- High
	Low probability	Very low	Low	Low	Medium	Medium- High
	Unlikely	Very low	Very low	Low	Medium	Medium
		Negligible	Insignificant	To be examined	Significant	Must be treated immediately
	Impact (Consequence)					

Risk Evaluation

The output of RA (ranked risks), will be utilized to determine the significance of each risk and decide whether it should be treated, transferred (e.g. through insurance), avoided (if possible), not acted upon, or be analyzed again. This is the process of Risk Evaluation. The decision is related to cost and benefit of adopting and implementing mitigation measures, while taking into consideration their effect on the operation and the strategic scope of the infrastructure.

Risk treatment options in the case of CI protection include avoiding the acknowledged risk by termination of the activity which generates it, removing the source of the risk, or adopting measures in order to make lower the likelihood and consequences of a disruption to operations and delivery of services.

In respect to the processes described above, the main difference between BIA and RA, is that BIA assesses the impact that specific critical operations of an infrastructures will have on the viability of the infrastructure if they stop working. This helps in the identification of the infrastructure's most critical operations and assets which will be the focus of the Business Continuity Plan. On the other hand, Risk Assessment identifies, quantifies and prioritizes risks/threats to provide an educated and targeted approach towards planning and implementing an adequate emergency response in case of specific threats.

The BIA and the RA should both be documented with periodical reviews and updated, to ensure continuous improvement and adaptability

Business Continuity Strategy

The design of the BC strategy based on the conclusions of the BIA and RA is the next immediate step towards the construction of the BCP. Its main objective is to mitigate the risks identified in the RA for the prioritized operations designated by the BIA through specific prevention measures, immediate and structured response to contain the damage of a threat and timely restoration of an infrastructure's critical functions. As part of an organizations BC Strategy the resources which will be needed to implement the plan (including personnel, facilities, equipment, vehicles, communications technology, IT support, budget and others) and will support the emergency response, should be identified and documented. In addition, the strategy describes under what circumstances a BCP shall be activated, and what the response and communications protocol entails. (ISO, 2014). A core concept for the design of the Strategy is the establishment of scenarios which must respond to the contingency situations that have been assumed in the preceding analysis. These include the identification of unavailability of existing Critical Infrastructure assets and resources, what may refer to as unavailability of locations (buildings, data centers, etc.), unavailability of human resources (personnel,), of supplies or loss of data. For each of the considered scenarios, one or more recovery alternatives should be set. For example, alternative buildings, mobile sites may be anticipated, alternative personnel and/or collaboration with other similar service providers might be agreed in advance, technology should be put in place for conservation of data (see operational requirements for business continuity). After the design and the establishment of the BC strategy, the plan should be tested at a pre-disaster stage in order to confirm its applicability and identify its weaknesses and strengths.

Periodically, but especially after an incident, the overall Strategy and Business Continuity Plan should be reviewed and compared against the documented response during past incidents. Accordingly, weaknesses, gaps and strengths should be identified the plan should be adjusted to raise the infrastructures adaptability and resilience to disruptive incidents.

All the actions and results of the implementation of a Business Continuity Plan and emergency response ought to be monitored and documented, to be later reported to the management for review and further improvement of the Plan. As a final step, based on the results of the management review the Business Continuity Plan may be improved and the overall Business Continuity policy and objectives of an infrastructure adjusted (BSI, 2014b).

Furthermore, the management review of an infrastructure's Business Continuity Plan is not the only way to reflect upon past response and existing policies and improve for future incidents. Other options include, internal auditing from a team formed of experts and certified professionals in the respective field and external auditing by an auditing firm which is fairly common practice (Center for Security Studies, 2018a).

In regard to further improving the knowledge of an organization about Business Continuity and enhancing its response capability, there also a few options available. A popular choice among practitioner are preparedness exercises which can be carried out through a scenario-based exercises (without specific roles), table-top exercises (with specific roles for the participants), simulated exercises by testing the existing plan in place, or realistic exercises which are similar to simulated exercises but on a bigger scale, also involving external stakeholders. Seminars and educational program before exercises can also prove useful as an introduction to the principles of Business Continuity and recent practices (Center for Security Studies, 2018a).

Finally, concluding from the presented chapter, a list of the essential information which should be included in a Business Continuity Plan, can be found in Appendix 3.

Operational Requirements for Business Continuity

On an operational level a Business Continuity Plan is structured to cover the following procedures:

Prevention: Prevention measures aim at mitigating the impact of a threat towards an infrastructure. They are based on the conclusions of the Risk Analysis and Risk Assessment (see respective section for details) which reveal the gaps regarding an infrastructure's protection. The adoption of these measures is decided during the design of the business continuity strategy and enforce the emergency response capability upon the occurrence of an incident, as well as to ensure the safety and security of the infrastructure's critical assets and operations (EU- CIRCLE consortium, 2017b). These measures include so-called "redundancy" measures such as:

- Alternative power solutions (e.g. backup generators, UPS, etc.)
- Alternative Data Center in a distant and safe location
- Backup information which supports the critical activity of the organization such as an alternative Control Center which can gather and manage information which supports the infrastructure's operations (e.g. surveillance material from cameras)
- Redundant computer systems and communication lines
- Emergency internal communications plan for general communication and fast notification of the top management, the management of the infrastructure and its personnel
- Emergency external communications plan for communication of an incident with other infrastructures, emergency responders (police, fire brigade, medical services etc.), municipality and others which can be affected or should be aware of the incident.
- Personnel and visitor's documented relocation policy including the assets needed to support the processes (vehicles and type, equipment, specific roles of personnel in support of the evacuation, a safe destination location)
- Alternative work/ operations locations
- Enhanced fire detection measures and firefighting support
- Intrusion detection and other incidents awareness measures such as surveillance systems and alarms
- Intrusion prevention measures such as fences, gates, traffic barriers and access control systems

Response: Response is as the word suggests, activated upon the detection of a disruptive incident. It is essentially the facilitation of the Business Continuity Plan. It is initiated by a specific group of people (e.g. Business Continuity Team) which as a first step emphasizes on the collection of data about the incident, the affected assets and the communication of the information across the organization. The activation of the Plan will depend upon the extent of the damage upon the systems, the criticality of the affected systems for the infrastructure, the safety of personnel and other visitors and the Recovery Time Objectives (RTO) dictated in the Plan (EU-CIRCLE consortium, 2017b). The response will include actions such as the coordination of the staff relocation, designation of alternative working locations, buildings, offices etc., internal and external communications management with continuous updates about the situation's status, awareness of people in need of medical attention, the security procedures, monitoring and reporting events (ISO, 2013).

Recovery: The first step of recovery is to restore basic resources such as power, water, communications and other utilities (The National Academies of Sciences, Engineering and Medicine [NASEM], 2009). Following that, emphasis should be given to recovery strategies for the infrastructure's identified critical processes, the reparation of damage and the restoration of the system's operational capacity to a certain acceptable level according to the BCP (EU- CIRCLE consortium, 2017b).

Restoration: Following the recovery effort, restoration is aimed towards the system's restore in its normal levels and way of operating. To ensure that this is achieved the capability and functionality of the system (infrastructure) must be tested and validated.

Validation of systems can be implemented in the following ways (EU- CIRCLE consortium, 2017b):

- Concurrent Processing: Involves running a system simultaneously at two separate locations concurrently, until it is validated that the recovered system is secured and runs as intended

- Data Testing Validation: The recovered data is tested and validated to ensure the complete recovery and restoration of data files and data base.
- Functionality Test Validation: All infrastructure's functionality must be tested to ensure that the system is ready to return to its normal operations.

During the design process of a Business Continuity Plan, for the recovery and restoration of critical operations specific accepted timeframes should be designated. This is defined as "Recovery Time Objective" (RTO) and indicates how quickly a process must be recovered after it fails. The time window of the RTO is smaller depending on the criticality of an operation or asset (Watters, 2014). Figure 6 presents the findings of research extracted from CI operators and demonstrates the RTO and possible disruptions to their critical operations along with the department in charge of the restoration, in the energy sector. A registry such as this must be included in a Business Continuity Plan as it can benefit an infrastructure's efficiency towards targeted and timely response due to successful prioritization during emergencies which will lead to avoidance of randomized and desperate actions.

SOLUTIONS AND RECOMMENDATIONS

Challenges to Critical Infrastructures can be posed from floods, earthquakes, storms, sabotage, cyberattacks, terrorist attacks, bombs loaded on drones and a variety of other threats. Analysis of these threats and solutions should be generated to cover the needs of all sectors of critical infrastructures. That is especially important when all the sectors of CIs are interconnected on CI's status affects the rest of its network.

The action of CI protection should be proactive and adaptive. However, risk cannot be nullified, thus there is no absolute measure which could ensure the total defense of infrastructures against threats, nor can threats be absolutely prevented. Therefore, a plan for the continuation of operations is more necessary than ever. Best practices, scientific research and existing literature can assist in the effort of those responsible for the protection of infrastructures to safeguard them from disasters. Through thorough analysis of the infrastructure systems, efficient plans and measures can generated, but a holistic understanding of the available methods is required.

Business Impact Analysis and Risk Assessment provide all the information needed for the prioritization of CIs' protection, their operations, critical assets, network of operations and result into a solid foundation for a successful Business Continuity Management which can be combined with technological solutions, to maximize the results of risk mitigation. It is an expensive and time-consuming process but if a crisis or disaster is realized it can have a higher economic impact. However, the selection of the mitigation measures which are to be adopted, will be based on the risk evaluation decided initiative to address specific risk, along with a cost-effectiveness analysis which will consider budget limitation for the implementation and sustainability of the measures.

The content of this chapter acts as a layout of theory and framework to be followed for understanding and implementing a BCP which will also compliment a wider risk management effort. The purpose of the framework is to enable users to follow a general approach which could be utilized for the purpose of both safety and security threats. However, it should not be assumed that all phenomena natural or man-made have the same characteristics and every type of infrastructure should be treated the same way. What the chapter proposes is that an integrated step-by-step approach/ framework can be used for the

Figure 6. Recovery Time Objectives and disruptions identification for Business Continuity (EU-CIRCLE consortium, 2017b, p. 35)

Please indicate Recovery Time Objectives (RTO) and possible disruptions				
CI sector	RTO	Department/Process	Essential Operations	Disruptions/Incidents
Energy	15 minutes	Maintenance traffic	Maintain operational process	Loss of electrical power
	3 hours	Technical	Loading/Unloading oil products	Pumping unit damage
	6 hours	HR	Provide additional personnel	Physical and intellectual fatigue of personnel during long time response
	Supply must be restored to 90% of the final clients that have been cut in maximum 5 days	Network control	Maintain supply	Supply to distribution networks
	Depending on each contract (confidential)	Network control	Maintain supply	Supply to industrial clients
	12 hours between the alert and the beginning of the intervention	Maintenance	Maintain supply or control by critical equipment availability	

analysis procedures of Business Continuity. Following the analysis, each threat and hazard which might affect an infrastructure could possibly require special attention towards each unique characteristic and some specific measures of mitigation and further analysis.

Moreover, the framework proposed can be benefited and enriched based on the users' needs, by examining more methods for risk assessment depending on whether the analysis is quantitative, qualitative or a mix of the two. IEC 31010:2009 or the updated IEC 31010:2019 Risk Management- Risk assessment techniques provide many alternatives. Standards such as the ones mentioned, can be combined with results for scientific research such the framework proposed by the EU- CIRCLE project, which focused on the strengthening of CIs' resilience and provide more insight on specific hazards.

An important aspect which should not be overlooked, is the communication of knowledge and expertise. The EU Directives have set a course of action, however in many cases there is still lack of awareness and a gap between the literature of risk management and the practitioners who are responsible for applying the theory during emergency operations. In a past vulnerability assessment workshop for the protection of public spaces (Center for Security Studies, 2019), it was observed that practitioners such as emergency first responders, felt the need to be informed about risk assessment as a structured process and admitted their unfamiliarity with using a specific risk assessment framework, although in practice they had already been implementing actions suggested by internationally accepted good emergency response practices and risk assessment frameworks. Additionally, a gap of communication, risk perception and information on security issues became visible between different security stakeholders (emergency responders, law enforcement agencies and local management) which could undermine the operability and response during emergencies between different agencies. As a response to the above, this chapter aims to raise awareness around issues surrounding the safety and security of CI's, their common themes of analysis and introduce a practical methodology for building a BCP.

FUTURE RESEARCH DIRECTIONS

Intense effort is currently undertaken in the EU for scientific and documented research on the protection of Critical Infrastructure involving not limited of research teams but also involving law enforcement agents, emergency first responders, critical infrastructure operators and European municipalities. There are many past and present EU initiatives such as the Horizon 2020 programme for research and innovation, or General Directorates (DGs) such as DG for Migration and Home Affairs (DG HOME), which fund many European projects with the aim to push research and innovation forward on subjects relevant to the protection and resilience of Critical Infrastructures which produce much valuable material such as theorical frameworks but also technological solutions. Particularly, the next research and innovation framework programme of Horizon Europe for 2021-2027 envisages to incorporate under Pillar II "Global challenges and European Industrial Competitiveness", a cluster for enhancement of civil security research. Investments in this Research and Innovation Actions will focus, among others, to the improved security and resilience of infrastructure with special focus on "resilience by design" and vital societal functions, increased cybersecurity and will further promote the collaboration of science, policy-makers, practitioners, technology providers and CI operators towards capacity building in the security sector. As a matter of fact, as the "Orientations towards the first Strategic Plan implementing the research and innovation framework programme Horizon Europe" (2019) underline, achievement of sovereignty in strategic technology areas and critical infrastructures is one of the main policy objectives and priorities of EU for the upcoming future.

Independently of the thematic aim of the projects, there is a common focus towards Risk Management and its subfields of Crisis Management, Disaster Management, Emergency Response, Safety and Security which are producing valuable material such as risk assessment frameworks and field-tested technological solutions but also result into networks for information, knowledge and expertise exchange.

Some successfully completed and currently ongoing EU projects focusing on the protection of Critical Infrastructures are:

- **EU-CIRCLE**: A pan-European framework for strengthening Critical Infrastructure resilience to climate change (European Commission, 2017)
- **ERNCIP**: European Reference Network for Critical Infrastructure Protection (Joint Research Center (JRC), 2018)
- **SATIE**: Security of Air Transport Infrastructure of Europe (European Commission, 2019c)
- **IMPROVER**: Improved risk evaluation and implementation of resilience concepts to critical infrastructure (European Commission, 2019b)
- **SecureGas**: Securing the European Gas Network (European Commission, 2019c)

- **InfraStress**: Improving resilience of sensitive industrial plants & infrastructures exposed to cyber-physical threats, by means of an open testbed stress-testing system (European Commission, 2019a)
- **STOP-IT**: Strategic, Tactical, Operational Protection of water Infrastructure against cyber-physical Threats (European Commission, 2019e)
- **DEFENDER-**: Defending the European Energy Infrastructures (European Commission, 2019f)

There is also plethora of publications by the ISO or other standardization organizations which can be followed as best practice for a Business Continuity, Emergency Response and Risk Management and other related disciplines and can be utilized to inform users as to what elements are necessary and should be incorporated into an organization's activity towards the protection and resilience of an infrastructure. Such guidelines include ISO 31000- Risk Management, ISO 27001- Risk Assessment Methodology, ISO 22301- Business Continuity, ISO 22320- Guidelines for incident management, or BS 11200- Crisis management. For a Business Continuity Strategy with focus on ICT (Information and Communications Technology) ISO 27031 (2011) can be utilized.

CONCLUSION

The issue of technical infrastructure protection has gathered a lot of attention has led to much initiative worldwide, both from a theoretical and technical research point of view, as well as legislation and standardization of best practice. This worldwide focus on CI protection, can be explained by three basic factors.

The first one is the necessities that the technical infrastructures produce for the daily functions of a nation in its interior and as a part of interactive worldwide system. It is clear that technical infrastructures satisfy all the basic and vital everyday operations, but also support the economy, the security and the health and overall societal well-being.

The second reason is the wide networks national and international that have been created between critical infrastructures due to urbanization, globalization, the information and communication systems, and the exchange of products and services between infrastructure which support their production and increase their efficiency. The interdependencies discussed in this chapter, can be beneficial towards the productivity of CIs but can also activate a chain of events spreading across systems, if an infrastructure is disrupted or is destroyed with devastating effects for a society and cross- border entities part of an interconnected network of productivity. Additionally, the make the infrastructures and attractive target for terrorists since they can attack an individual infrastructure with relatively low-cost means (e.g. Improvised Explosive Devices on drones) and amplify their damage by creating cascading effects.

The third reason is that the threats which may theoretically disrupt critical infrastructures are not just scenarios discussed on paper, but they are real and there are many case studies which can be analyzed. Critical Infrastructures have either experienced many natural hazards in the past or have constituted the stage for man-made threats whether intentional or not (e.g. the 9/11 terrorist attacks on the World Trade Center).

The safety and security of critical infrastructure cannot be guaranteed by taking up conventional security and safety measures without analysis of the infrastructure's external environment. That is because the value of critical infrastructures is not found in their real estate, but in what it contributes to the society and other infrastructures through its operations. Therefore, the need for securing CIs' integrity lies within the continuation of their activities, their ability to absorb the damage disruption, responds through adaptability to crises and resume their operations through a culture of risk management and Business Continuity.

This effort can be supported by adjustable documented analysis by following standard practice, which can be supported by the results of academic research for a more complete approach, expert opinion and examination of past events for deeper understanding of threats. Last but not least, CIs' protection must follow an approach of ecosystem analysis if it is to be truly effective and mitigate risks which might result into one-time crises but could terminate a CI's life cycle.

REFERENCES

Ένας χρόνος από την εθνική τραγωδία στο Μάτι - Μνημόσυνο για τους 102 νεκρούς [One year from the national tragedy of Mati- Memorial for the 102 dead] (2019, July 13). *CNN Greece*. Retrieved from: https://www.cnn.gr/news/ellada/story/185215/enas-xronos-apo-tin-ethniki-tragodia-sto-mati-mnimosyno-gia-toys-102-nekroys

Φωτιά Αττική. Σταδιακή αποκατάσταση της ηλεκτροδότησης στις πληγείσες περιοχές [Fire Attica: Gradual restoration of electricity in the affected areas] (2018, July 24). *CNN Greece*. Retrieved from https://www.cnn.gr/news/ellada/story/139950/fotia-attiki-stadiaki-apokatastasi-tis-ilektrodotisis-stis-pligeises-perioxes

Πουλιόπουλος, Γ. (2012, February 12). *Παραδόθηκαν στις φλόγες 45 κτίρια και καταστήματα του Κέντρου [45 buildings and retail shops in the Center of Athens surrendered to flames]*. To BHMA (TO VIMA). Retrieved from https://www.tovima.gr/2012/02/12/society/paradothikan-stis-floges-45-ktiria-kai-katastimata-toy-kentroy/

Administration of Barack Obama. (2013). *Directive on Critical Infrastructure Security and Resilience February 12- 2013 Presidential Policy Directive/PPD–21*. Retrieved from https://obamawhitehouse.archives.gov/the-press-office/2013/02/12/presidential-policy-directive-critical-infrastructure-security-and-resil

Arnold, D. R., & Wade, P. J. (2015). A definition of Systems Thinking: A Systems Approach. 2015 Conference of Systems Engineering Research. *Procedia Computer Science*, *44*, 669–678. doi:10.1016/j.procs.2015.03.050

Below, R. Wirtz., A., Sapir- Guha, D. (2009, October). Disaster Category Classification and peril Terminology for Operational Purposes, *Working Paper*. Retrieved from https://www.cred.be/downloadFile.php?file=sites/default/files/DisCatClass_264.pdf

Borodzicz, E. P. (2005). *Risk, Crisis and Security Management*. Chichester, West Sussex: John Wiley & Sons Ltd.

British Standards Institution. (2010). *Risk management- Risk assessment techniques (ISO 31010:2010)*. Retrieved from https://shop.bsigroup.com/ProductDetail?pid=000000000030183975

British Standards Institution. (2014a). *Adapting to Climate Change using your Business Continuity Management System*. Retrieved from https://www.bsigroup.com/localfiles/en-gb/iso-22301/resources/bsi-sustainability-report-adapting-to-climate-change-using-your-business-continuity-management-system-uk-en.pdf

British Standards Institution. (2014b). *Societal security — Business continuity management systems — Requirements (ISO 22301:2012)* Retrieved from https://shop.bsigroup.com/ProductDetail?pid=000000000030292502

British Standards Institution. (2014c). *Crisis management – Guidance and good practice* (BS 11200:2014). Retrieved from https://shop.bsigroup.com/ProductDetail/?pid=000000000030274343

British Standards Institution. (2015). *Societal security — Business continuity management systems — Guidelines for business impact analysis (BIA) (PD ISO/TS 22317:2015)*. Retrieved from https://shop.bsigroup.com/ProductDetail/?pid=000000000030299894

British Standards Institution. (2018a). *Risk management — Guidelines (BS ISO 31000:2018)*. Retrieved from https://shop.bsigroup.com/ProductDetail?pid=000000000030315447

British Standards Institution. (2018b). *Security and resilience- Vocabulary* (ISO 22300:2018). Retrieved from https://shop.bsigroup.com/ProductDetail/?pid=000000000030348153

Brussels explosions: What we know about airport and metro attacks. (2016, April 9). BBC NEWS. Retrieved from https://www.bbc.com/news/world-europe-35869985

Ceballos, G. P. J., Fuentes, T. F., Diaz, P. D., Sanchez, S. M., Llorente, M. C., & Sanz, G. J. E. (2005). 11 March 2004: The terrorist bomb explosions in Madrid, Spain – an analysis of the logistics, injuries sustained, and clinical management of casualties treated at the closest hospital., *Critical Care Journal, 9*(1), 104-111. Retrieved from https://www.ncbi.nlm.nih.gov/pmc/articles/PMC1065101/#

Center for Security Studies. (2018a, March). Εγχειρίδιο- Εκπαιδεύσεις για την Προστασία των Υποδομών Ζωτικής Σημασίας [*Manual- Training for the Protection of Critical Infrastructure*]. Retrieved from http://www.ciprotection.gr/index.php/el/

Center for Security Studies. (2018b, March). Στοχευμένες Δράσεις για την Αύξηση της Προστασίας των Εθνικά Χαρακτηρισμένων Ευρωπαϊκών Υποδομών Ζωτικής Σημασίας [Targeted actions for enhancing the protection of national characterized European critical infrastructure, WP 1.1 (V.2)], Οδικός Χάρτης Παραδοτέο 1 (E1) [Roadmap- Deliverable 1 (E1)]. SAE 050/2. Unpublished internal document.

Commission of the European Communities. (2006, December 12). *Communication from the commission on a European Programme for Critical Infrastructure Protection*. Retrieved from https://eur-lex.europa.eu/legal-content/EN/TXT/?uri=CELEX:52006DC0786

Commonwealth of Australia. (2015). *Critical Infrastructure Resilience Strategy: Plan*. Retrieved from https://www.tisn.gov.au/Documents/CriticalInfrastructureResilienceStrategyPlan.PDF

Council of the European Union. (2008, December 8). *COUNCIL DIRECTIVE 2008/114/EC of 8 December 2008 on the identification and designation of European critical infrastructures and the assessment of the need to improve their protection.* Retrieved from https://eur-lex.europa.eu/LexUriServ/LexUriServ.do?uri=OJ:L:2008:345:0075:0082:EN:PDF

Dumbrava, V., & Iacob, S. V. (2013). Using Probability- Impact Matrix in Analysis and Risk Assessment Projects. *Journal of Knowledge Management, Economics and Information Technologies.* Retrieved from https://pdfs.semanticscholar.org/e2b7/82cb13b324d156e1c1db08fc2f2c267e1247.pdf

Edwards, M. (2014). *Critical Infrastructure protection, NATO Science for Peace and Security Series - E: Human and Societal Dynamics NATO.* IOS Press. Retrieved from https://ebookcentral.proquest.com/lib/portsmouth-ebooks/detail.action?docID=1637644

ENISA. (2014, December). Methodologies for the identification of Critical Information Infrastructure assets and services. Retrieved from https://www.academia.edu/25419910/Methodologies_for_the_identification_of_Critical_Information_Infrastructure_assets_and_services_Guidelines_for_charting_electronic_data_communication_networks?auto=download

EU-CIRCLE consortium. (2016). *EU- CIRCLE-* A pan European Framework for strengthening Critical Infrastructure resilience to climate change. Retrieved from http://www.eu-circle.eu/wp-content/uploads/2017/01/D4.1_EU-CIRCLE_resilience_initialversion.pdf

EU-CIRCLE consortium. (2017a). *EU- CIRCLE- A pan- European Framework for strengthening Critical Infrastructure resilience to climate change.* Retrieved from http://www.eu-circle.eu/wp-content/uploads/2015/07/D3.1.pdf

EU-CIRCLE consortium. (2017b). EU-CIRCLE A pan- European Framework for strengthening Critical Infrastructure resilience to climate change. Retrieved from http://www.eu-circle.eu/wp-content/uploads/2015/07/D4.4.pdf

European Commission. (2017). EU-CIRLE- *A pan European framework for strengthening Critical Infrastructure resilience to climate change.* Retrieved from https://cordis.europa.eu/project/rcn/196896/factsheet/en

European Commission. (2019a). *InfraStress- Improving resilience of sensitive industrial plants & infrastructures exposed to cyber-physical threats, by means of an open testbed stress-testing system.* Retrieved from https://cordis.europa.eu/project/rcn/222602/factsheet/en

European Commission. (2019b). IMPROVER. *Improved risk evaluation and implementation of resilience concepts to critical infrastructure.* Retrieved from https://cordis.europa.eu/project/rcn/196889/factsheet/en

European Commission. (2019c). *SATIE. Security of Air Transport Infrastructure of Europe.* Retrieved from https://cordis.europa.eu/project/rcn/222594/factsheet/en

European Commission. (2019d). *SecureGas. Securing the European Gas Network.* Retrieved from https://cordis.europa.eu/project/rcn/222598/factsheet/en

European Commission. (2019e). *STOP-IT. Strategic, Tactical, Operational Protection of water Infrastructure against cyber-physical Threats.* Retrieved from https://cordis.europa.eu/project/rcn/210216/factsheet/en

European Commission. (2019f), September 16). *DEFENDER- Defending the European Energy Infrastructures.*

European Parliament, Policy Department for Citizens' Rights and Constitutional Affairs. (2018). *Member States' Preparedness for CBRN Threats- Terrorism.* Retrieved from http://www.europarl.europa.eu/RegData/etudes/STUD/2018/604960/IPOL_STU(2018)604960_EN.pdf

Gibbs, L. I., & Holloway, C. F. (2013, May). *Hurricane Sandy After Action. Report and Recommendations to Mayor Michael R. Bloomberg.* Retrieved from https://www1.nyc.gov/assets/housingrecovery/downloads/pdf/2017/sandy_aar_5-2-13.pdf

Homeland Security. (2013). NIPP 2013. *Partnering for Critical Infrastructure Security and Resilience.* Retrieved from: https://www.dhs.gov/sites/default/files/publications/national-infrastructure-protection-plan-2013-508.pdf

International Organization for Standardization. (2011, March). *Information technology – Security techniques - Guidelines for information and communication technology readiness for business continuity.* Retrieved from https://www.iso.org/standard/44374.html

International Organization for Standardization. (2012, December). *Societal security — Business continuity management systems — Guidance.* Retrieved from https://www.iso.org/standard/50050.html

Joint Research Center. (2018). JRC Technical Reports. *European Reference Network for Critical Infrastructure Protection: ERNCIP Handbook 2018 edition.* Retrieved from https://erncip-project.jrc.ec.europa.eu/sites/default/files/ERNCIP%20Handbook%202018.pdf

Labid, A., & Harris, M. J. (2014, July 8). Learning how to learn from failures: The Fukushima nuclear disaster. *Engineer Failure Analysis Journal Volume 47* P.A 117-128. Available from: https://www.sciencedirect.com/science/article/abs/pii/S1350630714002933

Lekkas, E., Carydis, P., & Lagouvardos, K. Mavroulis. S, Diakakis, M. … Papagiannaki, K. (2018). The July 2018 Attica (Central Greece) Wildfires- Scientific report (Version 1.1). *Newsletter of Environmental, Disaster, and Crisis Management Strategies.* Retrieved from https://www.researchgate.net/publication/326672342_The_July_2018_Attica_wildfires_Scientific_report_v11

McGuiness, D. (2017, April 27). How a cyber-attack transformed Estonia. *BBC News.* Retrieved from https://www.bbc.com/news/39655415

Morris, S. (2018, September 6). Novichok poisoning timeline: Q&A- Details released by police of the suspects' movements in Salisbury answer some questions but raise some more. *The Guardian.* Retrieved from https://www.theguardian.com/uk-news/2018/sep/06/novichok-poisoning-timeline-qa

Mueller, B., & Tsang, A. (2018, December 20). Gatwick Airport Shut Down by "Deliberate" Drone Incursions. *The New York Times.* Retrieved form https://www.nytimes.com/2018/12/20/world/europe/gatwick-airport-drones.html

North Atlantic Treaty Organization Strategic Communications Centre of Excellence. (n.d.). *Hybrid Threats: 2007 cyber-attacks on Estonia.* Retrieved from https://www.stratcomcoe.org/hybrid-threats-2007-cyber-attacks-estonia

Orientations towards the first Strategic Plan implementing the research and innovation framework programme Horizon Europe. (2019). *Co-design via web open consultation.* Retrieved from https://ec.europa.eu/research/pdf/horizon-europe/ec_rtd_orientations-towards-the-strategic-planning.pdf

Pursianen, C. (2018). The Crisis Management Cycle. New York: Routledge

Rinaldi, M. S., Peerenboom, J., & Terrence, K. K. (2001, December). *Complex Networks- Identifying, Understanding, and Analyzing Critical Infrastructure Interdependencies.* Retrieved from https://pdfs.semanticscholar.org/b1b7/d1e0bb39badc3592373427840a4039d9717d.pdf

Russian spy poisoning: What we know so far (2018, October 8). *BBC NEWS.* Retrieved from https://www.bbc.com/news/uk-43315636

Senate of the United States. (2001, October 24). *Uniting and Strengthening America by Providing Appropriate Tools Required to Intercept and Obstruct Terrorism (USA PATRIOT ACT) Act of 2001.* Retrieved from https://epic.org/privacy/terrorism/hr3162.pdf

Takabatake, T., Mall, M., Esteban, M., Nakamura, R., Kyaw, O. T., Ishii, H., ... Shibayama, T. (2018, November 9). Field Survey of 2018 Typhoon Jebi in Japan: Lessons for Disaster Risk Management. *Geosciences Journal, 8*(11). Retrieved from https://www.mdpi.com/2076-3263/8/11/412

The National Academies of Sciences, Engineering and Medicine. (2009). *A Guide to Planning Resources on Transportation and Hazards.* Washington, DC: The National Academies. Retrieved from https://www.nap.edu/catalog/23007/a-guide-to-planning-resources-on-transportation-and-hazards

United Nations Counter-Terrorism Centre, Counter Terrorism Committee Executive Directorate, Interpol. (2018). The protection of critical infrastructure against terrorist attacks: *Compendium of good practices.* Retrieved from https://www.un.org/sc/ctc/wp-content/uploads/2018/06/Compendium-CIP-final-version-120618_new_fonts_18_june_2018_optimized.pdf

United Nations General Assembly. (2016, December 1). *Report of the open-ended intergovernmental expert working group on indicators and terminology relating to disaster risk reduction.* Retrieved from https://www.preventionweb.net/files/50683_oiewgreportenglish.pdf

United Nations Office for Disaster Reduction- UNISDR. (2015). *Senday Framework for Disaster Risk Reduction 2015-2030.* Retrieved from https://www.unisdr.org/we/coordinate/sendai-framework

United Nations Office for Disaster Reduction- UNISDR. (2018). Words into Action Guidelines. Man-made and Technological Hazards. *Practical considerations for Addresing Man-made and Technological Hazards in Disaster Reduction.* Retrieved from https://www.preventionweb.net/publications/view/54012

United Nations Office of Counter- terrorism, United Nations Security Council, Interpol. (2018). *The protection of critical infrastructures against terrorist attacks: Compendium of good practices.* Retrieved from https://www.un.org/sc/ctc/wp-content/uploads/2018/06/Compendium-CIP-final-version-120618_new_fonts_18_june_2018_optimized.pdf

Watters J. (2014). Disaster Recovery, Crisis Response, and Business Continuity: A Management Desk Reference. Springer Science+Business.

ADDITIONAL READING

Gritzalis, D., Theocharidou., M., Stergiopoulos., G. (2019). *Critical Infrastructure Security and Resilience Theories, Methods, Tools and Technologies*. Springer.

Hedel, R., Boustras, G., Gkotsis, I., Vasiliadou, I., & Rathke, P. (2018). Assessment of the European Programme for Critical Infrastructure Protection in the surface transport sector. *International Journal of Critical Infrastructures, 14*(4), 2018. doi:10.1504/IJCIS.2018.095616

Joint Research Center. (2018, May 31). European reference Network for Critical Infrastructure Protection: ERNCIP Handbook 2018 edition. *JRC Technical Reports.* Publications Office of the European Union, Luxembourg.

Joint Research Center. (2019). *JRC Science for Policy Report. Recommendations for National Risk Assessment for Disaster Risk Management in EU*. Luxembourg: Publications Office of the European Union.

Labib, A. (2014). *Learning from failures. Decision Analysis of Major Disasters.* Oxford: Butterworth-Heinemann.

Lazari, A. (2014). *European Critical Infrastructure Protection.* Springer.

NCHRP. (2009). NCHRP Report 525. Surface Transportation Security Volume 15. Costing Asset Protection: An All Hazards Guide for Transportation Agencies (CAPTA). Transportation Research Board. Washington, D.C.

Rome, E., Theocharidou, M., & Wolthusen, S. (2015). Critical Information Infrastructures Security. In *Proceedings of the 10th International Conference, CRITIS 2015.* Springer. AG Switzerland doi:10.1007/978-3-319-07497-9

KEY TERMS AND DEFINITIONS

Assets: All the components (tangible and intangible) of an infrastructure which support its operation. Some examples of tanglible assets include staff, buildings, technical equipment, vehicles. Some examples of intangible assets include digitally stored data, financial assets, company reputation, network of operations etc.

Crisis: An unpredictable and unstable adverse situation which threatens an organization strategic objective

Disaster: An irreversible adverse situation as a result of inefficient response to emergencies or crises

Holistic Approach: The examination of an infrastructure or a phenomenon based on its function, its activities and components as a whole.

Resilience: The ability of an infrastructure to resist, respond and overcome adverse events

Risk Management: An ongoing process of activities aimed towards safeguarding an organization from uncertainties which can negatively impact its objectives

Stakeholders: All the parties (natural or legal entities) which are affected by, affect or participate in the activities of an infrastructure.

Threat: A cause of an incident/event (mostly man-made) with negative impact on the activities of an infrastructure

Vulnerability: Weaknesses of an infrastructure towards security and safety threats due to lack of measures, policies or strategy.

APPENDIX 1

Table 6. Registry of CIs' critical services by sector (EU- Circle consortium, 2017a, p.15-17)

Critical Infrastructure Sectors, Subsectors and Critical Services		
Critical Infrastructure Sector	**Critical Infrastructure Subsectors**	**Critical Services**
Energy	Oil	Exploration, Drilling, Extraction/Production, Refining/Processing, Transport, Storage, Distribution to users
	Gas	Exploration, Drilling, Extraction/Production, Fractioning, Treatment, Transport, Storage, Distribution to users
	Coal	Mining/Extraction, Treatment, Transport, Storage, Distribution to users, Gasification, Liquefaction
	Electricity	Generation, Cogeneration, Transmission, Distribution
	Renewables	Generation, Cogeneration, Transmission, Distribution
	District Heating	Transportation, Distribution
Transportation	Road	Ground passenger transportation, Road freight, transport including dangerous goods transport
	Rail	commuter passenger transport, long distance passenger transport, mass transit e.g. subways, trams etc., freight transport, including dangerous goods transport, Loading/unloading freight, Shunting and holding
	Aviation	passenger transport, loading and unloading of passengers, freight transport, postal transport, loading/unloading freight, air traffic control
	Maritime	freight/cargo transport, cargo loading/ unloading, cargo storage, passenger transport, embarking/ disembarking for passengers, vessel traffic control
	Inland waterway	freight/cargo transport, cargo loading/ unloading, cargo storage, passenger transport, embarking/ disembarking for passengers, nautical navigation
Water	Drinking Water	Abstraction, conveyance, storage, treatment, distribution, monitoring
	Wastewater	Collection, storage, treatment, discharge, monitoring
	Flood Water	Defense, conveyance, storage, discharge, monitoring
ICT	Telecommunication	Wired Communications, Wireless Communications, Internet, Information Services, Satellite Communications (SatCom), High Frequency Radio Communications (HF Com)
	Information	IT Products and Services, Internet Routing, Access and Connection Services
Chemical Industry	Basic	Manufacturing, Distribution, Storage and Warehousing, Chemicals Disposal
	Chemicals/Commodities	Manufacturing, Distribution, Storage and Warehousing, Chemicals Disposal
	Specialty/Fine Chemicals	Manufacturing, Distribution, Storage and Warehousing, Chemicals Disposal
	Consumer Chemicals	Manufacturing, Distribution, Storage and Warehousing, Chemicals Disposal
Public/Government	Fire and Rescue Services	Firefighting, water Pumping, Search and Rescue, first aid, sheltering, evacuation
	Emergency Medical Services	First aid, triage, Prehospital (on scene) treatment, Patient transport
	Military	Emergency transport, Logistics, Advance Field hospitals
	Law Enforcement	Traffic management, property protection, Jail, detention facility management, Coordination of operations, Crowd management, Forced Evacuation
	Public Services	Public administration
	Public Health and Healthcare	Primary care including health promotion, disease prevention, health maintenance, diagnosis and treatment of acute and chronic illnesses, and disease-management. Secondary care by specialists including acute care such as surgery Tertiary care which includes advanced medical investigation and treatment often using highly specialized equipment and expertise Pharmaceutical care Public health care (e.g. community health, management of disease outbreaks/pandemics etc.)

APPENDIX 2

Table 7. Threat registry for critical infrastructure (Center for Security Studies, 2018; Below, Wirtz, & Guha-Sapir, 2009)

Threat Type	Threat Category	Threat Sub-Category	Danger	Incident
Man-made	Organized and non-organized criminal activity	Terrorism-domestic and international (by conventional means)	Bomb /explosives placement	Bombings (via timer or remote control) / High Power explosives
				Bombs in abandoned objects
				Bomb in vehicle
				IED carried by UAV (Drone)
			Armed Attacks	Attempted murder or manslaughter
				Launch of anti-tank or other type of missile
			Vehicles	Land or air vehicle- ramming attack
				Vehicle transferring explosive/ CBRN/ illegal material, other dangerous material or firearms
			Theft	Armed robbery inside or outside an installation, vehicle theft, theft of equipment or product
				Armed robbery or theft in a storage area or counting house
			Kidnapping/ Highjack	Kidnapping / captivity of company personnel or customers/ visitors
				Hijack of vehicle or infrastructure / station
			Arson	Incendiary device in a facility or in vehicles outside installation (IIDs)
			Sabotage	Sabotage of vehicles, sabotage of production processes, sabotage or destruction of transportation network
		Technological attacks - Cyber-threats	Tampering	Unauthorized access / entry to controlled areas
				Unauthorized data alteration
			Hacking -Cybercrime	Cyber Attack / cybercrime (hacking, cracking, phishing)
				Robbery, intimidation/ threat, interception of sensitive / confidential/ personal data
			Communication loss	Monitoring of communications
				Interference or interruption to communications
			Viruses	Technological viruses
			System failure	Network malfunction
				Power outage
		Terrorism by unconventional means	Threat by Chemical, Biological, Radiological, Nuclear and Explosive (CBRN(E)) matter	Unconventional weapons (CBRN) in facilities / stations / air management systems
				Combination of CBRN materials with conventional explosives / usage of high impact explosives
				Dirty Bombs (with radioactive material)
				Asphyxiating gases or suffocating gases, caustic gases, acid or corrosive gases, irritant gases, toxic gases, explosive gases and fumigants (e.g. smoke grenades) (Chemical)
				Nuclear bombs
				Bacteria, viruses, toxins (Organic)
				Hazardous mail e.g. letters/ parcel with contagious/ poisonous or explosive material
		Anarchism	Attacks	Armed attacks / robberies in vehicles or storage areas or counting houses
				Vandalism and seize of facilities
			Bombing	Low impact Improvised explosive devices (IEDs)
			Arson	Improvised incendiary devices (IIDs)
			Sabotage	Sabotage of installations
				Prevention of transportation
			Hijack	Hijacked vehicle or infrastructures / station
			Bombing	Low impact Improvised explosive devices (IEDs)
		Organized and non-organized crime	Kidnapping	Kidnapping / captivity of company personnel, visitors or customers
			Hijack	Vehicle hijack/ captivity of passengers
			Homicide	Homicide of organization's staff
			Disturbance	Disruption/ turmoil in the operator's facilities
			Theft	Theft / theft of company fund / customer/ visitor robbery
			Smuggling	Smuggling of illicit goods/ materials
			Damage (direct / indirect) to infrastructure	Argument between gangs
				Injuries
				Sabotage / arson (IIDs)
		Antisocial behavior	Damage to infrastructure	Damage to property (IIDs, IEDs etc.)
				Vandalism to (night) patrol
			Suicide	Suicide attempt of passenger at transportation station
			Violence	Fight between team supporters or other groups
			Fraud - Threats	Pranks about bombs placement

continued on following page

Table 7. Continued

Threat Type	Threat Category	Threat Sub-Category	Danger	Incident
	Mass demonstrations / strikes (as a means of protest)	Demonstrations / public gatherings / strikes which turn violent	Hijack	Seizure of facilities / stations / mobile assets
				Preventing access to station/streets
				Preventing start and general operation of facility
			Violence	Sabotage / vandalism / arson
	Accident / random events	Environmental Accidents	Pollution	Air pollution with chemical / radioactive /biological substances (explosive, flammable, toxic, radioactive or corrosive) due to atmospheric dispersion / ignition
				Water pollution from chemical / biological / radioactive substances (explosive, flammable, toxic, radioactive or corrosive), due to leakage / ignition
				Pollution of the soil by chemical / radioactive / biological substances (explosive, flammable, toxic, radioactive or corrosive), due to leakage / ignition
		Technological Accidents	Fire	Urban fires of large scale (except industrial installations)
			Explosions	Malfunctions and destruction in industrial premises
				Release and spread of hazardous materials, gas, liquids (oil, radioactive energy,
			Damage in systems	Damage to electricity production plant (power failure)
				Damage to energy grid
				Damage to communications network
				Damage to water supply network
		Transportation accidents	Car accidents	Train accident / collision
				Fall aircraft
				Underground railway accidents, elevators, escalator, train, passenger accident
		Infrastructure collapse	Natural disasters	Natural disaster or fire
			Faulty infrastructure	Manufacturing fault
				Non-compliance with safety rules
				Insufficient or inadequate infrastructure against natural hazards
			Error	Human error
			Deliberate damage	Deliberate damage by infrastructure staff
	Other	Abandoned objects (hazardous materials)	Presence of hazardous materials	Unattended: Phosphate fertilizers, solvents, pesticides, flammable materials, fuel, explosives, weapons and ammunition
		Insufficient resources	Lack of staff	Absence of staff
		Insufficient resources	Lack of equipment	Fatigue / personal disease
				Equipment failure
			Lack of resources	Insufficient resources
		Panic without good reason (e.g. dissemination of false information)	Wounding	Trampling
			Wear and tear of facility	Destruction of operator's property
				Equipment abuse
		Panic due to emergency (e.g., fire, earthquake)	Rescue prevention	Sabotage of rescue teams
				Trapping
				Preventing the evacuation process
Natural Hazards		Meteorological Phenomena	Malfunctions, Damages, injuries due to weather conditions	Heat Waves, thunderstorms, heavy snowfall, snowstorms, ice, extreme winds, hail, hurricanes
		Geophysical Phenomena		Earthquakes, landslides, volcanic eruptions, mass movement
		Hydrological Phenomena		Floods (small scale), Floods (large scale), moving liquid (liquid) masses
		Climatological Phenomena		Extreme Temperatures (cold waves, heat waves, frost)
		Biological Phenomena		Infectious diseases, epidemics
		Extra- terrestrial	Malfunctions of equipment, damages, injuries	Asteroids/ Meteorites collision. space weather (e.g. geomagnetic storm)
		Fires (depending on the asset at risk)	Fire	Aerial/ Submarine

APPENDIX 3

A Business Continuity Plan should contain at least the following information contributing to the preparedness and response during emergencies:

Infrastructure Information

- Type of infrastructure
- Location
- Main activity/service
- Business Continuity Plan editor
- Contact details

Roles

- List of persons with key roles and the description of their responsibilities and expected action during an emergency

(e.g. Business Continuity Team manager, Communications Team, IT team, Security Team, Emergency Reponse Team, Personnel, etc.)

Communications

- Important Contacts (from inside the organization e.g. management, security, medical support, etc.)
- Communications Team Contact information
- Other important Contacts (external responders)
- Communications protocol for emergencies (basic steps), including informing management, employees on site, employees away from site, visitors, emergency response stakeholders.
- List of means for communication
- List of infrastructure's personnel

Response Protocol

- Basic Action for management during emergencies
- Basic Action for employees
- Basic Action for visitors
- Description of general response in case of fire, flood, loss of power, loss of communications, loss of information systems etc. (including evacuation of site)
- Description of transportation assets, routes, destinations/ safe places
- Alternative workspaces
- Description of personnel relocation process

Continuation of Business Protocol

- Relocation process and site of key personnel
- Description of immediate action in regard to documented critical assets and processes
- Location of alternative distant workplaces
- Backup equipment and systems (e.g. alternative data center, alternative communications, backup generators etc.)

Description of Critical Activities

- Description of Critical Operations
- Ranking of Operations for recovery including their RTO
- Basic action for the protection, recovery and restoration of each operation

Description of Critical Technical Systems and Assets

- Ranking of technical systems and assets for recovery (including human assets) and their respective RTO
- Basic action to ensure the protection, recovery and restoration of each critical system and assets

Basic Layout of the Infrastructure

- Scheme of the layout of the infrastructure
- Key locations
- Key emergency equipment
- Safety pathways

Compilation of References

Abbasi, T., Pasman, H. J., & Abbas, S. A. (2010). A scheme for the classification of explosions in the chemical process industry. *Journal of Hazardous Materials, 174*(1-3), 270–280. doi:10.1016/j.jhazmat.2009.09.047 PMID:19857922

ABC. (2003, February 21). Man Who Threatened Melbourne Water Supply Bailed. Retrieved from http://www.abc.net.au/news/2003-02-21/man-who-threatened-melbourne-water-supply-bailed/2689584

Abdalla, R. M., & Niall, K. M. (2010). *Location-Based Critical Infrastructure Interdependency (LBCII). Tech. Rep. TR 2009-130.* Toronto, Canada: Defence Research and Development Canada.

Abrams, M., & Weiss, J. (2008). *Malicious Control System Cyber Security Attack Case Study - Maroochy Water Services, Australia.* Retrieved from http://csrc.nist.gov/groups/SMA/fisma/ics/documents/Maroochy-Water-Services-Case-Study_report.pdf

Act No. 240/2000 Coll. on Crisis Management as amended. (2000).

Act No. 45/2011 Coll. on the Critical Infrastructure as amended. (2011).

Addai, E. A. (2016). *Investigation of explosion characteristics of multiphase fuel mixtures with air.* Powell, Wyoming: Western Engineering, Inc.

Adger, W. N. (2006). Vulnerability. *Global Environmental Change, 16*(3), 268–281. doi:10.1016/j.gloenvcha.2006.02.006

Adger, W. N., Hughes, T.P., Folke, C., Carpenter, S.R., & Rockstrom, J. (2005). Social-ecological resilience to coastal disasters. *Science, 309*(5737), 1036–1039. doi:10.1126cience.1112122 PMID:16099974

Administration of Barack Obama. (2013). *Directive on Critical Infrastructure Security and Resilience February 12- 2013 Presidential Policy Directive/PPD–21.* Retrieved from https://obamawhitehouse.archives.gov/the-press-office/2013/02/12/presidential-policy-directive-critical-infrastructure-security-and-resil

Advisory Committee on the Safety of Nuclear Installations (ACSNI). (1993). *ACSNI study group on human factors. United Kingdom.* London, United Kingdom: HM Stationery Office.

AEPA. (2008). *Asphalt pavements in tunnels.* Retrieved from https://eapa.org/wp-content/uploads/2018/07/asphalt_pavements_tunnelsMay2008-1.pdf

Agent Based Evacuation Simulation. (n.d.). Retrieved from https://www.thunderheadeng.com/pathfinder/

Alcaraz, C., & Zeadally, S. (2015). Critical Infrastructure Protection: Requirements and Challenges for the 21st Century. *International Journal of Critical Infrastructure Protection, 8*(1), 53–66. doi:10.1016/j.ijcip.2014.12.002

Altman, P., Cunningham, J., Dahnesha, U., Ballard, M., Thompson, J., & Marsh, F. (2006). *Disturbance of Cerebral Function in People Exposed to Drinking Water Contaminated With Aluminium Sulphate: Retrospective Study of the Camelford Water Incident*. UK: Retrieved from Oxford. Retrieved from http://discovery.ucl.ac.uk/2010/1/807.pdf

ANAKAN. (2019). Retrieved from http://anakan.cz/

Andersson, A., O'Connor, A., & Karoumi, R. (2015). Passive and Adaptive Damping Systems for Vibration Mitigation and Increased Fatigue Service Life of a Tied Arch Railway Bridge. *Computer-Aided Civil and Infrastructure Engineering*, *30*(9), 748–757. doi:10.1111/mice.12116

ARAMIS. (2004). Accidental risk assessment methodology for industries in the framework of the SEVESO II directive, User Guide.

Arnold, D. R., & Wade, P. J. (2015). A definition of Systems Thinking: A Systems Approach. 2015 Conference of Systems Engineering Research. *Procedia Computer Science*, *44*, 669–678. doi:10.1016/j.procs.2015.03.050

ASMR. (2014). Type plan for Large-Scale Oil and Petroleum Products Disruption. *Administration of State Material Reserves*. Retrieved from http://www.sshr.cz/pro-verejnou-spravu/ropna_bezpecnost/ropna_bezpecnost/Typovy%20plan.pdf

ASTRA. (2008). *Lüftung der Strassentunnel, Systemwahl, Dimensionierung und Ausstattung, ASTRA 13001* (2.03). Retrieved from www.astra.admin.ch

ASTRA. (2010). *Sicherheitsanforderungen an Tunnel im Nationalstrassennetz, ASTRA 74001* (01.08.2010 V1.02). Retrieved from www.astra.admin.ch

Atwan, A. B. (2007). *The Secret History of Al-Qaida*. London: Saqi Books.

Auld, H., & MacIver, D. (2007). Changing weather patterns, uncertainty and infrastructure risks: emerging adaptation requirements. Environment Canada.

Babrauskas, V. (2003). *Ignition handbook principles and applications to fire safety engineering, fire investigation, risk management and forensic science*. Issaquah, WA: Fire Science Publ.

Babrauskas, V. (2003). The effects of FR agents on polymer performance. In V. Babrauskas & S. J. Grayson (Eds.), *Heat release in fires* (pp. 423–446). London, UK: Interscience Communications.

Babrauskas, V., & Parker, W. J. (1987). Ignitability measurements with the cone calorimeter. *Fire and Materials*, *11*(1), 31–43. doi:10.1002/fam.810110103

Babrauskas, V., & Peacock, D. (1992). Heat release rate: The single most important variable in fire hazard. *Fire Safety Journal*, *18*(3), 255–272. doi:10.1016/0379-7112(92)90019-9

Bajpai, S., & Gupta, P. J. (2005). Site security for chemical process industries. *Journal of Loss Prevention in the Process Industries*, *18*(4-6), 301-309.

Bakand, S., Hayes, A., & Dechsakuthorn, F. (2012). Nanoparticles. A review of particle toxikology following inhalation exposure. *Inhalation Toxicology*, *24*(2), 125–135. doi:10.3109/08958378.2010.642021 PMID:22260506

Baker, G. H. (2005). *A vulnerability assessment methodology for critical infrastructure sites*. Retrieved from http://works.bepress.com/george_h_baker/2/

Balmer, C. (2004, April 29). Italian Court Acquits Nine in Alleged Plot Against US Embassy. Boston. Retrieved from http://www.boston.com/news/world/europe/articles/2004/04/29/italian_court_acquits_nine_in_alleged_plot_against_us_embassy/

Bařinová, E. (2018). *Application of Certified Methodology, for Providing Personal Protective Equipment, in a Nanoparticle Environment at Selected Operatoíons* [Diploma thesis]. VSB-Technical university of Ostrava, Ostrava, Czech Republic.

Bařinová, E., Klouda, K., Lustyk, K., Nechvátal, M., & Senčík, J. (2018). Presentation and analysis of results from field measurement of nanoparticles in a machine plant. *Paper presented at XVIII. International conference: Occupational health and safety*. Academic Press.

Barker, K., & Haimes, Y. Y. (2009). Assessing uncertainty in extreme events: Applications to risk-based decision making in interdependent infrastructure sectors. *Reliability Engineering & System Safety, 94*(4), 819–829. doi:10.1016/j.ress.2008.09.008

Bartknecht, W., & Zwahlen, G. (2013). *Explosionsschutz*. Berlin: Springer Berlin.

Bátrlová, K., Klouda, K., Kubátová, H., Roupcová, P., & Nechvátal, M. (2019). Monitoring of the occurrence of nanoparticles in the working environment focusing on agricultural activities. *Josra*. Retrieved from https://www.bozpinfo.cz/josra/monitorovani-vyskytu-nanocastic-v-prostredi-prvotni-zavery-mereni

Baus, O., & Bouchard, S. (2014). Moving from virtual reality exposure-based therapy to augmented reality exposure-based therapy: A review. *Frontiers in Human Neuroscience, 8*(MAR). doi:10.3389/fnhum.2014.00112 PMID:24624073

BBC. (2005). Massive Power Outage in Indonesia. Retrieved from http://news.bbc.co.uk/2/hi/asia-pacific/4162902.stm

BBC. (2017). Who was hit by the NHS cyber-attack? Retrieved from https://www.bbc.com/news/health-39904851

BBC. (2017). Wiggins and Froome medical records released by Russian hackers. Retrieved from http://www.bbc.co.uk/news/world-37369705

Bea, R. G., Holdsworth, R. D., & Smith, C. (1996). Human and Organization Factors in the Safety of Offshore Platforms. *Paper presented at the International Workshop on Human Factors in Offshore Operations*. Academic Press.

Beering, P. S. (2002). Threats on Tap: Understanding the Terrorist Threat to Water. *Journal of Water Resources Planning and Management, 128*(3), 163–167. doi:10.1061/(ASCE)0733-9496(2002)128:3(163)

Belan, L. (2015). *Safety management – security management*. Zilina, Slovak Republic: EDIS.

Belan, L. (2015). *Security management. Security and risk management. (Bezpečnostný manažment. Bezpečnosť a manažérstvo rizika)*. Žilina, Slovakia: EDIS University of Žilina.

Below, R. Wirtz., A., Sapir- Guha, D. (2009, October). Disaster Category Classification and peril Terminology for Operational Purposes, *Working Paper*. Retrieved from https://www.cred.be/downloadFile.php?file=sites/default/files/DisCatClass_264.pdf

BenOSH. (n.d.). *Socio-economic costs of accidents at work and work-related ill health*. Retrieved from http://ec.europa.eu/social

Berger, F., Bernatíková, Š., Přichystalová, R., Schreiberová, L., Šigutová, L., Vavrečková, K. & Tymráková, L. (2019). *Professional exposure and health risks of nanostructured materials and possibilities of their prevention at workplaces across sectors*. Academic Press.

Bernatik, A., Zimmerman, W., Pitt, M., Strizik, M., Nevrly, V., & Zelinger, Z. (2008). Modelling accidental releases of dangerous gases into the lower troposphere from mobile sources. *Process Safety and Environmental Protection, 86*(3B3), 198–207. doi:10.1016/j.psep.2007.12.002

Berzi, L. (1996). *Teória policajno-bezpečnostných služieb [Theory of police security services]*. Bratislava, Slovakia: Academy of the Police Force in Bratislava.

Bird, F. E., & Germain, G. L. (Eds.). (1992). *Practical Loss Control Leadership*. Loganville, GA: International Loss Control Institute.

Birkett, D., & Mala-Jetmarova, H. (2012). Are Risk Mitigation Strategies of Water Critical Infrastructure Adequate within a European Environment of 21st Century. *Paper presented at the 10th International Conference on Hydroinformatics HIC2012*. Academic Press.

Birkett, D. (2017). Water Critical Infrastructure Security and Its Dependencies. *Journal of Terrorism Research*, *8*(2), 1–21. doi:10.15664/jtr.1289

Birkett, D., & Mala-Jetmarova, H. (2014). Plan, Prepare and Safeguard: Water Critical Infrastructure Protection in Australia. In R. M. Clark & S. Hakim (Eds.), *Securing Water and Wastewater Systems, Protecting Critical Infrastructure 2*. Switzerland: Springer International Publishing. doi:10.1007/978-3-319-01092-2_14

Birkett, D., Truscott, J., Mala-Jetmarova, H., & Barton, A. F. (2011). Vulnerability of Water and Wastewater Infrastructure and its Protection from Acts of Terrorism: A Business Perspective. In R. M. Clark, S. Hakim, & A. Ostfeld (Eds.), *Handbook of Water and Wastewater Systems Protection, Series Protecting Critical Infrastructure* (pp. 457–483). New York, USA: Springer. doi:10.1007/978-1-4614-0189-6_23

Boček, J. (2012). *Uplatnění nanotechnologií v zemědělství*. Jihočeská univerzita in České Budějovice.

Bogdevičius, M., Prentkovskis, O., & Vladimirov, O. (2004). Engineering solutions of traffic safety problems of road transport. *Transport*, *19*(1), 43–50. doi:10.3846/16484142.2004.9637952

Boile, M. (2000). Intermodal transportation network analysis - a GIS application. In *Proceedings of the 10th Mediterranean Electrotechnical Conference - MELECON* (vol. 2, pp.660-663). Academic Press.

Bologna, S. (2016). *Guidelines for Critical Infrastructure Resilience Evaluation*. Roma, Italy: Italian Association of Critical Infrastructures' Experts.

Borodzicz, E. P. (2005). *Risk, Crisis and Security Management*. Chichester, West Sussex: John Wiley & Sons Ltd.

Bosch C., & van den Weterings R. (2005) *Methods for the calculation of the physical effects—due to releases of hazardous materials (liquids and gases)—yellow book*. CPR 14 E, The Hague.

Bragança, G., Siakas, K., & Anastasiadis, T. (2019). Internet of Things in the Context of Industry 4.0: An Overview. *International Journal of Entrepreneurial Knowledge*, *7*(1), 4–19.

Bragança, S., Costa, E., Castellucci, I., & Arezes, P. M. (2019). A Brief Overview of the Use of Collaborative Robots in Industry 4.0: Human Role and Safety. In *Occupational and Environmental Safety and Health* (pp. 641–650). Cham: Springer. doi:10.1007/978-3-030-14730-3_68

Brandt, P. T., & Sandler, T. (2010). What do Transnational Terrorists Target? Has It Changed? Are We Safer? *The Journal of Conflict Resolution*, *54*(2), 214–236. doi:10.1177/0022002709355437

Brekelmans, J., & Bosh, R. (2003). Summary of Large Scale Fire Tests in the RUNEHAMAR Tunnel in Norway. Retrieved from http://www.vlada.cz/cz/media-centrum/aktualne/audit-narodni-bezpecnosti-151410/

British Standard Institute (BSI). (2014). *BS11200: Crisis Management – guidance and good practice*. BSI.

British Standards Institution. (2010). *Risk management- Risk assessment techniques (ISO 31010:2010)*. Retrieved from https://shop.bsigroup.com/ProductDetail?pid=000000000030183975

British Standards Institution. (2014a). *Adapting to Climate Change using your Business Continuity Management System.* Retrieved from https://www.bsigroup.com/localfiles/en-gb/iso-22301/resources/bsi-sustainability-report-adapting-to-climate-change-using-your-business-continuity-management-system-uk-en.pdf

British Standards Institution. (2014b). *Societal security — Business continuity management systems — Requirements (ISO 22301:2012)* Retrieved from https://shop.bsigroup.com/ProductDetail?pid=000000000030292502

British Standards Institution. (2014c). *Crisis management – Guidance and good practice* (BS 11200:2014). Retrieved from https://shop.bsigroup.com/ProductDetail/?pid=000000000030274343

British Standards Institution. (2015). *Societal security — Business continuity management systems — Guidelines for business impact analysis (BIA) (PD ISO/TS 22317:2015).* Retrieved from https://shop.bsigroup.com/ProductDetail/?pid=000000000030299894

British Standards Institution. (2018a). *Risk management — Guidelines (BS ISO 31000:2018).* Retrieved from https://shop.bsigroup.com/ProductDetail?pid=000000000030315447

British Standards Institution. (2018b). *Security and resilience- Vocabulary* (ISO 22300:2018). Retrieved from https://shop.bsigroup.com/ProductDetail/?pid=000000000030348153

Brockett, J. (2015, July 13). UK Water Networks Vulnerable to Terrorist Attack. Utility Week. Retrieved from http://utilityweek.co.uk/news/uk-water-networks-%E2%80%98vulnerable-to-terrorist-attack/1150512#.VkC-c8tOcyU

Brooks, N., Adger, N. W., & Kelly, M. P. (2005). The determinants of vulnerability and adaptive capacity at the national level and the implications for adaptation. *Global Enviro. Change. Part A, 15*(2), 151–163. doi:10.1016/j.gloenvcha.2004.12.006

Brown, D.F., Dunn, W.E., & Policastro, A.J. (2000). *A National Risk Assessment for Selected Hazardous Materials in Transportation.* Argonne National Laboratory.

Brown, G., Carlyle, M., Salmeron, J., & Wood, K. (2005). Analyzing the Vulnerability of Critical Infrastructure to Attack and Planning Defenses. *Paper presented at the INFORMS Tutorials in Operations Research, Institute for Operations Research and the Management Sciences.* Academic Press. 10.1287/educ.1053.0018

Brown, T., Beyeler, W., & Barton, D. (2004). Assessing Infrastructure Interdependencies: The Challenge of Risk Analysis for Complex Adaptive Systems. *International Journal of Critical Infrastructures, 1*(1), 108–117. doi:10.1504/IJCIS.2004.003800

Bruneau, M., Chang, S. E., Eguchi, R. T., Lee, G. C., O'Rourke, T. D., Reinhorn, A. M., ... von Winterfeldt, D. (2003). A framework to quantitatively assess and enhance the seismic resilience of communities. *Earthquake Spectra, 19*(4), 733–752. doi:10.1193/1.1623497

Brunner, E.M. & Suter, M. (2008). *International CIIP Handbook.* Academic Press.

Brussels explosions: What we know about airport and metro attacks. (2016, April 9). BBC NEWS. Retrieved from https://www.bbc.com/news/world-europe-35869985

Bryan, B. A., & Crossman, N. D. (2008). Systematic regional planning for multiple objective natural resource management. *Journal of Environmental Management, 88*(4), 1175–1189. doi:10.1016/j.jenvman.2007.06.003 PMID:17643737

Brzicová, T. (2019). *Nanoparticle safety management to occupational safety* [Doctoral dissertation]. VSB-Technical university of Ostrava, Ostrava, Czech Republic.

Bundesamt für Bevölkerungsschutz und Katastrophenhilfe. (2008). *Schutz Kritischer Infrastruktur: Risikomanagement im Krankenhaus.* Retrieved from https://www.google.com/url?sa=t&rct=j&q=&esrc=s&source=web&cd=1&cad=rja&uact=8&ved =2ahUKEwiC9oTthv3kAhVDb1AKHUiECG8QFjAAegQIAh AB&url=https%3A%2F%2Fwww.bbk.bund.de%2FDE%2FAufgabenundAusstattung %2FKritischeInfrastrukturen%2FPationenKritis%2FSchutz_KRITIS_Ri

Bundesamt für Sicherheit in der Informationstechnik. (2013). *Schutz Kritischer Infrastrukturen: Risikoanalyse Krankenhaus-IT. Bonn.* Retrieved from https://www.google.com/url?sa=t&rct=j&q=&esrc=s&source=web&cd=1&ved= 2ahUKEwjrxtyQhv3kAhUQJlAKHTkcC3wQFjAAegQIABAC&url=https% 3A%2F%2Fwww.bsi.bund.de%2FSharedDocs%2FDownloads%2FDE%2FBSI %2FPublikationen%2FBroschueren%2FRisikoanalyseKrankenhaus.pdf%3F__blob%3

Burby, R. J., Deyle, R. E., Godschalk, D. R., & Olshansky, R. B. (2000). Creating hazard resilient communities through land-use planning. *Natural Hazards Review*, *2*(1), 99–106. doi:10.1061/(ASCE)1527-6988(2000)1:2(99)

Burke, D. A. (1999). Towards a Game Theory Model of Information Warfare [PhD thesis]. Air Force Institute of Technology.

Burton, I., Huq, S., Lim, B., Pilifosova, O., & Schipper, E. L. (2002). From impacts assessment to adaptation priorities: The shaping of adaptation policy. *Climate Policy*, *2*(2-3), 145–159. doi:10.3763/cpol.2002.0217

Bussey, M., Carter, R. W. B., Keys, N., Carter, J., Mangoyana, R., Matthews, J., ... Smith, T. F. (2011). Framing Adaptive Capacity through a History-Futures Lens: Lessons from the South East Queensland Climate Adaptation Research Initiative. *Futures*, *44*(4), 385–397. doi:10.1016/j.futures.2011.12.002

Cagno, V., Andreozzi, P., D'Alicarnasso, M., Silva, P. J., Mueller, M., Galloux, M., ... & Weber, J. (2018). Broad-spectrum non-toxic antiviral nanoparticles with a virucidal inhibition mechanism. *Naturematerial*, *17*, 195-203. Retrieved from: https://www.nature.com/articles/nmat5053?fbclid=IwAR0O0PjOFI5sWG8NpeML3JfsFSfYSZ9QTPsTsyg_xebnTQXR-6RolzsTXKfE

Cain, J. (2001). *Planning improvements in natural resources management. Guidelines for using Bayesian networks to support the planning and management of development programmes in the water sector and beyond.* UK: Centre for Ecology and Hydrology Wallingford.

Cambridge. (2019). Dictionary. Retrieved from https://dictionary.cambridge.org/

Cameron, C. (2002, July 30). Feds Arrest al-Queda Suspects with Plans to Poison Water Supplies. Fox News. Retrieved from http://www.foxnews.com/story/2002/07/30/feds-arrest-al-qaeda-suspects-with-plans-to-poison-water-supplies/

Canadian Wood Council. (1996). Fire Safety Design in Buildings. A reference for applying the National Building Code of Canada fire safety requirements in building design Fire Safety Design in Buildings.

Carmigniani, J., Furht, B., Anisetti, M., Ceravolo, P., Damiani, E., & Ivkovic, M. (2011). Augmented reality technologies, systems and applications. *Multimedia Tools and Applications*, *51*(1), 341–377. doi:10.100711042-010-0660-6

Carvel, R. O., & Torero, J. L. (2006). *The contribution of asphalt road surfaces to fire risk in tunnel fires: Preliminary findings.* Retrieved from https://www.era.lib.ed.ac.uk/bitstream/handle/1842/895/Carvel%20Torero%20Hong%20Kon;jsessionid=F1F323A8F25A2DE73E34343958E27CC7?sequence=1

Carvel, R. (2019). A review of tunnel fire research from Edinburgh. *Fire Safety Journal*, *105*(1), 300–306. doi:10.1016/j.firesaf.2016.02.004

CCPS. (1989). *Guidelines for Chemical Process Quantitative Risk Analysis - CPQRA.* New York: Center for Chemical Process Safety of the American Institute of Chemical Engineers.

CCPS. (2001). *Layer of Protection Analysis: Simplified Process Risk Assessment.* New York: American Institute of Chemical Engineers CCPS.

Ceballos, G. P. J., Fuentes, T. F., Diaz, P. D., Sanchez, S. M., Llorente, M. C., & Sanz, G. J. E. (2005). 11 March 2004: The terrorist bomb explosions in Madrid, Spain – an analysis of the logistics, injuries sustained, and clinical management of casualties treated at the closest hospital., *Critical Care Journal, 9*(1), 104-111. Retrieved from https://www.ncbi.nlm.nih.gov/pmc/articles/PMC1065101/#

CEI. (1994). Manual – Dow's Chemical Exposure Index (1st ed.). New York: AIChE.

Center for Chemical Process Safety (AIChE). (1992). Guidelines for Hazard Evaluation Procedures.

Center for Security Studies. (2018a, March). Εγχειρίδιο- Εκπαιδεύσεις για την Προστασία των Υποδομών Ζωτικής Σημασίας [*Manual- Training for the Protection of Critical Infrastructure*]. Retrieved from http://www.ciprotection.gr/index.php/el/

Center for Security Studies. (2018b, March). Στοχευμένες Δράσεις για την Αύξηση της Προστασίας των Εθνικά Χαρακτηρισμένων Ευρωπαϊκών Υποδομών Ζωτικής Σημασίας [Targeted actions for enhancing the protection of national characterized European critical infrastructure, WP 1.1 (V.2)], Οδικός Χάρτης Παραδοτέο 1 (Ε1) [Roadmap- Deliverable 1 (E1)]. SAE 050/2. Unpublished internal document.

Certified methodology for providing personal protective equipment in workplaces with a presence of nanoparticles: (focused on employers in need of protective equipment with respiratory protection). (2016). Prague, Czech Republic: Occupational Safety Research Institute.

CEU. (2008). *Council Directive 2008/114/EC of 8 December 2008 on the identification and designation of European critical infrastructures and the assessment of the need to improve their protection.* Brussels: Council of the European Union.

Chao, E. L., & Henshaw, J. L. (2002). *Job Hazard Analysis*. Occupational Safety and Health Administration, US Department of Labor. Retrieved from https://www.osha.gov/Publications/osha3071.pdf

Charniak, E. (1991). Bayesian networks without tears. *AI Magazine, 12*, 50–63.

Chen, H., Wood, M. D., Linstead, C., & Maltby, E. (2011). Uncertainty analysis in a GIS-based multi-criteria analysis tool for river catchment management. *Environmental Modelling & Software, 26*(4), 395–405. doi:10.1016/j.envsoft.2010.09.005

Christou, M.D., Struckl, M., & Biermann T. (2006). Land Use Planning Guidelines in the context of Article 12 of the Seveso II Directive 96/82/EC as amended by Directive 105/2003/EC.; JRC.

Chromečka, O. (2019) *Measurement of dust particles less than 10 μm in working atmosphere.* (Bachelor thesis), VSB-Technical university of Ostrava, Ostrava, Czech Republic.

Chung, P. W. H., & Jefferson, M. (1998). The integration of accident databases with computer tools in the chemical industry. *Computers & Chemical Engineering, 22*(Supplement 1), 729–732. doi:10.1016/S0098-1354(98)00135-5

CI. (2000). *Annexe n° 2 à la circulaire interministérielle n° 2000- 63 du 25 août 2000 relative à la sécurité dans les tunnels du réseau routier national—Instruction technique relative aux dispositions de sécurité dans les nouveaux tunnels routiers (conception et exploitation).*

CIPRNET. (2019). Retrieved from https://cordis.europa.eu/project/rcn/107425/factsheet/en

Clark, R. & Hakim, S. (2014). Securing Water and Wastewater Systems. Springer.

CLC/TS 50398 Alarm systems. Combined and integrated alarm systems. General requirements. (2009).

Cohen, F. (2010). What Makes Critical Infrastructures Critical? *International Journal of Critical Infrastructure Protection, 3*(2), 53–54. doi:10.1016/j.ijcip.2010.06.002

Coleman, K. (2005). *Protecting the Water Supply From Terrorism.* Directions Magazine. Retrieved from http://www.directionsmag.com/entry/protecting-the-water-supply-from-terrorism/123563

Colim, A., Costa, S., Cardoso, A., Arezes, P., & Silva, C. (2019, July). Robots and Human Interaction in a Furniture Manufacturing Industry-Risk Assessment. In *Proceedings of the International Conference on Applied Human Factors and Ergonomics* (pp. 81-90). Springer.

Coll, S. (2008). *The Bin Ladens - The Story of a Family and its Fortune.* London: Penguin books.

Collins, M., Petit, F., Buehring, W., Fisher, R., & Whitfield, R. (2011). Protective measures and vulnerability indices for the Enhanced Critical Infrastructure Protection Programme. In International Journal of Critical Infrastructures (vol. 7, pp. 200-219). Retrieved from doi:10.1504/IJCIS.2011.042976

Comite Europeen de Normalisation. (2002). *EN 1991-1-2 Eurocode 1: Actions on structures—Part 1-2: General Actions—Actions on structures exposed to fire.*

Comite Europeen de Normalisation. (2005). *Standard CEN/TR 12101-5* Smoke and heat control systems - Part 5: Guidelines on functional recommendations and calculation methods for smoke and heat exhaust ventilation systems.

Comite Europeen de Normalisation. (2009). *EN 12591 Bitumen and bituminous binders – Specifications for paving grade bitumens.*

Commission of The European Communities. (2005). *Green Paper on a European programme for critical infrastructure protection.* Retrieved from https://eur-lex.europa.eu/legal-content/EN/TXT/?uri=CELEX:52005DC0576

Commission of the European Communities. (2006). *Communication from the Commission on a European Programme for Critical Infrastructure Protection.* Retrieved from http://europa.eu/legislation_summaries/justice_freedom_security/fight_against_terrorism/l33260_en.htm

Commission of the European Communities. (2006, December 12). *Communication from the commission on a European Programme for Critical Infrastructure Protection.* Retrieved from https://eur-lex.europa.eu/legal-content/EN/TXT/?uri=CELEX:52006DC0786

Commonwealth of Australia, (2011). *National guidelines for protecting critical infrastructure from terrorism.*

Commonwealth of Australia. (2010). *Critical infrastructure resilience strategy.* Retrieved from http://www.tisn.gov.au/

Commonwealth of Australia. (2015). *Critical Infrastructure Resilience Strategy: Plan.* Retrieved from https://www.tisn.gov.au/Documents/CriticalInfrastructureResilienceStrategyPlan.PDF

Control banding . (2015) ACS. Retrieved from https://www.acs.org/content/acs/en/chemical-safety/hazard-assessment/ways-to-conduct/control-banding.html

Control banding. Canadian centre for occupational health and safety . (2017). Retrieved from: https://www.ccohs.ca/oshanswers/chemicals/control_banding.html

Cooke, R. M., & Goossens, L. H. J. (2004). Expert judgment elicitation for risk assessments of critical infrastructures. *Journal of Risk Research*, *7*(6), 643–656. doi:10.1080/1366987042000192237

Cooper, A. (2013, April 9). Man's al-Qaeda Link, Court Reporting. The Melbourne Age, Australia.

Copeland, C. (2010). *Terrorism and Security Issues Facing the Water Infrastructure Sector. CRS Report for Congress.* Washington, USA: Congressional Research Service.

Council Directive. (2001). Council directive 2001/14/EC on the allocation of railway infrastructure capacity and the levying of charges for the use of railway infrastructure and safety certification.

Council Directive. (2008). Council directive 2008/114 / EC of 8 December 2008 on the Identification and design of European critical infrastructures and the assessment.

Council of the European Union. (2008, December 8). *COUNCIL DIRECTIVE 2008/114/EC of 8 December 2008 on the identification and designation of European critical infrastructures and the assessment of the need to improve their protection.* Retrieved from https://eur-lex.europa.eu/LexUriServ/LexUriServ.do?uri=OJ:L:2008:345:0075:0082:EN:PDF

Court-Young, H. C. (2003). *Understanding Water and Terrorism* (September 2003 ed.). Denver, CO: Burg Young Publishing LLC.

Covert, A. J. (2008). Water: Vital to Life, Vulnerable to Terrorism [PhD] Air University.

Crnčevič, D. (2015). Transport Risk Analysis. In Transportation Systems and Engineering, Concepts, Methodologies, Tools, and Applications. Hershey, PA: IGI Global. doi:10.4018/978-1-4666-8473-7.ch001

CTIF. (2018). *Extreme Fire Behavior*: Understanding the Hazard. Retrieved from https://www.ctif.org/news/extreme-fire-behavior-understanding-hazard

Current control banding approaches for nanotechnologies . (2008). Safenano. Retrieved from https://www.safenano.org/knowledgebase/guidance/banding/

Cutter, S. L., Barnes, L., Berry, M., Burton, C., Evans, E., Tate, E., & Webb, J. (2008). A place-based model for understanding community resilience to natural disasters. *Global Environmental Change, 18*(4), 598–606. doi:10.1016/j.gloenvcha.2008.07.013

Czech Office for Standards, Metrology and Testing. (2010). ČSN 73 0804 Fire safety of constructions – Production structures.

Czech Office for Standards, Metrology and Testing. (2012). *Low-voltage electrical installations. Part 5-52: Selection and erection of electrical equipment – wiring systems* (CSN 33 2000-5-52:2012).

Czech Office for Standards, Metrology and Testing. (2013). ČSN 737507 The designing of road tunnels.

Dado, M., & Zahradnik, M. (2007). *Intelligent Transport Technologies and Services*. Zilina, Slovak Republic: University of Zilina.

Danihelka, P. (2016). *Legislation of nanomaterials within the EU, environmental protection.* Retrieved from https://www.enviprofi.cz/33/legislativa-nanomaterialu-v-ramci-eu-ochrana-zivotniho-prostredi-uniqueidmRRWSbk196FNf8-jVUh4EpGY5V1LxZIbDfbtq18xpwk/

Das, P., Barua, S., Sarkar, S., Chatterjee, S. K., Mukherjee, S., Goswami, L., ... Bhattacharya, S. S. (2018). Mechanisms of toxicity and transformation of silver nanoparticles: Inclusive assessment in earthworm-microbe-soil-plant system. *Geoderma, 314*, 73–84. doi:10.1016/j.geoderma.2017.11.008

Data Protection Authority. (2019). *Data Protection Authority.* Retrieved from https://www.dpa.gr/portal/page?_pageid=33,40911&_dad=portal&_schema=PORTAL

De Porcellinis, S., Oliva, G., Panzieri, S., & Setola, R. (2009, March). A holistic-reductionistic approach for modeling interdependencies. In *Proceedings of the International Conference on Critical Infrastructure Protection* (pp. 215-227). Springer. 10.1007/978-3-642-04798-5_15

De Rademaeker, E., Suter, G., Pasman, H. J., & Fabiano, B. (2014). A review of the past, present and future of the European loss prevention and safety promotion in the process industries. *Process Safety and Environmental Protection, 92*(4), 280–291. doi:10.1016/j.psep.2014.03.007

Deloitte Advisory. (2012). *Methodology to ensure of critical infrastructure protection in the area of electricity generation, transmission and distribution.* Prague: Deloitte Advisory. (in Czech)

Deloitte. (2016). *Cyber crisis management: Readiness, response, and recovery.* Retrieved from https://www.google.com/url?sa=t&rct=j&q=&esrc=s&source=web&cd=16&cad= rja&uact=8&ved=2ahUKEwij0amRn_3lAhXISxUIHeu5AWAQFjAPegQICRAC&url= https%3A%2F%2Fwww2.deloitte.com%2Fcontent%2Fdam%2FDeloitte%2Fde%2FDocuments%2Frisk%2FDeloitte-Cyber-crisis-management-Rea

DEMASST. (2019). Factsheet. Retrieved from https://cordis.europa.eu/project/rcn/91165/factsheet/en

DHS. (2013). *National Infrastructure Protection Plan: Partnering for Critical Infrastructure Security and Resilience.* Washington, DC: U.S. Department of Homeland Security.

Di Giorgio, A., & Liberati, F. (2012). A Bayesian Network-Based Approach to the Critical Infrastructure Interdependencies Analysis. *IEEE Systems Journal, 6*(3), 510–519. doi:10.1109/JSYST.2012.2190695

DMRB. (1999). *Highway structures: Design (substructures and special structures) material. Special structures. Design of road tunnels (BD 78/99).* London: The Highways Agency.

Dolnak, I. (2008). *Security of information systems in railway transport* [Dissertation thesis]. University of Zilina.

Drdla, P. (2018). *Passenger Transport of Regional and Supra-regional Importance.* Pardubice, Czech Republic: University of Pardubice. (in Czech)

Dreazen, Y. J. (2001, December 28). 'Backflow' Water-Line Attack Feared - Terrorists Could Reverse Flow in System to Introduce Toxins, Water Security. *The Wall Street Journal.* Retrieved from https://cryptome.org/backflow-panic.htm

DSB. (2003). *The role and Status of DoD Red Teaming Activities.* Retrieved from http://www.fas.org/irp/agency/dod/dsb/redteam.pdf

Dudenhoeffer, D. D., Permann, M. R., & Manic, M. (2006). CIMS: A Framework for Infrastructure Interdependency Modelling and Analysis. In *Proc. of the 38th Winter Simulation Conference* (pp. 478-485). Academic Press. 10.1109/WSC.2006.323119

Dugal, J. (1999). Guadalajara Gas Explosion Disaster. *Disaster Recovery Journal, 5*(3). Retrieved from http://www.drj.com/drworld/content/w2_028.htm

Dumbrava, V., & Iacob, S. V. (2013). Using Probability- Impact Matrix in Analysis and Risk Assessment Projects. *Journal of Knowledge Management, Economics and Information Technologies.* Retrieved from https://pdfs.semanticscholar.org/e2b7/82cb13b324d156e1c1db08fc2f2c267e1247.pdf

Dunn, W. N. (2001). Using the method of context validation to mitigate type III Errors in environmental policy analysis. In M. Hisschemoller, R. Hoppe, W. N. Dunn, & J. Ravetz (Eds.), *Knowledge, power and participation in environmental policy* (pp. 417–436). New Brunswick, London: Transaction Publishers.

Dvorak, Z., Cizlak, M., Leitner, B., Sousek, R., & Sventekova, E. (2010). *Risk management in railway transport.* Pardubice, Czech Republic: Jan Perner Institute. (in Slovak)

Dvorak, Z., Sventekova, E., Rehak, D., & Cekerevac, Z. (2017). Assessment of Critical Infrastructure Elements in Transport. *Procedia Engineering, 187*, 548–555.

Dyllick, T., & Hockerts, K. (2002). Beyond the Business Case for Corporate Sustainability. *Business Strategy and the Environment*, *11*(2), 130–141. doi:10.1002/bse.323

E.C. transport Themes. (2011). Roadmap to a Single European Transport Area - Towards a competitive and resource efficient transport system.

EC. (2010). *Commission staff working paper: Risk assessment and mapping guidelines for disaster management.* Brussels: European Commission.

EC. (2013). *Commission staff working paper on a new approach to the European programme for critical infrastructure protection - making European critical infrastructures more secure.* Brussels: European Commission.

Eckhoff, R. (2008). *Dust explosions in the process industries.* Amsterdam: Gulf Professional Pub.

ECTRI. (2019). Retrieved from https://www.ectri.org/

Edwards, M. (2014). *Critical Infrastructure protection, NATO Science for Peace and Security Series - E: Human and Societal Dynamics NATO.* IOS Press. Retrieved from https://ebookcentral.proquest.com/lib/portsmouth-ebooks/detail.action?docID=1637644

EFFECTS. (2018). Software EFFECTS, Gexcon. TNO. Retrieved from https://www.gexcon.com/products-services/EFFECTS/31/en

Egan, M. J. (2007). Anticipating future vulnerability: Defining characteristics of increasingly critical infrastructure-like systems. *J. Counting. Crisis Manage.*, *15*(1), 4–17. doi:10.1111/j.1468-5973.2007.00500.x

EGIG. (2018). *Gas Pipeline Incidents: 10th Report of the European Gas Pipeline Incident Data Group (period 1970 – 2016).* Groningen: European Gas Pipeline Incident Data Group.

eMars. (2019). Major Accident Hazards Bureau (MAHB), eMARS Data. Retrieved from http://mahbsrv.jrc.it

EN 15602 (2008). *Security service providers. Terminology.*

EN 16082 (2011). *Airport and aviation security services.*

EN 16747 (2015). *Maritime and port security services.*

EN 16763 (2017). *Services for fire safety systems and security systems.*

EN 50130-4 (2011). *Alarm systems. Electromagnetic compatibility. Product family standard: Immunity requirements for components of fire, intruder, hold up, CCTV, access control and social alarm systems.*

EN 50131–1 (2006). *Alarm systems. Intrusion and hold-up systems. Part 1: System requirements.*

EN 50131-7 (2010). *Alarm systems. Intrusion and hold-up systems. Part 7: Application guidelines.*

EN 50131-8 (2019). *Alarm systems. Intrusion and hold-up systems. Part 8: Security fog devices.*

EN 50134-1 (2002). *Alarm systems. Social alarm systems. Part 1: System requirements.*

EN 50136-1 (2012). *Alarm systems. Alarm transmission systems and equipment. Part 1: General requirements for alarm transmission systems.*

EN 54-1 (2011). *Fire detection and fire alarm systems. Introduction.*

EN 54-2 (1997). *Fire detection and fire alarm systems. Control and indicating equipment.*

EN 60839-11-1 (2013). *Alarm and electronic security systems - Part 11-1: Electronic access control systems - System and components requirements.*

EN 60839-11-2 (2014). *Alarm and electronic security systems - Part 11-2: Electronic access control systems - Application guidelines.*

EN 62676-1-1 (2014). *Video surveillance systems for use in security applications. System requirements. General.*

Enders, W., & Sandler, T. (1993). The Effectiveness of Anti-Terrorism Policies: A Vector-Autoregression-Intervention Analysis. *The American Political Science Review, 87*(4), 829–844. doi:10.2307/2938817

Engelbrecht, H., Lindeman, R. W., & Hoermann, S. (2019). A SWOT Analysis of the Field of Virtual Reality for Firefighter Training. *Frontiers in Robotics and AI, 6*(October), 1–14. doi:10.3389/frobt.2019.00101

ENISA. (2014, December). Methodologies for the identification of Critical Information Infrastructure assets and services. Retrieved from https://www.academia.edu/25419910/Methodologies_for_the_identification_of_Critical_Information_Infrastructure_assets_and_services_Guidelines_for_charting_electronic_data_communication_networks?auto=download

ENISA. (2016). *Good Practice Guide on Vulnerability Disclosure. From challenges to recommendations.* Retrieved from https://www.enisa.europa.eu/publications/vulnerability-disclosure

ENISA. (2016). *Securing Hospitals: A research study and blueprint. Independent Security Evaluators.* Retrieved from https://www.securityevaluators.com/wp- content/uploads/2017/07/securing_hospitals.pdf

ENISA. (2016). *Smart Hospitals: Security and Resilience for Smart Health Service and Infrastructures.* Retrieved from https://www.enisa.europa.eu/publications/cyber-security-and -resilience-for-smart-hospitals

ENISA. (2019). *Greek National Cyber Security Strategy*. Retrieved from https://www.enisa.europa.eu/topics/national-cyber-security-strategies/ncss-map/ national-cyber-security-strategies-interactive-map/strategies/national-cyber-security-strategy-greece/view

ERACHAIR. (2019). Retrieved from http://www.erachair.uniza.sk/

EU ERA. (2019). Retrieved from https://www.era.europa.eu/

EU. (2008). *Council Directive 2008/114/EC.* Retrieved from https://eur-lex.europa.eu/legal-content/EN/TXT/?uri=uriserv%3AOJ.L_.2008.345.01.0075.01.ENG

EU. (2013). *Decision No 1082/2013/EU of The European Parliament and of the Council of 22 October 2013 on serious cross-border threats to health and repealing Decision No 2119/98/EC.* Retrieved from https://ec.europa.eu/health/sites/health/files/preparedness_response

EU. (2016). *Directive (EU) 2016/1148 of The European Parliament and of the Council of 6 July 2016 concerning measures for a high common level of security of network and information systems across the Union.* Retrieved from https://eur-lex.europa.eu/legal-content/EN/TXT

EU. (2016). *Regulation (EU) 2016/679 of the European Parliament and of The Council of 27 April 2016 on the protection of natural persons with regard to the processing of personal data and on the free movement of such data, and repealing Directive 95/46/EC (GDPR).* Retrieved from https://eur-lex.europa.eu/legal-content/EN/TXT/PDF/?uri=CELEX:32016R0679&from=EN

EU. (2017). *Cybersecurity Act.* Retrieved from https://eur-lex.europa.eu/legal-content/EN/TXT/?uri=COM:2017:0477:FIN

EU. (2017). *Regulation (EU) 2017/746.* Retrieved from https://eur-lex.europa.eu/legal-content/EN/TXT/?uri=CELEX:32017R0746

EU. (n.d.). REGULATION (EU) 2017/745. Retrieved from https://eur-lex.europa.eu/legal-content/EN/TXT/?uri=CELEX:32017R0745

EU-CIRCLE consortium. (2016). *EU- CIRCLE-* A pan European Framework for strengthening Critical Infrastructure resilience to climate change. Retrieved from http://www.eu-circle.eu/wp-content/uploads/2017/01/D4.1_EU-CIRCLE_resilience_initialversion.pdf

EU-CIRCLE consortium. (2017a). *EU- CIRCLE- A pan- European Framework for strengthening Critical Infrastructure resilience to climate change.* Retrieved from http://www.eu-circle.eu/wp-content/uploads/2015/07/D3.1.pdf

EU-CIRCLE consortium. (2017b). EU-CIRCLE A pan- European Framework for strengthening Critical Infrastructure resilience to climate change. Retrieved from http://www.eu-circle.eu/wp-content/uploads/2015/07/D4.4.pdf

Eur-Lex. (2019). Retrieved from https://eur-lex.europa.eu/LexUriServ/LexUriServ.do?uri=OJ:L:2008:345:0075:0082:EN:PDF

European Commission (2018). Commission communication in the framework of the implementation of Directive 2014/34/EU of the European Parliament and of the Council on the harmonisation of the laws of the Member States relating to equipment and protective systems intended for use in potentially explosive atmospheres. *Official Journal of the European Union*, 61.

European Commission. (2006). *Communication from the Commission on a European Programme for Critical Infrastructure Protection.* Brussels.

European Commission. (2008). *Council Directive 2008/114/EC on the Identification and Designation of European Critical Infrastructures and the Assessment of the Need to Improve their Protection.* Brussels.

European Commission. (2014). Communication from the Commission to the European Parliament, the Council, the European Economic and Social Committee and the Committee of the Regions on an EU Strategic Framework on Health and Safety at Work 2014–2020.

European Commission. (2017). EU-CIRLE- *A pan European framework for strengthening Critical Infrastructure resilience to climate change.* Retrieved from https://cordis.europa.eu/project/rcn/196896/factsheet/en

European Commission. (2019a). *InfraStress- Improving resilience of sensitive industrial plants & infrastructures exposed to cyber-physical threats, by means of an open testbed stress-testing system.* Retrieved from https://cordis.europa.eu/project/rcn/222602/factsheet/en

European Commission. (2019b). IMPROVER. *Improved risk evaluation and implementation of resilience concepts to critical infrastructure.* Retrieved from https://cordis.europa.eu/project/rcn/196889/factsheet/en

European Commission. (2019c). *SATIE. Security of Air Transport Infrastructure of Europe.* Retrieved from https://cordis.europa.eu/project/rcn/222594/factsheet/en

European Commission. (2019d). *SecureGas. Securing the European Gas Network.* Retrieved from https://cordis.europa.eu/project/rcn/222598/factsheet/en

European Commission. (2019e). *STOP-IT. Strategic, Tactical, Operational Protection of water Infrastructure against cyber-physical Threats.* Retrieved from https://cordis.europa.eu/project/rcn/210216/factsheet/en

European Commission. (2019f), September 16). *DEFENDER- Defending the European Energy Infrastructures.*

European Committee for Standardization. (2004). *Protective clothing for use against solid particulates - Part 1: Performance requirements for chemical protective clothing providing protection to the full body against airborne solid particulates (type 5 clothing)* (EN ISO 13982-1:2004/A1:2010).

European Committee for Standardization. (2005). *Protective clothing against liquid chemicals - Performance requirements for chemical protective clothing offering limited protective performance against liquid chemicals (Type 6 and Type PB [6] equipment)* (EN 13034:2005+A1:2009).

European Committee for Standardization. (2005). *Protective clothing against liquid chemicals - performance requirements for clothing with liquid-tight (Type 3) or spray-tight (Type 4) connections, including items providing protection to parts of the body only (Types PB [3] and PB [4])* (EN 14605:2005+A1:2009).

European Committee for Standardization. (2011). *Common test methods for cables under fire conditions.Heat release and smoke production measurement on cables during flame spread test. Test apparatus, procedures, results* (EN 50399:2011/A1:2016).

European Committee for Standardization. (2015). *Method of test for resistance to fire of unprotected small cables for use in emergency circuits* (EN 50200:2015).

European Committee for Standardization. (2015). *Protective clothing against dangerous solid, liquid and gaseous chemicals, including liquid and solid aerosols - Part 1: Performance requirements for Type 1 (gas-tight) chemical protective suits* (EN 943-1:2015+A1:2019).

European Committee for Standardization. (2016). *Common test methods for cables under fire conditions. Heat release and smoke production measurement on cables during flame spread test. Test apparatus, procedures, results* (EN 50399:2011+A1:2016).

European Committee for Standardization. (2018). *Fire classification of construction products and building elements - Part 1: Classification using data from reaction to fire tests* (EN 13501-1:2018).

European Committee for Standardization. (2018). *Fire classification of construction products and building elements - Part 6: Classification using data from reaction to fire tests on power, control and communication cables* (EN 13501-6:2018).

European Committee for Standardization. (2019). *Protective clothing against dangerous solid, liquid and gaseous chemicals, including liquid and solid aerosols - Part 2: Performance requirements for Type 1 (gas-tight) chemical protective suits for emergency teams (ET)* (EN 943-2:2019).

European Parliament and the Council. (2004). *Directive 2004/54/EC of the European Parliament and of the Council of 29 April 2004 on minimum safety requirements for tunnels in the Trans-European Road Network* (Directive 2004/54/EC).

European Parliament, Policy Department for Citizens' Rights and Constitutional Affairs. (2018). *Member States' Preparedness for CBRN Threats- Terrorism.* Retrieved from http://www.europarl.europa.eu/RegData/etudes/STUD/2018/604960/IPOL_STU(2018)604960_EN.pdf

Europol. (2019). *Terrorism Situation and Trend Report 2019.*

Evaluation study. (2019). European Union. Retrieved from https://ec.europa.eu/info/law/better-regulation/initiatives/ares-2018-1378074/public-consultation_en

Ex Guide. (2018) IEC System for Certification to Standards relating to Equipment for use in Explosive Atmospheres - An Informative Guide comparing various elements of both IECEx and ATEX. Geneva: IEC.

Examples of nanotechnology misuse . (2015). Retrieved from http://www.seminarky.cz/Priklady-moznosti-zneuziti-nanotechnologie-30621

Ezell, B. C. (2007). Infrastructure Vulnerability Assessment Model (I-VAM). *Risk Analysis, 27*(3), 571–583. doi:10.1111/j.1539-6924.2007.00907.x PMID:17640208

Fanfarová, A., & Mariš, L. (2017). Simulation tool for fire and rescue services. *Procedia Engineering, 192*, 160–165.

FDS-SMV. (2019). Fire Dynamics Simulator. Retrieved from https://pages.nist.gov/fds-smv/

FEI. (1994). Manual – Dow's Fire & Explosion Index, Hazard Classification Guide (7th ed.). Academic Press.

Fenton, N., & Neil, M. (2013). *Risk Assessment and Decision Analysis with Bayesian Networks*. London: CRC Press.

Filipová, Z., Kukutschová, J., & Mašláň, M. (2012). *Risk of nanomaterials*. Olomouc, Czech Republic: University of Palacký.

FLAIM systems. (2019). *Virtual Reality for Firefighting*. Retrieved from https://www.flaimsystems.com/flaim-trainer/flaim-trainer-options/

Folke, C. (2006). Resilience: The emergence of a perspective for social-ecological systems analyses. *Global Environmental Change, 16*(3), 253–267. doi:10.1016/j.gloenvcha.2006.04.002

Folwarczny, L., & Pokorný, J. (2006). *Evacuation of persons*. Ostrava: The Association of fire and safety engineering.

Ford, J. D., Smit, B., & Wandel, J. (2006). Vulnerability to climate change in the Arctic: A case study from Arctic Bay, Canada. *Global Environmental Change, 16*(2), 145–160. doi:10.1016/j.gloenvcha.2005.11.007

FORTESS. (2019). Retrieved from https://cordis.europa.eu/project/rcn/185488/factsheet/en

Freeman, P. K. (2003). Natural Hazard Risk and Privatization. In A. Kreimer, M. Arnold, and A. Carlin (Eds.), Building Safer Cities: The Future of Disaster Risk (pp. 33-44). Academic Press.

Freitas, L.C. (2016). Manual de segurança e saúde do trabalho. Lisboa, Portugal: Sílabo.

Friedlingstein, P., & Solomon, S. (2005). Contributions of past and present human generations to committed warming caused by carbon dioxide. *Proceedings of the National Academy of Sciences of the United States of America, 102*(31), 10832–10836. doi:10.1073/pnas.0504755102 PMID:16037209

Friend, M. A., & Kohn, J. P. (2003). Fundamentals of occupational safety and health. UK: The Scarecrow Press, Inc.

Frišhansová, L., & Klouda, K. (2016). *Ambivalence of nanoparticles*. Retrieved from http://www.odpadoveforum.cz/TVIP2017/prispevky/220.pdf

Frišhansová, L., & Klouda, K. (2017). Ambivalence of nanoparticles. In Sborník příspěvků ze symposia Týden vědy a inovací pro praxi a životní prostředí. Academic Press.

Frišhansová, L., Klouda, K., Nechvátal, M., Roupcová, P., & Barták, P. (2018). Analysis of unique operation for processing of sandstone from the view of released nanoparticles. *Paper presented at Sborník přednášek XVIII. ročníku mezinárodní konference. Ostravice VŠB-TU 2018*. Academic Press.

Gaddis, T. (1998). *Virtual reality in the school. Virtual reality and Education Laboratory*. East Carolina University.

Gangopadhyay, R. K. (2005a). *Study of some accidents in chemical industries – Lessons learnt*. Unpublished doctoral dissertation, Birla Institute of Technology, Meshra, India.

Gangopadhyay, R. K., & Das, S. K. (2009a). Fire during transfer of LPG from the road tanker to Horton sphere – rupture of the hose attached to the manifold. In I. Sogaard & H. Krogh (Ed.), Fire Safety (pp. 113-120). Nova Science Publishers, Inc.

Gangopadhyay, R. K., & Das, S. K. (2009b). Fire accident at a refinery in India: The causes and the lesson learnt. In I. Sogaard & H. Krogh (Ed.), Fire Safety (pp. 121-126). Nova Science Publishers, Inc.

Gangopadhyay, R. K., & Das, S. K. (2007a). Chlorine emission during the chlorination of water case study. *Environmental Monitoring and Assessment, 125*(1-3), 197–200. doi:10.100710661-006-9252-3 PMID:16897510

Gangopadhyay, R. K., & Das, S. K. (2007b). Accident due to release of hydrogen sulphide in a manufacturing process of cobalt sulphide – case study. *Environmental Monitoring and Assessment, 129*(1-3), 133–135. doi:10.100710661-006-9347-x PMID:17057980

Gangopadhyay, R. K., & Das, S. K. (2007c). Lesson learned – fire due to leakage of ethanol during transfer. *Process Safety Progress, 26*(3), 235–236. doi:10.1002/prs.10197

Gangopadhyay, R. K., & Das, S. K. (2007d). Accidental emissions of sulphur oxides from the stack of a sulphuric acid plant – case study. *Process Safety Progress, 26*(4), 283–288. doi:10.1002/prs.10199

Gangopadhyay, R. K., & Das, S. K. (2008a). Ammonia leakage from refrigeration plant and the management practice. *Process Safety Progress, 27*(1), 15–20. doi:10.1002/prs.10208

Gangopadhyay, R. K., & Das, S. K. (2008b). Lessons learned from a fuming sulfuric acid tank overflow incident. *Journal of Chemical Health and Safety, 15*(5), 13–15. doi:10.1016/j.jchas.2008.02.002

Gangopadhyay, R. K., Das, S. K., & Mukherjee, M. (2005b). Chlorine leakage from bonnet of a valve in a bullet – a case study. *Journal of Loss Prevention in the Process Industries, 18*(4-6), 526–530. doi:10.1016/j.jlp.2005.07.008

Garcia, M. L. (2001). *The Design and Evaluation of Physical Protection Systems*. USA: Elsevier.

Giaccone, M. (2010). European Foundation for the Improvement of Living and Working Conditions. Health and safety at work in SMEs: Strategies for employee information and consultation. Retrieved from: https://www.eurofound.europa.eu/publications/report/2010/health-and-safety-at-work-in-smes-strategies-for-employee-information-and-consultation

Giannopoulos, G., Filippini, R., & Schimmer, M. (2012). *Risk assessment methodologies for critical infrastructure protection. Part I: A state of the art*. Ispra: European Commission, Joint Research Centre; doi:10.2788/22260

Giannopoulos, G., Filippini, R., & Schimmer, M. (2012). *Risk assessment methodologies for Critical Infrastructure Protection. Part I: State of the art*. Luxembourg: Publications Office of the European Union.

Gibbs, L. I., & Holloway, C. F. (2013, May). *Hurricane Sandy After Action. Report and Recommendations to Mayor Michael R. Bloomberg*. Retrieved from https://www1.nyc.gov/assets/housingrecovery/downloads/pdf/2017/sandy_aar_5-2-13.pdf

Gilbert, P. H., Isenberg, J., Baecher, G. B., Papay, L. T., Spielvogel, L. G., Woodard, J. B., & Badolato, E. V. (2003). Infrastructure Issues for Cities - Countering Terrorist Threat. *Journal of Infrastructure Systems, 9*(1), 44–54. doi:10.1061/(ASCE)1076-0342(2003)9:1(44)

Gillette, J., Fisher, R., Peerenboom, J., & Whitfield, R. (2002). Analysing Water/Wastewater Infrastructure Interdependencies. *Paper presented at the 6th International Conference on Probabilistic Safety Assessment and Management (PSAM6)*. Retrieved from http://www.ipd.anl.gov/anlpubs/2002/03/42598.pdf

Gleick, P. H. (2006). Water and Terrorism. *Water Policy,* (8), 481-503.

Gleick, P. H. (2006). *Water Conflict Chronology*. Retrieved from http://citeseerx.ist.psu.edu/viewdoc/download?rep=rep1&type=pdf&doi=10.1.1.204.9178

Gleick, P. H., & Heberger, M. (2014). *Water Conflict Chronology* (1597264210). World water. Retrieved from http://www.worldwater.org/wp-content/uploads/sites/22/2013/07/ww8-red- water-conflict-chronology-2014.pdf

Gleick, P. H. (1993). Water and Conflict: Fresh Water Resources and International Security. *International Security*, *18*(1), 79–112. doi:10.2307/2539033

Gleick, P. H. (1996). Basic Water Requirements for Human Activities: Meeting Basic Needs. *Water International*, *21*(2), 83–92. doi:10.1080/02508069608686494

Glendon, A. I., Clarke, S. G., & McKenna, E. F. (Eds.). (2006). *Human Safety and Risk Managemen*. London, New york: Taylor & Francis Publishing.

Godschalk, D. R. (2003). Urban hazard mitigation: Creating resilient cities. *Natural Hazards Review*, *4*(3), 136–143. doi:10.1061/(ASCE)1527-6988(2003)4:3(136)

Gonzalez, J. J., Sarriegi, J. M., & Gurrutxaga, A. (2006). A Framework for Conceptualizing Social Engineering Attacks. In J. L'opez (Ed.), *CRITIS 2006* (pp. 79–90). Springer. doi:10.1007/11962977_7

Government of Canada, (2009). National strategy for critical infrastructure.

Government of the Czech Republic. (1997). Act no. 13/1997 Coll., on land roads, as amended. Retrieved from https://www.zakonyprolidi.cz/cs/1997-13

Government Regulation. (2010). Coll. on Criteria for Determination of the Critical Infrastructure Element.

Grant, G. B., & Drysdale, D. D. (1997). Estimating heat release rates from large-scale tunnel fires. [). London, UK: International Association for Fire Safety Science.]. *Proceedings of Fire Safety Science*, *5*, 1213–1224. doi:10.3801/IAFSS.FSS.5-1213

Großhans, H. (2019). *Simulation of Explosion Hazards in Powder Flows*. Braunschweig: Physikalisch-Technische Bundesanstalt (PTB). doi:10.7795/210.20190521P

Guay, B. (2019, July 23). Probe Opened in France Over Radioactive Water Rumours, Water contamination. *AFP*. Retrieved from https://www.afp.com/en/news/826/probe-opened-france-over- radioactive-water-rumours-doc-1j12mc2

Hamins, A.P., Bryner, N.P., Jones, A.W., & Koepke, G.H. (2015). *Research Roadmap for Smart Fire Fighting Summary Report Research Roadmap for Smart Fire Fighting*. NIST and The Fire Protection Research Foundation. Retrieved from: www.nfpa.org/SmartFireFighting

Hartin, E. Compartment Fire Behaviour Training: Reading the Fire. Retrieved http://cfbt-us.com/wordpress/?p=38

Hattwig, M., & Steen, H. (2008). *Handbook of Explosion Prevention and Protection*. Weinheim: Wiley-VCH.

Hauptmanns, U. (2015). *Process and Plant Safety*. Berlin, Heidelberg: Springer Berlin Heidelberg.

He, W., Liu, Y., Wamer, W. G., & Yin, J. (2014). Electron spin resonance spectroscopy for the study of nanomaterial-mediated generation of reactive oxygen species. *Journal of Food and Drug Analysis*. *22*(1), 49-63. Retrieved from: https://linkinghub.elsevier.com/retrieve/pii/S1021949814000052

Health and Safety Executive (HSE). (2019). Common Topic 4: Safety Culture. Retrieved from http://www.hse.gov.uk/humanfactors/topics/common4.pdf

Healthcare and Public Health Sector Coordinating Councils. (2017). *Health Industry Cybersecurity Practices: managing threats and protecting patients.* Retrieved from https://www.phe.gov/Preparedness/planning/405d/Documents/HICP-Main-508.pdf

Heisler, A. (2018). Seven Critical Risks Impacting the Energy Industry. *Risk & Insurance*. Retrieved from https://riskandinsurance.com/7-risks-impacting-energy-industry

Hei, X. D. X. (2013). Conclusion and Future Directions. In *Security for Wireless Implantable Medical Devices*. SpringerBriefs in Computer Science. doi:10.1007/978-1-4614-7153-0_5

Hellenic National Defence General Staff. (2019). *Hellenic National Defence General Staff*. Retrieved from http://www.geetha.mil.gr/en/hndgs-en/history-en.html

Henning, T. (2006). *Explosible region, warning sign*. Wikimedia. Retrieved from https://commons.wikimedia.org/wiki/File:D-W021_Warnung_vor_explosionsfaehiger_Atmosphaere_ty.svg

Heuvel, S., Zwaan, L., Dam, L. V., Oude-Hengel, K. M., Eekhout, I., van Emmerik, M. L., . . . Wilhelm, C. (2017). *Estimating the costs of work-related accidents and ill-health: An analysis of European data sources*. European Agency for Safety and Health at Work (EU-OSHA). Retrieved from https://osha.europa.eu/pt/themes/good-osh-is-good-for-business

HIPAA. (2018). Healthcare Data Breach Statistics. Retrieved from https://www.hipaajournal.com/healthcare-data-breach-statistics/

HIPAA. (2018). Largest Healthcare Data Breaches of 2018. Retrieved from https://www.hipaajournal.com/largest-healthcare-data-breaches-of-2018/

Hirschler, M. M. (2003). Heat release from plastic materials. In V. Babrauskas & S. J. Grayson (Eds.), *Heat release in fires* (pp. 375–422). London, UK: Interscience Communications.

Hofreiter, L., Lovecek, T., & Velas, A. (2005). Zásady a princípy analýzy rizík v oblasti fyzickej a objektovej bezpečnosti [Principles of risk analysis in the area of physical and object security]. National Security Agency of Slovak Republic.

Hofreiter, L. (2003). Nové determinanty ochrany objektov [New determinants of object protection]. In *Proceedings of the Solving crisis situations in a specific environment*. (vol. 8, pp. 155-160). Žilina, Slovakia: EDIS University of Žilina.

Hofreiter, L. (2004). *Security and security risks and threats*. University of Zilina. (in Slovak)

Höj, N. P. (2005). *Fire in tunnels, Thematic network, Technical Report - Part 2 Fire Safety Design - Road Tunnels*. Brussels: European thematic network on fire in tunnels.

Høj, N. P. (2004). Guidelines for fire safe design compared fire safety features for road tunnels. In *Proceedings of the 1st International Symposium on Safe & Reliable Tunnels* (Vol. 133). Academic Press.

Hollnagel, E. (2005). Human reliability assessment in context. *Nuclear Engineering and Technology*, *37*(2), 159–166.

Holt, A. S. J. (2001). *Principles of Construction Safety*. London: Blackwell Science Ltd. doi:10.1002/9780470690529

Homeland Security. (2013). NIPP 2013. *Partnering for Critical Infrastructure Security and Resilience*. Retrieved from: https://www.dhs.gov/sites/default/files/publications/national-infrastructure-protection-plan-2013-508.pdf

Homer-Dixon, T. (2013). *The upside of down: Catastrophe, creativity and the renewal of civilization*. Toronto, Canada: Knopf.

Hosser, D. (2013). *Leitfaden Ingenieurmethoden des Brandschutzes*. Retrieved from http://www.kd-brandschutz.de/files/downloads/Leitfaden2013.pdf

Howie, L. (Ed.). (2009). *Terrorism, the Worker and the City - Simulations and Security in a Time of Terror*. Farnham, UK: Gower.

Hruza, P., Sousek, R., & Szabo, S. (2014). Cyber-attacks and attack protection. In *World Multi-Conference on Systemics* (Vol. 18, pp. 170–174). Orlando, FL: Cybernetics, and Informatics.

HSE. (2005). *Inspector tool kit. Human factors in the management of major accidents hazards – introduction to human factors.* Retrieved from http://www.hse.gov.uk/humanfactors/topics/humanfail.html

Huang, Z., He, K., Song, Z., Zeng, G., Chen, A., Yuan, L., ... Chen, G. (2018). Antioxidative response of Phanerochaete chrysosporium against silver nanoparticle-induced toxicity and its potential mechanism. *Chemosphere*, *211*, 573–583. doi:10.1016/j.chemosphere.2018.07.192 PMID:30092538

Huijben, J. W., Fournier, P., & Rigter, B. P. (2006). *Aanbevelingen ventilatie van verkeerstunnels.* Utrecht: Ministerie van Verkeer en Waterstaat, Rijkswaterstaat, Bouwdienst.

Hurley, M. (2015). SFPE handbook of fire protection engineering. New York, NY: Springer Science+Business Media.

HZS CR. (2017). *Polygon – Klecovy trenazer.* Retrieved from https://www.hzscr.cz/clanek/trenazery-polygon-klecovy-trenazer.aspx

HZS CR. (2019). Hasičský záchranný sbor České republiky. Retrieved from https://www.hzscr.cz

HZS CR. (2019). *Statistical Yearbooks.* Retrieved from https://www.hzscr.cz/hasicien/article/statistical-yearbooks.aspx

IDEKO Project. (2019). Retrieved from http://oblast.cdv.cz/cz/O37/user/project/detail/2

IEC 31010. (2019). *Risk management -- Risk assessment techniques.* Geneva: International Organization for Standardization.

IEC 60812. (2018). *Failure modes and effects analysis (FMEA and FMECA).* Geneva: International Electrotechnical Commission.

IEC 61025. (2006). *Fault Tree Analysis.* Geneva: International Electrotechnical Commission.

IEC 61882. (2016). *Hazard and operability studies (HAZOP studies) - Application guide.* Geneva: International Electrotechnical Commission.

IEC 62502. (2010). *Analysis techniques for dependability – Event tree analysis.* Geneva: International Electrotechnical Commission.

IISD. (2019). International Institute for Sustainable Development, Business Strategy for Sustainable Development. Retrieved from https://www.iisd.org/business/pdf/business_strategy.pdf

ILO. (2019). International Labour Organization. Retrieved from https://www.ilo.org/global/lang--en/index.htm

Independent Security Evaluators. (2016). *Securing Hospitals - A research study and blueprinT.* Retrieved from https://www.securityevaluators.com/wp-content/uploads/2017/07/securing_hospitals.pdf

INE. (2019). Instituto Nacional de Estatística, Statistics Portugal. Retrieved from https://www.ine.pt/

INFOSEC. (2019). Hospital Security. Retrieved from https://resources.infosecinstitute.com/category/healthcare-information-security/security-awareness-for-healthcare-professionals/hospital-security/

INFRARISK. (2019). Retrieved from https://cordis.europa.eu/project/rcn/110820/factsheet/en

INTACT. (2019). Retrieved from https://cordis.europa.eu/project/rcn/185476/factsheet/en

International Organization for Standardization. (2011, March). *Information technology – Security techniques - Guidelines for information and communication technology readiness for business continuity.* Retrieved from https://www.iso.org/standard/44374.html

International Organization for Standardization. (2012, December). *Societal security — Business continuity management systems — Guidance*. Retrieved from https://www.iso.org/standard/50050.html

International Organization for Standardization. (2015). *Reaction-to-fire tests - Heat release, smoke production and mass loss rate - Part 1: Heat release rate (cone calorimeter method) and smoke production rate (dynamic measurement)* (ISO 5660-1:2015).

International Organization for Standardization. (2016). *Reaction to fire tests - Room corner test for wall and ceiling lining products - Part 1: Test method for a small room configuration* (ISO 9705:2016).

IRM. (2016). *About Risk Management*, Institute of risk management: Enterprise risk magazine. Retrieved from https://www.theirm.org/the-risk-profession/risk-management.aspx

Iromuanya, C., Hrgiss, K., & Howard C. (2015). *Critical risk path method.* Hershey, PA: IGI global. doi:10.4018/978-1-4666-8473-7.ch028

ISO 31000 (2018). *Risk management. Guidelines.*

ISO. (2009). ISO/IEC 31010:2009. Risk Assessment Techniques. International Organization for Standardization. Geneva-Switzerland.

ISO. (2009). ISO/IEC Guide 73:2009. Risk Management – Vocabulary. International Organization for Standardization. Geneva-Switzerland.

ISO. (2018). 31000:2018. Risk Management – Guidelines. ISO/TC, 262. International Organization for Standardization. Geneva, Switzerland.

ISO. (2018). 45001:2018. Occupational Health and Safety Management Systems – Requirements with Guidance for Use. International Organization for Standardization.

ISO. (2018). ISO 31000:2018 Risk management. Retrieved from https://www.iso.org/iso-31000-risk-management.html

ISO. (2019). International Organization for Standardization. Retrieved from https://www.iso.org/iso-45001-occupational-health-and-safety.html

ISO. (2019). ISO 31010:2019. Risk management - Risk assessment techniques.

Izuakor, C., & White, R. (2016). Critical Infrastructure Asset Identification: Policy, Methodology and Gap Analysis. In *Proceedings of the 10th International Conference on Critical Infrastructure Protection (ICCIP)*, (pp. 27-41). Academic Press. 10.1007/978-3-319-48737-3_2

Jackson, J.H., Mountcastle, R., & Charles, E. (2005). *The Counter-Terrorist Handbook.* London: Michael O'Mara Books Ltd.

Jedermann-Verl. (2007). *Safety characteristics determination and rating*. Heidelberg.

Jenkins, B. M. (1975). *Will Terrorists Go Nuclear.* Retrieved from http://www.defence.org.cn/aspnet/vip-usa/UploadFiles/2008-10/200810132327510156.pdf

Jiang, M., Zhou, G., & Zhang, Q. (2018). Fire-fighting Training System Based on Virtual reality. *IOP Conference Series: Earth and Environmental Science, 170*(4). 10.1088/1755-1315/170/4/042113

Jofré, C., Romero, J., & Rueda, R. (2010). *Contribution of concrete pavements to the safety of tunnels in case of fire.* Eupave. Retrieved from https://www.eupave.eu/wp-content/uploads/eupave-safety-of-tunnels-in-case-of-fire.pdf

Joint Research Center. (2018). JRC Technical Reports. *European Reference Network for Critical Infrastructure Protection: ERNCIP Handbook 2018 edition*. Retrieved from https://erncip-project.jrc.ec.europa.eu/sites/default/files/ERNCIP%20 Handbook%202018.pdf

Joschek, H. I. (1981). Risk assessment in the chemical industry. In *Proceedings of the International topical meeting on probabilistic risk assessment* (pp. 122 – 131). New York: American Nuclear Society.

Kadri, F., Babiga, B., & Châtelet, E. (2014). The impact of natural disasters on critical infrastructures. *Journal of Homeland Security and Emergency Management, 11*(2), 217–241.

Kahani, M. & Beadle, H. W. P. (1997). Comparing Immersive and Non-Immersive Virtual Reality User Interfaces for Management of Telecommunication for Management of Telecommunication Networks. Retrieved from https://www.researchgate.net/publication/228742087

Kalil, J. M., & Berns, D. (2004). *Drinking Supply: Terrorist Had Eyes on Water*. Defendyourh2o. Retrieved from http://www.defendyourh2o.com/pdfs/DRINKING%20SUPPLY.pdf

Kampova, K., & Makka, K. (2018). Economic aspects of the risk impact on the fuel distribution enterprises. In *Proceedings of the International Conference Transport Means* (pp. 231-235). Academic Press.

Kasa, J. (2012). The issue of flammable bitumen surfaces (asphalts) on roads in tunnels in the CR. Czech Technical University in Prague. Retrieved from http://people.fsv.cvut.cz/~wald/edu/134SEP_Seminar_IBS/2012/06_SP12_Kasa_Zivice.pdf

Katsumiti, A., Thornley, A. J., Arostegui, I., Reip, P., Valsami-Jones, E., Tetley, T. D., & Cajaraville, M. P. (2018). Cytoxicity and cellular mechanisms of toxicity of CuO NPs in mussel cells in vitro and comparative sensitivity with human cells. *Toxicology In Vitro, 48*, 146–158. doi:10.1016/j.tiv.2018.01.013 PMID:29408664

Kelemen, M., & Jevcak, J. (2018). Security Management Education and Training of Critical Infrastructure Sectors' Experts. In *Proceedings of* NTAD 2018 - 13th International Scientific Conference - New Trends in Aviation Development (pp. 72–75).

Kersten H., Klett G. (2017). Business Continuity und IT-Notfall management.

Khamar. (2016). *CFR 1910.119 14 Elements of Process Safety Management*. Retrieved from https://commons.wikimedia.org/w/index.php?curid=49203116

Khan, F. I., & Abbasi, S. A. (1999). Major accidents in process industries and analysis of causes and consequences. *Journal of Loss Prevention in the Process Industries, 12*(5), 361–378. doi:10.1016/S0950-4230(98)00062-X

Khan, F. I., & Amoyette, P. R. (2003). How to make inherent safety practice a reality. *Canadian Journal of Chemical Engineering, 81*(1), 2–16. doi:10.1002/cjce.5450810101

Khan, F., Rathnayaka, S., & Ahmed, S. (2015). Methods and models in process safety and risk management: Past, present and future. *Process Safety and Environmental Protection, 98*, 116–147. doi:10.1016/j.psep.2015.07.005

Kidam, K., & Hurme, M. (2013). Analysis of equipment failures as contributors to chemical process accidents. *Process Safety and Environmental Protection, 91*(1-2), 61–78. doi:10.1016/j.psep.2012.02.001

Kirwan, B. (1994). *A guide to practical human reliability assessment*. London: Taylor & Francis Publishing.

KISDIS. (2019). Retrieved from https://starfos.tacr.cz/en/project/VG20122015070

Kletz, T. A. (1991). *Plant Design for Safety – A User-Friendly Approach*. New York: Hemisphere.

Kletz, T. A. (1993). *Lessons from disaster: How organizations have no memory and accidents recur.* U.K.: Gulf Professional Publishing.

Kletz, T. A. (2001). *Learning from accidents.* Oxford: Gulf Professional Publishing.

Klouda, K., Frišhansová, L., & Senčík, J. (2016). Nanoparticles, nanotechnologies and nanoproducts and their connection with occupational health and safety. In *Occupational health and safety & life quality 2016*. Academic Press.

Knight, A. M., & Kevin, W. (2010). *A Journey Not a Destination.* Moscow: Risk Management, Presentation to the RusRisk.

Koch, R. (2008). *The 80/20 Principle: The Secret to Achieving More with Less.* London: Crown Publishing Group.

Kokkala, M. A., Thomas, P. H., & Karlsson, B. (1993). Rate of heat release and ignitability indices for surface linings. *Fire and Materials, 17*(5), 209–216. doi:10.1002/fam.810170503

Kolářová, L. (2014). *Introduction to nanosciences and nanotechnologies.* Retrieved from: http://mofychem.upol.cz/KA4/Nanotechnologie.pdf

Koncek, R. (2013). *Characteristics and shielding of ionizing radiation.* Brno, Czech Republic: Brno University of Technology.

Korshed, P., Li, L., Liu, Z., Mironov, A., & Wang, T. (2018). Antibacterial mechanisms of a novel type picosecond laser-generated silver-titanium nanoparticles and their toxicity to human cells. *International Journal of Nanomedicine, 13*, 89–101. doi:10.2147/IJN.S140222 PMID:29317818

Kovács, T. (2015). *Ergonomics.* Budapest: Óbuda University.

KPMG. (2015). *Health care and cyber security: increasing threats require increased capabilities.* Retrieved from https://assets.kpmg/content/dam/kpmg/pdf/2015/09/cyber-health-care-survey-kpmg-2015.pdf

Krause, T., Wu, J., & Markus, D. (2019). *Ringversuche im Bereich des Explosionsschutzes - Ergebnisse und Erkenntnisse aus dem Programm "Explosion Pressure".* In Proceedings of the 15th BAM-PTB. Academic Press. doi:10.7795/210.20190521D

Krause, U. (2009). *Fires in silos.* Weinheim: Wiley-VCH. doi:10.1002/9783527623822

Krietsch, A. (2015). *Untersuchung der Brand- und Explosionsgefahren von Nanostäuben.* Magdeburg: Otto-von-Guericke-Universität.

Kroll, D. (2008, September 22). Testing the Waters: Improving Water Quality and Security Via On-Line Monitoring. *Paper presented at the ICMA.* Academic Press. Retrieved from http://hachhst.com/wp-content/uploads/2010/07/Presentation_Water-Security-and-Quality.pdf

Kroll, D. (2010a). *Aqua ut a Telum "Water as a Weapon."* Retrieved from http://hachhst.com/wp-content/uploads/2010/07/White-Paper_Water-as-a-weapon.pdf

Kroll, D. (2010b). *A Reinterpretation of the 2002 Attempted Water Terror Attack on the U.S. Embassy - Don't Underestimate the Enemy.* Hach. Retrieved from http://hachhst.com/wp- content/uploads/2010/07/White-Paper_-WATER-TERROR-ATTACK1.pdf

Kroll, D. J. (2013). The Terrorist Threat to Water and Technology's Role in Safeguarding Supplies. In *Proceedings of the 45th Session of the International Seminars on Nuclear War and Planetary Emergencies.* Academic Press. 10.1142/9789814531788_0030

Kroll, D. J., King, K., Engelhardt, T., Gibson, M., Craig, K., & Securities, H. H. (2010). *Terrorism Vulnerabilities to the Water Supply and the Role of the Consumer: A Water Security White Paper*. Waterworld. Retrieved from http://www.waterworld.com/articles/2010/03/terrorism- vulnerabilities-to-the-water-supply-and-the-role-of-the-consumer.html

Kroll, D., & King, K. (2006). Laboratory and Flow Loop Validation and Testing of the Operational Effectiveness of an On-Line Security Platform for the Water Distribution System. *Paper presented at the 8th Annual Water Distribution Systems Analysis Symposium*. Academic Press.

Kroll, D., King, K., & Klein, G. (2009). Real World Experiences With Real-Time On-Line Monitoring for Security and Quality, Detecting and Responding to Events. *Paper presented at the International Workshop on Water and Wastewater Security Incidents*. Hach. Retrieved from http://hachhst.com/wp-content/uploads/2010/07/Presentation_Real-World-events.pdf

Kroll, D. (2006). *Securing Our Water Supply: Protecting a Vulnerable Resource* (S. Hill, Ed.). Tulsa, OK: PennWell Corporation.

Kudrna, J., Bebčák, P., Bebčáková, K., Šperka, P., & Stoklásek, S. (2018). *Asphalt road surfaces in tunnels*. Retrieved from http://www.silnice-zeleznice.cz/clanek/asfaltove-vozovky-v-tunelech/

Labid, A., & Harris, M. J. (2014, July 8). Learning how to learn from failures: The Fukushima nuclear disaster. *Engineer Failure Analysis Journal Volume 47* P.A 117-128. Available from: https://www.sciencedirect.com/science/article/abs/pii/S1350630714002933

LaFree, G. (2017, May 22). 6 Reasons Why Stopping Worldwide Terrorism is so Challenging. The Conversation. Retrieved from https://theconversation.com/6-reasons-why- stopping-worldwide-terrorism-is-so-challenging-70626

Larena, G. I. (2017). *Antimicrobial activity of graphene and its viability.* Retrieved from http://diposit.ub.edu/dspace/handle/2445/112186

Lauterwasser, C. (2016) *Opportunities and risks of Nanotechnologies* OECD. Retrieved from https://www.oecd.org/science/nanosafety/44108334.pdf

Lazarova, D. (2015, May 27). Problem With Contaminated Tap Water in Prague 6 Still Unresolved. Radio Prague International. Retrieved from http://www.radio.cz/en/section/news/problem-with-contaminated-tap-water-in-prague-6-still-unresolved

Leão, C. P., Costa, S., Costa, N., & Arezes, P. (2018, October). Capturing the ups and downs of accidents' figures–the Portuguese case study. In *Proceedings of the International Conference on Human Systems Engineering and Design: Future Trends and Applications* (pp. 675-681). Springer.

Lee, J.-J. M. (2009). A War for Water. *Paper presented at the World Environmental and Water Resources Congress 2009*. Academic Press. Retrieved from http://link.aip.org/link/?ASC/342/393/1

Lee, E. (2009). *Homeland Security and Private Sector Business: Corporations' Role in Critical Infrastructure Protection*. Boca Raton, FL: Auerbach Publications.

Lees, F. (2005). *Lees´ Loss Prevention in the Process Industries, Hazard identification, assessment and control* (S. Mannan, Ed.) (3rd ed.). Elsevier.

Lees, R. E., & Laudry, B. R. (1989). Increasing the understanding of industrial accidents: An analysis of potential major injury records. *Canadian Journal of Public Health*, *80*(6), 423–426. PMID:2611739

Lekkas, E., Carydis, P., & Lagouvardos, K. Mavroulis. S, Diakakis, M. … Papagiannaki, K. (2018). The July 2018 Attica (Central Greece) Wildfires- Scientific report (Version 1.1). *Newsletter of Environmental, Disaster, and Crisis Management Strategies*. Retrieved from https://www.researchgate.net/publication/326672342_The_July_2018_Attica_wildfires_Scientific_report_v11

Li, J., Li, Y. F., Cheng, C. H., & Chow, W. K. (2019). A study on the effects of the slope on the critical velocity for longitudinal ventilation in tilted tunnels. *Tunnelling and Underground Space Technology*, *89*, 262–267. doi:10.1016/j.tust.2019.04.015

Lima, T. M. (2004). Trabalho e Risco no Sector da Construção Civil em Portugal: Desafios a uma cultura de prevenção. *Oficina do CES*, *211*, 1–13.

Lincon, R. S. (1960). Human factors in the attainment of reliability. *Transactions on Reliability*, *9*, 97–103.

Li, X., Zhou, Z., Deng, X., & You, Z. (2017). Flame Resistance of Asphalt Mixtures with Flame Retardants through a Comprehensive Testing Program. *Journal of Materials in Civil Engineering*, *29*(4), 04016266. doi:10.1061/(ASCE)MT.1943-5533.0001788

Li, Y. Z., & Ingason, H. (2018). Overview of research on fire safety in underground road and railway tunnels. *Tunnelling and Underground Space Technology*, *81*, 568–589. doi:10.1016/j.tust.2018.08.013

Locken. (2017, January 9). Understanding the Terrorist Threat to our Water Sector. Security Newsdesk. Retrieved from https://securitynewsdesk.com/locken-understanding-terrorist-threat-water-sector/

Lorenzo, D. K. (1990). *A guide to reducing human errors, improving human performance in the chemical industry*. Washington, DC: The Chemical Manufacturers' Association, Inc.

Lovecek, T., & Reitspis, J. (2011). *Projektovanie a hodnotenie systémov ochrany objektov [Design and evaluation of physical protection systems]*. Slovakia: University of Žilina.

Lovecek, T., Ristvej, J., & Simak, L. (2010). Critical Infrastructure Protection Systems Effectiveness Evaluation. *Journal of Homeland Security and Emergency Management*, *7*(1), 34. doi:10.2202/1547-7355.1613

Lovecek, T., Velas, A., Kampova, K., Maris, L., & Mozer, V. (2013). Cumulative Probability of Detecting an Intruder by Alarm Systems. In *Proceedings of the IEEE International Carnahan Conference on Security Technology*. IEEE Press. 10.1109/CCST.2013.6922037

Lovreglio, R. (2018). A Review of Augmented Reality Applications for Building Evacuation. Retrieved from http://arxiv.org/abs/1804.04186

Löwe, K., & Kariuki, S. G. (2004). Berücksichtigung des menschen beim design verfahrenstechnischer anlagen. Fortschritt-Berichte, 22(16), 88-103.

Lustyk, K. (2018). *Exposure to Fine and Ultrafine Particles during Smoking and Comparison of Used Measurement Techniques* [Diploma thesis]. Vysoká škola báňská – Technická univerzita Ostrava.

Mach, V., & Boros, M. (2017). Perimeter protection elements testing for burglar resistance. In Key Engineering Materials (vol. 755, pp. 292-299). Retrieved from doi:10.4028/www.scientific.net/KEM.755.292

Maiolo, M., & Pantusa, D. (2018). Infrastructure Vulnerability Index of Drinking Water Systems to Terrorist Attacks. Academic Press. doi:10.1080/23311916.2018.1456710

Mannan, M. S., O'Connor, T. M., & Keren, N. (2009). Patterns and trends in injuries due to chemical based on OSHA occupational injury and illness statistics. *Journal of Hazardous Materials, 163*(1), 349–356. doi:10.1016/j.jhazmat.2008.06.121 PMID:18718716

Mannan, S., & Lees, F. (2005). *Lee's loss prevention in the process industries*. Burlington, MA: Elsevier Butterworth-Heinemann.

Mantovani, F., Castelnuovo, G., Gaggioli, A., & Riva, G. (2003). Virtual reality training for health-care professionals. *Cyberpsychology & Behavior, 6*(4), 389–395. doi:10.1089/109493103322278772 PMID:14511451

Margaritis, D., Anagnostopoulou, A., Tromaras, A., & Boile, M. (2016). Electric commercial vehicles: Practical perspectives and future research directions. *Research in Transportation Business & Management, 18*, 4–10. doi:10.1016/j.rtbm.2016.01.005

Marsh, R. T. (1997). *Critical Foundations: Protecting America's Infrastructure*. President's Commission on Critical Infrastructure Protection.

Martin, L. (2019). Hackers scramble patient files in Melbourne heart clinic cyber attack. *The Guardian*. Retrieved from https://www.theguardian.com/technology/2019/feb/21/hackers-scramble-patient-files-in-melbourne-heart-clinic-cyber-attack

Martin, G. M. P. (2017). Cybersecurity and healthcare: How safe are we? *BMJ (Clinical Research Ed.), 358*(j3179). PMID:28684400

Martirosov, S. & Kopeček, P. (2017). *Virtual Reality and its Influence on Training and Education - Literature Review*. doi:. doi:10.2507/28th.daaam.proceedings

Masarik, I. (2003). *Plastics and its fire hazard*. Ostrava, Czech Republic: SPBI.

Masarik, I., Dvorak, I., & Charvatova, V. (1999). *Investigation of combustion products toxicity*. Prague, Czech Republic: Technical Institute of Fire Protection.

Mathams, R. H. (1982). *Sub Rosa - Memoirs of an Australian Intelligence Analyst*. Sydney, Australia: George Allen & Unwin.

May, I., & Deckker, E. (2009). Reducing the risk of failure by better training and education. *Engineering Failure Analysis, 16*(4), 1153–1162. doi:10.1016/j.engfailanal.2008.07.006

McGuiness, D. (2017, April 27). How a cyber-attack transformed Estonia. *BBC News*. Retrieved from https://www.bbc.com/news/39655415

McGuire, E. G. (1996). Knowledge representation and construction in hypermedia and environments. *Telematics and Informatics, 13*(4), 251–260. doi:10.1016/S0736-5853(96)00025-1

MD CR. (2019). Retrieved from https://www.mdcr.cz/?lang=en-GB

MDV SR. (2019). Retrieved from https://www.mindop.sk/en

Medinsky, M. A., & Bond, J. A. (2001). Sites and mechanisms for uptake of gases and vapors in the respirátory tract. *Toxicology, 160*(1-3), 165–172. doi:10.1016/S0300-483X(00)00448-0 PMID:11246136

Meinhardt, P. L. (2006). Medical Preparedness for Acts of Water Terrorism. An On-Line Readiness Guide. American College of Preventative Medicine.

Meinhardt, P. L. (2005). Water and Bioterrorism: Preparing for the Potential Threat to U.S. Water Supplies and Public Health. *Annual Review of Public Health*, *24*(26), 213–237. doi:10.1146/annurev.publhealth.24.100901.140910 PMID:15760287

Meister, D. (1962). The problem of human-initiated failures. In Proceedings 8th National System Reliability and Quality Control (pp. 234-239). Academic Press.

Methods and way of risk assessment in the workplace. Documentation OSH. (2018) Retreived from https://www.dokumentacebozp.cz/aktuality/metody-hodnoceni-rizik-bozp/

Meye, T. (2016). *Untersuchungen zu Gasdetonationen in Kapillaren für die Mikroreaktionstechnik.* Magdeburg: Otto-von-Guericke-Universität.

Michalik, D. (2010). *Safety culture. Methodological manual.* Prague, Czech Republic: Occupational Safety Research Institute. (in Czech)

Miguel, A. S. (2014). *Manual de Higiene e Segurança no Trabalho.* Porto, Portugal: Porto Editora.

Mikkola, E., & Wichman, I. S. (1989). On the thermal ignition of combustible materials. *Fire and Materials*, *14*(3), 87–96. doi:10.1002/fam.810140303

Mikušová, M., & Horváthová, P. (2019). Prepared for a crisis? Basic elements of crisis management in an organisation. *Economic Research-Ekonomska Istraživanja*, *32*(1), 1844–1868. doi:10.1080/1331677X.2019.1640625

Milgram, P., Takemura, H., Utsumi, A., & Kishino, F. (1995). Augmented reality: a class of displays on the reality-virtuality continuum. In *Proceedings of Society of Photo-Optical Instrumentation Engineers: Telemanipulator and Telepresence Technologies*. Academic Press. doi:10.1117/12.197321

Ministry of Business, Innovation & Employment. (2017). *Verification method: Framework for fire safety design* (C/VM2:2017).

Ministry of Industry (India). Department of Industrial Development, (1987). Calcium Carbide Rules. Retrieved from https://dipp.gov.in/sites/default/files/Calcium_carbide_rules_1987_0.pdf

Ministry of Transport, the Land roads division. (2013). *A methodological instruction, Ventilation of road tunnels, Selection of the system, designing, operation, the quality assurance of ventilation systems of road tunnels.*

Ministry of Transport, the Road infrastructure division. (2010). Tests of fire safety devices for road tunnels.

Miniter, R. (2011). *Mastermind, The many Faces of the 9/11 Architect, Khalid Shaikh Mohammad.* New York: Penguin Group.

MIT. (2014a). Type plan for large-scale electricity supply disruption. *Ministry of Industry and Trade*. Retrieved June 10, 2019, from http://download.mpo.cz/get/26093/58202/615552/priloha007.doc

MIT. (2014b). Type plan for large-scale gas supply disruptions. *Ministry of Industry and Trade*. Retrieved June 10, 2019, from http://download.mpo.cz/get/26093/58202/615554/priloha005.doc

MIT. (2014c). Type plan for large-scale disruption of thermal energy supply. *Ministry of Industry and Trade*. Retrieved June 10, 2019, from http://download.mpo.cz/get/26093/58202/615556/priloha003.doc

Mleziva, J., & Snuparek, J. (2000). *Polymers: production, structure, properties and using*. Praha, Czech Republic: Sobotales.

MODSAFE. (2019). Retrieved from https://cordis.europa.eu/project/rcn/92875/factsheet/en

Moktadir, M. A., Ali, S. M., Kusi-Sarpong, S., & Shaikh, M. A. A. (2018). Assessing challenges for implementing Industry 4.0: Implications for process safety and environmental protection. *Process Safety and Environmental Protection, 117*, 730–741. doi:10.1016/j.psep.2018.04.020

Molnárné, M., Schendler, T., & Schröder, V. (2003). *Explosionsbereiche von Gasgemischen*. Bremerhaven: Wirtschafts-verl. NW.

Moridpour, s., Pour, A.T., & Saghapour, T. (2019). *Big Data Analytics in Traffic and Transportation Engineering. Emerging Research and Opportunities*. Hershey, PA: IGI Global.

Morris, S. (2012). The Camelford Poisoning: Blackwater, A Driver's Mistake. Retrieved from https://encrypted-tbn0.gstatic.com/images?q=tbn:ANd9GcQwx- FrraixjDahj7ftjHqWCBRBVF1WH0Aq-l0fE-qlEjIIZMnD

Morris, S. (2018, September 6). Novichok poisoning timeline: Q&A- Details released by police of the suspects' movements in Salisbury answer some questions but raise some more. *The Guardian*. Retrieved from https://www.theguardian.com/uk-news/2018/sep/06/novichok-poisoning-timeline-qa

Motaki, K. (2016). *Risk Analysis and Risk Management in Critical Infrastructures* (Master Thesis). Piraeus: University of Piraeus.

Mueller, B., & Tsang, A. (2018, December 20). Gatwick Airport Shut Down by "Deliberate" Drone Incursions. *The New York Times*. Retrieved form https://www.nytimes.com/2018/12/20/world/europe/gatwick-airport-drones.html

MZP. (2007). Methodological guideline of the Department of Environmental Risks of the Ministry of the Environment for the procedure of preparation of the document "Analysis and evaluation of major accident risks" pursuant to Act No. 59/2006 Coll., On prevention of major accidents. *Bulletin of the Ministry of the Environment, 3*, 1–15.

Nandavar, A., Petzold, F., Nassif, J., & Schubert, G. (2018). Interactive virtual reality tool for bim based on ifc: Development of openbim and game engine-based layout planning tool. In Proceedings of the *23rd International Conference on Computer-Aided Architectural Design Research in Asia: Learning, Prototyping and Adapting CAADRIA 2018* (Vol. 1, pp. 453–462). Academic Press.

Nando (New Approach Notified and Designated Organisations) Information System. (2019). Retrieved from https://ec.europa.eu/growth/tools-databases/nando/index.cfm

Nanoparticles and their toxicity. Chemical papers. (2015). Retrieved from: http://ww.chemicke-listy.cz/docs/full/2015_06_444-450.pdf

National Cybersecurity Agency of France (ANSSI). (2018). *EBIOS Risk Manager – The method*. Retrieved from https://www.ssi.gouv.fr/en/guide/ebios-risk-manager-the-method/

Nel, A. E., Mädler, L., Velegol, D., Xia, T., Hoek, E. M. V., Somasundaran, P., ... Thompson, M. (2009). Understanding biophysicochemical interactions at the nano-bio interface. *Nature Materials, 8*(7), 543–557. doi:10.1038/nmat2442 PMID:19525947

NFPA. (2017). *FACT SHEET Smart Firefighting We're revolutionizing fire safety*. Retrieved from https://www.nfpa.org/-/media/Files/Code-or-topic-fact-sheets/SmartFFFactSheet.ashx

NFPA. (2017). NFPA 502, Standard for road tunnels, bridges, and other limited access highways (National Fire Protection Association, Technical Committee on Road Tunnel and Highway Fire Protection).

NIS. (2019). *NIS*. Retrieved from http://www.nis.gr/portal/page/portal/NIS/

NIST-Blog. (2019). Making fighting fires virtual reality more real reality. Retrieved from https://www.nist.gov/blogs/taking-measure/making-fighting-fires-virtual-reality-more-real-reality

Noble, T., & Schrembi, J. (1984, September 17). Geelong Man Charged over Poisoning Threat. *The Melbourne Age*. Retrieved from https://news.google.com/newspapers?nid=1300&dat=19840917&id=7lVVAAAAIBAJ&sjid=n5UDAAAAIBAJ&pg=4756,7124&hl=en

North Atlantic Treaty Organization Strategic Communications Centre of Excellence. (n.d.). *Hybrid Threats: 2007 cyber-attacks on Estonia*. Retrieved from https://www.stratcomcoe.org/hybrid-threats-2007-cyber-attacks-estonia

Norwegian Public Roads Administration Standard Road Tunnels 03.04. (2004). Norwegian Public Roads Administration Road Tunnels.

NV. (2006). Government decree no. 344/2006 Col. On minimum safety requirements for tunnels in the road network. Retrieved from https://www.slov-lex.sk/pravne-predpisy/SK/ZZ/2006/344/20060601

NV. (2009). Government decree no. 264/2009 Col., on safety requirements for road tunnels longer than 500 meters. Retrieved from https://www.zakonyprolidi.cz/cs/2009-264/zneni-20090901

Nyikes, Z. (2017). *Digital Competence and the Safety Awareness Base in the Assessments Results of the Missle-East-European Generation*. Tirgu-Mures, Romania: INTER-ENG.

O'Connor, A., & Eichinger, E. M. (2007). Site-specific traffic load modeling for bridge assessment. *Proceedings of the Institution of Civil Engineers: Bridge Engineering*, *160*(4), 185–194.

OECD. (2003). OECD Guiding Principles for Chemical Accident Prevention, Preparedness and Response - Guidance for Industry (Including Management and Labor), Public Authorities, Communities, and other Stakeholders.

OFFSITE. (2017). 27χρονος επιτέθηκε σε γιατρούς στο Νοσοκομείο Λάρνακας. Retrieved from https://www.offsite.com.cy/articles/eidiseis/topika/231890-27hronos-epitethike-se-giatroys-sto-nosokomeio-larnakas

Orientations towards the first Strategic Plan implementing the research and innovation framework programme Horizon Europe. (2019). *Co-design via web open consultation*. Retrieved from https://ec.europa.eu/research/pdf/horizon-europe/ec_rtd_orientations-towards-the-strategic-planning.pdf

OSCE. (2012). Retrieved from https://www.osce.org/secretariat/99852?download=true

OSHA. (2013). *Occupational Safety and Health Standards 29 CFR 1910.119*. United States Department of Labor Occupational Safety and Health Administration. OSHA.

Ostman, B. L. A. (2003). Smoke and soot. In V. Babrauskas & S. J. Grayson (Eds.), *Heat release in fires* (pp. 233–250). London, UK: Interscience Communications.

Ouyang, M. (2014). Review on the Modeling and Simulation of Interdependent Critical Infrastructure Systems. *Reliability Engineering & System Safety*, *121*, 43–60. doi:10.1016/j.ress.2013.06.040

Pacinda, S. (2010). Network Analysis and KARS. *The Science for Population Protection*, *2*(1), 75–96.

Pamphlet 64, 5th Edition. Emergency response plans for chlorine facilities. (2000). The Chlorine Institute, Inc.

Parker, W. J. (2003). Prediction of the heat release rate from basic measurements. In V. Babrauskas & S. J. Grayson (Eds.), *Heat release in fires* (pp. 333–356). London, UK: Interscience Communications.

Perlroth, N., & Sanger, D. E. (2017). *The New York Times*. Retrieved from https://www.nytimes.com/2017/05/12/world/europe/uk-national-health-service-cyberattack.html

Peters, D., Kandola, K., Elmendorf, A. E., & Chellaraj, G. (1999). *Health expenditures, services, and outcomes in Africa: basic data and cross-national comparisons, 1990-1996 (English)*. Washington, D.C.: The World Bank. doi:10.1596/0-8213-4438-2

Petrova, S., Soudek, P., & Vanek, T. (2015). Flame retardants, their use and environmental impact. *Chemical Papers*, *109*(9), 679–686.

PIARC Committee on Road Tunnels. (1999). *Fire and smoke control in road tunnels*. La Défense.

PIARC Committee on Road Tunnels. (2019). World Road Association. Retrieved from https://www.piarc.org/en/

Pimentel, D., Houser, J., Preiss, E., White, O., Fang, H., Mesnick, L., ... Alpert, S. (1997). Water Resources: Agriculture, the Environment, and Society. *Bioscience*, *47*(2), 97–106. doi:10.2307/1313020

PLASA. (2019). Retrieved from https://cordis.europa.eu/project/rcn/207498/factsheet/en

Plos, V., Sousek, R., & Szabo, S. (2016). Risk-based indicators implementation and usage. In *Proceedings* the 20th World Multi-Conference on Systemics, Cybernetics and Informatics (Vol. *2*, pp. 235–237). Academic Press.

POEDIN. (2018). Κέντρα Υγείας σε Αποδιοργάνωση. *POEDIN.* Retrieved from https://www.poedhn.gr/deltia-typoy/item/3413-kentra-ygeias-se-apodiorganosi-viaiopragies-se-varos-iatrikoy-kai-nosileftikoy-prosopikoy--klopes-sto-kentro-ygeias-salaminas-epithesi-apo-omada-roma-se-giatro-tou-ky-lygouriou-pou-efimereve-sto-tep-tou-gnnafpl

Pokorny, J., Malerova, L., Wojnarova, J., Tomaskova, M., & Gondek, H. (2018). Perspective of the use of road asphalt surfaces in tunnel construction. In *Proceedings of the 22nd International Scientific on Conference Transport Means 2018* (pp. 290-296). Academic Press. Retrieved from https://www.scopus.com/inward/record.uri?eid=2-s2.0-85055566722&partnerID=40&md5=b5478e2d8440ed6be2fca65db63ed873

Pokorny, J., Mozer, V., Malerova, L., Dlouha, D., & Wilkinson, P. (2019). A simplified method for establishing safe available evacuation time based on a descending smoke layer. *Communications - Scientific Letters of the University of Zilina*, *20*(2), 28-34.

Pokorny, J. & Gondek, H. (2016). Comparison of theoretical method of the gas flow in corridors with experimental measurement in real scale. *Acta Montanistica Slovaca*, *21*(2), 146–153.

Porod, C., Collins, M., & Petit, F. (2014). Water Treatment Dependencies. *The CIP Report, 13*(2), 10-13. Retrieved from http://cip.gmu.edu

Postranecky, M., & Svitek, M. (2018). *Cities in future*. NADATUR. (in Czech)

PPJK. (2018). The policy of the quality of land roads. Retrieved from http://www.pjpk.cz/

Prague Daily Monitor (PDM). (2015a, June 8). Prague to Vaccinate Children From Districts With Contaminated Water. Retrieved from http://praguemonitor.com/2015/06/08/prague-vaccinate- children-districts-contaminated-water

Prague Daily Monitor (PDM). (2015b, May 28). Prague Town Hall Files Complaint Over Contaminated Drinking Water. Retrieved from http://www.praguemonitor.com/2015/05/28/prague- town-hall-files-complaint-over-contaminated-drinking-water

Prentkovskis, O., Sokolovskij, E., & Bartulis, V. (2010). Investigating traffic accidents: A Collision of two motor vehicles. *Transport*, *25*(2), 105–115. doi:10.3846/transport.2010.14

Pursianen, C. (2018). The Crisis Management Cycle. New York: Routledge

Qi, R., Prem, K. P., Ng, D., Rana, M. A., Yun, G., & Mannan, M. S. (2012). Challenges and needs for process safety in the new millennium. *Process Safety and Environmental Protection*, *90*(2), 91–100. doi:10.1016/j.psep.2011.08.002

Qiu, J., Yang, T., Wang, X., Wang, L., & Zhang, G. (2019). Review of the flame retardancy on highway tunnel asphalt pavement. *Construction & Building Materials*, *195*, 468–482. doi:10.1016/j.conbuildmat.2018.11.034

Quayle, M. (2006). Transport. In Purchasing and Supply Chain Management: Strategies and Realities. Hershey, PA: IGI Global.

RABT. (2006). *Richtlinien für die Ausstattung und den Betrieb von Straßentunneln (RABT 2006)*. Köln: Forschungsgesellschaft für Straßen- und Verkehrswesen e.V.

Racek, J. (2009). *Nuclear installations*. Brno, Czech Republic: Novpress.

Rådemar, D., Blixt, D., Debrouwere, B., Melin, B. G., & Purchase, A. (2018). Practicalities and Limitations of Coupling FDS with Evacuation Software. In *SFPE 12th International Conference on Performance-Based Codes and Fire Safety Design Methods*. Academic Press. Retrieved from https://c.ymcdn.com/sites/www.sfpe.org/resource/resmgr/2018_Conference_&_Expo/PBD/Program/Hawaii_Program.pdf%0Ahttps://www.sfpe.org/events/EventDetails.aspx?id=901118&group=

Rahim, E. (2016). The Benefits of Risk Management Planning. Retrieved from https://pmcenter.bellevue.edu/2016/06/19/the-benefits-of-risk-management-planning/

RAIN. (2019). Retrieved from https://cordis.europa.eu/project/rcn/185513/factsheet/en

Ramsay, J., Denny, F., Szirotnyak, K., Thomas, J., Corneliuson, E., & Paxton, K. L. (2006). Identifying nursing hazards in the emergency department: A new approach to nursing job hazard analysis. *Journal of Safety Research*, *37*(1), 63–74. doi:10.1016/j.jsr.2005.10.018 PMID:16490215

RASEN. (2019). Retrieved from https://cordis.europa.eu/project/rcn/105547/factsheet/en

Rasmussen, J., Nixon, P., & Warner, F. (1990). Human error and the problem of causality in analysis of accidents. *Philosophical Transactions of the Royal Society of London. Biological Sciences, (B),* 1241-1327 & *Human Factors in Hazardous Situations*, 449-462.

Reason, J. (1997). *Managing the Risks of Organizational Accidents*. Ashgate Publishing.

Rehak, D. (2012). Introduction to risk management. In L. Lukas (Ed), Security technologies, systems and management II (pp. 74-95). Zlin: VeRBuM. (in Czech)

Rehak, D., Cigler, J., Nemec, P., & Hadacek, L. (2013). *Critical Infrastructure in the Energy Sector: Identification, Assessment and Protection*. Ostrava: Association of Fire and Safety Engineering. (in Czech)

Rehak, D., Danihelka, P., & Bernatik, A. (2014a). Criteria Risk Analysis of Facilities for Electricity Generation and Transmission. In R. D. J. M. Steenbergen, P. H. A. J. M. van Gelder, S. Miraglia, & A. C. W. M. Vrouwenvelder (Eds.), *Safety, Reliability and Risk Analysis: Beyond the Horizon: ESREL 2013* (pp. 2073–2080). Boca Raton, FL: CRC Press.

Rehak, D., Hromada, M., & Novotny, P. (2016). European Critical Infrastructure Risk and Safety Management: Directive Implementation in Practice. *Chemical Engineering Transactions*, *48*, 943–948. doi:10.3303/CET1648158

Rehak, D., Markuci, J., Hromada, M., & Barcova, K. (2016). Quantitative evaluation of the synergistic effects of failures in a critical infrastructure system. *International Journal of Critical Infrastructure Protection, 14*, 3–17. doi:10.1016/j.ijcip.2016.06.002

Rehak, D., & Senovsky, P. (2014b). Preference Risk Assessment of Electric Power Critical Infrastructure. *Chemical Engineering Transactions*, *36*, 469–474. doi:10.3303/CET1436079

Rehak, D., Senovsky, P., Hromada, M., & Lovecek, T. (2019). Complex Approach to Assessing Resilience of Critical Infrastructure Elements. *International Journal of Critical Infrastructure Protection*, *25*, 125–138. doi:10.1016/j.ijcip.2019.03.003

Rehak, D., Senovsky, P., Hromada, M., Lovecek, T., & Novotny, P. (2018). Cascading Impact Assessment in a Critical Infrastructure System. *International Journal of Critical Infrastructure Protection*, *22*, 125–138. doi:10.1016/j.ijcip.2018.06.004

Reid, S. (2007, December 14). A Lethal Cover Up: Britain's Worst Water poisoning Scandal. Daily Mail. Retrieved from http://www.dailymail.co.uk/news/article-502442/A-lethal-cover- Britains-worst-water-poisoning-scandal.html

Reniers, G., & Amyotte, P. (2012). Prevention in the chemical and process industries: Future directions. *Journal of Loss Prevention in the Process Industries*, *25*(1), 227–231. doi:10.1016/j.jlp.2011.06.016

RESIST. (2019). Retrieved from https://cordis.europa.eu/project/rcn/215997/factsheet/en

Rimac, I., Šimun, M., & Dimter, S. (2014). Comparison of pavement structures in tunnel. *Elektronički časopis građevinskog fakulteta Osijek*, 12–18. doi:10.13167/2014.8.2

Rinaldi, M. S., Peerenboom, J., & Terrence, K. K. (2001, December). *Complex Networks- Identifying, Understanding, and Analyzing Critical Infrastructure Interdependencies.* Retrieved from https://pdfs.semanticscholar.org/b1b7/d1e0bb39badc3592373427840a4039d9717d.pdf

Rinaldi, S. M., Peerenboom, J. P., & Kelly, T. K. (2001). Identifying, Understanding and Analyzing Critical Infrastructure Interdependencies. *IEEE Control Systems Magazine*, *21*(6), 11–25. doi:10.1109/37.969131

Rodgers, G., Lee, E., Swepston, L., & Van Daele, J. (2009). The International Labour Organization and the quest for social justice, 1919-2009. *Book Samples*, 53. Retrieved from https://digitalcommons.ilr.cornell.edu/books/53

Rodrigues, C. (2006). *Higiene e Segurança do Trabalho – Manual Técnico do Formador*. Braga, Portugal: Nufec - Núcleo de Formação, Estudos e Consultoria.

Ronchi, E., Ph, D., & Bari, P. (2015). Evacuation modelling and Virtual Reality for fire safety engineering.

Rose, M. (2012, March 14). Camelford Water Poisoning: Authority 'Gambled With Lives.' BBC. Retrieved from http://www.bbc.co.uk/news/uk-england-cornwall-17367243

Roupcová, P. (2018). *Monitoring of ecotoxicity of carbon based nanoparticles* [Doctoral dissertation]. VSB-Technical university of Ostrava, Ostrava, Czech Republic.

Royal, M. (2019). Heads-Up / Hands-Free Firefighting Solutions: Requirements, State of Technology Overview and Market Characterization. Defense research and development Canada.

Russian spy poisoning: What we know so far (2018, October 8). *BBC NEWS*. Retrieved from https://www.bbc.com/news/uk-43315636

RVA. (2016). *DEMA's model for risk and vulnerability analysis (the RVA Model). Birkerød.* Danish Emergency Management Agency.

RVS. (2008). *RVS 09.02.31 Tunnel Ventilation—Basic Principles*. Austrian Research Association for Roads, Rail and Transport.

SAFE-10-T. (2019). Retrieved from https://cordis.europa.eu/project/rcn/209711/factsheet/en

SAFECARE project. (2018). *Grant Agreement Number 787005, European Commission H2020.* Retrieved from https://www.safecare-project.eu/

SAFURE. (2019). Retrieved from https://cordis.europa.eu/project/rcn/194149/factsheet/en

Salim, R., & Johansson, J. (2018). Automation decisions in investment projects: A study in Swedish wood products industry. *Procedia Manufacturing*, *25*, 255–262. doi:10.1016/j.promfg.2018.06.081

Salminen, S., & Tallberg, T. (1996). Human errors in fatal and serious occupational accidents in Findland. *Ergonomics*, *39*(7), 980–988. doi:10.1080/00140139608964518 PMID:8690011

Salvi, O., Corden, C., Cherrier, V., Kreissig, J., Calero, J., & Mazri, Ch. (2017). Analysis and summary of Member States' reports on the implementation of Directive 96/82/EC on the control of major accident hazards involving dangerous substances, Amec Foster Wheeler, Directorate-General for Environment (European Commission), EU-VRI, Final report – Study. doi:10.2779/2037

Saroha, A. K. (2006). Safe handling of chlorine. *Journal of Chemical Health and Safety*, *13*(2), 5–11. doi:10.1016/j.chs.2005.02.005

Schüllerová, B., Adamec, V., & Benco, V. (2019). Current approaches of risk assessment of nanomaterials. *Paper presented at XIX. International conference: Occupational safety and health,* Ostravice, *2019*. Academic Press.

Scudamore, M. J., Briggs, P. J., & Prager, F. H. (1991). Cone calorimetry – a review of tests carried out on plastics for the association of plastics manufacturers in Europe. *Fire and Materials*, *15*(2), 65–84. doi:10.1002/fam.810150205

Senate of the United States. (2001, October 24). *Uniting and Strengthening America by Providing Appropriate Tools Required to Intercept and Obstruct Terrorism (USA PATRIOT ACT) Act of 2001*. Retrieved from https://epic.org/privacy/terrorism/hr3162.pdf

SERON. (2019). Retrieved from https://cordis.europa.eu/project/rcn/92516/factsheet/en

Seveso. (2012). Council Directive 2012/18/EU of 4 July 2012 on the control of major-accident hazards involving dangerous substances, Official Journal of the European Communities (SEVESO Directive III).

Silva, C. J. (2011, August 23). More on the Attempted Water-Poisoning in Spain. Gates of Vienna. Retrieved from http://gatesofvienna.blogspot.cz/2011/08/more-on-attempted-water- poisoning-in.html

Simak, L. (2015). *Crisis management in public administration (Krízový manažment vo verejnej správe)*. Žilina, Slovakia: University of Žilina.

Sirovátka, J. (2018). *Measurement of dust Particles of less than 10 μm in Ambient Air in Transport with regard to Particle Phytotoxicity* [Diploma thesis]. VSB-TUO, Ostrava, Czech Republic.

Siser, A., Maris, L., Rehak, D., & Pellowski, W. (2018). The use of expert judgement as the method. to obtain delay time values of passive barriers in the context of the physical protection system. In *Proceedings of the IEEE International Carnahan Conference on Security Technology* (pp. 126-130). IEEE. 10.1109/CCST.2018.8585718

Skřehot, P., & Rupová, M. (2011). Nanosafety. Praha, Czech: Occupational Safety Research Institute.

Slovak Office for Standards, Metrology and Testing. (2008). STN 737507 Designing road tunnels.

SMART RAIL. (2019). Retrieved from https://cordis.europa.eu/project/rcn/100584/factsheet/en

Smith, S., & Ericson, E. (2009). Using immersive game-based virtual reality to teach fire-safety skills to children. *Virtual Reality, 13*(2), 87–99. doi:10.100710055-009-0113-6

Smolka, J. (2020). Vliv proudění vzduchu na průběh požáru. Unpublished doctoral dissertation, Technical University of Ostrava, Czech Republic.

Smolka, J., Kempna, K., Hošťálková, M., & Hozjan, T. (n.d.). Guidance on Fire-fighting and Bio-Based Materials.

Snow, J. (1857). Cholera, and the Water Supply in the South Districts of London. *British Medical Journal, 4*(42), 864–865. doi:10.1136/bmj.s4-1.42.864

Solon, A. H. A. (2017). Petya ransomware cyber attack who what why how. *The Guardian*. Retrieved from https://www.theguardian.com/technology/2017/jun/27/petya-ransomware-cyber-attack-who-what-why-how

Soltes, V., Kubas, J., & Stofkova, Z. (2018). The safety of citizens in road transport as a factor of quality of life. In *Proceedings of the International Conference Transport Means* (pp. 370-374). Academic Press.

Spear-Cole, R. (2019, June 12). Thames Water: No Water Supply for 100,000 Properties and Schools Across Hampton, Twickenham and West London Due to Burst Pipes. *The Evening Standard*. Retrieved from https://www.standard.co.uk/news/london/100000-properties-left-without- water-in-south-and-west-london-after-pipe-bursts-in-hampton-a4165441.html#spark_wn=1

Spearpoint, M. J., & Quintiere, J. G. (2001). Predicting the piloted ignition of wood in the cone calorimeter using an integral model - effect of species, grain orientation and heat flux. *Fire Safety Journal, 36*(4), 391–415. doi:10.1016/S0379-7112(00)00055-2

Spellman, F. R. (2003). *Disinfection of wastewater. Handbook of water and wastewater treatment plant operations*. London: CRC Publication. doi:10.1201/9780203489833

Stahmer, K. W. (2019). Explosion parameters of aluminium dust in different volumes: the limits of the cube law. *Chemical Engineering Transactions, 75*.

Stanleigh, M. (2016). *Risk Management: The What, Why, and How, Business Improvement Architects*. Retrieved from https://bia.ca/risk-management-the-what-why-and-how/

STERSST. (2019). Retrieved from https://cordis.europa.eu/project/rcn/110339/factsheet/en

Stoller, J., & Dubec, B. (2017). Designing protective structure using FEM simulations. In *Proceedings of the International Conference on Military Technologies (ICMT)* (pp. 236-241). Academic Press. Retrieved from 10.1109/MILTECHS.2017.7988762

Sulleyman, A. (2017). *NHS cyber attack: Why stolen medical information is so much more valuable than financial data*. The Independent. Retrieved from https://www.independent.co.uk/life-style/gadgets-and-tech/news/nhs-cyber-attack-medical-data-records-stolen-why-so-valuable-to-sell-financial-a7733171.html

Sullivan, J. K. (2011). *Water Sector Interdependencies - Summary Report*. Retrieved from http://www.wef.org/uploadedFiles/Access_Water_Knowledge/Water_Security_and_Emergency_Res ponse/Final_WEF_Summary_WSI.pdf

Swain, A. D., & Guttmann, H. E. (1983). *Handbook of human reliability analysis with emphasis on nuclear power plant applications*. U.S. Nuclear Regulatory Commission. doi:10.2172/5752058

Szakali, M., & Szűcs, E. (2017). The defence planning model formation and development, Védelmi tervezésimodellek kialakulása és fejlődése. *Hadmérnök, 12*(1), 24–40.

Szakali, M., & Szűcs, E. (2018). The price of the security, probably effects of the defence budget expansion, A biztonságára, avagy a védelmiköltségvetésemelésénakhatásai. *Hadmérnök*, *13*(1), 314–325.

Takabatake, T., Mall, M., Esteban, M., Nakamura, R., Kyaw, O. T., Ishii, H., ... Shibayama, T. (2018, November 9). Field Survey of 2018 Typhoon Jebi in Japan: Lessons for Disaster Risk Management. *Geosciences Journal*, *8*(11). Retrieved from https://www.mdpi.com/2076-3263/8/11/412

Tallo, A., Rak, R., & Turecek, J. (2006). *Moderné technológie ochrany osôb a majetku [Modern technologies of protection of persons and property]*. Bratislava, Slovakia: Academy of the Police Force in Bratislava.

Tang, J., Wu, Y., Esquivel-Elizondo, S., Sørensen, S. J., & Rittmann, B. E. (2018). How Microbial Aggregates Protect against Nanoparticle Toxicity. *Trends in biology, 36*(11), 1171-1182. Retrieved from: https://www.sciencedirect.com/science/article/abs/pii/S0167779918301732?fbclid=IwAR3eWRkxiAW0sQSzXJ7gwYKBOVaJv_ukliP85j-WVUz-i4XeFfae8pXFt9gI

Tenschert, J. (2016). NANOaers - Fate of aerosolized nanoparticles: the influence of surface active substances on lung deposition and respiratory effects. *Nanopartikel.* Retrieved from https://nanopartikel.info/en/projects/era-net-siinn/nanoaers-en

Tewarson, A. (2002). Generation of heat and chemical compounds in fires. In P. J. DiNenno (Ed.), *The SFPE Handbook of Fire Protection Engineering* (pp. 618–697). Quincy, US: National Fire Protection Association.

The EP and Council. (2004). Directive of the EP and of the Council 2004/54/ES on minimum safety requirements for tunnels of the Trans-European Road Network. Retrieved from https://eur-lex.europa.eu/legal-content/CS/TXT/PDF/?uri=CELEX:32004L0054&from=CS

The Guardian. (2019). Violence in the NHS: staff face routine assault and intimidation. Retrieved from https://www.theguardian.com/society/2019/sep/04/violence-nhs-staff-face-routine-assault-intimidation

The National Academies of Sciences, Engineering and Medicine. (2009). *A Guide to Planning Resources on Transportation and Hazards*. Washington, DC: The National Academies. Retrieved from https://www.nap.edu/catalog/23007/a-guide-to-planning-resources-on-transportation-and-hazards

The precautionary principle . (2000) EUR-Lex: Acess to European Union law. Retreived from: https://eur-lex.europa.eu/legal-content/CS/TXT/?uri=LEGISSUM:l32042

The Regulation of the EP and of the Council. (2011). *Regulation of the European Parliament and of the Council (EU) no. 305/2011 dated 9th March 2011, which defines harmonized conditions for the introduction of building products to the market and which revokes the Directive of the Council 89/106/EEC. Text significant for the EEC.*

The Times. (2010). Woman kills 3 in hospital shooting spree. Retrieved from https://www.thetimes.co.uk/article/woman-kills-3-in-hospital-shooting-spree-n9q9nbhws9b

The White House. (2013). *Presidential Policy Directive - Critical Infrastructure Security and Resilience*. Washington, DC: The White House.

Theocharidou, M., & Giannopoulos, G. (2015). *Risk assessment methodologies for critical infrastructure protection. Part II: A new approach.* Ispra: European Commission, Joint Research Centre; doi:10.2788/621843

Tichy, L. (2019). Energy infrastructure as a target of terrorist attacks from the Islamic state in Iraq and Syria. *International Journal of Critical Infrastructure Protection*, *25*, 1–13. doi:10.1016/j.ijcip.2019.01.003

Tixier, J., Dusserre, G., Salvi, O., & Gaston, D. (2002). Review of 62 risk analysis methodologies of industrial plants. *Journal of Loss Prevention in the Process Industries*, *15*(4), 291–303. doi:10.1016/S0950-4230(02)00008-6

TNO. (1997). Methods for the calculation of physical effects due to releases of hazardous materials (liquids and gases) (3rd ed.). Academic Press.

TNO. (2005). Guidelines for Quantitative Risk Assessment. Academic Press.

TP 049. (2018). *TP 049 Technical conditions for the ventilation of road tunnels*. Bratislava: Ministry of Transport, construction and regional development of the Slovak Republic road transport and land roads division.

TP 13. (2015). *TP 13 Technical conditions, Fire safety of road tunnels*. Bratislava: Ministry of transport, construction and regional development of the Slovak Republic Road transport and land roads division.

Tran, H. C. (2003). Experiment data on wood materials. In V. Babrauskas & S. J. Grayson (Eds.), *Heat release in fires* (pp. 357–372). London, UK: Interscience Communications.

Transportgeography. (2019). Retrieved from https://transportgeography.org/?page_id=247

Trucco, P., Cagno, E., & De Ambroggi, M. (2012). Dynamic Functional Modeling of Vulnerability and Interoperability of Critical Infrastructures. *Reliability Engineering & System Safety*, *105*, 51–63. doi:10.1016/j.ress.2011.12.003

Truscott, J. (2015). The Art of Crisis Leadership: Incident Management in the Digital Age. Mission Mode. Retrieved from http://www.missionmode.com/wp-content/uploads/2014/08/The-Art- of-Crisis-Leadership-Incident-Mgmnt-In-Digital-Age.pdf

Tschirschwitz, R. (2015). *Entwicklung von Bestimmungsverfahren für Explosionskenngrößen von Gasen und Dämpfen für nichtatmosphärische Bedingungen*. Magdeburg: Otto-von-Guericke-Universität.

Tůma, M. (2004). Are nanotechnologies the salvation of humanity with safety risks?

UIC. (2019). Retrieved from https://uic.org/

UITP. (2019). Retrieved from https://www.uitp.org/

UN Water. (2013, May 8). What is Water Security? Infographic. Retrieved from https://www.unwater.org/publications/water-security-infographic/

UNECE. (2019). Retrieved from http://www.unece.org/trans/theme_infrastructure.html

United Nations Counter-Terrorism Centre, Counter Terrorism Committee Executive Directorate, Interpol. (2018). The protection of critical infrastructure against terrorist attacks: *Compendium of good practices*. Retrieved from https://www.un.org/sc/ctc/wp-content/uploads/2018/06/Compendium-CIP-final-version-120618_new_fonts_18_june_2018_optimized.pdf

United Nations General Assembly. (2016, December 1). *Report of the open-ended intergovernmental expert working group on indicators and terminology relating to disaster risk reduction*. Retrieved from https://www.preventionweb.net/files/50683_oiewgreportenglish.pdf

United Nations Office for Disaster Reduction- UNISDR. (2015). *Senday Framework for Disaster Risk Reduction 2015-2030*. Retrieved from https://www.unisdr.org/we/coordinate/sendai-framework

United Nations Office for Disaster Reduction- UNISDR. (2018). Words into Action Guidelines. Man-made and Technological Hazards. *Practical considerations for Addresing Man-made and Technological Hazards in Disaster Reduction*. Retrieved from https://www.preventionweb.net/publications/view/54012

United Nations Office of Counter- terrorism, United Nations Security Council, Interpol. (2018). *The protection of critical infrastructures against terrorist attacks: Compendium of good practices.* Retrieved from https://www.un.org/sc/ctc/wp-content/uploads/2018/06/Compendium-CIP-final-version-120618_new_fonts_18_june_2018_optimized.pdf

Unity Technologies. (n.d.). Unity Manual. Retrieved from https://docs.unity3d.com/540/Documentation/Manual/UnityManual.html

Updyke, L. J. (1982). Method for calculating water distribution in a chlorine condensing system. *Paper presented at the 25th Chlorine Plant Operations*. Academic Press.

US Army Center for Health Promotion and Preventive Medicine (USACHPPM). (2019). *Countering Terrorism of Water Supplies.*

Van Leuven, L. J. (2011). Water/Wastewater Infrastructure Security: Threats and Vulnerabilities. In R.M. Clark, S. Hakim, & A. Ostfield (Ed.), Handbook of Water and Wastewater Protection (Vol. 1, pp. 27-46). New York, USA: Springer.

VDI 2263. (2018). *Staubbrände und Staubexplosionen - Gefahren - Beurteilung – Schutzmaßnahmen.* VDI-Gesellschaft Energie und Umwelt.

Veilleux, J., & Dinar, S. (2018, May 8). New Global Analysis Finds Water-Related Terrorism Is On the Rise. New security beat. Retrieved from https://www.newsecuritybeat.org/2018/05/global- analysis-finds-water-related-terrorism-rise/

Velas, A., Zvakova, Z., & Svetlik, J. (2016). Education and lifelong learning opportunities in the private security services in Slovak republic. In *EDULEARN16: 8th international conference on Education and new learning technologies* (pp. 6517-6523). Barcelona, Spain: IATED Academy.

Velazquez, J. (2016) *Firefighters*. Retrieved from https://pixabay.com/photos/fire-volunteer-firefighters-rescue-2263406

Vempati, S., & Uyar, T. (2014). Fluorescence from graphene oxide and the influence of ionic, π–π interactions and hetero-interfaces: Electron or energy transfer dynamics. *Journal of Physical Chemistry Chemical Physics, 16*(39), 21183–21203. doi:10.1039/C4CP03317E PMID:25197977

Vidrikova, D., Boc, K., Dvorak, Z., & Rehak, D. (2017). *Critical Infrastructure and Integrated Protection.* Ostrava: Association of Fire and Safety Engineering.

Virtual Reality for Firefighter Services. (2019). Retrieved from https://www.ludus-vr.com/en/portfolio/firefighter-services/

Wang, B., Li, H., Rezgui, Y., Bradley, A., & Ong, H. N. (2014). BIM based virtual environment for fire emergency evacuation. *The Scientific World Journal*, (August). doi:10.1155/2014/589016 PMID:25197704

Wang, S., Hong, L., & Chen, X. (2012). Vulnerability Analysis of Interdependent Infrastructure Systems: A Methodological Framework. *Physica A, 391*(11), 3323–3335. doi:10.1016/j.physa.2011.12.043

Watters J. (2014). Disaster Recovery, Crisis Response, and Business Continuity: A Management Desk Reference. Springer Science+Business.

Watts. (2010). Stop Backflow News! Retrieved from https://media.wattswater.com/F-SBN.pdf

Weiss, J. (2011, November 21). Hackers 'Hit' US Water Treatment Systems. *BBC News Technology*. Retrieved from http://www.bbc.co.uk/news/technology-15817335

Wells, W. D. (2005). Naval Postgraduate School. *Security*, (September).

Willoughby, I. (2015, May 26). Anger After Slow Announcement of Water Contamination Leaves Scores Sick in Prague. Retrieved from http://www.radio.cz/en/section/curraffrs/anger-after- slow-announcement-of-water-contamination-leaves-scores-sick-in-prague

Wood, J., & Van-Slyke, S. (2018). *Terrorism and New Ideologies*. Control Risks. Retrieved from https://www.controlrisks.com/-/media/3c6cfbc84397463bbe955f1c844cf78e.ashx

WoodwardB. (2019) RiVR. Retrieved from https://rivr.uk/

World Health Organisation. (2019). *Health Systems*. Retrieved from http://www.euro.who.int/en/health-topics/Health-systems/pages/health-systems

World Health Organisation. (2019). Hospitals. Retrieved from https://www.who.int/hospitals/en/

World Health Organization. (2015). *Hospital Safety Index: Guide for Evaluators*. Retrieved from https://www.google.com/url?sa=t&rct=j&q=&esrc=s&source=web&cd=1&cad=rja&uact=8&ved=2ahUKEwi_tdq7n__kAhVE2aQKHVZfBbkQFjAAegQIAxAC&url=https%3A%2F%2Fwww.who.int%2Fhac%2Ftechguidance%2Fhospital_safety_index_evaluators.pdf&usg=AOvVaw3Jb3x3xUgBh-IK84EtnKD8

Xu, Q., Majlingova, A., Zachar, M., Jin, C., & Jiang, Y. (2012). Correlation analysis of cone calorimetry test data assessment of the procedure with tests of different polymers. *Journal of Thermal Analysis and Calorimetry*, *110*(1), 65–70. doi:10.100710973-011-2059-7

Yan, L., Gu, Z., & Zhao, Y. (2013). Chemical Mechanisms of the Toxicological Properties of Nanomaterials: Generation of Intracellular Reactive Oxygen Species. *Chemistry - An Asian Journal*, 8(10), 2342-2353. Retreived from: http://doi.wiley.com/10.1002/asia.201300542

Yang, D., Li, P., Duan, H., Yang, C., Du, T., & Zhang, Z. (2019). Multiple patterns of heat and mass flow induced by the competition of forced longitudinal ventilation and stack effect in sloping tunnels. *International Journal of Thermal Sciences*, *138*, 35–46. doi:10.1016/j.ijthermalsci.2018.12.018

Young, W., Sobhani, A., Lenné, M., & Sarvi, M. (2014). Simulation of Safety: A Review of the State of the Art in Road Safety Simulation Modeling. *Accident; Analysis and Prevention*, *66*, 89–103. doi:10.1016/j.aap.2014.01.008 PMID:24531111

Zagorecki, A., Ristvej, J., & Klupa, K. (2015). Analytics for Protecting Critical Infrastructure. *Communications Scientific Letters of the University of Zilina*, *17*(1), 111–115.

Zalk, D. M., & Nelson, D. I. (2003). *History and evolution of control banding*. Retrieved from https://www.researchgate.net/publication/5502902_History_and_Evolution_of_Control_Banding_A_Review

Zemanova, E., Klouda, K., & Zeman, K. (2011). C60 fullerene derivative: Influence of nanoparticle size on toxicity and radioprotectivity of water soluble fullerene derivative. In *Proceedings of 3rd International Conference NANOCON 2011* (*Vol. 1*, pp. 1-10). Tanger.

Zhang, H. D., & Zheng, X. P. (2012). Characteristics of hazardous chemical accidents in China: A statistical investigation. *Journal of Loss Prevention in the Process Industries*, *25*(4), 686–693. doi:10.1016/j.jlp.2012.03.001

Zhang, J. (2008). *Study of polyamide 6-based nanocomposites*. New York: Polytechnic University.

Zio, E. (2016). Challenges in the vulnerability and risk analysis of critical infrastructures, *Reliability Engineering and System Safety, 152*, 137-150

Zio, E. (2016). Critical Infrastructures Vulnerability and Risk Analysis. *European Journal for Security Research, 1*(2), 97–114. doi:10.100741125-016-0004-2

Zohar, D. (1980). Safety climate in industrial organization: Theoretical and applied implications. *The Journal of Applied Psychology, 65*(1), 96–102. doi:10.1037/0021-9010.65.1.96 PMID:7364709

Zvakova, Z., Velas, A., & Mach, V. (2018). Security in the transport of valuables and cash. In *Proceedings of the International Conference Transport Means* (pp. 1209-1214). Academic Press.

Ένας χρόνος από την εθνική τραγωδία στο Μάτι - Μνημόσυνο για τους 102 νεκρούς [One year from the national tragedy of Mati- Memorial for the 102 dead] (2019, July 13). *CNN Greece*. Retrieved from: https://www.cnn.gr/news/ellada/story/185215/enas-xronos-apo-tin-ethniki-tragodia-sto-mati-mnimosyno-gia-toys-102-nekroys

Πουλιόπουλος, Γ. (2012, February 12). *Παραδόθηκαν στις φλόγες 45 κτίρια και καταστήματα του Κέντρου [45 buildings and retail shops in the Center of Athens surrendered to flames].* Το ΒΗΜΑ (TO VIMA). Retrieved from https://www.tovima.gr/2012/02/12/society/paradothikan-stis-floges-45-ktiria-kai-katastimata-toy-kentroy/

Φωτιά Αττική. Σταδιακή αποκατάσταση της ηλεκτροδότησης στις πληγείσες περιοχές [Fire Attica: Gradual restoration of electricity in the affected areas] (2018, July 24). *CNN Greece*. Retrieved from https://www.cnn.gr/news/ellada/story/139950/fotia-attiki-stadiaki-apokatastasi-tis-ilektrodotisis-stis-pligeises-perioxes

About the Contributors

David Rehak is a PhD in Processes Modelling and Simulation of Troops and Population Protection at the University of Defence. He is currently an Associate Professor at the VSB – Technical University of Ostrava, Faculty of Safety Engineering. His scientific and research work is aimed at critical infrastructure resilience and protection, risk management, and civil protection. He is a member of the International Association of Critical Infrastructure Protection Professionals (IACIPP), a member of the Czech Technology Platform Energy Security (TPEB), a reviewer of the "International Journal of Critical Infrastructure Protection" (IJCIP), a reviewer of the "Journal of Infrastructure Systems" (JIS) and the editor-in-chief of the "Transactions of the VSB-Technical University of Ostrava, Safety Engineering Series" (TSES).

* * *

Janette Alba obtained her degree in Chemistry from the University of Guanajuato (Mexico) in 2012 with the thesis entitled "Extraction of Hg (II) from HCl medium with resins impregnated with ionic liquids". Later she obtained her PhD degree from the University of Guanajuato in 2017 with the thesis entitled "Removal of toxic metals with encapsulated extractants". Currently works for the company Analitek in the area of development of projects for the industry.

Konstantinos Apostolou (MSc) is a research associate at the Center for Security Studies (KE.ME.A) based in Athens (Greece), currently participating in private research/studies and EU funded projects/studies, all relative to Risk Management and Risk Assessment, Operator Security Plans, Critical Infrastructure Protection, Disaster and Emergency Management, and Physical Security. In 2018, Konstantinos acquired his master's degree (MSc) in Risk, Crisis and Resilience Management from the University of Portsmouth - UK. His research consists of topics such as risk management standards and practice comparison, enterprise risk management (ERM), safety risk, crisis management and disaster analysis. Konstantinos has also acquired a bachelor's degree (BSc) in International, European and Regional Studies from Panteion University of Social and Political Sciences (Athens, GR). In 2013-2014 during his internship at KE.ME.A, he participated in studies relevant to risk management, security threats, critical infrastructure and security plan composition.

Ales Bernatik is a PhD in Methodology for Industrial Risk Assessment and graduated in Environmental Protection. He is currently a Professor at the Faculty of Safety Engineering, VSB-Technical University of Ostrava. His scientific and research work is aimed at risk analysis of industrial accidents and major-accident prevention. He is a member of the European Technology Platform on Industrial Safety, a member of the Association for Fire & Safety Engineering, a reviewer of international journals, etc.

Gargi Bhattacharjee is a Lecturer at the University of Calcutta and Prasanta Chandra Mahalanobis MahaVidyalaya, Department of Human Development, passed Bachelor of Science, Bachelor of Masters and Ph. D.in Human Development from University of Calcutta; working on Human factors in chemical industries, measures safety knowledge of workers, safety knowledge on LPG of truck drivers and auto drivers. Published 7 international and 1 national journal articles related to process safety and deliver lectures more than 10 conferences about organizational health and safety. He is a Life Member of the Indian Science Congress.

David Birkett has worked across the three tiers of government, and in the public and private sector in the water, power, mining, oil, gas and transport industries. He has specialised in the risk and emergency management, disaster mitigation planning and business continuity, with the emergency response aspects of managing public and private sector assets. Dave has applied his expertise with counter-intelligence in Defence, Australia, to advise organisations in the prevention and control of emergencies and crises. Dave's developed expertise is in the design and delivery of training in emergency response, and crisis management to middle and senior managers in the oil & gas, power, utilities, chemical and other high-risk industries. Experienced in leading and coaching crisis and emergency management teams in their roles and responsibilities during major emergencies. Other activities include competency assessments, delivering training in the form of presentations on various subjects involved in crisis and emergency response, such as, HR, communications, command and control, information handling, emergency control centre dynamics, roles & responsibilities, logistic and resource allocation.

Nélson Costa is an Ergonomics and Human Factors Professor at the University of Minho, Portugal, where he is a researcher at the Centro ALGORITMI in the "Industrial Engineering and Management (IEM)" research line, Ergonomics and Human Factors research group. He authored several papers in international peer-reviewed journals, more than 30 papers in international conference proceedings and 10 book chapters. Nélson Costa has been a member/co-coordinator of several international and national externally funded research projects in the domains of Ergonomics, Occupational Safety and Hygiene. He was the supervisor of more than 20 MSc. theses (concluded) in several Portuguese universities and of 1 concluded PhD thesis. He currently supervises 2 ongoing PhD projects. He has been collaborating with 3 international peer-reviewed scientific journals in the domain of ergonomics and occupational safety, as a reviewer. He is currently the associate editor of the "International Journal of Occupational and Environmental Safety" (IJOOES), an interdisciplinary journal with peer-review. He is also a member of the Organizing Committee of the International Symposium on Occupational Safety and Hygiene (SHO) since 2008 and of the International Symposium on Project Approaches in Engineering Education (PAEE) since 2016. Besides his appointment at the University of Minho, he has been a regular invited Professor at the Faculty of Medicine of the University of Coimbra, in Portugal.

Susana da Costa is an Invited Assistant Professor at the University of Minho (Portugal), where she also is a researcher at the Centro ALGORITMI in the "Industrial Engineering and Management (IEM)" research line, Ergonomics and Human Factors research group. Susana Costa holds a PhD in Industrial and Systems Engineering from the University of Minho and a master's degree in Biomedical Engineering from the same teaching institution. She is an occupational safety and hygiene specialist since 2010 and a certified internal auditor of integrated management systems for quality, environment and safety since 2015. Susana previously integrated the INNOVCAR project and is currently a fellow researcher at the EasyRide project, both resulting from a partnership between the University of Minho and Bosch. Susana has collaborated with several institutions and companies in the scope of professional training and consulting in occupational ergonomics, as well as occupational safety and hygiene. She is part of the International Scientific Committee of the Occupational Safety and Hygiene (SHO) Symposium since 2016 and a Member of the Advisory Board of the Conference on Safety Management and Human factors (SMHF) since 2017. She is the author and reviewer of several articles in magazines, symposia and conferences in the area of occupational safety, hygiene, and ergonomics.

Eleni Darra (MSc) (female) received a degree in the Department of Technology Education and Digital Systems from the University of Piraeus, Greece, in 2005, and a Master of Science (M.Sc.) in Network Oriented Systems from the University of Piraeus, Greece, in 2008. Her job experience includes her involvement as an instructional designer, tester of websites and IT support in Exodus Ltd.(2005), as an adjunct lecturer of Computer Science at the higher education (2008) and as assistant and advisor of commission and the member states in the ENISA (2015) in their dialogue with the industry to address security related problems and, when called upon, in the technical preparatory work for updating and developing Community legislation in the field of network and information security. In the same context, she has participated in promoting best practices, contributing to awareness raising actions in the areas of smart infrastructures, intelligent public transport, national cyber security strategies, economic impact of cyber-security incidents on critical information infrastructures. Her experience also lies in the areas of intrusion detection systems in wireless sensor networks, computer networks security, mobile communications security and privacy, mobile ad hoc and sensor network security and privacy, cloud computing security, user modeling. She has authored or co-authored several journal publications, book chapters and conference proceedings publications and she has participated in funded national and international R&D projects in these areas. She has been actively involved in local organizing committee of international scientific conferences co-organized by the research group of systems security information and communication of the University of Piraeus. Since 2018, she is a research associate at the Centre for Security Studies (KEMEA), of the Hellenic Ministry of Citizen Protection.

Sudip Kumar Das, Dr., Professor, and Head, Chemical Engineering Department, University of Calcutta, passed B. Tech. from the University of Calcutta and M. Tech. and Ph.D. from Indian Institute of Technology, Kharagpur, working on the different fields of engineering like process safety, ANN application, CFD, single and multiphase flow in pipeline and piping components, fluidization and inverse fluidization, wastewater technology, etc. He has published 190 papers in the above mentioned fields, guided 17 PhDs, working 15 students for their PhDs and 37 M.Tech. students, and delivered more than 100 invited lectures and chair in the technical sessions in international/national conferences. He is a member of different professional societies like the IEEE, the Indian Institute of Chemical Engineers, the Institution of Engineers (India), the Indian Association for Air Pollution Control, the Indian Chemical Society, the Millennium Institute of Energy and Environment Management, and the International Congress of Chemistry and Environment.

Janka Dibdiakova graduated as a PhD. in the Wood Engineering – Chemical Processing of Wood from the Faculty of Wood Sciences and Technology of the Technical University in Zvolen. She is currently Researcher at the Division of Forest and Forest Resources Wood Technology of the Norwegian Institute of Bioeconomy Research. She is engaged in the wood products, wood materials and its protection against atmospheric effects, wood decaying fungi, wood-destroying insects, heat, flame and fire, mainly by the flame retardants. She has been member of the scientific board of the Advances in Fire and Safety Engineering international scientific conference since 2012.

Zdenek Dvorak completed his university education in the field of Military Transport in 1986. Until 1987 he worked at the Regional Administration of Military Transport in Hradec Kralove. Since 1988, he worked as a researcher at the Research Institute in Zilina. In 1993 he successfully defended his dissertation and became a PhD. Since 1996 he has worked as an assistant professor at the Military Faculty in Zilina. In 2000, after a successful habilitation procedure, he became associate professor. Since 2009 he has been Head of Research Department of Crisis Management. In 2011 he successfully inaugurated and was appointed professor. In 2015-2019 he was dean of the Faculty of Security Engineering, University of Zilina. He is currently professor, guarantor of the study program Security and Protection of Critical Infrastructure.

George Eftychidis has a University degree in Forestry and Environmental Protection from the Aristotelian University of Thessaloniki (Gr) since 1987. For years, he has worked in the private ICT sector on public and private contracts dealing with security, environmental and civil protection applications. In addition, he has participated in several European and National R&D projects and he cooperated with EC services as a reviewer and expert in policy making missions. His R&D topics of interest refer to modeling and simulating natural hazards and assessing related risks, using information technology. He contributed to the implementation of GIS and web applications for assessing forest fire behaviour, propagation rate and growth patterns using properly developed simulation tools. Such tools are currently used operationally by public services in EU countries. Eftychidis contributed also to the development of forest fuel and risk analysis maps at the local, regional, national and EU level. During the last decades, he gained extensive experience in managing and coordinating national and European R&D projects. Furthermore, Eftychidis cooperated with several public EU organizations regarding the analysis of needs and requirements of end users in security, critical infrastructure operators, civil protection authorities and environmental organizations and he contributed to disseminate results of research projects to the respective groups.

Dieter Gabel is a scientific coworker at the Otto von Guericke University since 2000; Faculty of Process & Systems Engineering, Department of Plant Systems Engineering and Safety.

Ilias Gkotsis (MSc) is a Mechanical and Aeronautics Engineer with an MSc degree in Energy Production and Management. Currently, he is a PhD candidate in the Dep. of Transportation Planning & Engineering of NTUA, through which he has also participated in several transportation research projects and studies (e.g. traffic modelling, transport network management, road transport emissions, etc.). From 2012, he is heavily involved (implementation, management and coordination) in EU and Nationally Funded R&D projects (H2020, FP7, DG ECHO, CIPS, ISF, NSRF, etc.), regarding security, critical infrastructure protection, disaster resilience, civil protection, crisis management, border surveillance, ICT and Unmanned Aerial Systems for which he is a certified pilot by HCAA, as an associate researcher at the Center for Security Studies (KEMEA), of the Hellenic Ministry of Citizen Protection.

Tomaž Hozjan is an Associate professor of civil engineering at the University of Ljubljana, Slovenia, Faculty of civil and geodetic engineering. His research involves mechanics of structures, numerical methods for the analysis of structures, composite structures, thermodynamics, transfer of moisture in porous materials such as concrete and wood, coupled problems in concrete and timber in fire conditions and fire safety of buildings. So far, he has published over 80 journal and conference papers. Up to now he was a twice guest lecturer at foreign universities. In 2014 he visited the Technical University of Denmark (DTU) and in 2017 University of Boras, Sweden and has been an invited speaker at several European universities and international conferences. He has participated in many COST actions. He was the leader of the working group during the action COST FP 1404. He is also a member of the RILEM Technical committee 256-SPF, Slovenian association of mechanics and Slovenian association for fire safety, which is a member of CFPA Europe and CFPA International.

Martin Hromada graduated master education in Tomas Bata University in Zlin (TBU) in the study programme Security Technologies, Systems and Management in 2008. His dissertation thesis: "Technological aspects of critical infrastructure protection in SR" was defended in 2011. In 2017, he defended his habilitation thesis at the Faculty of Safety Engineering, VŠB - TU in Ostrava, and received the academic title of associate professor in Safety and Fire Protection. In the years 2010 - 2015 he worked in Deloitte Advisory Ltd. as a consultant and lead of the security research projects, Critical Infrastructure Protection in the Electricity Production, Transmission and Distribution Sector, and Actual Cyber Threats of Czech Republic and their Elimination. He currently works as an associated professor at the Department of Security Engineering, Faculty of Applied Informatics, Tomas Bata University in Zlin. He is a project manager and leads the security research project RESILIENCE 2015 Dynamic Resilience Evaluation of Interrelated Critical Infrastructure Subsystems. Within the scientific-research, activities are mostly dedicated to the protection and resilience of critical (information) infrastructure issues and assessment of the physical protection systems functionality. He is a member of significant national expert groups of the Czech and Slovak Republic Ministry of Interior and the European institutions such as ERNCIP and CIWIN. He regularly participated in practical exercises relating to the protection of energy infrastructures (ARGOS project Bucharest, Bulgaria, Krizovany Slovakia, NATO CoE, Vilnius, Lithuania).

Danai Kazantzidou-Firtinidou is a Civil Engineer with MSc in Earthquake Engineering. She currently works as a Research Associate at the Center for Security Studies (Ministry of Citizen Protection) in Greece, where she is involved in projects related with civil protection and critical infrastructures, holding the position of Officer at the Coordination Center for Critical Infrastructures Protection. Her expertise is in seismic risk assessment, having 10 years experience with participation in several European and national funded programs in Greece and other European countries, focusing on the scientific support of civil protection and risk management. More specifically, she has participated in disaster management planning and evacuation exercises, trials for the development of technological tools supporting crisis management, technical post-earthquake visits, studies of seismic hazard, vulnerability, risk and loss estimation and has contributed as national expert in risk assessment projects. She has more than 15 scientific publications in peer reviewed journals and international conferences. She is currently enrolled as PhD candidate at National Kapodistrian University of Athens conducting research in seismic risk.

Kamila Kempna is a Czech researcher, engineer, lecturer. Co-establisher of the Czech Fire Protection Association Majaczech as a member of CFPA-Europe with a goal to increase fire prevention by producing national and international training, guidelines, and safety days for all, firefighters, engineers, and civilians in Europe. Her research is focused on disaster resiliency, fire science, and risk management. She is employed as a researcher at VSB-Technical University of Ostrava and Occupational Safety Research Institute.

Karel Klouda, Doc. Ing. et Ing., Ph.D., MBA. graduated from the Institute of Chemical Technology in Prague, majoring in chemical technology. From 1976 to 1987 he worked at the Nuclear Research Institute Řež, Department of Fluoride Chemistry, and from 1987 to 1992 at the Czechoslovak Atomic Energy Commission in Prague. In 1993 he joined the State Office for Nuclear Safety, where he still works. In 2009 he defended his habilitation thesis on Underground constructions at VŠB - TU of Ostrava; their benefits and risks. Since 2007 he has been teaching at the Faculty of Safety Engineering of VŠB - TU of Ostrava, where he is a guarantor of CBRN substances and industrial toxicology. He is the author of more than 50 scientific papers, both contributions in the proceedings of international conferences, as well as journal articles and reports of the category "Special Facts" for the Ministry of the Interior of the Czech Republic.

Tünde Kovács is an Associate professor in the Óbuda University, Donát Bánki Faculty of Mechanical and Safety Engineering, Materials Technology Department. Her interests focus on material science, composites, metal alloys testing, manufacturing, and innovations.

Bohus Leitner is an Associate professor in the Faculty of Security Engineering, University of Žilina (FSE UNIZA). He is a co-guarantor of the accredited study program Security and Protection of Critical Infrastructure and currently he holds the position as Head of Department of technical sciences and informatics at FSE UNIZA. His scientific activity is focuses on the technical aspects of the resilience of critical infrastructure elements, reliability and safety of technical systems and complex systems analysis. He is a member of the International Association of Critical Infrastructure Protection Professionals (IACIPP).

Tomáš Loveček, Prof. Ing., PhD., is currently vice-dean for science and research. He is the chairman of study program committee 8.3.1 People and property protection in the study program Security Management. In his research activities since 2003, deals with issues of design and evaluation security systems of state strategic objects (e.g. critical infrastructure objects, SEVESO companies, special importance objects, other important objects, etc.). Among its main tasks include partial research modeling and simulation security systems of objects and evaluation their effectiveness, efficiency, reliability and quality. In the years 2013-2014 he was a member of the working group of Ministry of Economy of Slovak Republic for preparation methodical guidelines for critical infrastructure operators. He also participates in departmental roles of the Nuclear Regulatory Authority of Slovak Republic, in the context of the policy of physical protection of nuclear material and nuclear facilities International Atomic Energy in Austria and tasks of the National Security Authority in the context of making risk analysis of classified information protection.

Vasiliki Mantzana (PhD) (female) holds a PhD degree in Healthcare Information Systems (HIS) from Brunel University in West London. She is a research associate at the Centre for Security Studies (KEMEA), of the Hellenic Ministry of Citizen Protection, teaches at both undergraduate and postgraduate programmes and has published books and research papers in internationally refereed journals (CAIS,

IJTM, EJIS) as well as international conferences (ECIS, HICSS, AMCIS). In terms of applied research and collaboration with industry, she has close relationships with industry and has experience in the analysis, design, development and management of international and national projects in the areas of Healthcare Information Systems, e-health, e-government and e-learning. She has participated in research and development projects funded by the European Commission, the UK government, the Greek Ministry of Health and Social Solidarity and Cyprus. More specifically, from 2005 she has participated in several projects, such as "Virtual Institute for Electronic Government Research (VIEGO)," "Older Employees Training on Information and Communication Technologies (REFOCUS)," "Supporting computing studies at the University of Piraeus," "Advancing the skills of professionals in the healthcare sector, using digital systems," "Greek Classification/coding of diseases, diagnoses and medical processes" as well as the "National electronic referral and prescribing systems for Greece."

Ladislav Mariš, PhD, is a university teacher at the University of Žilina, Faculty of Security Engineering, Slovakia. He secures subjects: video surveillance systems and technical drawing for security management. In his research area, he focuses on security management, security assessment, security design measures, VSS solutions (CCTV), and other interesting security areas.

Jozef Martinka graduated as a PhD. in the Fire Protection and Safety Engineering at the Technical University in Zvolen. He is currently an Associate Professor and head of the Fire Engineering Department of the Institute of Integrated Safety of the Faculty of Materials Science and Technology in Trnava of the Slovak University of Technology in Bratislava. He is engaged in the occupational safety and health, fire protection engineering, fire hazard of materials, fire investigation and environmental safety. He is member of the "Wood Research" scientific journal editorial board and reviewer of many international scientific journals indexed in the Web of Science Core Collection (e.g. the Journal of Thermal Analysis and Calorimetry or the Energies).

Lenka Môcová is a language lecturer in security and safety programs at the Faculty of Security Engineering of the University of Žilina. Her scientific work focuses on the linguistics aspects of security research and academic writing for PhD. students. Besides this, she is a PhD.student at the Catholic University of Ružomberok, where she studies the interdisciplinary problem of the image of Slovakia in the British media.

Zoltán Nyikes, BSc, is an IT engineer and MSc is a safety and security engineer, PhD, professional soldier, CIS officer of the Hungarian Defence Forces Command in Budapest. His responsibilities include infocommunications technology, logistics and information security. his interest focus information security and physical defense.

Jiri Pokorny graduated from the Faculty of Mining and Geology of VSB – Technical University of Ostrava, specialization fire protection in 1992. In 2002, he finished the doctoral program Fire protection and Industrial Safety. In 2013, he graduated from CEVRO Institut in Prague, specialization Master of Public Administration. In 2017 he finished habilitation. For more than 20 years, he has been engaged in fire prevention and population protection. He has worked in various positions within professional firefighters. Since September 2015 has worked as an academic employee in VSB – Technical University of Ostrava, Faculty of Safety Engineering. He is a member of the Subcommission SK1 Designing fire safety constructions and SK4 Fire engineering Technical Standardization Commission TNK 27 Fire Safety of Buildings, Czech Association of Firefighting Officers, Association of Fire and Safety engineering and editorial boards magazines.

Petra Roupcova, Ing., PhD, obtained her degree in Occupational Health and Safety at VSB- Technical University of Ostrava (Czech Republic) in 2015 with the thesis entitled "Assessment and proposal of increasing the OSH level in the case of service and maintenance workers." Later, she obtained her PhD degree from VSB-Technical University of Ostrava in 2018 with the thesis entitled "Monitoring of ecotoxicity of carbon based nanomaterials." Currently works as an academic staff at VSB-Technical University of Ostrava. She mainly focuses on nanomaterials safety.

Jan Smolka works as a researcher in the Faculty of Safety Engineering, VSB-Technical University of Ostrava and Occupational Safety Research Institute (VÚBP). With a goal to improve prevention and awareness of safety in Europe and the Czech Republic, he is a vice-chairman of the Czech Fire Protection Association – Majaczech as a member of the CFPA-Europe where he participates in training and guidelines development. He was also an active member of COST Action FP1404, where he led a task group focused on fighting fires in bio-based buildings.

Miguel Vidueira is a Technical Manager at CEPREVEN, the Spanish Fire Protection Association. With 17 years of experience, mainly in the field of inspection and commissioning of fire protection systems. Former chairman of the CFPA-E Guidelines Commission (2014-2019). Member of the Experts Groups of Insurance Europe Prevention Forum in the areas of sprinkler systems, water mist, and sandwich panels.

Vladimir Vlcek graduated from the Machinery High School with specialization on Nuclear Power Engineering and then he finished Technical University in Ostrava, Department of Fire Safety in 1984. Since 1984 he has worked in the Fire Rescue Service in the Czech Republic in various positions. Nowadays, he has more than 35 years of practical experience. Since 1990 he has worked in various management positions, and from 1993-1995 he worked as a Chief of Czech Fire Service. Currently, he is the Chief Fire Officer of FRB MSR. In the year 2017 he was appointed by the President of the Czech Republic to the rank of Brigadier General. Vladimir Vlcek has practical experience in management of large-scale incidents - for example in floods 2007, railway accidents, explosion and obviously large-scale fires. He is a Chief of USAR - Urban Search and Rescue Team (reclassified by INSARAG), Chief of WASAR – Water Search and Rescue Team (certified by EU), HCP – High Capacity Pumping Team, all those teams are ready for international deployment. Vladimir Vlcek has taught for more than 25 years as an associated professor at the Technical University of Ostrava, subjects of organization and managing system, management of fire units as well as classes for the master's degree – Crisis and Emergency Management in Large Scale Disasters. In 2006 he finished his PhD. He is a President of the Czech Association of Fire Officers (NGO), which is an organization is a member of FEU (Federation of European Union Fire Officers Associations), where Vladimir represents CAFO since 1995. For last couple of years, he is one of the classifiers within INSARAG for USAR classifications/reclassifications and he permanently presents at INSARAG USAR Team Leaders Meetings. Within EU, Vladimir is an expert of EU Civil Protection and participated on some of international missions like Japan (TL of EUCP Team), Jordan (TL of EUCP Team), Namibia, etc.

Neil Walker is a Co-founder and Director of the International Association of Critical Infrastructure Protection Professionals (IACIPP).

Mike Zeegers finished 18 years of service with the police, where he has been designing and implementing protective and security measures for companies operating in the chemical industry, oil and gas companies, transport and banking chains for over 30 years. For more than 10 years he has been cooperating with the European Commission as a Security Risk Advisor. From 2010 to 2016 he was a Chairman and board member in the European Energy Information Security Network. Since 2017, he has worked as a Senior Security Consultant at Security Risk Watch.

Index

G

H

I

L

M

N

O

P

Q

R

S

T

U

V

W

Printed in the United States
By Bookmasters